Gernot Beer (ed.)

Numerical Simulation in Tunnelling

Springer-Verlag Wien GmbH

Univ. -Prof. Dipl. Ing. Dr. Gernot Beer
Institut für Baustatik, Technische Universität Graz, Graz, Austria

This work is sponsored by
Fonds zur Förderung der wissenschaftlichen Forschung, Wien

Originally published by Springer-Verlag Wien New York in 2003

Printed on acid-free and chlorine-free bleached paper
SPIN: 10913369

With numerous, partly coloured Figures

CIP data applied for

ISBN 978-3-211-00515-6 ISBN 978-3-7091-6099-2 (eBook)
DOI 10.1007/978-3-7091-6099-2

Preface

For many years the Austrian tunneling industry has demanded that research is urgently required to establish a theoretical basis for the New Austrian Tunneling Method and to assist site engineers in the often difficult day to day decisions. In particular it was felt that numerical models need to be improved considerably in order to be able to act as useful tools in predicting soil/rock mass behavior during tunneling. The required improvement not only refers to the quality of the models but also to their ease of use. As long as an experienced modeler is required to spend days in preparing the input and in interpreting the results the models will not be useful at the tunnel site.

It is heartening therefore that a group of scientists in Austria has come together to attempt to tackle this challenging task. The initiative has been supported in a number of ways by the Austrian tunneling industry. All Austrian companies associated with tunneling sent representatives to the management advisory board, which ensured that the research carried out in the project, was of benefit to the industry. The Austrian Geomechanics Society sponsored the project with a considerable amount, which was mainly used to cover site costs. HL-AG and ÖSAG, as well as the joint ventures allowed access to tunnel sites thereby making it possible to test new developments.

I believe that the Austrian tunneling community can be happy with the outcome of the research initiative and this can be seen as a start in the right direction. I believe that the results reported here are significant. There are several examples where new findings are applied already in practice. This relates for example to the method of predicting the degree of utilization of the shotcrete lining during tunnel construction, which has already been applied on several sites in Austria and the new apparatus for geological survey, which I believe has great potential for commercialization and export to other countries.

However, there is a lot of work left to be done. For example, since the FWF mainly funds fundamental research, there have been insufficient resources for the dissemination and exploitation. It is important that the findings of the research reported here are exploited commercially. An important beginning has already been made with respect to the geological survey equipment but much more remains to be done.

Last but not least this work has to be made known widely within the tunneling community to designers, contractors as well as to clients, who should demand the application of this new technology in new projects. I hope that the dissemination of research results will be partly achieved by this book and I wish the research group all the best in its future activities.

Vienna, January 2003

G.M. Vavrovsky
(President of the Austrian Society for Geomechanics (ÖGG) and the Austrian National Committee of the International Tunnel Association)

Editors preface

The work reported herein was performed within the joint research initiative Numerical Simulation in Tunnelling which was established and financially supported by the Austrian National Science Fund (FWF). The FWF also provided financial support to enable the printing of this book.

As the coordinator of the research initiative, it was a pleasure for me to be associated with a group of outstanding Austrian scientists. I particularly enjoyed the workshops and meetings we had during the five years, where many innovative ideas were presented and discussed. It was also a pleasure to edit the chapters of this book, as I believe they are of very high quality. My thanks go to all who participated in the task of writing them.

However, the new technology and knowledge developed would be of limited use if it were not transferred to the construction site. The cooperation of the tunnelling industry proved crucial here. The guidance given by the management advisory board was invaluable to ensure that the focus of the research effort was maintained.

This advisory board consisted of the following institutions:

ILF consulting engineers, Innsbruck (Dipl.-Ing. Dr. Rudolf Pöttler)
Österreichische Bundesbahnen, Planung und Engineering, Wien
(Dipl.-Ing. Dr. Hans Wehr)
Geoconsult ZT GmbH, Salzburg (Univ.-Prof. Dipl.-Ing. Dr. Johann Golser)
Tauernplan Consulting GmbH, Salzburg (Dipl.-Ing. Hans Döpper)
VOEST-ALPINE Bergtechnik GmbH, Zeltweg
(Dipl.-Ing. Dr. Karl-Heinz Gehring)
Eisenbahn Hochleistungsstrecken AG, Wien
(Dipl.-Ing. Dr. Georg Michael Vavrovsky)
ÖSTU Stettin Hoch-und Tiefbau GmbH, Leoben (Ing. Peter Schwab)
Ingenieurgemeinschaft für Geotechnik und Tunnelbau, Salzburg
(Dipl.-Ing. Nejad Ayaydin)
D2 Consult, Linz (Dipl.-Ing. Dr. Harald Wagner)
Stmk. Landesregierung, Fachabteilung 18a, Graz (HR Dr. Wolfgang Gobiet)
Porr-Asdag Tunnelbau GmbH, Wien (Dipl.-Ing. Dr. Harald Lauffer)
Beton- und Monierbau, Innsbruck (Dipl.-Ing. Josef Arnold)
Ingenieurbüro Laabmayr, Salzburg (Dipl.-Ing. Manfred Eder)
Silicon Graphics GmbH, Wiener Neudorf (Dipl.-Ing. Heinz Moser)

However, the industry not only gave us advice but also actively participated in the development, evaluation and exploitation of the developments. I would like to mention some of the companies and individuals that participated in the various sub-projects and therefore made contributions to the research reported in this book:

For the sub-project "site investigation" (Chap. 3):

- The access to the Sieberg tunnel for the application of numerical modelling was granted by ARGE ÖSTU-Stettin/Haider and HL-AG (client)
- The access to the tunnel Spital, for the application of numerical modelling, was granted by ARGE Poor-Asdag, Östu-Stettin, Illbau, Hinteregger, Ast and ÖSAG

For the sub-project "shotcrete" (Chap. 5, Chap. 10):

- Adiabatic temperature tests of shotcrete cement were carried out by Dr. Rudolf Röck (Schretter and Cie, Tirol)
- Experimental work on determination of evolution of strength and stiffness of shotcrete was carried out by Dr. Rupert Friedle and Ing. Erwin Schön (Lafarge Centre Europe Central, Prüfanstalt Mannersdorf)
- Measurement data and geometrical dimensions of the investigated tunnel site LT22 Lainzer tunnel and helpful comments and suggestions were provided by Dr. Georg-Michael Vavrovsky (HL-AG), Ing. Johann Lemmerer (HL-AG), DI Oliver Wagner (HL-AG), Dr. Peter Schubert (iC Consulenten), Prof. Lutz Sparowitz (TU Graz), Dr. Helmut Hartl (TU Graz)

For the sub-project "damage tensor" (Chap. 7):

- The *Land Salzburg* supported the research project with a special measurement program at the Schönberg tunnel on the B 311 provincial highway. The help of Dipl.-Ing. Friedrich Herold-Wildfeldner is gratefully acknowledged. These measurements formed the basis for the numerical models used by Dipl.-Ing. Manfred Eder, of the Ingenieurbüro Laabmayr, Salzburg
- The design for the new *Music Theater Linz* (consultant Dipl.Ing. Franz Laabmayr, Salzburg), was based on the numerical model developed in the sub-project.

For the sub-project "large deformations" (Chap. 6, Chap. 9):

- Geological data and geometrical dimensions of the investigated underground station Taborstraße and helpful comments were provided by Dr. Lothar Martak (MA 29) and DI Helmut Liebsch (Wiener Linien)
- Helpful comments on the tunnel analysis considering fire load were given by Ing. Johann Lemmerer and DI Oliver Wagner (HL-AG)

For the sub-project "compressed air" (Chap. 11):

- The access to the tunnel Siegaue was granted and measurement data were provided by Dr.-Ing. Jens Gattermann (Ed. Züblin AG).
- The access to Lainzer tunnel was granted and shotcrete panels for laboratory experiments were provided by Dipl.-Ing. Christian Haberland (Porr-Asdag Tunnelbau GmbH).
- Material for the laboratory experiments was provided by Ing. Karl Böhm (ALWAG Tunnelausbau)

The Austrian Society of Geomechanics (ÖGG), whose members include representatives of the Austrian tunnelling industry, supported the project and also made financial contributions towards it. The chairman of the ÖGG, Dr. Georg-Michael Vavrovsky, participated in some workshops and meetings and actively initiated industry interaction. Last, but not least, I would like to thank Dipl.Ing. Dr.techn. Thomas Reichl for converting Word documents to Latex, for the initial preparation of the cover design and for assisting in the layout of the book. I hope that readers will find the information contained in this book useful and that the new developments reported here will assist in making the difficult job of tunnelling easier.

Graz, January 2003 *Gernot Beer*

List of contributors

Gernot Beer
Institute for Structural Analysis,
Graz University of Technology,
Lessingstrasse 25, 8010 Graz,
Austria
beer@ifb.tu-graz.ac.at

Zhangzhi Cen
Department of Engineering
Mechanics,
Tsinghua University,
Beijing 100084,
P.R. China
demczz@mail.tsinghua.edu.cn

Alfred Fasching
3G Gruppe Geotechnik Graz
ZT GmbH,
Elisabethstrasse 22/II, 8010 Graz,
Austria
fasching@3-g.at

Andreas Gaich
3G Software & Measurement GmbH,
Plüddemanngasse 77, 8010 Graz,
Austria
a.gaich@3gsm.at

Jens Gattermann
Ed. Züblin AG,
Albstadtweg 5, 70567 Stuttgart,
Germany

Harald Golser
GEOCONSULT ZT GmbH /
GEOCONSULT Consulting
Engineers
Sterneckstrasse 55, 5020 Salzburg,
Austria
harald.golser@geoconsult.at

Christian Hellmich
Institute for Strength of Materials,
Vienna University of Technology,
Karlsplatz 13, 1040 Vienna, Austria

Günter Hofstetter
Institute for Structural Analysis and
Strength of Materials,
University of Innsbruck,
Technikerstrasse 13, 6020 Innsbruck,
Austria
guenter.hofstetter@uibk.ac.at

Franz Laabmayr
Laabmayr and Partner Consulting
Engineers Salzburg,
Preishartlweg 4, 5020 Salzburg,
Austria

Roman Lackner
Institute for Strength of Materials,
Vienna University of Technology,
Karlsplatz 13, 1040 Vienna, Austria
roman.lackner@tuwien.ac.at

Jürgen Macht
Institute for Strength of Materials,
Vienna University of Technology,
Karlsplatz 13, 1040 Vienna, Austria
juergen.macht@tuwien.ac.at

Herbert A. Mang
Institute for Strength of Materials,
Vienna University of Technology,
Karlsplatz 13, 1040 Vienna, Austria
herbert.mang@tuwien.ac.at

Bernhard Mathis†
Institute for Structural Analysis,
Graz University of Technology,
Lessingstrasse 25, 8010 Graz,
Austria

Gerhard Öttl
Institute for Structural Analysis and Strength of Materials,
University of Innsbruck,
Technikerstrasse 13, 6020 Innsbruck, Austria
gerhard.oettl@uibk.ac.at

Gernot Opriessnig
Institute for Structural Analysis,
Graz University of Technology,
Lessingstrasse 25, 8010 Graz,
Austria
opriessnig@ifb.tu-graz.ac.at

Bernhard Pichler
Institute for Strength of Materials,
Vienna University of Technology,
Karlsplatz 13, 1040 Vienna, Austria
bernhard.pichler@tuwien.ac.at

Christian Pichler
Institute for Strength of Materials,
Vienna University of Technology,
Karlsplatz 13, 1040 Vienna, Austria
christian.pichler@tuwien.ac.at

Thomas Reichl
Institute for Structural Analysis,
Graz University of Technology,
Lessingstrasse 25, 8010 Graz,
Austria
reichl@ifb.tu-graz.ac.at

Yannick Scheid
Institute for Soil Mechanics and Foundation Engineering,
Graz University of Technology,
Rechbauerstrasse 12, 8010 Graz,
Austria
scheid@ibg.tu-graz.ac.at

Wulf Schubert
Institute for Rock Mechanics and Tunnelling,
Graz University of Technology,
Rechbauerstrasse 12, 8010 Graz,
Austria
schubert@fmt.tu-graz.ac.at

Hartmut Schuller
Garber and Dalmatiner Consulting Engineers Graz,
Leechgasse 37, 8010 Graz, Austria,
formerly Institute for Soil Mechanics and Foundation Engineering,
Computational Geotechnics Group,
Graz University of Technology,
Rechbauerstrasse 12, 8010 Graz,
Austria
hartmut.schuller@utanet.at

Helmut F. Schweiger
Institute for Soil Mechanics and Foundation Engineering,
Computational Geotechnics Group,
Graz University of Technology,
Rechbauerstrasse 12, 8010 Graz,
Austria
helmut.schweiger@tugraz.at

Stephan Semprich
Institute for Soil Mechanics and Foundation Engineering,
Graz University of Technology,
Rechbauerstrasse 12, 8010 Graz,
Austria
semprich@ibg.tu-graz.ac.at

Yvonne Spira
Institute for Strength of Materials,
Vienna University of Technology,
Karlsplatz 13, 1040 Vienna, Austria
yvonne.spira@tuwien.ac.at

Rudolf F. Stark
Institute for Structural Analysis and Strength of Materials,
University of Innsbruck,
Technikerstrasse 13, 6020 Innsbruck, Austria
rudolf.stark@uibk.ac.at

Robert Stelzer
Institute for Structural Analysis and Strength of Materials,
University of Innsbruck,
Technikerstrasse 13, 6020 Innsbruck, Austria
robert.stelzer@uibk.ac.at

Gunter Swoboda
Institute for Structural Analysis and Strength of Materials,
University of Innsbruck,
Technikerstrasse 13, 6020 Innsbruck, Austria
gunter.swoboda@uibk.ac.at

Zhihai Xiang
Department of Engineering Mechanics,
Tsinghua University,
Beijing 100084, P. R. China
xiangzh@mails.tsinghua.edu.cn

Qiang Yang
Department of Hydraulic Engineering,
Tsinghua University,
Beijing 100084, P. R. China
yangq@mail.tsinghua.edu.cn

Dean Zhao
Department of Civil Engineering,
Lanzhou Railway Institute,
Lanzhou, P. R. China
zhao_dean@sina.com

Contents

Chapter 1
Introduction

Tunnelling plays an important role in the construction industry. Great challenges await us in the new millennium. For example the construction of tunnels through the Alps in order to allow the environmentally friendly transport of goods from the northern to the southern part of Europe involve high overburdens and difficult ground conditions. The planned *City of Tomorrow* will involve underground space for the transport of people and goods as well as for housing. Some very large projects which are in the concept stage are a tunnel through the Andes in South America with extremely challenging ground conditions and undersea tunnel projects to connect Korea and Japan and Russia and America. In addition there will be increased use of underground space for the disposal of hazardous and nuclear waste.

Austrian scientists and engineers have made significant contributions to the development of tunnel construction technology. In 1965, a new concept of tunnelling was developed by Professor Rabcewicz, which is now known as the *New Austrian Tunnelling Method* (NATM). In this method the ground support is adjusted in such a way that the load carrying capacity of the ground is mobilized as much as possible. This means that a ground arch is formed and only a thin shotcrete shell suffices for the support, instead of the thick concrete shells used in the old Austrian tunnelling method. The NATM has been used successfully for many projects around the world. Only in a few very isolated cases (Jubilee extension of the London underground, Munich underground, Galgenberg tunnel) problems were experienced with the method.

However, further development and economic success of modern tunnelling methods, which are still partly based on empirical assumptions, can only be assured if it is possible to establish a more solid and sound scientific basis for them. This relates in particular to the application of modern computer based data collection, simulation and visualisation methods. To address this issue the Austrian science fund established a joint research initiative (JRI) "Numerical Simulation in Tunnelling" in spring 1997. The project aimed to bundle existing world class research activities in order to achieve a major breakthrough. Eight research groups from three Universities in Austria worked together for five years. The budget for the JRI was more than € two million.

The main aims of the Joint Research Initiative "Numerical Simulation in Tunnelling" were:

- development of the scientific and technological basis for numerical simulation in tunnelling

- development of on site devices for data collection, for example for the automatic geological data acquisition using stereophotogrammetric techniques
- substantial improvement and further development of computer-aided simulation models in conjunction with *in situ* measurements
- further and new developments of simulation and visualization software including their application in tunnelling

The purpose of this work was to achieve a good correlation between numerical prognosis and the actual *in situ* behaviour. The technical aim of the JRI was to develop and apply numerical simulation tools to estimate the required tunnel support, the required construction measures and their impact on the environment at the planning stage and during construction. This results in an improvement of the safety and economy of tunnels. Different aspects are considered, from the acquisition of data to the development of efficient algorithms for the numerical simulation of tunnelling in highly deformable soil. In addition to basic research, which represented the central element of the JRI, a close collaboration with the Austrian tunnelling industry was practised.

The purpose of this book is to disseminate the findings of the JRI. The sequence of the various chapters of the book are meant to reflect the various stages of a numerical simulation depicted in Fig. 1.

The results of numerical simulations are greatly influenced by geological conditions. In the numerical model geological features are considered in the generation of meshes and in the definition of material properties. Geological features are difficult to determine since they are locked inside the ground. Common methods to determine geological features are exploration drilling, ground probing radar and seismic methods. The method used and developed further here is the geological survey of the tunnel face after each excavation stage. Here significant advances were made in technology. In Chap. 3, a new computer based system for the acquisition of geological data is presented which represents a significant improvement over the old method, whereby a geologist made hand sketches standing in front of the unsecured tunnel face. The procedure developed not only makes the data collection safer but allows a more accurate interpretation and prediction to be made since the geological data of tunnel faces maybe superimposed digitally to make a 3-D model of the geology.

From information about geology, the tunnel cross-section, the construction type and sequence a mesh has to be constructed for the numerical simulation. This pre-processing forms an important part of the numerical simulation and is with present systems rather cumbersome and time consuming. Here the aim of further development is to make the procedure completely automatic so that the user need not be involved in the detailed mesh generation. In such a system only the tunnel shape and construction sequence need to be input by the user or read from a CAD file. Using this input together with

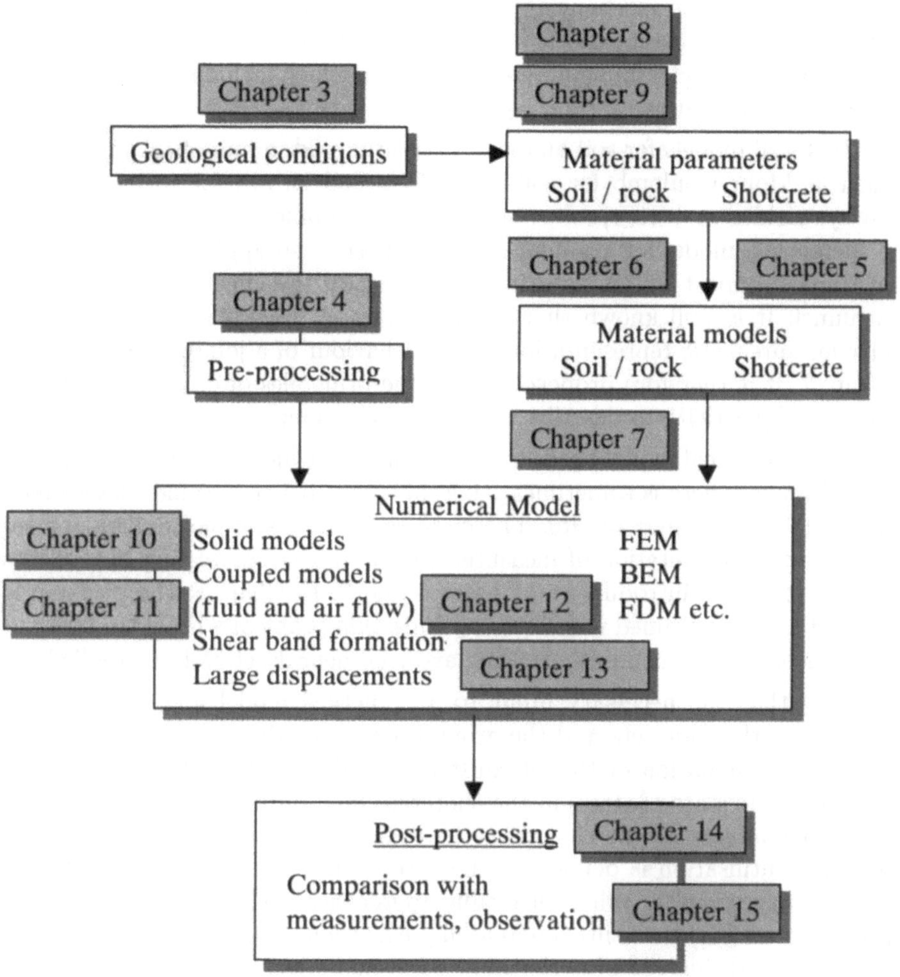

Fig. 1. Flow of numerical simulation and relevant chapters in the book

geological information supplied digitally by the newly developed acquisition system described above the program generates a mesh for the simulation software. This pre-processor is described in Chap. 4.

The next important input items are the specification of the material models used for describing the shotcrete and the soil/rock behaviour. Shotcrete is an integral part of the NATM since its deformation behaviour will influence the distribution of load between the rock/soil mass and the shotcrete lining. Sometimes large (controlled) deformations have to occur in the shotcrete lining in order to allow the ground arch to fully develop. Shotcrete is sprayed onto the tunnel walls and sets while being loaded. It therefore exhibits complex thermo-chemo-mechanical behaviour. In previous models for shotcrete many of the important aspects of material behaviour were not considered. In Chap. 5 a new model for shotcrete is presented which includes the description of all these influences.

Material models for soil and rock currently used in numerical simulations such as Mohr Coulomb for soils and Hoek and Brown for rocks have not always yielded realistic results and further development is required. In Chap. 6 therefore a model for granular soil is proposed and applied to the analysis of tunnel support by jet grouting and to study the effect of a fire load in a tunnel. It is well known that the principles of visco-plasticity sometimes give an unrealistic representation of the behaviour of a jointed rock mass. In Chap. 7, it is therefore proposed to use the principles of damage mechanics instead of plasticity to describe rock mass behaviour.

Any numerical model needs input parameters which are difficult to determine because there is sometimes a big difference between values determined in the laboratory and *in situ*. Therefore to get realistic parameters one relies on the interpretation of measurement data. Two interesting approaches to this topic are introduced in Chaps. 8 and 9. In Chap. 8, the problem is treated as a constrained non-linear optimisation problem whereas in Chap. 9 the concept of soft computing, using artificial neural networks is applied.

Now that the necessary input to a numerical model and the material models for the shotcrete and the ground have been discussed we proceed to the actual simulation of tunnel construction. Chap. 10 deals with the prediction of the state of stress in the shotcrete shell during tunnel advance. Of particular interest is the change in the degree of utilisation with time. The degree of utilisation is defined as the ratio of the stress in the shell to the compressive strength. When this value approaches 1.0 (or 100 % of the compressive strength) then this is potentially hazardous. An interesting approach has been used. Instead of performing a full numerical analysis including the ground and the shotcrete, which would have been rather time consuming only the shotcrete was considered and the interpolated measured displacements applied as a boundary condition. This concept reminds one of the early attempts at numerical simulation in tunnelling by Pacher described in the next chapter where the non-linear load-deformation behaviour of the support and the ground were considered separately. Simplification of the model allows the results to be obtained in such a short time that this tool can be used on site as a warning system to warn tunnel engineers of an unacceptably high degree of utilisation.

Often fluid is present in the ground and this affects the tunnel construction because there may be water inflow from the tunnel face and from cracks in the lining. Compressed air is often used to prevent water inflow and to reduce surface settlements. Chaps. 11 and 12 deal with this topic. Chap. 11 is concerned with the determination of the volume of air required to ensure that water is prevented from entering the tunnel and to ensure that settlements are reduced for shallow tunnels. It has been found, that especially for long tunnels most of the air loss is through cracks in the concrete shell. Therefore first experiments are carried out to study the flow of air and water in the ground if air pressure is applied at a crack in the concrete shell. Next, some numerical calculations are carried out which result in recommendations on formulae that can be used to calculate the requirement of compressed air. In Chap. 12, a sophisticated numerical model is developed, that allows to consider the movement of water and air through the soil skeleton in interaction with its deformation. The model has been tested on a number of examples and has been shown to give realistic results.

Once the analysis has been performed it is necessary to display the results in such a way that they can be easily understood and interpreted. Chap. 14 therefore presents novel display methods which allow three-dimensional time dependent results to be visualised in an effective way and compared with observation.

A final chapter deals with the testing of existing modelling software at tunnel sites. The application on site places very high demands on any modelling software since results are required in a very short time. It is also possible to immediately check the accuracy of solutions by comparing computed and observed values. Chap. 15 therefore deals in some detail with the comparison of simulation results with observation and measurements.

Chapter 2
Tunnelling - the need for technological development and innovation

Wulf Schubert[1] and Gernot Beer[2]

[1] Institute for Rock Mechanics and Tunnelling, Graz University of Technology
[2] Institute for Structural Analysis, Graz University of Technology

1 Introduction

Tunnelling at a larger scale started in the 19th century with the development of the railway systems. In the beginning the practices used in mining for excavation and support were adopted also for the traffic tunnels. In contrast to tunnelling, in mining only a rather small portion of the underground structures is designed for permanent use. The rock mass conditions in mines are usually well known due to the long experience gained during the lifetime of a mine. This experience of excavation under the same or similar conditions explains also why in mining the layout of the excavations and the required support is mainly determined in an empirical way.

With the increase in the construction of traffic routes and hydropower plants, tunnels and other underground structures had to be built in areas, where no or little experience with the rock mass behaviour was available. This soon created a demand for the development of investigation and laboratory testing, as well as for consistent design methods. Primary supports up to the first half of the 20th century usually consisted of timber while for the permanent linings dressed-stone, stone or brick masonry was used. Due to the multi-drift methods employed at the time, and the use of supports, which were not well adapted to the task, a loosening of the rock mass was unavoidable. The amount of timber propping also blocked much of the tunnel profile, hindering the supply and the construction of the permanent linings. With the development of rock bolts and shotcrete, which started in the twenties of the last century, many of the problems in tunnelling could be solved. In particular, the amount of load on the supports which was activated by the loosening of the rock mass could be minimized.

The enormous boom in tunnelling in the 19th century and at the beginning of the 20th century also lead to an increase in the efforts to establish a theoretical background to supplement or replace the classification methods used at that time. In [1] and [2] the development of the stresses around a tunnel was investigated and some fundamental investigations performed. Due to the limited possibilities, in particular related to computing capacity, these workers used homogeneous models. In [3] and [4] also relevant research was conducted in 1938, which helped the understanding of the development of the state of stress around tunnels. Fenner also described the relationship between tunnel wall displacements and the required support in order to ensure stability. The well-known Fenner-Pacher curve, which is based on this research is still widely discussed in the tunnelling community.

The state of the art in tunnelling up to the mid forties of the 20th century has been summarized in [5]. In his work Rabcewicz also discusses the shortcomings of the theoretical approaches used at that time, namely the fact that they neglect the heterogeneity of the rock mass, that they do not consider the influence of time and the effect of the presence of water. Taking in consideration the complex nature of the rock mass, he proposed a systematic observation the rock mass behaviour in order to develop appropriate design methods for various geological conditions.

In the second half of the 20th century the availability of computers made more advanced methods for tunnel design available to the tunnel engineer. This relates in particular to numerical methods such as the finite element, boundary element and distinct element method. These methods have been enthusiastically applied, but because of several limitations of these methods, the results of those computations did very often not realistically reflect actual behaviour. At the same time classification systems by Bieniawski and Barton [6] [7] were applied more frequently for tunnel design. These purely empirical systems are very attractive for tunnelling engineers, as with only a few parameters the construction method and the support requirements can be determined. The major drawback of classification methods is the fact, that these schemes are based on experience in certain rock mass conditions, and thus cannot be generalized. In addition these methods create the impression, that tunnel engineering is as simple as the classification schemes themselves. Major setbacks, failures, and financial losses are often the consequence of such simplifications. The application of these oversimplified methods does not promote the concepts of a more rigorous engineering analysis, leading to a rather limited learning effect. In Europe, and especially in Austria, the observation of the rock mass behaviour during tunnelling has reached a high standard. Unfortunately the huge amount of data collected is sometimes not evaluated systematically. Nevertheless the continuous observation during construction and the evaluation of the collected data has led to improvements with respect to safety and economy of tunnel construction.

2 Need for further development

Today numerical methods are widely used for tunnel design. However, problems with available models still exist in terms of the limited existence of realistic material models and the fact that post failure behaviour is not considered. Often users apply numerical models without critically examining the results. Misleading and unrealistic results of numerical models have led in the past to a certain frustration on the side of the users and on the side of the code developers. A general lack of communication between developers and users could be observed in the last decades.

Quite a number of unrealistic modelling assumptions are still widely used. Some of those are listed here:

- Unrealistic homogenisation assumptions for blocky and jointed rock mass
- Post failure behaviour is not modelled realistically
- The use of visco-plastic material models for a jointed rock mass
- Assumption of elastic behaviour for shotcrete, which has a pronounced non-linear and time dependent behaviour
- Unrealistic modelling of grouted bolts
- The assumption of small displacements, when displacements are rather large
- Time dependent behaviour is not modelled

It is well known, that a realistic simulation of tunnelling requires three-dimensional models. This has been shown again in a recent dissertation [9]. With the increasing power of computers and more user-friendly and efficient software more realistic three-dimensional computations become feasible also for the everyday use at the construction site. However, a precondition for realistic models and results is a high quality of the input. In too many cases the results of geological investigations are patchy and are not presented in a form, which can be used directly in a computer-model. In addition field and laboratory tests in many cases do not consider the special requirements of the numerical models. No consistent rules or guidelines exist, when to use which of the many numerical models available today. This sometimes leads to the misuse of models, in many cases often simply because the user prefers to use a certain commercial computer program, he is familiar with.

In order to increase the reliability of computer predictions it is also important to calibrate the models with real data. This stresses the importance of an appropriate data collection procedure on site and of a close co-operation between geologists, site engineers and modellers. New methods in evaluating and processing site data in order to allow a direct input into the model have to be developed and applied.

3 New developments and innovation

The problems with numerical simulation mentioned above have led to the establishment of the joint research effort, the results of which are reported in this book.

An improvement of the geological data collection on site and of improving the geological input into numerical models was obtained by applying photogrammetric principles using digital line cameras and the associated software. With this system, geological data are available in digital form and may be directly used by numerical simulation software. The issue of realistic modelling of shotcrete was addressed by the development of a sophisticated non-linear thermo-chemical-mechanical model that considers all aspects of shotcrete behaviour. The problem of determining reliable input data has been addressed and sophisticated parameter identification methods have been developed. The problem of the interaction of soil, water and air was tackled from the theoretical and the experimental side. A new approach to the modelling of shear bands, a form of failure especially common in shallow tunnels, is presented and will allow a more realistic design. The sometimes unrealistic models of jointed rock using visco-plastic or elasto-plastic theory were replaced with models based on the concept of damage mechanics and a more realistic prediction was achieved for some cases. Last but not least, the presentation of the computation results in a way, that also non-specialists are able to easily comprehend essential aspects, is of great importance. In this context advanced scientific visualisation plays an important role. Here new concepts of visualisation are presented.

It is quite clear, that with the uncertainties inherent in each geological-geotechnical model, an accurate a-priori numerical model will still be possible only in cases, where extensive investigation and experience exists. The continuous updating of structural models, adjustments to the material models, and improvements in the rock mass characterization will be necessary to achieve a further improvement of tunnelling practice. However, we believe that with the research work presented here an important step in the right direction has been made. The observation of the actual rock mass behaviour during tunnelling will also in future remain an essential task to verify design assumptions, refine models, and build a database, probably leading to an expert system with an extensive numerical component.

References

[1] Wiesmann E. (1909) Ein Beitrag zur Frage der Gebirgs- und Gesteinsfestigkeit. *Schweizerische Bauzeitung*, Bd. LIII, Nr. 13

[2] Leon A., Willheim F. (1910, 1912) Über die Zerstörung in tunnelartig gelochten Gesteinen. *Österreichische Wochenschrift für den öffentlichen Baudienst, Amtliches Fachblatt*, published by k.k. Ministerium der öffentlichen Arbeiten, XVI. Jahrgang, Heft 44. und Heft 16.

[3] Kühn P. (1931) Betrachtungen über die Gebirgsdruckfrage. *Glückauf*, Berg- und Hüttenmännische Zeitschrift, 67. Jahrg., Nr. 48

[4] Fenner R. (1938) Untersuchungen zur Erkenntnis des Gebirgsdrucks. *Dissertation*, Verlag Glückauf G.m.b. H., Essen

[5] Rabcewicz L. v. (1944) Gebirgsdruck und Tunnelbau. Springer Verlag Wien

[6] Bieniawski Z.T. (1974) Geomechanics classification of rock masses and its application in tunnelling. *Advances in rock mechanics 2 (A)*, 27-32. Washington, D.C.: Nat. Acad.Sci.

[7] Barton N., Lien R., Lunde, J. (1974) Engineering classification of rock masses for the design of tunnel support. *Rock Mechanics* 6 (4): 189-236, Vienna: Springer.

[8] Pacher F. (1964) Deformationsmessungen im Versuchsstollen als Mittel zur Erforschung des Gebirgsverhaltens und zur Bemessung des Ausbaues. *Felsmechanik*, Suppl. I

[9] Golser H. (2001) The Application of Finite Element and Boundary Element Methods in Tunelling. *Doctoral thesis*, Graz University of Technology

Chapter 3
Improved site investigation
Acquisition of geotechnical rock mass parameters based on 3D computer vision

Andreas Gaich[1], Alfred Fasching[2], and Wulf Schubert[3]

[1] 3G Software & Measurement GmbH
[2] 3G Gruppe Geotechnik Graz
[3] Institute for Rock Mechanics and Tunnelling, Graz University of Technology

Abstract. Acquisition and evaluation of geotechnical data are integrated parts of subsurface and surface construction works. Geotechnical data serve as input for decision making processes during all phases of projects, ranging from feasibility studies to construction and maintenance. The present system of data acquisition, specifically applied during underground construction works, has a number of constraints. Sampling bias may be caused by the "human factor" of individual capabilities, inaccessibility of the rock exposure and time limitations. In most cases data are irrecoverable when excavation proceeds or support has to be applied. Data processing and evaluation is time consuming so that input data for numerical calculations cannot be provided on a daily basis. To overcome the listed shortcomings a digital stereoscopic colour imaging system has been developed which enables the evaluation of a large number of geotechnical data by interactive two and three dimensional image analysis. Among others the data can be used for the innovative modelling of the rock mass structure, for the provision of geometrical input data for numerical simulations performed on site as well as for a descriptive visualisation of complex structural conditions. The developed hardware and software components have been tested in different environments and on different rock mass types to investigate their general suitability and effectiveness. It was found that digital stereoscopic imaging and image evaluation are suitable for a cpmprehensive and reproducible documentation of the structural inventory of rock surfaces, and are most effective for acquisition of geotechnical data.

1 Introduction

1.1 Motivation

Acquisition and evaluation of geotechnical data are integrated parts of subsurface and surface construction works, prior to, as well as during their execution. Decisions about feasibility, design and construction method depend to a certain extent on the predictions made on the basis of acquired data. Evidently the quality of data has a direct influence on its interpretation and consequently on predicting engineering consequences. Specifically assessments concerning the distribution of different rock mass types, defined by distinguishable properties like lithology, rock mass structure, tectonic influences, ground water conditions and mechanical and rheological parameters will always be affected by uncertainties until construction. These uncertainties with repsect to the geological appraisement consequently lead to an increase of uncertainties within the geotechnical design, all the more when numerical modelling is based on those uncertain model parameters [30].

The acquisition of geotechnical data from a rock surface, no matter if it is a natural or artificial one, is always influenced by three limitations, which are the (1) *human factor* (2) *accessibility* and (3) *time.*

(1) Presently no data acquisition system exists which has the versatile capabilities of the human being. The ability to view and move simultaneously to survey a rock mass in three dimensions from a range of distances, locations and angles, to touch and probe the rock exposure, to make comparisons with features observed elsewhere instantaneously, are not achievable by an automatic data collection system. But it has also to be considered that humans are susceptible to subjectivity and personal bias. Humans may have different education, degrees of experience and motivation; they work relatively slowly and can get bored or tired, particularly when working under adverse environmental conditions [23].
(2) Excluding flat rock surfaces, like foundation areas, accessibility is mainly limited to a strip of some 2 m height as a maximum. The geologist without auxiliary equipment can cover this area at the bottom of a rock surface, steeply inclined to the vertical. Within this area, data can be collected which are relevant for the description of the rock mass concerning rock type, rock mass structure, weathering etc. Data from the remaining part of the rock exposure can be collected by visual inspection only.
(3) With the exception of natural rock surfaces, uninfluenced by or prior to excavation activity, the time available for data acquisition is rather limited. This limitation may be caused by the total available time for data collection within a project or by the working cycle of rock excavation for tunnelling or slope excavation.

The limitations mentioned above, result in the collection of incomplete and highly biased data that influence all further evaluations and conclusions. This can only be overcome by using an alternative data acquisition and evaluation system, which can at least handle the major part of the required data and provide information of much higher quality. An important aspect, which has to be considered when looking for an alternative data acquisition system, is the chance for acceptance and integration into a routine working cycle at a construction site. Any alternative from the "human" data acquisition system is less versatile and disturbs the working process more. Over and above an extra system causes additional cost. Therefore an alternative system has to be quick enough and the benefit of the resulting high quality data must exceed the expenses.

2 Geotechnical data acquisition

2.1 Relevant parameters for rock mass characterisation

The purpose of data, collected during the geological/geotechnical survey is to describe the encountered rock mass in a comprehensive way. The data should allow for the classification of the rock mass concerning lithology, structural conditions and rock mass quality, in combination with results of geotechnical monitoring programs. From these data the rock mass behaviour should be derivable, utilising the data in combination with evaluation programs for displacement monitoring data [35] or for prediction of displacements in tunnelling [32]. Rock mass modelling and numerical simulation, based on such data, should result in the most realistic results concerning the interaction between rock mass and construction works.

Geotechnical relevant parameters, particularly important for rock mass modelling are discontinuity related data such as orientation, surface geometry, spacing, frequency, size, aperture / filling width, termination index or the georeferenced position, i.e. in relation to a world co-ordinate system. The data are completed with information concerning distribution of rock types, weathering conditions and specific local phenomena like karst. All these data can be acquired by visual methods when measuring of orientations and distances within a known 3D co-ordinate system are provided.

Definitions on the parameters vary slightly. The following are given according to Priest [23]

Discontinuity: A discontinuity refers to cracks and fractures in rock along with which there has been extremely little or no movement (otherwise it is spoken of a *fault*). The size for a discontinuity ranges from a few millimetres up to several hundred metres. At a tunnel face a discontinuity is represented by its *trace* in the cutting plane defined by the tunnel face.

Discontinuity set: Grouping of discontinuities based on certain attributes like genetic origin, orientation, type of fracturing, or lithology.

Discontinuity orientation or *Orientation*: Spatial direction of a discontinuity, described by the two angles, *dip* and *dip direction.* The dip is the acute angle measured in vertical plane between a given line and the horizontal plane. The dip direction refers to the geographical azimuth, measured in clockwise rotation from north of the vertical plane containing the given line.

Discontinuity Gap Length or *Discontinuity bridge*: distance between two discontinuities in main direction.

Discontinuity spacing: represents the distance between two discontinuities measured perpendicular to the main direction them.

Termination Index: It distinguishes whether the endpoints of a discontinuity are visible or not.

Trace Length: is the distance between the starting point and the endpoint of a discontinuity projected onto the cutting plane.

Co-ordinates: Location of a point in relation to a defined co-ordinate frame.

Area: An area contains features with common attributes. Since areas are not measured in present practice they are specified in relation to some reference area, e.g. the tunnel face.

As an example of data to be evaluated from a jointed rock mass, a listing of parameters, required for preparation of numerical models (e.g. UDEC [40]) is shown in Fig. 1.

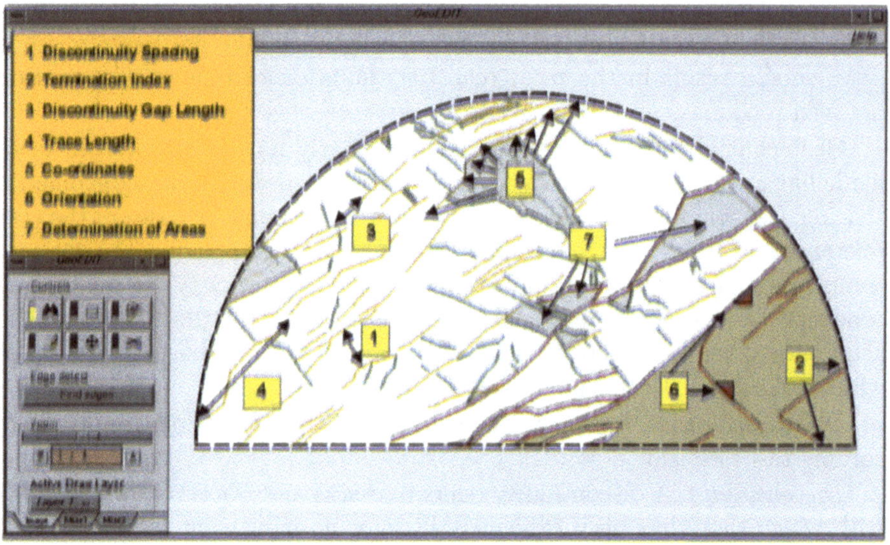

Fig. 1. Selection of parameters, required for the preparation of numerical models for jointed rock mass.

A computer vision system cannot acquire all geotechnical parameters usually needed for a comprehensive description of a rock mass. Certain magnitudes rely on a first direct observation by an experienced geologist, e.g. just to "feel" the constitution of filling material. But many of the parameters used on a regular basis in order to describe the actual rock mass situation are geometric, thus derivable from images. Table 1 outlines those parameters that can be computed by a computer vision system.

Summarising, a vision-based approach allows to determine all those geotechnical parameters that are currently identified by visual inspection and/or estimation. Those parameters which cannot be determined by am image need be collected by the geologist on site.

However, the regular data acquisition at the tunnel site (several times per day) can be done using a computer vision concept since it acquires all

Table 1. Selected parameter of a geological tunnel documentation, current methods of data acquisition and potentially collectable data by a 3D computer vision system (marked with a '*').

Parameter		**Method of data acquisition**
Rock type / Kind of rock mass / Rock mass type		Visual identification based on macroscopic features
Discontinuities	*Type**	Visual identification
	*Orientation**	Compass readings within the accessible area
	*Shape / Curvature**	Visual classification
	*Roughness**	Visual classification in combination with tactile inspection within accessible areas
	*Persistence**	Visual classification, supported by isolated measurements within accessible areas
	*Spacing**	Visual classification, supported by isolated measurements within accessible areas
	Filling / Thickness	Visual classification, supported by isolated measurements and tactile assessment within accessible areas
	*Termination**	Visual classification
	*Aperture**	Visual estimation, supported by isolated measurements within accessible areas
Jointing of rock mass	*Shape of joint bodies*	Visual classification
	Dilation of rock mass	Visual and manual classification
	*Degree of fracturing**	Visual classification
Water, encountered during excavation		Visual estimation
Water pressure inside the rock mass		Visual estimation
Degree of weathering*		Visual classification
Behaviour of rock mass	*Behaviour during excavation*	Visual classification
	Shape / location of overbreak	Visual estimation
	Rock Mass Bond / Interlocking	Visual and manual classification

geometric features of the visible rock mass and conditions usually do not change rapidly. In addition important changes in the rock mass or upcoming faults can be determined roughly in advance, e.g. by the interpretation of discontinuity measured orientations [35].

2.2 Present methods of geotechnical data acquisition

Presently the documentation of geological conditions during construction is performed using a combination of manual and visual methods. Depending on the type of work, the documented area normally consists of a rock face (foundation area, cut slope, pit wall, tunnel face) or of a wall, crown and invert area of an underground opening (circumferential mapping).

Figure 2 shows excellent working conditions for the geologist during excavation works of a tunnel but still outlines one major problem: without additional support he cannot reach areas above approximately 2 m off the ground. Besides, normal working conditions are characterised by the interference of the documentation work with other construction activities such as excavation, mucking, support installation or drilling and charging for the next blast. Consequently the documentation work has to be performed while machinery like boomers, excavators or muck loading and hauling equipment are operating.

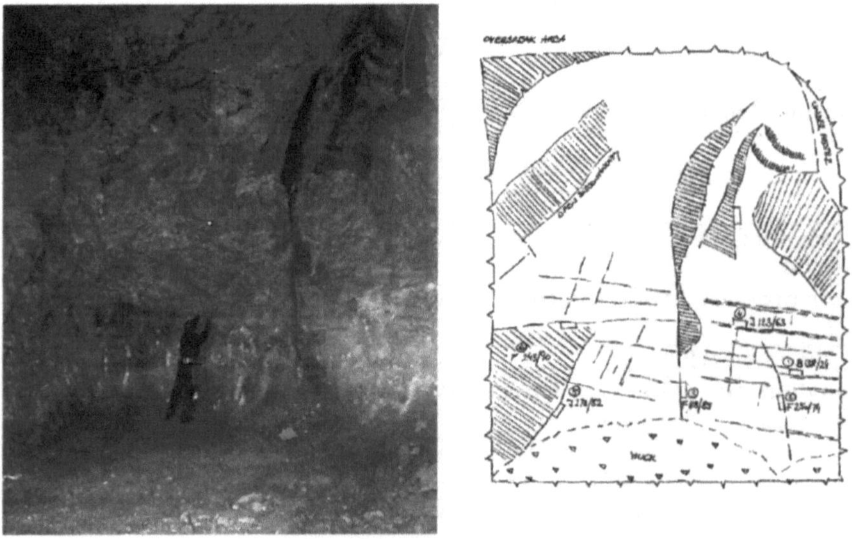

(a) (b)

Fig. 2. Present practice of acquiring discontinuity orientation data at a tunnel face (a). Picture shows an approx. 7.5 m high face of a drift tunnel in the RHI Magnesite Mine in Breitenau, Austria. Resulting manual sketch with six measured discontinuity orientations (b). Hatched areas represent slickensided fault planes, lines represent discontinuity traces.

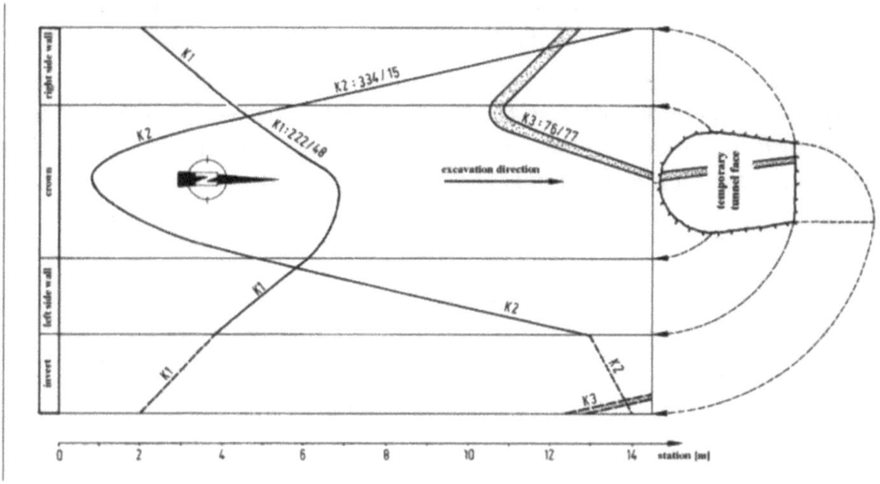

Fig. 3. Schematic sketch of circumferential mapping of a tunnel (modified from Wittke [38]).

When unstable rock mass conditions are encountered the geologist has no opportunity at all to enter the unsupported area. In that case a direct collection of data is impossible.

For documentation a combination of parameters is selected, with the intent to ensure a comprehensive characterisation of the geotechnical conditions of a specific project. From the documented parameters and information the rock mass behaviour should be derivable.

To facilitate this process and to ensure the completeness of the documentation the use of project tailored data sheets as shown in Fig. 4 proved to be helpful.

A standardised parameter selection, suitable for all possible geological environments is judged to be unpractical as the variability of geological features is too large. Such standardised parameter combinations are used frequently for quantitative rock mass classification (e.g., Q-System, RMR System) [1][17][3]. The application of such project-independent parameter combinations may imply that certain features, governing the rock mass behaviour remain unaccounted or features become under-/overestimated in general [8][29].

Depending on the occurrence of the individual features, discontinuity patterns, distribution of different lithological units or rock mass types, location and shape of overbreak, locations of water inflow, are all drawn schematically by hand. Fig. 2b shows the result of a mapping of the tunnel face shown in Fig. 2a. A schematic example of a circumferential mapping is shown in Fig. 3.

Instead of paper and pencil electronic plane tables may be used [39] which facilitate further data processing but do not improve the quality of data.

In addition, images taken by a conventional (digital) camera, may supply additional visual information to the face mapping. However such images cannot be used for further evaluation as the images are influenced by distortion and no orientation information is given for interpretation.

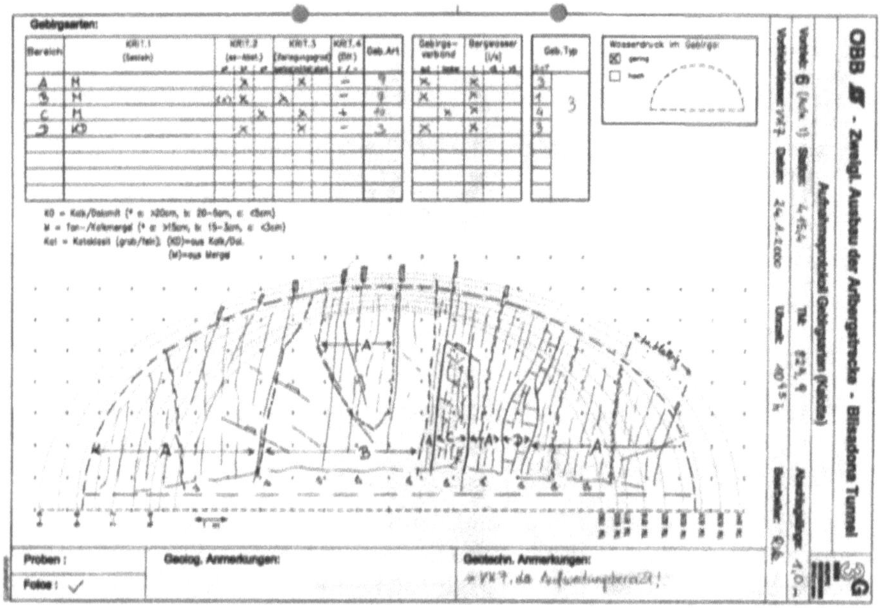

Fig. 4. Documentation of rock mass types based on project tailored criteria. Double track extension of Arlberg Railway Line – Blisadona Tunnel (courtesy of 3G GmbH).

Innovative "state of the art" documentation systems are based on specifically selected criteria. In the given example of a geological documentation of a railway tunnel (see Fig. 4) rock type, spacing of bedding planes, degree of fracturing, aperture, rock mass bond, and water inflow have been chosen as relevant input parameters. In combination with the interpretation of the geotechnical monitoring results, the data are used for decisions on round length and support. Such systems have to be designed and fitted to each new individual project.

In all cases each criterion or parameter has been subdivided in several subcategories. An identification of those subcategories is required to allocate them to distinct portions of the rock face.

Provided that access to the rock surface is possible, discontinuity orientation data are the only ones that are measured, while all other information is obtained by estimation. Predominantly though, the parameters are determined visually.

Visual estimation and the classification of qualitative as well as quantitative data highly depends on the competency and experience of the operating geologist. Even more so as complex coherences can only be recognised and assessed when experience from similar situations already exists [25].

2.3 Image based methods

Basic considerations. Imaging and image evaluation techniques have been part of geological analytical work for quite a time. Known as "remote sensing" it is used for identification of large scale lineaments, so called photo lineaments, large scale discontinuity patterns (e.g., master joint systems), evaluation of morphological features like slope instabilities, and demarcation of different lithological units, especially in areas where visibility is not hampered by vegetation (e.g., arid areas).

The images are made from air - or spacecrafts, normally with a setup which allows for stereoscopic inspection. Evaluation is frequently performed in an interactive way, with the possibility to use automatic systems [26].

In contrast the application of terrestrial imaging techniques for medium to small scale features has not become a practical application until now. However, it's application for documentation of rock surfaces has a number of advantages in comparison with the graphical documentation, such as

- Objectivity of images,
- Reduction of time required for field survey,
- Data acquisition in inaccessible areas,
- Acquisition of statistically sufficient data quantities,
- Processing and evaluation of data in the office without time constrains,
- Permanent record for review purposes.

Application for geological problems.

2D - structure analysis on the basis of standard images. The possibilities to evaluate geological features, especially structural parameters on the basis of standard two-dimensional images or videos have been investigated by a number of authors. Main topics have been discontinuity parameters like, discontinuity identification, 2D orientation, persistence, spacing, and surface properties, as described by Tsoutrelis [36], or Cravero [6]. In addition planar density, frequency, type of discontinuity termination and spatial correlation were also analysed, Crosta [7]. Identification of individual blocks within a muck pile and analysis of block sizes, shapes as well as volume were performed, investigating the applicability of image analysis for blasting design and quarry management (Franklin [13]).

Common properties of these methods are the use of digitised monochromatic or panchromatic images, image-processing steps like thresholding, dilation, and gradient calculations and the manual digitising of extractable features.

Research on methods of automatic image analysis for detection of discontinuity traces has been performed, using edge detection methods. Gradient filter operations and computation of a topographic value for each image point were applied on digitised monochrome rock mass images with low resolution. Thereby the association of image points to a joint trace was investigated. The outcome is that a number of external influences like changing lighting conditions, different lithologies as well as different image resolutions require more robust algorithms. Fundamental feasibility of automated structural analysis under standardised imaging conditions has been demonstrated for evaluation of core photos.

3D – structure analysis on the basis of stereoscopic Imaging. In comparison with two-dimensional imaging the application of stereoscopic imaging comprises additional important features, these are:

- Accurate determination of the space arrangement of all features in a local or world co-ordinate system,
- Possibility of unbiased data evaluation for all structural data,
- Reproducible documentation of temporary outcrop conditions, including all geometrical information like orientation and scale,
- Option to switch between two- and three-dimensional perspective to facilitate the image analysis,
- High information content for demonstration purposes to external reviews.

The system of stereoscopic imaging for geological problems has been applied first in the 1960's. At this time it was tried to utilise images, taken by photo-theodolites (Linkwitz [20]) or photogrammetric cameras (Rengers [28], Vieten [37]) for determination of discontinuity orientations.

For stereographic set-up two photogrammetric cameras were used, mounted on a tripod with a base-length of 120 cm [28]. This arrangement allowed a distance to the relevant object of about 6 m to 25 m. For larger distances the use of photo-theodolites is recommended. The images were oriented by observing control points attached to the rock face of interest. The evaluation was performed on negatives using a stereo-autograph or stereo-comparator. The achieved accuracy of discontinuity orientation measurements were analysed by comparison with field measurements. An average error of 3 degrees and a maximum error of 5 degrees for three measurements were determined.

Vieten uses basically the same approach as described by Rengers, without information about the achieved accuracy. Based on a camera base-distance of 120 cm (ref. Fig. 5) a maximum distance to the rock face of 40 m, and minimum size of discontinuity surfaces to be measured of 25 cm^2 is specified. For large-scale imaging, covering distances between 20 m and 2 000 m, the use of a single camera with variable base-distance is recommended.

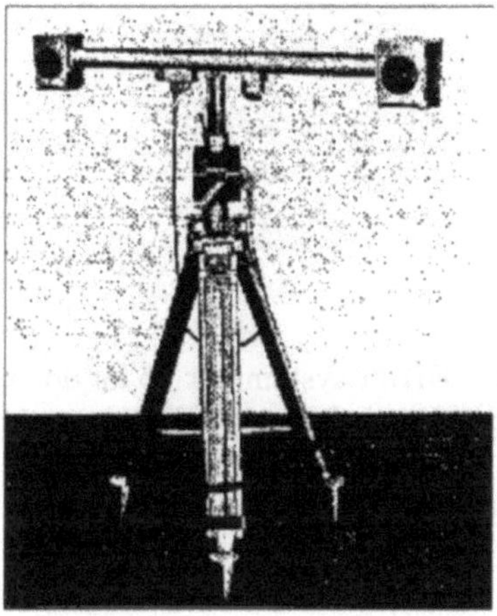

Fig. 5. Stereometric camera SMK 120, Carl Zeiss Company , camera base-distance 120 cm.

Some ten years later, for detailed studies of rock structure a set-up of two single reflex cameras with a base of about 20 cm, housed in a box, was used for deep-mine rock structure studies [18]. The interactive evaluation of fracture type (e.g., natural or mining induced), frequency of particular discontinuity sets, discontinuity orientations, in-filling thickness, persistence, and surface properties, was performed on black and white positives using a stereoscope and a digitiser.

The general applicability of stereo imaging systems was demonstrated and the achieved qualities by the above mentioned methods, checked by compass comparison measurement, was stated to be satisfactory, sometimes even more representative as the bias of compass readings, caused by the smallness of mapped areas, could be avoided. But there are no indications that the described systems have been investigated, developed or used further as no literature could be found, referring to these sources.

For geometric reconstruction of the tunnel circumference before and after application of shotcrete a stereoscopic set-up with two digital panchromatic perspective cameras on a revolvable rig has been developed recently (base distance 1.7 m) [40]. The DIBIT system (Digitales Bildmeßsystem für den Tunnelbau) is shown in Fig. 6, together with application examples of proposed geological tunnel face documentation.

(a)

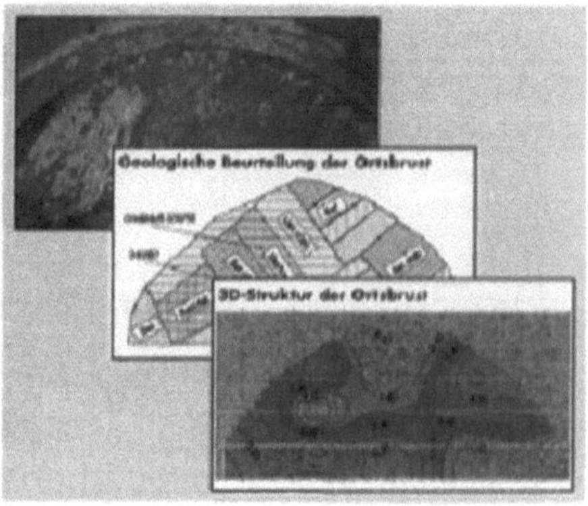

(b)

Fig. 6. Image a) shows the DIBIT system, set up for documentation of tunnel walls. In b) examples for geological evaluation of DIBIT tunnel face images are displayed, where the face image is not derived from the cameras of the stereographic set-up as greyscale images are acquired presently. Images taken from the dibit Homepage [40].

It is basically possible to document the tunnel face for *geotechnical purposes* with this system, which has not been applied as a routine till now, but this has three disadvantages (it must be stated that the system was not designed for that): (i) the system does not provide colour information up to now, which is an important element for geological evaluation, (ii) several images of the tunnel face have to be made and merged to achieve a sufficient high geometric resolution, and (iii) the system does not provide the possibility to inspect and evaluate the individual scenes in 3D.

2.4 Data acquisition by computer vision

Computer vision can be understood as connecting a computer with (digital) cameras and processing the images on it. Three-dimensional computer vision (3D vision) is inspired by human perception and tries either to reconstruct objects from images or to understand the contents of a scene [34]. The image quality is an essential design criterion for a computer vision system.

The basic concept is to generate two images of the rock face in proper quality and to use the combination of automatic processing and the highly sophisticated human visual system together with the knowledge of an expert on rock mass conditions. The result are geometric parameters that allow an abstract description of a rock mass situation which is to be used for cyclic decisions concerning support and further outcrop at a tunnel site. Figure 7 gives an overview on such a system.

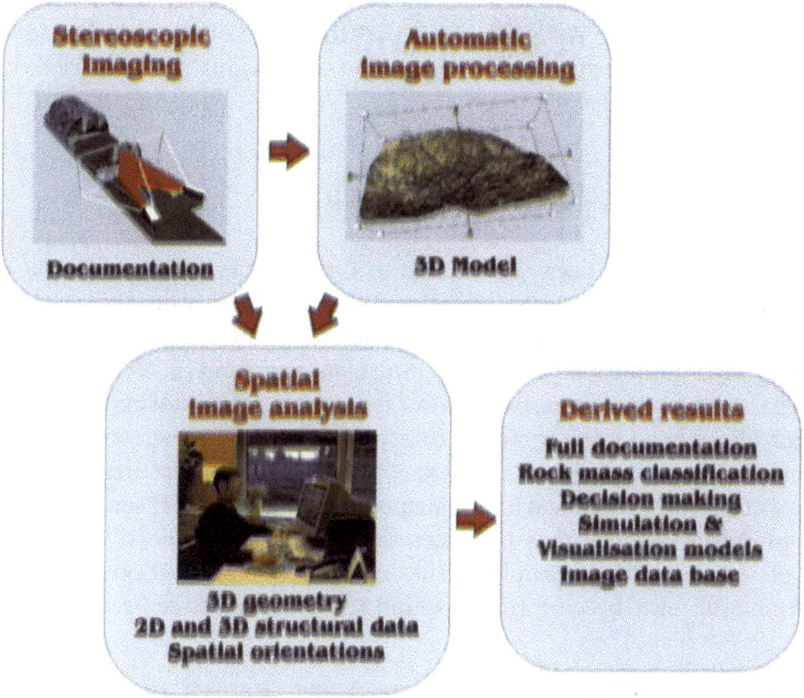

Fig. 7. Concept for data acquisition using a computer vision system. An imaging system and software components are used to derive geometric parameters of the rock mass used for decisions on the tunnel site, numerical simulations and visualisation purposes.

Requirements to a computer vision system. A computer vision system for the tunnel site has the following requirements (the order does not reflect importance):

- Instant image processing directly at the site is required which brings the need for digital images.
- Image resolution of about 2 - 3 mm on the rock surface (object space) to identify smaller structures which requires image sizes beyond customary digital cameras (e.g. 4000 x 3000 pixels).
- Colour information to identify lithological boundaries or weathering.
- A stereoscopic camera configuration, which is the basis for metrics, i.e. at least two images with two different viewpoints showing the rock face are needed.
- Geometric accuracy at least in the cm-range.
- A geo-referencing mechanism, that brings a relation between the images and the surroundings.
- Data acquisition within minutes.

These challenging specifications cannot be fulfilled using standard off-the-shelf cameras. In order to get a solution at reasonable cost the specifications led to the design and construction of a proprietary imaging system based on panoramic line-scan cameras.

3 The panoramic line-scan camera

Panoramic cameras have been used since the 19th century and received their impetus primarily to be used as aerial cameras, i.e. for use in aircraft [33] in order to overcome the limits of a single lens field of view and therefore to combine high resolution and wider swath in one camera. The cameras were mechanically quite expensive and worked mostly with rotating prisms or lenses. For image processing a precise rotation around the perspective centre was required in order to calibrate the camera and to allow measurements [5].

Substituting the mechanical film-transporting unit by a CCD sensor (charge coupled device: an unit that produces electrical charge depending on the amount of incoming light) simplifies the mechanical construction although the rotation still must be performed around the perspective centre.

Fig. 8. Example for a 360° panoramic image taken in an indoor environment using a rotating line-scan camera with the sensor oriented vertically. Note the varying camera perspective resulting in horizontal edges warped to arcs.

The properties of a panoramic image are (cf. Fig. 8):

- The field of view is extended to 360°.
- The scanning camera requires a static scene.
- The image is composed by several (thousands) image columns, each individually exposed.
- Every single image column has its own exterior camera orientation.
- Due to the rotation the perspective of the cameras changes during the scan. Level edges that perpendicular to the sensor are warped to arcs whereas parallel edges remain rectilinear (cf. Fig. 8).
- Every image column is individually exposed which requires a continuous illumination during the scan or synchronised flashlight (stroboscope).

- Without doing a pre-scan camera settings, like exposure time or focus cannot be found automatically.

Since panoramic line-scan cameras are not really off-the-shelf products and the cameras have be suitable for the rough tunnel construction site conditions, specially designed cameras were implemented.

The realised imaging system consists of two chromatic CCD line-scan cameras with 6000 x 3 sensor elements each. The cameras are mounted on rotation units controlled by stepper engines. The camera settings and the imaging process are controlled via PC or notebook (see Fig. 9)

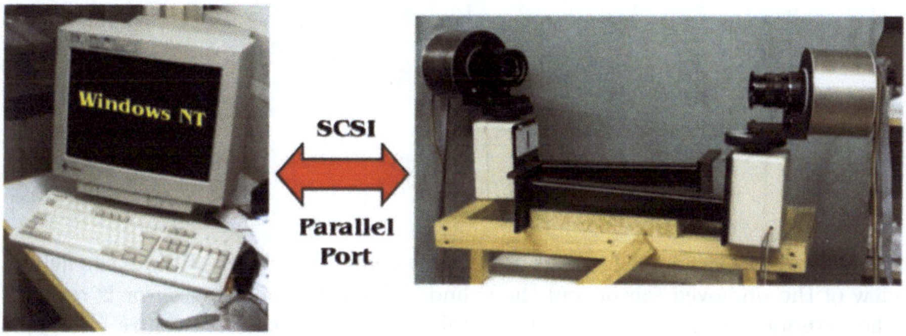

Fig. 9. The specially constructed CCD line-scan cameras mounted on rotation units. The communication channels to the hosting PC are SCSI for the data transfer and the parallel port for control signals.

The use of linear CCD sensors is not very common outside classical applications, like in flatbed scanners, in remote sensing aircrafts, or industrial inspection [14][21]. Besides, the derivation of measurements from images taken with line-scan cameras is less applied.

3.1 Imaging with a panoramic line-scan camera

The image acquisition with a rotating line-scan camera bears certain properties and discrepancies from a conventional frame camera. Fig. 10 depicts the basic geometric conditions of a single image column. A point $P(X, Y, Z)$ in object space is mapped onto the sensor point $P(\phi,y)$ passing the perspective centre $O(X, Y, Z)$. It also outlines the small field of view of single image column taken with an unmoved sensor. In order to get a two-dimensional image the sensor is rotated, juxtaposing the single image columns taking along the motion. Assuming an ideal rotation, the resulting images "resides" geometrically on a cylinder which explains the perspective differing from perspective frame camera.

As the camera rotates during the image acquisition, each of the image columns has its own exterior orientation. Therefore commonly used principles

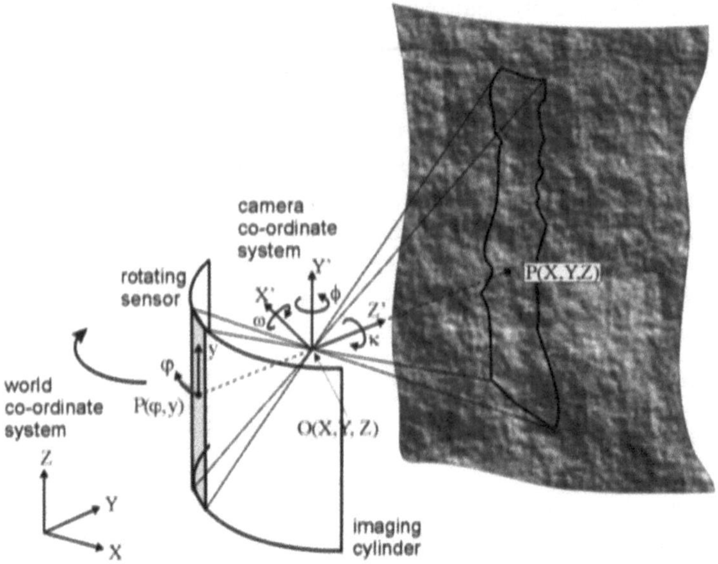

Fig. 10. The sensor model and co-ordinate systems. The sketch outlines the field of view of the unmoved sensor and the cylindrical image when the sensor is rotated. The exterior camera orientation is described by the perspective centre O(X,Y,Z) and the angles of rotation (ω, ϕ, κ)

to recover the exterior orientation from an image that contains control points with known co-ordinates in the object space cannot be applied. But as the rotation axis was assumed to pass the perspective centre, the cylindrical image can be transformed into a planar one that follows the perspective of a single lens camera. This transformation is denoted as a "warp" operation. Case [5] described the mathematical relationship for the warp operation.

After applying the warp operation (cf. Fig. 11) the image can be treated as emerging from a conventional single lens frame camera, thus all well established algorithms and procedures for camera orientation can be used.

The comprehensive sensor model for the line-scan camera is given in [15].

3.2 The panoramic line-scan camera at the tunnel site

The advantage of a full panoramic image can be observed when looking at the basic geometric conditions on a tunnel site, outlined in Fig. 12.

Retroreflective targets are commonly used on tunnel sites in order to regularly measure the displacements of the tunnel shell. These targets are mounted in sections every 5 - 25 m depending on the actual rock mass conditions, 5 to 7 targets per section. These targets can be used by the computer vision system as so-called *control points* to determine the position and orien-

(a) (b)

Fig. 11. Effect of the image warp. The panoramic image (a) is transformed onto an image plane resulting in a central perspective image (b). Note that the straightened lamp in the upper part of the image. If the rotation was performed around the perspective centre image (b) has a unique exterior orientation whereas in image (a) every image column has its own orientation.

tation of the system in relation to the tunnel. With a panoramic camera it is possible to acquire all the targets and the tunnel face with a single image.

As three-dimensional measurements from the images are needed, a stereoscopic arrangement on top of a car was used to test the behaviour of the imaging system (see Fig. 13). The car allowed to position the imaging equipment quickly on site.

In order to record the displacement monitoring targets bheind the car, small lamps are mounted on top of the cameras. The lamps rotate with the cameras ensuring a good visibility of the retro-reflective monitoring targets which are situated in the otherwise dark area. Figure 14 shows an example of an image viewing behind the car. For this purpose lamps are rotated together with the cameras. The lighting and the image resolution is sufficient to identify displacement monitoring target at least to a distance of 60 m. This ensures that at least three targets are visible within an image at the tunnel site which is needed to determine the exterior camera orientation from a single central-perspective image [33].

The time needed to take an image is a crucial part at the tunnel site. The reflection characteristics of the tunnel's surface are poor that means light falling onto the rock mass or shotcrete surface is reflectedl ess, thus a powerful lighting equipment is required.

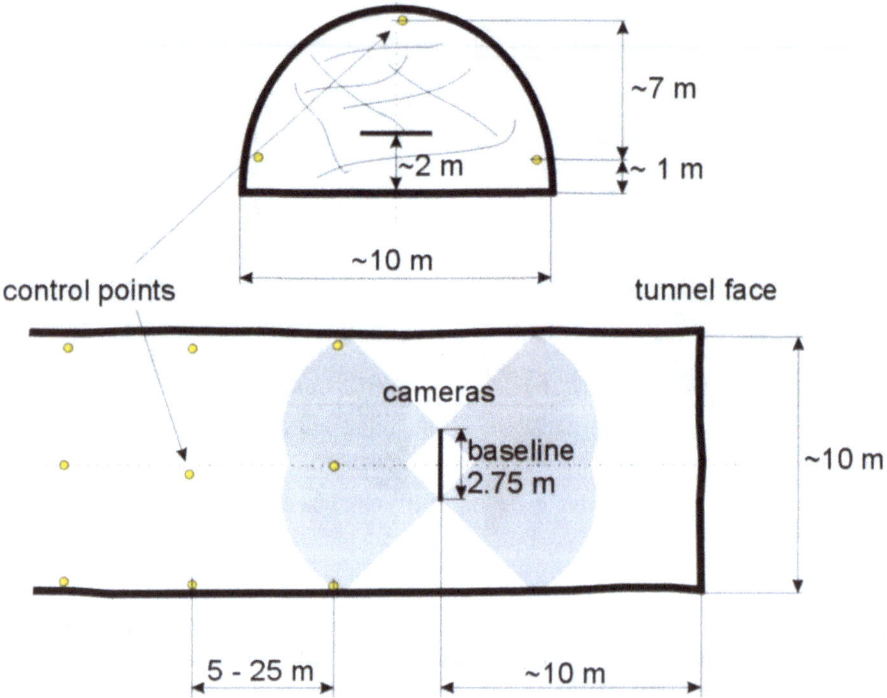

Fig. 12. Arrangement at the tunnel site. The tunnel face is in front of the cameras whereas the control points are behind them.

(a) (b)

Fig. 13. Rotating line-scan cameras mounted on top of a vehicle (a). The image was taken at the magnesite mine testing site (see Fig. 2). The imaging principle is illustrated in (b) – the cameras scan independently but simultaneously the rock mass.

Fig. 14. Image taken at the Breitenau magnesite mine testing site (cf. Fig. 2 and Fig. 13).

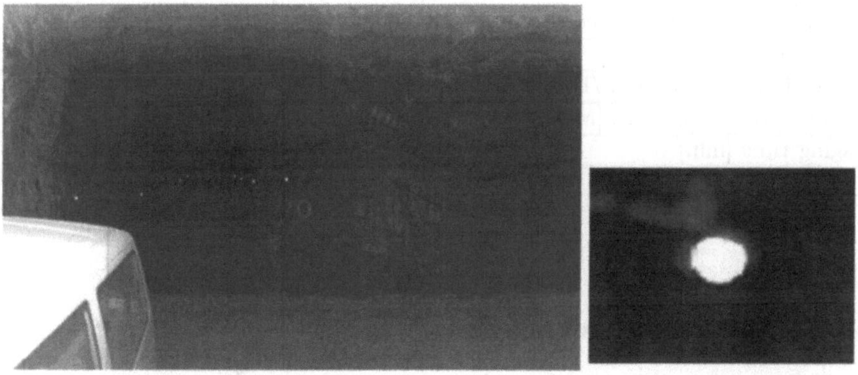

(a) (b)

Fig. 15. Image of displacement monitoring targets from the same standpoint as Fig. 13 (a). The targets are mounted for testing purposes at distances from 5 to 60 m. The only light source were small halogen lamps on top of the cameras. The image detail (b) shows that the resolution is sufficient for easily detecting the targets within the images.

The line-scan cameras need a constant lighting during the scanning process. The exposure time needed for a single image column is a time-critical component for the whole imaging process, since several thousands of such column need to be taken. Metal-halide lamps are a good choice when CCD cameras are used because the spectral distribution of the sensor's sensitivity fits well to the lamp. In the actual application, two metal-halide lamps with a power consumption of 250 W each were used which makes a bright impression for the human observer.

The implemented cameras allow to merge adjacent sensor elements electronically (usually denoted as hardware binning) which doubles the area of a single sensor element. This bisects the exposure time needed for a single image column but also halves the geometric resolution of the image. The doubling of the sensor area requires larger angular steps during the camera rotation resulting in a quarter of the image size and imaging time.

Tab. 2 shows typical values for the imaging with the panoramic line-scan cameras and the influence of hardware binning.

Table 2. Key values for the imaging system. Data base on an object distance of about 10 m and an observed tunnel face of about 10 m x 8 m. The middle column refers to typical values

Parameter	**Value**		
Angular coverage [°]	270		
Total image size [pixel]	20000 x 6000	10000 x 3000	5000 x 1500
Image size of the tunnel face region [pixel]	8000 x 6000	4000 x 3000	2000 x 1500
Image scale [1]	180		
Geometric resolution [mm/Pixel]	2	4	8
Uncompressed image size [MB]	343	86	21
Imaging time [min]	16	4	1

3.3 Measurement from images

In order to measure from the images the perspective geometry of the cameras, the so-called *internal orientation* must be known. The internal orientation includes the focal length, the geometric distortion of the used lens, and the position of the principal point [33]. Once these parameters are known metric measurements can be taken within the scope of two different co-ordinate systems:

Camera co-ordinate system (no geo-reference): The imaging system itself defines the co-ordinates, i.e. measurements do not relate to the surroundings. But relative measurements within a single image pair are valid and have physical reality. This is the simpler configuration, which requires no other positioning system. But it is sufficient for many geotechnical analyses, like the evaluation of a single rock wall that uses measurements of structures only in relation to each other. Since the measurements are based on panoramic images the field of view might be 120° without any problems, thus long drawn-out rock walls can be acquired.

Object co-ordinate system (geo-referenced): Structures are measured in relation to a given (external) object co-ordinate system. To achieve this, the so-called exterior orientations of the cameras within the object space must be determined. The exterior orientation of a camera is defined by the location of the centre of projection and the pose of the optical axis. By observing at least three non co-planar control points (points with known co-ordinates in the object space) the exterior orientation can be determined. This process was adapted for panoramic line-scan cameras [5][15]. In the case of tunnel construction sites, control points already exist. Since commonly applied reflection targets are used for displacement monitoring it is quite natural to use them to determine the camera orientation.

Using the parameters of the interior and exterior camera orientation the three-dimensional reconstruction of surface points becomes possible. The points are connected to a surface description and combined with geometrical structures originating from rock mass discontinuities, which finally leads to three-dimensionally defined geotechnical parameters.

4 Image analysis

4.1 Interactive structural analysis

A single rock mass image from a stereoscopic pair is used to interactively annotate geotechnically relevant structures identified within the rock mass. This approach was plurally used in the past in order to digitise structural (2D) information from conventional photographs, e.g. [18][13][7][36].

The major problem in automatic identification of discontinuities originates from changing rock mass conditions and the task to determine the geotechnically relevant (or significant) structures. Fully automatic approaches [27][11] do not deliver satisfactory results at present. Too many parameters in a chain of processing steps influence the final results and have to be set individually. Current analysis tools are not robust to be adaptive to varying rock mass conditions. Therefore an interactive analysis performed by an experienced geologist is proposed. Using software that allows the handling of the images, traces of discontinuities can be marked. These structures are displayed as graphical overlays on the original images. Internally they are grouped to manage discontinuity sets and are stored in a hierarchical manner, which simplifies the computation of statistical values. Fig. 16 shows a tunnel face and the 2D structural map of the main joints grouped into four discontinuity sets. It took an experienced geologist about 5 minutes to come to this result, which is now metrically correct and based on a reproducible 3D model.

4.2 Stereoscopic vision

A major support for the interactive analysis is the use of stereoscopic vision that allows to inspect the images three-dimensionally on the computer. The

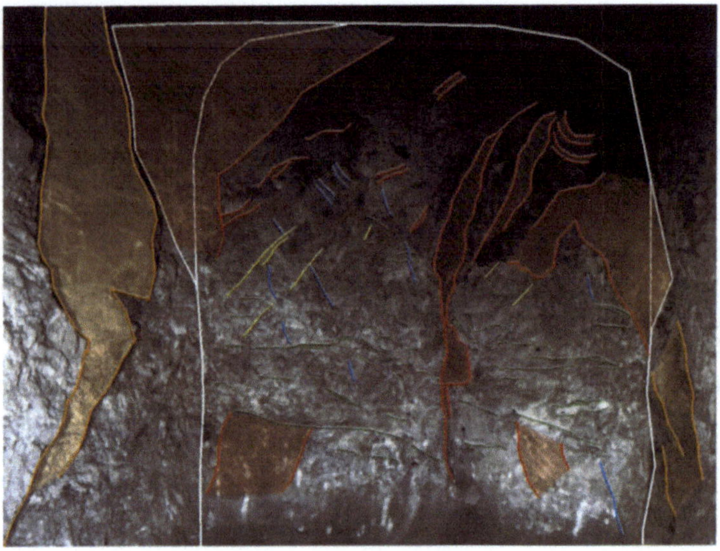

Fig. 16. Image taken in a drift tunnel in the RHI mine in Breitenau, Austria (red: fault planes, green: foliation planes, blue and yellow: joint planes, orange: fault planes in the sidewall areas, white: cross section with over break areas).

two images of the rock mass exposure are simultaneously displayed and separated for each eye by means of graphic's hardware (e.g. synchronised shutter glasses) which enables three-dimensional perception [12]. This allows a better assessment of ambiguous regions. In the working process the mode is switched between two-dimensional annotation and three-dimensional vision.

4.3 Automatic surface reconstruction

Surface reconstruction allows recovering the three-dimensional shape of the rock mass exposure. Starting from the stereoscopic image pair, three steps are performed (cf. Fig. 17):

Matching is the process of identifying corresponding points within the image pair. This task is an important basis for the quality of the reconstructed surface: the denser and more accurate the point correspondences are detected, the more reliable are measurements derived from the surface model.

Three-dimensional point reconstruction refers to the computation of the three-dimensional object co-ordinates of a surface point based on the corresponding image points and the orientation of the cameras. The results is an unorganised set of 3D surface points (point cloud)

Mesh generation ensures a connection between the single points of a point cloud. Often used are triangulated irregular networks (TIN) like the Delaunay

Fig. 17. Principle of stereo reconstruction. From the two images and known camera parameters a 3D surface of the rock face can be computed (middle) and aligned with one of the images (bottom).

Triangulation [9], which results in a surface description exclusively composed by triangles.

4.4 Fusion of results

The resulting structural maps from the interactive analysis can be geometrically aligned with the automatically computed surface reconstructions if both results base on the very same images. With that 2D structural maps become 3D existence and therefore enable the derivation of 3D magnitudes of the structures, like the true length of discontinuities or distances between them.

The resulting geotechnical surface models consists of 3D points, their connection among each other and the 3D structural data. The models are handled in a standard graphics file format called virtual reality modelling language (VRML), which is virtually supported by any Internet browser. This allows easily interchanging such models using the World Wide Web. In Fig. 18 the steps of surface reconstruction from the 3D point cloud to the textured surface model are displayed.

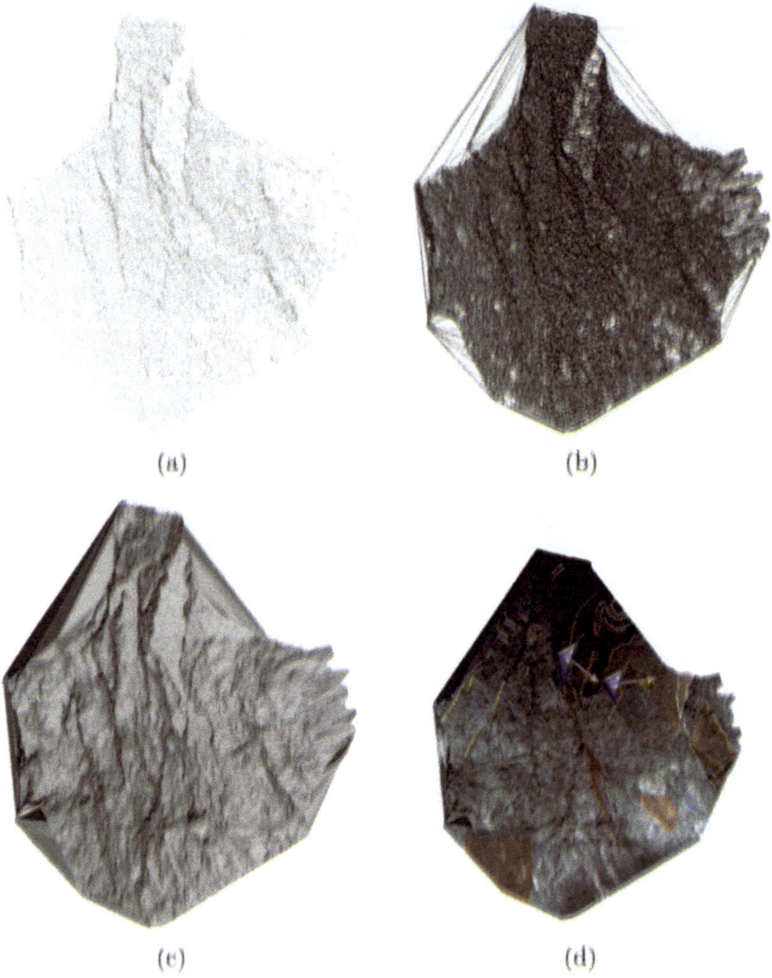
(a) (b)
(c) (d)

Fig. 18. Surface reconstruction and display of results: a) 3D point cloud, b) triangulated surface model, c) shaded surface model, d) textured surface reconstruction.

4.5 Discontinuity orientations

The results so far enable 3D point measurements either in relation to the camera co-ordinate system or to an external object co-ordinate system. The measurement of spatial orientation information can be derived from single 3D point measurements where a spatial triangle is determined by three surface points. If the surface points are chosen to lie on a discontinuity then the normal vector to this surface triangle represents the orientation of this discontinuity. This principle was already used in geotechnical analyses based on conventional photographs [20][28].

Another way is to use the stereoscopic vision system. A virtual triangle can be navigated through three-dimensionally perceived rock mass by means of a 3D input device, like a space ball. This triangle is positioned until it matches the discontinuity and the (also displayed) normal vector represents the discontinuity orientation (Fig. 18d).

5 Applications

5.1 General modelling of jointed rock mass

Stereoscopic rock mass images contain high quality information on the discontinuity systems. Therefore the elaboration of discontinuity models based on these images and data is an obvious field of application. The complexity of jointed rock mass is a field of worldwide research activities. It is still very difficult or even impossible to model discontinuity networks as encountered in nature. Therefore software for generating three-dimensional artificial discontinuity systems, based on investigation data, has been developed (Fracman [41]). Modules for modelling discontinuity networks are integrated parts in software for numerical calculations of jointed rock mass as well (UDEC [19]). Collecting and evaluating data of discontinuities from stereoscopic images provides two major advantages. First, it can improve the input data for geometrical modelling of discontinuity systems. Second, by routine application of stereoscopic imaging, e.g. during tunnel excavation, a high density of in-situ geometric data can be acquired. Based on these data the modelling of actual discontinuity networks, closer to their natural properties as well as the evaluation of the influence of discontinuities on the excavation geometry (overbreak) is possible. The acquired models can be used directly for the preparation of numerical models as well as for the verification of artificially generated discontinuity models [10]. An example of a discontinuity model, based on rock mass images, is shown in Fig. 19.

5.2 Data used for numerical modelling

An evaluation of selected discontinuity parameter (trace length, normal spacing, gap length) has been performed with discontinuity traces shown in Fig. 20b. The traces result from an interactive image mapping of a limestone cliff (Fig. 20a). The data were used for generating a UDEC model (Fig. 20c). The selected window has a size of about 17.5 m^2. Traces of bedding planes (276 samples) and traces of a joint set (289 samples) are marked. Both discontinuity sets are oriented normal to the image plane. The results of parameter evaluation for discontinuity trace length and normal spacing are displayed in Fig. 21a and Fig. 21b as well as in Tab. 3 and Tab. 4. The acquired input parameters for the automated generation of the discontinuity mesh of UDEC result in a discontinuity pattern which reflects the general layout of the two discontinuity systems satisfactorily, displaying the local concentration of parallel traces as well as the open joint pattern.

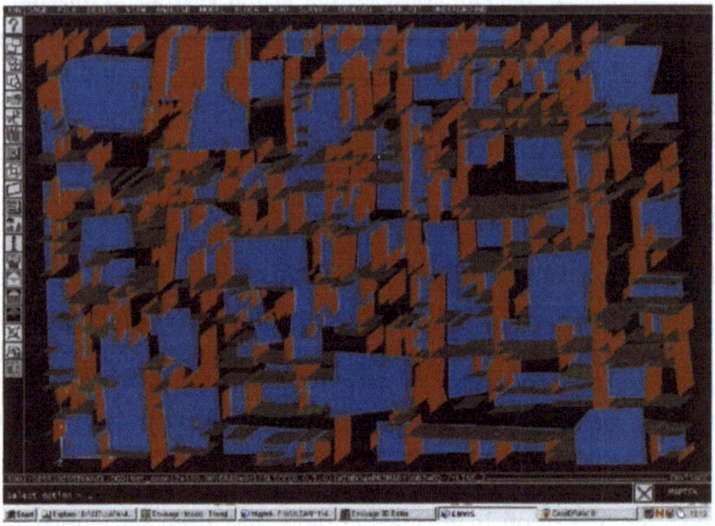

Fig. 19. Artificial discontinuity system (green: bedding planes, red and blue: joint sets), modelled on the basis of rock mass image evaluation (Fig. 18) using the software VULCAN [42].

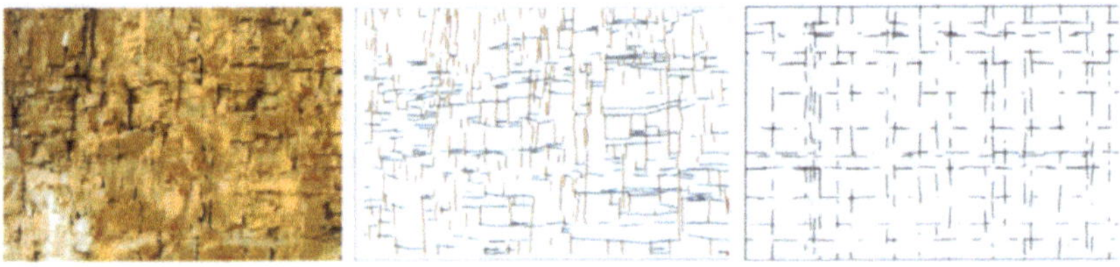

(a) (b) (c)

Fig. 20. UDEC model, generated on the basis of a detailed image analysis of discontinuity parameters (discontinuity trace length, spacing, gap length); a) original image, b) result of detailed interactive image mapping, c) UDEC mesh based on statistical evaluation of discontinuity data showing joints for block generation.

Table 3. Results of discontinuity trace length analysis

	Horizontal set	**Vertical set**
Mean value	0.21 m	0.23 m
Standard deviation	0.18 m	0.20 m
Minimum value	0.03 m	0.03 m
Maximum value	1.48 m	1.58 m
Number of samples	276	289

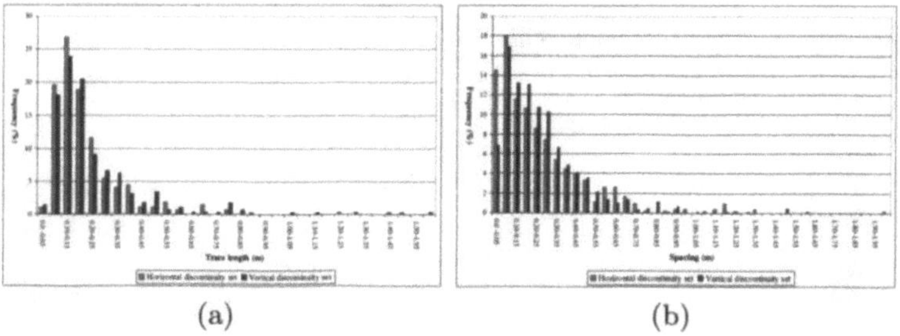

Fig. 21. a) Discontinuity trace length histograms, b) Discontinuity spacing histograms

Table 4. Results of discontinuity spacing analysis

	Horizontal set	Vertical set
Mean value	0.25 m	0.26 m
Standard deviation	0.24 m	0.22 m
Minimum value	0.01 m	0.03 m
Maximum value	1.97 m	1.59 m
Number of samples	538	614
Number of sampling lines	50	35
Total length of sampling lines	134 m	157 m

5.3 Other test sites

The imaging system was tested in several environments, among others a rock wall in a limestone quarry was recorded and thoroughly evaluated. Fig. 22 shows one image with the structural map overlaid. The test showed the importance of colour as lithological boundaries can be identified clearly.

Another test site were the air - raid protection tunnels in the Grazer Schlossberg (Fig. 23) where capacious testing in a subsurface environment was possible and a subsurface quarry in South-Styria (Fig. 24) that allow testing of different lighting sources and conditions.

6 Discussion and outlook

For rock mass modelling high-resolution stereoscopic images as well as data derived from their evaluation are enormous improvements as the accuracy of evaluation as well as of interpretation increases. The recoverable objective documentation of the majority of essential structural rock mass parameters allows for a flexible use of the data. In dependency from the question formulation and urgency the exactness of evaluation can be selected, always with the option to go back to the initial data. Rock mass modelling in combination with image information increases the content of information to the

Fig. 22. Structural map evaluated by an experienced geologist. The image was taken at a limestone quarry north of Graz, Austria, the resolution is about 4 mm/pixel, the image size about 4000 x 3000 pixel.

viewer as major features of a jointed rock mass are easy to recognise visually. The transportation of information between different parties involved is improved as the professional filter, implied by the mapping person, can be moved to the background when required. Imaging provides the possibility to trace structures by successive documentation of rock faces, as excavation works proceeds, independent if this concerns tunnels or slopes. With the accuracy, contained in scaled and oriented images, it provides information for a better understanding of the observed rock mass portion. Therefore the quality of interpretation as well as extrapolation can be increased. Critical evaluation of numerical models can only be performed if a realistic data source is available. Parameter modifications are justifiable only when the raw data are of high quality and when they can serve as starting point for simplifications required by the modelling software as well as for checking the plausibility of results. In order to have at least a better control on geometrical parameters, data acquisition based on stereoscopic imaging may reduce the total amount of uncertainties, that still exist with repect to material properties, material laws, failure modes, and to by the concepts of numerical models on their own. As demonstrated the possibility of a detailed evaluation of discontinuity parameters improves the accuracy of statistically generated geometrical models for numerical simulations. The general subjective impression of rock mass quality, supplemented by a few measurements cannot guarantee the correctness of input data for any further evaluation. Therefore the provision

Fig. 23. Imaging inside the Grazer Schlossberg.

Fig. 24. Subsurface quarry in the south of Styria.

of detailed data should be possible as a routine on a daily basis, and should not be restricted to exceptional situations or projects.

6.1 Next steps

The prototype from the computer vision system, i.e. the imaging system and software components showed up two things: (i) this kind of data acquisition, documentation and measurement highly improves the actual data acquisition process by its better data quality (ii) the effort getting these data is reduced since the process of acquisition and analysis is separated.

The prototype is to be redesigned and brought into a condition to be suitable for the daily use on site. For this purpose a commercial camera system was tested that has come onto market in 2000. The camera follows the same principle as those built for the prototype system, but it is faster and lighter which make a car carrying the system no longer a requirement. The imaging system therefore is more flexible and can be incorporated into the working cycle on site easier. Fig. 25 shows an image from a site test taken at the Unterwald tunnel which is part of the railway track Amstetten-Tarvis (Many thanks to ÖBB, PAT, ÖSTU Stettin, Geodata).

A single camera is used that resides on top of tripod, where also a laptop is mounted. No external power cabling is needed. The overall weight of the system is less that 10 kg which makes it easy to be carried by single person. Image data acquisition (2 sequential images) took 10 minutes in front of the tunnel face which includes the information for orienting the camera with respect to the tunnel.

Results of the image evaluation are shown in Fig. 26 and Fig. 27. Figure 26 shows the main structures, lithological units, and rock mass conditions. The area based assessment of rock mass conditions highly supports the geologist. Figure 27 shows a three-diemensional surface reconstruction with the main structures overlaid.

In order to support the regular decision making process on site another step is required: an easy to use tool that allows the combination of subsequent surface reconstructions. With it appropriate structures can be connected and brought into a model (see Fig. 28), which improves the actual rock mass description.

7 Final remarks

The project was performed at the Institutes for Rock Mechanics and Tunnelling and for Computer Graphics and Vision both from the University of Technology in Graz. This interdisciplinary co-operation showed to be a synergetic and productive environment for the development of new concepts in the objective field of improving current geological survey.

Fig. 25. Single panoramic camera system tested at the Unterwalder tunnel site. The resulting image of the tunnel face is about 6000 x 5000 pixels, revealing a resolution of about 2 mm/pixel.

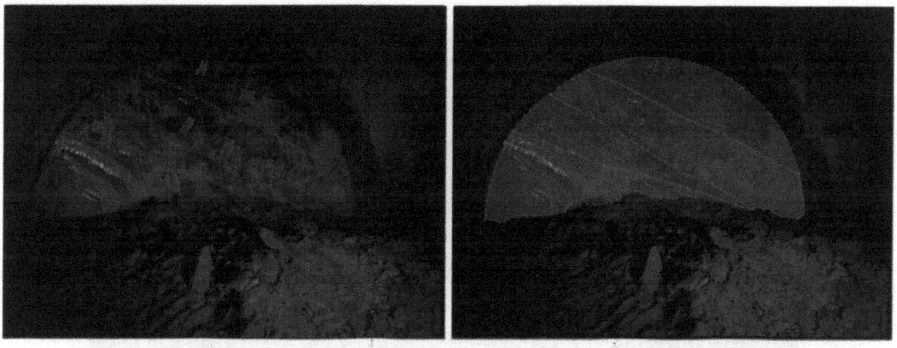

(a) (b)

Fig. 26. Structural analysis at station 738. Main structures are identified (a) and segmented into areas (b).

The project led to three Master's theses and two PhD theses, several contributions and presentations at conferences held in different countries, e.g. in Austria, Germany, Switzerland, Finland, Australia which gave positive echoes to the results of the project.

Fig. 27. Two views of the reconstructed tunnel face at station 738 with structural map overlaid.

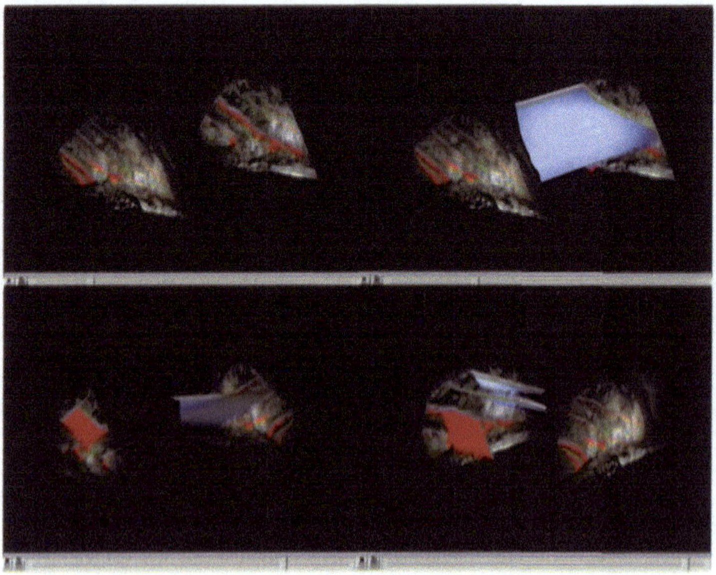

Fig. 28. Four images from a movie that shows the interconnection between two subsequent tunnel face evaluations. The images were taken at the Unterwald tunnel site. Thanks to R. Fuchs for the model and the animation.

References

[1] Barton N., Lien R., Lunde, J. (1983) Engineering classification of rock mass for the design of tunnel support. NGI Publication 106, *Rock Mech*, **6**, No. 4: 189-236

[2] Benosman R., Kang S. B. (eds.) (2001) *Panoramic Vision: Sensors, Theory and Applications*. Springer Verlag, Heidelberg

[3] Bienniawski Z. T. (1974) Engineering classification of jointed rock masses. *Trans. S. Afr. Inst. Civ. Engrs.*, **15**: 335-344

[4] Brown E. T. (ed.) (1981) *Rock characterisation, testing and monitoring*. ISRM Suggested Methods. Pergamon Press

[5] Case J. B. (1967) The analytical reduction of panoramic and strip photography, *Photogrammetria*: 127-141

[6] Cravero M. and Iabichino G. (1993) Characterization of joint patterns from trace maps. In: *Eurock '93*, Ribeiro e Sousa and Grossmann (eds.), Balkema

[7] Crosta G. (1997) Evaluating Rock Mass Geometry From Photogrammetric Images. *Rock Mechanics and Rock Engineering*, **30**(1): 35-38

[8] Daller J., Riedmüller G., Schubert W. (1994) Zur Problematik der Gebirgsklassifizierung im Tunnelbau. *Felsbau – Rock Soil Engng.* **6/94**: 443-447

[9] Delaunay B. (1934) *Sur la sphére vide*. Izvestia Akademia Nauk SSSR, VII Seria, Otdelenie Matematicheskii i Estestvennyka Nauk, 7: 793-800

[10] Dershowitz W., Lee, G. Geier, J. Foxford T., LaPointe P., Thomas A. (1996) *Fracman – Interactive discrete feature data analysis, geometric modelling, and exploration simulation*. User Documentation, Version 2.5, Golder Associates Inc.

[11] Fasching A. (2001) *Improvement of Acquisition Methods for Geotechnical Data*, Phd thesis, Riedmüller, G., Schubert, W. and Semprich, S. (eds.) Gruppe Geotechnik Graz, Schriftenreihe, Heft 12b, Technische Universität, Graz

[12] Foley J.D., van Dam A., Feiner S.K., Hughes J.F. (1991) *Computer graphics, principles and practice*, 2nd ed. Reading. MA: Addison-Wesley

[13] Franklin J.A., Maerz N.H. and Bennet C.P. (1988) Rock mass charactersation using photoanalysis. *Int. Journal of Mining and Geological Engineering* 6: 97-112

[14] Gaich, A., Sehovic A., Gruber M. (1998) System zur Erzeugung hochauflösender Stereobilder für die Nahbereichs-Objektrekonstruktion, *Vermessung und Geoinformation*, No. 4/98: 194-201

[15] Gaich A. (2001) *Panoramic Vision for Geotechnical Analyses in Tunnelling*, Phd thesis, Riedmüller, G., Schubert, W. and Semprich, S. (eds.) Gruppe Geotechnik Graz, Schriftenreihe, Heft 12a, Technische Universität, Graz

[16] Goodman R.E. (1989) *Introduction to rock mechanics*. 2nd Edition. John Wiley

[17] Grimstad E., Barton N. (1993) Updating of the Q-System for NMT. *Proc. Int. Symp. on Sprayed Shotcrete*, Fagernes

[18] Hagan T.O. (1980) A Case for Terrestrial Photogrammetry in Deep-Mine Rock Structure Studies. *Int. Journal of Rock Mechanics and Mining Sciences* 17: 191-198

[19] ITASCA Consulting Group. UDEC 3.00 user's manual, Minnesota, USA

[20] Linkwitz K. (1963) Terrestrisch-photogrammetrische Kluftmessung. *Rock Mechanics and Engineering Geology* I:152-159

[21] Maresch M. (1998) *Linear CCD array based recording of buildings for digital models*, Dissertation, OCG Schriftenreihe Vol. 109. Wien: Österreichische Computer Gesellschaft

[22] Mauldon M. (1998) Estimating mean fracture trace length and density from observations in convex windows. *Rock Mech. Rock Engng.* Vol. 31: 201-216

[23] Priest S. D. (1993) *Discontinuity Analysis for Rock Engineering*, London: Chapman and Hall

[24] Price N. J., Cosgrove J. W. (1990) *Analysis of geological structures.* Cambridge: University Press

[25] Raab T. (1994) . *Über Tendenzen der Fehleinschätzungen einiger ingenieurgeologischer Parameter.* Diploma thesis. Graz: University of Technology

[26] Raghavan V., Masumoto S., Koike K. and Nagano S. (1995) Automatic lineament extraction from digital images using a segment tracing and rotation transformation approach. *Computers and Geosciences*, Vol. 21, No. 4, 555-591

[27] Reid T.R., Harrison J.P. (2000) A semi automated methodology for discontinuity trace detection in digital images of rock mass exposures. *Int. Journal of Rock Mechanics and Mining Sciences* 37: 1073-1089

[28] Rengers N. (1967) Terrestrial Photogrammetry: A Valuable Tool for Engineering Geological Purposes. *Rock Mechanics and Engineering Geology V*: 150-154

[29] Riedmüller G., Schubert, W. (1999) Critical comments on quantitative rock mass classifications. *Felsbau - Rock Soil Engng.* 3/99: 164-167

[30] Schubert P. (1992) Die Ungewißheit bei der Standsicherheitsanalyse von Felsbauwerken. *Felsbau* Jhg. 10, Nr. 4: 191-195

[31] Schubert W., Steindorfer A. (1997) Application of new Methods of Monitoring Data Analysis for short term Prediction in Tunnelling, *Proc. of the World Tunnel Congress 97 – Tunnels for People*, Rotterdam: Balkema: 65-70

[32] Sellner P. (2000) *Prediction of displacements in tunnelling.* Phd thesis. Riedmüller, G., Schubert, W. and Semprich, S. (eds.) Gruppe Geotechnik Graz, Schriftenreihe, Heft 9. Technische Universität, Graz

[33] Slama Ch. C. (ed.) (1980) *Manual of Photogrammetry*, Fourth Edition, Falls Church, VA: American Society of Photogrammetry

[34] Sonka M., Hlavac V., Boyle, R. (1998) *Image Processing, Analysis, and Machine Vision*, 2nd ed., Pacific Grove, CA: Brooks/Cole Publishing

[35] Steindorfer A. (1998) *Short Term Prediction of Rock Mass Behaviour in Tunnelling by Advanced Analysis of Displacement Monitoring Data.* Phd thesis, Riedmüller, G., Schubert, W. and Semprich, S. (eds.) Gruppe Geotechnik Graz, Schriftenreihe, Heft 1, Technische Universität, Graz

[36] Tsoutrelis C.E., Exadactylos G.E., Kapenis A.P. (1990) Study of the rock mass discontinuity system using photoanalysis. In Rossmanith (ed.), *Proc. int. conf. on Mechanics of Jointed and Faulted Rock.* Rotterdam: Balkema

[37] Vieten W. (1970) Die Ermittlung tektonischer Gefügedaten aus stereophotogrammetrischen Bruchwandaufnahmen. *Clausthaler Tektonische Hefte.* Heft 10, 319-336

[38] Wittke W. (1984) *Felsmechanik – Grundlagen für wirtschaftliches Bauen im Fels.* Springer Verlag

[39] Condor Eartch Technologies Inc., http://www.condorearth.com.

[40] DIBIT tunnel scanner. http://dib.joanneum.ac.at/bv_home.html.

[41] Golder Associates Inc. 2000. http://fracman.golder.com/

[42] MAPTEK Inc.: http://www.maptek.com/vulcan/vulcan.htm

Chapter 4
Advanced pre-processing methods

Thomas Reichl, Bernhard Mathis†, and Gernot Beer

Institute for Structural Analysis, Graz University of Technology

† Bernhard Mathis tragically passed away in an avalanche accident in 2001

Abstract. A user friendly computer program is presented for the automatic generation of finite and boundary element meshes. The program is designed to be used by professionals that are relatively inexperienced in the use of numerical simulation software.

1 Introduction

Mesh generation is one of the most time consuming tasks in the numerical simulation in tunnelling. However, most of the steps in the generation of meshes from information about the tunnel shape, the construction sequence/method and the geological information are fairly methodical. There is no need to require the analyst to perform these steps as they can be made by the computer.

In numerical simulation there are basically two main approaches: one based on volume, the other on boundary discretisation. In the boundary element method (BEM) for example it is sufficient to only describe the boundary of the problem with elements, whereas in the finite element method (FEM) the volume of the soil/rock mass has to be described. In numerical simulation meshes have a dual purpose, they are used to describe the geometry of the problem and used for the numerical analysis. For example in the BEM isoparametric linear or parabolic elements describe the shape of the excavation as well as the variation of surface tractions and displacement. Inside the rock mass no approximation is required since exact solutions of the governing equations are used. In the FEM elements are used also to describe the variation of displacements inside the ground.

In a numerical simulation for tunnelling not only the mesh has to be specified but also the construction/excavation sequence. This information is then used to identify elements that have to be removed or added in order to simulate the excavation of material and the installation of support.

2 General approach

The approach selected here considers the special requirements for tunnelling and therefore differs form other general purpose pre-processors. The generation of meshes proceeds in the following steps:

(1) Input of the tunnel axes. If more than one tunnel axis is input then an intersection is automatically generated. Axes may be curved.
(2) Input of tunnel cross-section. The cross section of a tunnel may be read from a file in AUTOCAD DXF format or may be input by specifying centres of circles with radius and start/end angle. Cross-sections are fixed at a specific point on the tunnel axis.
(3) Input of excavation/construction sequence. An excavation/construction sequence is specified either via the mouse or a file.
(4) Generation of a 2-D BEM mesh. This mesh merely consists of linear or quadratic curves that describe the tunnel shape. This mesh can be used for the elastic stress analysis, assuming plane strain conditions.
(5) Generation of a 2-D FEM mesh. A square box is placed around the tunnel and a mesh is generated that fills the rock/soil mass. This would allow the non-linear analysis of tunnel excavation.
(6) Extrusion to a 3-D BEM mesh. A 3-D mesh which consists of linear or quadratic surface elements is generated by extruding the 2-D mesh along the tunnel axis. If two tunnels intersect then an intersection curve is generated. This mesh allows a three-dimensional elastic analysis to be carried out.
(7) Generation of a 3-D FEM mesh. A box is placed around the tunnel and a volume mesh is generated in such a way that element boundaries coincide with the boundaries of the tunnel. This mesh allows the 3-D nonlinear analysis of a tunnel.

3 Input of geometry

The pre-processor allows to input the tunnel geometry by hand or by reading from a DXF-File, which can be produced by most CAD programs. The tunnel geometry consists of two basic elements - axes and cross sections.

3.1 Input of axes

Figure 1 shows the dialog for the input of axes. It allows to either define a new axis or copy/delete existing axes. An axis consists of elements which may be straight lines or arcs. The small drawing on the left side of the dialog shows all elements that have been currently defined for the axis. The list of elements on the right describes the properties of the currently defined elements. Selecting one of the elements in this list will colour the element in the sketch in red. Adding new elements can be achieved with the dialog shown in Fig. 2. There are three possibilities for the definition, one for a line and two for arcs. Axes are defined as a polyline, so the end point of the previous element is the start point of the next element. The dialog automatically fills in the end point of the previous element as the starting point for the next element.

It is also possible to import the information about axes from a DXF file.

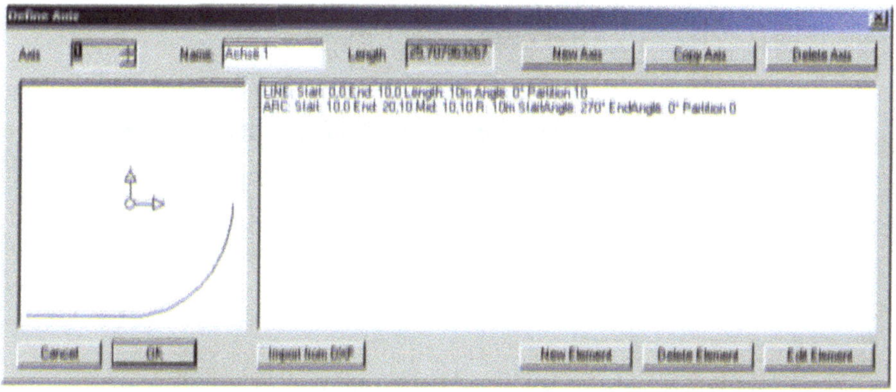

Fig. 1. Dialog for input of axes

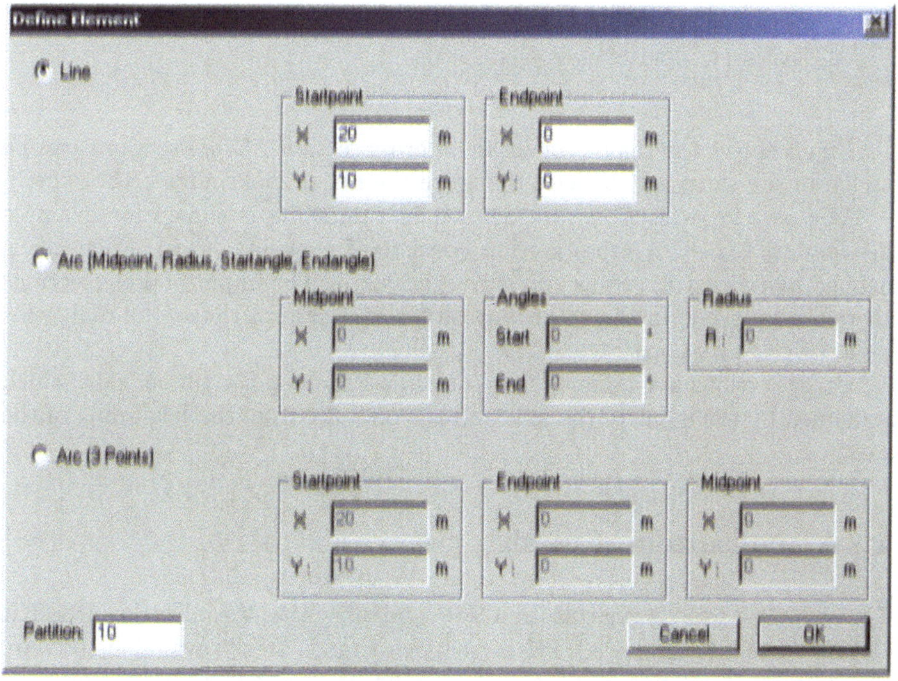

Fig. 2. Dialog for input of elements

3.2 Input of cross sections

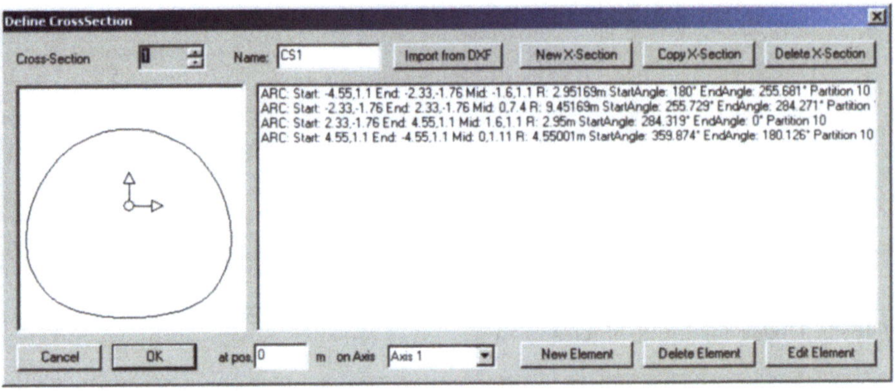

Fig. 3. Dialog for input of cross sections

Fig. 3 shows the dialog for input of cross sections. Cross sections can be added in the same way as axes therefore the dialog looks almost the same.

The way cross sections are defined in tunnelling is shown for example in Chap. 6 Fig. 3. A cross section consists of elements (arcs). Each arc is defined by its centre, radius and start -end angle. The imput of cross sections therefore proceeds in the same way as for axes. Fig. 3 shows the dialog for input of cross sections.

Cross sections are assigned to a certain position on the tunnel axis, which is defined by the name of the axis and the distance from the beginning of the axis.

4 Visualisation of mesh

The pre-processor is written in C++ and uses the Visualization ToolKit (VTK) from Kitware [1]. VTK is an open source, freely available software system for 3D computer graphics, image processing, and visualization used by thousands of researchers and developers around the world. VTK uses OpenGL and provides the possibility of rotating, zooming and selection of elements. It allows also to use many visualisation techniques which are

- light sources,
- perspective view,
- realistic textures.

Figure 4 shows an example of the visualisation using a light source.

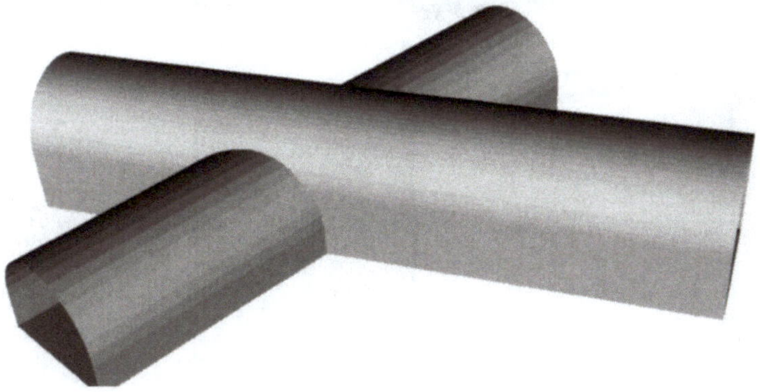

Fig. 4. Shading of tunnelsurface using light source

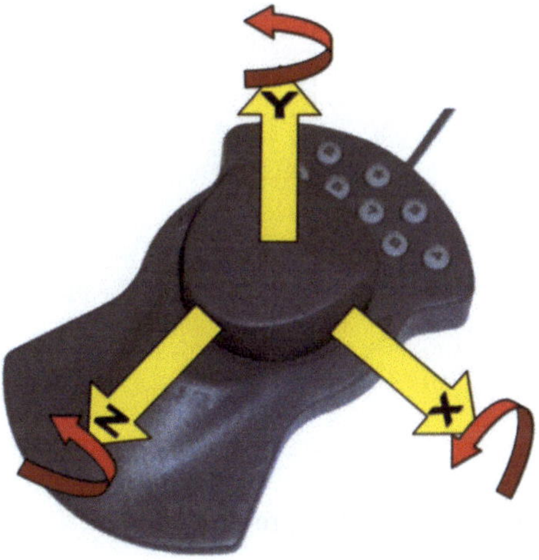

Fig. 5. Space mouse

4.1 Navigation

Most of the users will use the mouse for user interaction. Navigation with a mouse is implemented in VTK and works quite well. Special navigation tools such as a space mouse shown in Fig. 5 must be implemented manually by deriving from the VTK Interactor class. A space mouse supports 3 movement directions and 3 rotation directions with which the user can manipulate the camera.

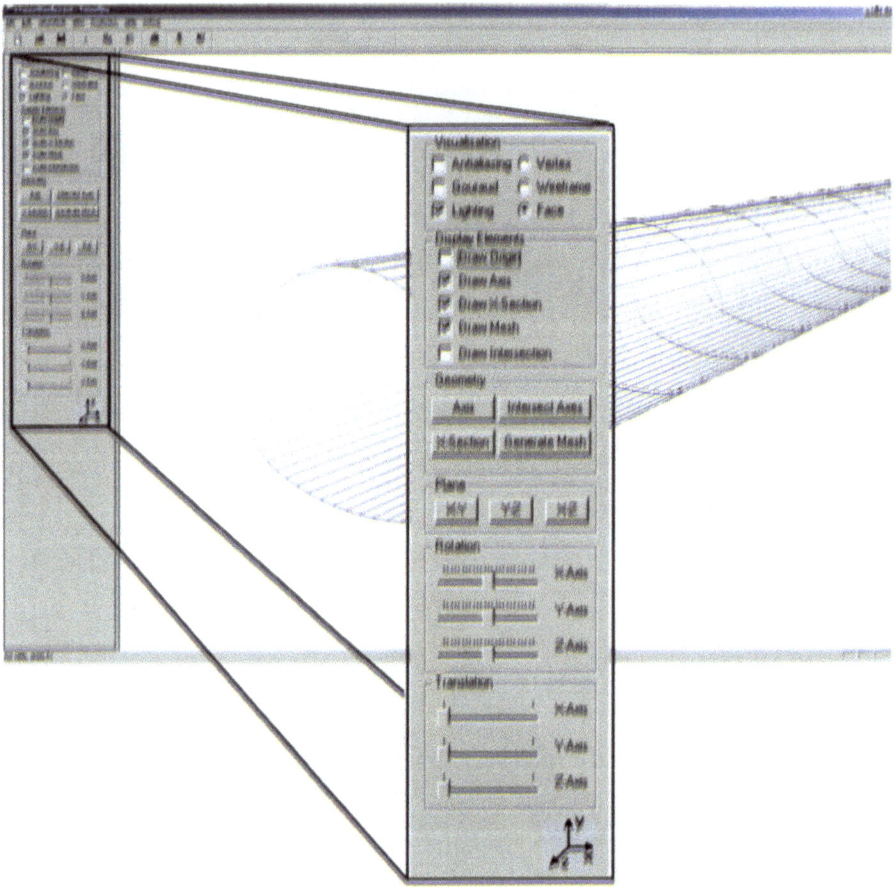

Fig. 6. Graphical user interface of the pre-processor

4.2 Graphical user interface

Figure 6 shows the graphical user interface of the developed pre-processor. The main window shows all geometric elements and on the left side all important user interactions are gathered.

4.3 Selection of elements

VTK supports selecting of elements in the mesh. In this way any element may be selected for example for the definition of material properties or for selecting elements to be excavated.

5 Mesh generation

As outlined above the generation of meshes proceeds in two steps. First a boundary element mesh is generated and then a finite element mesh. It should be mentioned at this point that as the capabilities of the method to deal with sequential excavation and nonlinear material behaviour [2] is extended the use of the BEM will increase in the future. There are distinct advantages in terms of ease of mesh generation and accuracy of the results [3].

5.1 Generation of BEM mesh

Two-dimensional BEM mesh. The generation of a mesh suitable for a 2-D BEM analysis is fairly straightforward and consists of subdividing the cross section into straight line or curved (second order curve) boundary elements. The number and type of elements can be determined by the user in the input dialog of elements shown in Fig. 2 but some default parameters can be used. This mesh forms the basis for the 3-D boundary element mesh.

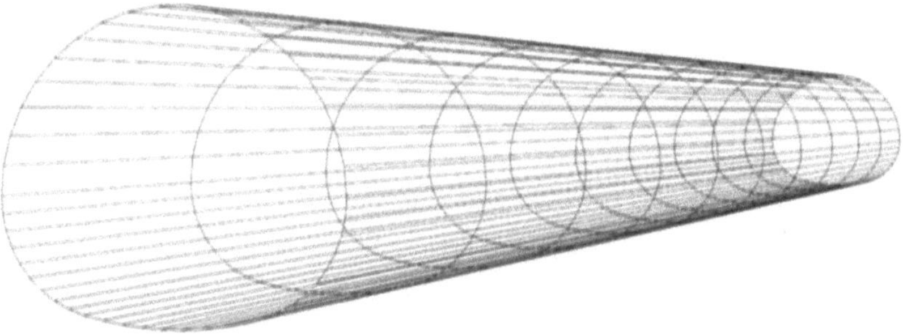

Fig. 7. Tunnel of Undergroundstation "Taborstraße"

Three-dimensional BEM mesh. The 3-D BEM mesh is generated by extruding the 2-D BEM mesh along the tunnel axis as shown in Fig. 7. The subdivision along the axis can be defined by the user and will be determined by the size of each excavation step. For each element of axes and cross sections it is possible to define the number of partitions. The partition fixes the number of slices for this element, so it is quite easy to vary the resolution of the mesh. The definition for the tunnel shown in Fig. 7 is done only by the definition of a line for the axis and the definition of the cross section which was imported from DXF-File created with AutoCAD 2000.

Unless infinite boundary elements are used the 3-D BEM requires the definition of a two-dimensional mesh which is identical to a 2-D FEM mesh except that 3 coordinates (x, y, z) are involved in the definition of nodes. These meshes are either required at the end of the tunnel or when the cross

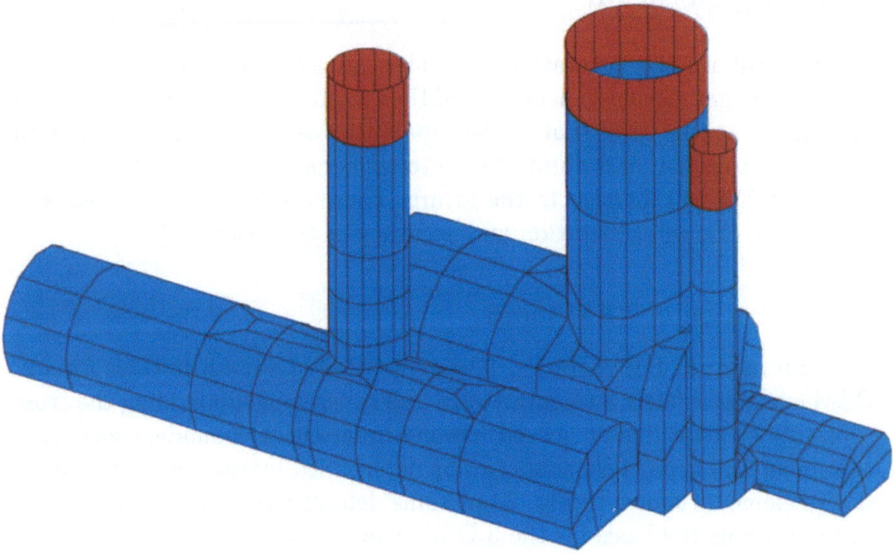

Fig. 8. Three-dimensional boundary element mesh of cavern CERN

section of the tunnel changes for example when modelling underground stations or caverns (see for example Fig. 8). The generation of this mesh is discussed in Sect. 5.2.

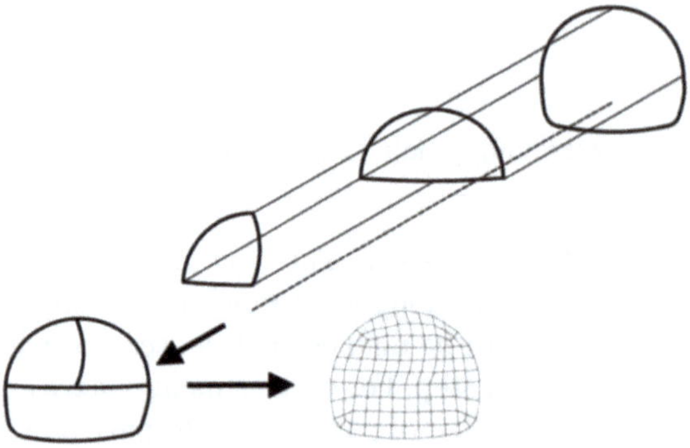

Fig. 9. Copying slices to master slice, mesh generation

Sequential excavation with the BEM. To model sequential excavation with the BEM we have to specify sliced regions. The main idea to get a 3D mesh is to divide the axes of the tunnel into slices. Each slice has an individual geometry or geology (Fig. 9). Slices are copied into one master slice in which all geometrical and geological information is stored. The master slice will be meshed in 2 dimensions with the new decomposition and paving algorithm (Fig. 9).

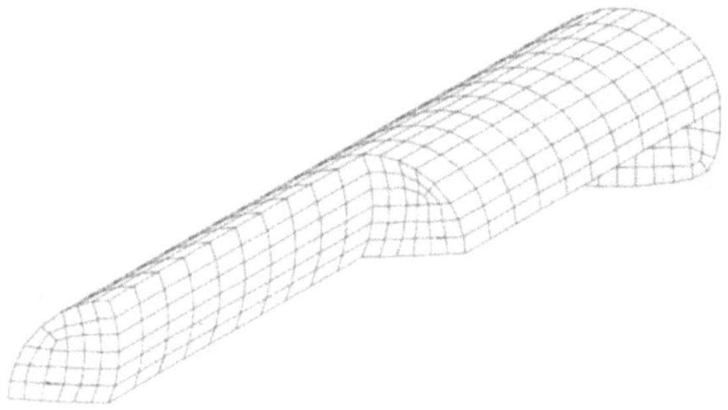

Fig. 10. 3D-boundary mesh of the tunnel

The last step is the generation of the sequential excavation regions by a 3D extrapolation of the master slice along the axis shown in Fig. 10.

Fig. 11. Intersection of axes

Intersection of axis. If there are two or more tunnel axes an intersection curve has to be calculated. As this curve is in general not linear or parabolic, a polygon is used to describe the intersection curve. The points of this line are calculated as intersections of horizontal lines in the direction of the tunnel axes. From each subdivision point of the cross section a line parallel to the axis is intersected with the line parallel to the second axis. A picture of an automatically generated intersection is shown in Fig. 11.

5.2 2D mesh generator

Here the generation of two-dimensional finite element meshes or three-dimensional boundary element meshes is discussed. The mesh generator in 2D is an paving algorithm for quadrilateral meshes developed in [4]. The three steps in meshing a geometry are

- topology decomposition,
- geometry decomposition,
- meshing simple regions.

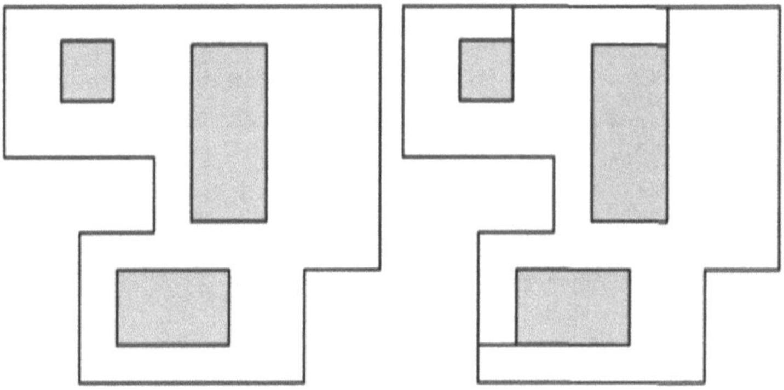

Fig. 12. Topology decomposition

Topology decomposition. If an geometry consists of more then one closed polygon and these polygons are not connected with each other then a number of free polygons exist. To mesh this geometry in a efficient way the polygons have to be connected using the topology decomposition as explained in Fig. 12.

Geometry decomposition. After the geometry has reduced to a topology without free polygons the domain has to be decomposed into a number of simple regions. Simple regions are regions such as triangles and quadrangles as shown in Fig. 14. These types of regions are easy to fill with quadrilaterals.

Meshing of simple regions. When all the regions of the geometry are of type simple, then the meshing process can be started. Due the meshing of the simple regions a few corrections have to be performed to increase the quality of the mesh. This corrections are smoothing and seaming functions. The smoothing process is split in two different algorithms. The first one is the boundary node smoother which is an modified isoparametric smoother

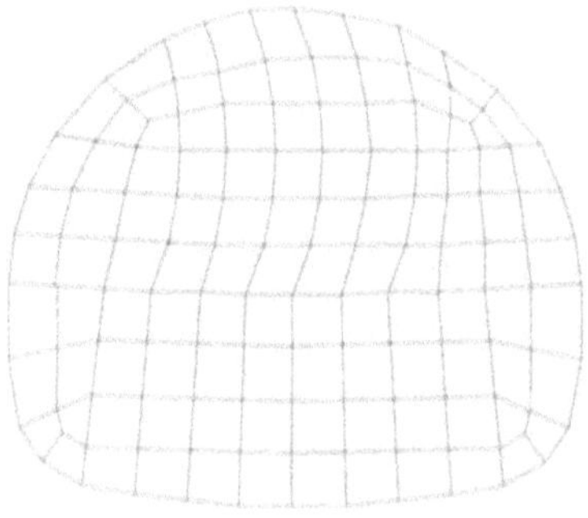

Fig. 13. Example of a two-dimensional mesh

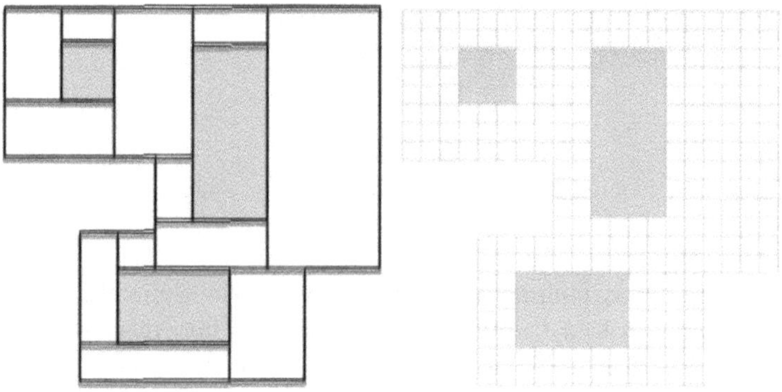

Fig. 14. Boundary decomposition, mesh

[6]. The interior node smoother is an length weighted Laplacian smoother [5]. To increase the calculation speed every mesh smoothing or mesh seaming step is checked if it is required. That means that the program tries to get the best mesh which has the lowest calculation time. An example of a generated mesh is shown in Fig. 13. In the end of the meshing algorithm the mesh of each region has to be connected to each other (Fig. 14).

5.3 Generation of a three-dimensional finite element mesh

For a three-dimensional finite element analysis of tunnel excavation a box is placed around the tunnel and filled with solid elements. The extent of the box is chosen in such a way that the influence of artificial boundary conditions that are applied at the edges of the box is minimised. The basic idea of the mesh generation method is to start with a box divided into a regular mesh of elements as shown in Fig. 15 and then to modify the mesh so that element boundaries coincide with the excavation boundaries. The latter can be achieved by either moving element nodes or by subdividing the elements as shown in Fig. 16.

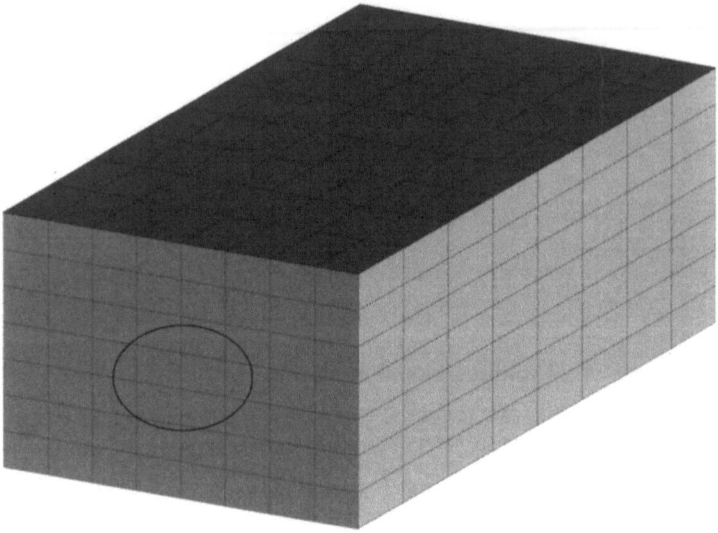

Fig. 15. Three-dimensional finite element mesh: box of solid elements

The decision on if an element node is moved or if the element is subdivided is made depending on how the boundary element surface intersects the solid finite element. The aim is to achieve a mesh with elements which are not highly distorted. A check is made after the generation to ensure that highly distorted elements are avoided. The advantage of this mesh generation scheme is that it is fairly general with respect to the excavation geometries that can be handled and with respect to transparency of the mesh generation since one starts with a boundary element mesh.

6 Summary

A new (fully automatic) mesh generation program has been developed especially for tunnelling applications. The advantage of such special purpose generators is that they can be made completely automatic, with the requirement of minimum user intervention. The user need only specify the input that is absolutely necessary with the computer taking over the tedious task of generating a mesh. The method proposed here for 3-D finite element mesh generation where on proceeds in stages by first generating a boundary element mesh and then a finite element mesh gives the mesh generation more transparency since one can see the tunnel geometry before it is hidden by solid finite elements. It is hoped that with this mesh generator the numerical simulation of tunnel construction with finite and boundary elements will become a routine application.

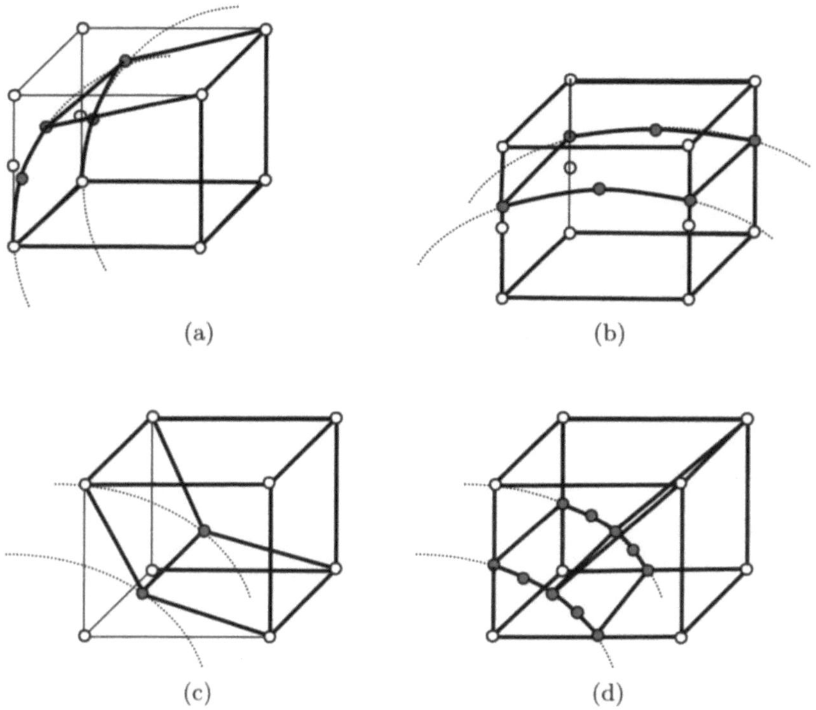

Fig. 16. Examples of changing the regular mesh so that element boundaries coincide with excavation boundaries (shown here by dotted lines) Moved or newly generated nodes are depicted by full circles. In (a) and (c) element nodes are moved and in (b) and (d) elements are subdivided.

References

[1] Kitware, Inc. 469 Clifton Corporate Parkway Clifton Park, New York 12065 USA http://public.kitware.com/VTK/

[2] Dünser C. (2001) Simulation of sequential tunnel excavation with the boundary element method. *Doctoral Thesis*, Institute for Structural Analysis, Graz University of Technology

[3] Beer, G. (2001) Programming the Boundary Element Method. Wiley, UK

[4] Blaker T.D., Stephenson M.B. (1991) Paving: a new approach to automated quadrilateral mesh generation. *International Journal for Numerical Methods in Engineering*, **32**: 811-847

[5] Herrmann L.R. (1976) Laplacian-isoparametric grid generation scheme. *Journal of the Engineering Mechanics Division ASCE*, **102**, EM 5: 749-756

[6] Jones R.E. (1979) QMESH: A self-organizing mesh generation program. SLA-73-1088, Sandia National Laboratories

Chapter 5
Thermochemomechanical material model for shotcrete

Christian Hellmich, Roman Lackner, and Herbert Mang

Institute for Strength of Materials, Vienna University of Technology

Abstract. Employing a thermodynamic framework, thermochemomechanical couplings for shotcrete are treated in this chapter. A material model based on multi-surface thermochemoplasticity is presented. It accounts for hydration kinetics, chemomechanical couplings related to strength growth, stiffness properties, and to autogeneous shrinkage in early-age shotcrete. Creep is modeled by means of two mechanisms: stress-induced water movement in the capillary pores of shotcrete, and a relaxation mechanism in the nanopores of the cement gel. The underlying material functions are intrinsic, i.e., independent of field and boundary conditions. They are determined from standard material tests. As for the numerical treatment of the constitutive equations of the material model, an extended form of the return map algorithm is presented. Microcracking is considered by means of a Drucker-Prager failure surface for the compressive load regime and by means of Rankine surfaces for tensile brittle failure.

1 Introduction

The continuous success of the New Austrian Tunneling Method (NATM) since the 1950s has increased the demand for an appropriate characterization of the material properties of shotcrete. It is needed to meet the challenge of estimating the forces acting in the shotcrete tunnel shell. A central concern in material modeling of shotcrete is the evolution of its mechanical properties which vary both in space and time. Different material models, based, e.g., on elasto-viscous formulations originating from mining science [25], or on rate flow theories [27], and ranging to enhanced plasticity-based models [30][23], have concentrated on the evolution *in time* by introducing the material stiffness and strength as prescribed functions of time. However, the hydration of shotcrete resulting in material stiffness and strength, interacts with the temperature distribution in the structure. The time-dependent evolutions of mechanical properties may depend significantly on the current field and boundary conditions in the laboratory and at the tunnel site, respectively. Depending on the deviations of these conditions in the field from the ones in the laboratory, the results obtained from analyses based on such models, calibrated by means of data from the laboratory, may deviate significantly from the real situation at the tunnel site.

A remedy was proposed by Ulm and Coussy [36] for early-age concrete in the framework of thermodynamics of reactive porous media [9]. Their concept was successfully applied to shotcrete [15]: In thermodynamics of continua, the considered solid is assumed to be composed of elementary systems. Their magnitude allows the material to be treated as homogeneous. The relevant scale for shotcrete is about 5 to 10 cm. This is because the size of the elementary inhomogeneity of the material is considerably smaller than this scale (10 nm to 10 mm), whereas the structural dimensions are significantly larger than this scale (0.3 m to 30 m). As is common in continuum mechanics, the considered solid is thought to be composed of infinitesimal representative volume elements dV. In the sense of a macroscopic theory, the properties of the elementary system are assigned to the volume elements. The phenomena in the microstructure of the material are represented by means of state variables and energetically conjugated thermodynamic forces, both attached to the elementary system. Cross-effects between phenomena on the microlevel and their macroscopic consequences can be determined from standard laboratory tests (at the macroscopic scale of engineering material description). Employing this strategy, the *intrinsic* material behavior, defined as *independent of field and boundary conditions*, can be determined. In this framework, material strength, e.g., is not a function of time, but is related to its origin on the microlevel, the formation of calcium silicate hydrates. In the same framework, material functions relating autogeneous (chemical) shrinkage strains, elastic stiffness, and creep compliance to the degree of hydration have been derived [11][28].

Use of relatively complex material laws for large-scale applications requires a proper numerical treatment of the underlying constitutive equations. As for the so-called thermochemical heat evolution problem in early-age concrete-shotcrete, reference is made to [14]. General aspects of single or multi-surface chemoplasticity have been described in [37] and [15], respectively. The algorithmic basis for the implementation of creep laws is documented in [28] and [22]. All of the algorithmically relevant references contain extensions or derivates of classical return map algorithms for elastoplasticity developed by Simo and co-workers in the 1980's and early 1990's (see, e.g., [29]).

This chapter concentrates on the model formulation, experimental remarks, and algorithmic aspects. Applications to tunnel engineering problems are addressed in Chap. 6 and Chap. 10.

2 Material model for shotcrete

2.1 State variables

Chemical and physical phenomena on the microlevel of the material are described, in a macroscopic context, by state variables. In the material model according to [11][15][28], two external (macroscopically measurable) variables ($\boldsymbol{\varepsilon}$ and T) and six internal variables (ξ, $\boldsymbol{\varepsilon}^p$, $\boldsymbol{\chi}$, $\boldsymbol{\varepsilon}^v$, $\boldsymbol{\varepsilon}^f$, and γ) are used:

- $\boldsymbol{\varepsilon}$ denotes the strain tensor.
- T stands for the absolute temperature.
- ξ is the degree of hydration, describing the state of this phenomena. Hydration is the chemical reaction between cement and free water. The reaction products are termed hydrates. ξ is the ratio between the current specific mass of water bound in hydrates, m, and the respective mass at complete hydration, m_∞.
- $\boldsymbol{\varepsilon}^p$ and $\boldsymbol{\chi}$ denote the tensor of plastic strains and the vector of hardening variables, respectively. $\boldsymbol{\varepsilon}^p$ represents deformations and $\boldsymbol{\chi}$ microstructural changes because of microcracking.
- According to [26][38][33], the reason for short-term creep[1] is stress-induced diffusion of water associated to the capillary pores of concrete. The size of these pores is in the range of micrometers. The resulting viscous strains are denoted by $\boldsymbol{\varepsilon}^v$.
- According to Wittmann [38], long-term or flow creep follows from dislocation-like processes in the nanopores of cement gel. Thus, the observation scale of this phenomena is 1000 times smaller than the one concerning short-term creep. The corresponding macroscopic flow strains are denoted by $\boldsymbol{\varepsilon}^f$. The internal variable γ, called viscous slip [33], represents microstructural changes resulting from the dislocation-like phenomena.

2.2 Field equations

Field equations are containing spatial derivatives. The physical space is described by means of two field equations. The first one is the equilibrium equation,

$$\mathrm{div}\boldsymbol{\sigma} + \mathbf{k} = \mathbf{0}, \tag{1}$$

with $\mathbf{k}$ denoting the vector of the volume-force density and $\boldsymbol{\sigma}$ standing for the (macroscopic) stress tensor. The second field equation is the first law of thermodynamics. In the absence of volume heat sources and accounting only for relevant terms [35], this law is given as

$$C\dot{T} - \ell_\xi \dot{\xi} = -\mathrm{div}\mathbf{q}, \tag{2}$$

with C as the heat capacity per unit volume, ℓ_ξ as the latent heat of hydration per unit volume, and $\mathbf{q}$ as the heat flow vector (positive for an efflux). Equation (2) states that the change of internal energy, $C\dot{T} - \ell_\xi\dot{\xi}$, equals the external heat supply, $-\mathrm{div}\mathbf{q}$.

---

[1] Two regimes of creep kinetics suggest the distinction of two types of creep of concrete [34][2]. They reflect two distinct phenomena in the material. In the following, these phenomena will be referred to as short-term creep and long-term creep. These terms are introduced with respect to the different creep kinetics and do not *a priori* imply the existence of characteristic times for these phenomena. More specifically, the second phenomenon does not exhibit an asymptotic behavior [4][33].

2.3 Heat conduction law

The heat flux $\mathbf{q}$ is related to T via Fourier's linear (isotropic) heat conduction law,

$$\mathbf{q} = -k \, \text{grad} T, \tag{3}$$

with k as the thermal conductivity.

2.4 Constitutive equations

State equations. State equations are relations between thermodynamic forces and state variables. The state equation for the stresses is formulated in infinitesimal (incremental) form, reading [28]

$$d\boldsymbol{\sigma} = \mathbf{C}(\xi) : [d\boldsymbol{\varepsilon} - d\boldsymbol{\varepsilon}^p - \mathbf{1}d\varepsilon^s(\xi) - \mathbf{1}\alpha_T dT - d\boldsymbol{\varepsilon}^v - d\boldsymbol{\varepsilon}^f], \tag{4}$$

with $\mathbf{C}(\xi)$ as the elasticity tensor for the case of material isotropy, depending on Young's modulus $E(\xi)$ and on a (constant) Poisson's ratio ν, and $\mathbf{1}\varepsilon^s(\xi)$ as the strains caused by autogeneous (chemical) shrinkage. These strains result from volume reduction of the formed hydration products with respect to the reactants water and clinker (Le Chatelier contraction), leading to tension in the liquid phase (see, e.g., [2]). $\mathbf{1}$ is the second-order volumetric unity tensor. α_T is the coefficient of thermal dilation, which is assumed to be constant.

Evolution equations. The set of governing equations is completed by the evolution equations for the internal state variables.

As for $\boldsymbol{\varepsilon}^p$ and $\boldsymbol{\chi}$, the concept of multi-surface chemoplasticity is applied [15][36]: Admissible stress states are defined by N loading functions f_α,

$$\boldsymbol{\sigma} \in C_E \quad \Leftrightarrow \quad f_\alpha = f_\alpha(\boldsymbol{\sigma}, \boldsymbol{\zeta}(\boldsymbol{\chi}, \xi)) \leq 0 \quad \forall \quad \alpha \in [1, 2, ...N] \, . \tag{5}$$

It is seen that the hardening force $\boldsymbol{\zeta}$, which is associated with the size of the yield surface in the stress space, is not only dependent on $\boldsymbol{\chi}$ but also on ξ (chemoplastic hardening). Assuming, for the sake of simplicity, associative hardening plasticity, the flow and the hardening rules read [18]

$$d\boldsymbol{\varepsilon}^p = \sum_{\alpha \in J_{act}} d\lambda_\alpha \frac{\partial f_\alpha}{\partial \boldsymbol{\sigma}}, \quad d\boldsymbol{\chi} = \sum_{\alpha \in J_{act}} d\lambda_\alpha \frac{\partial f_\alpha}{\partial \boldsymbol{\zeta}} \, , \tag{6}$$

where $d\lambda_\alpha$ are the plastic multipliers; J_{act} denotes the set of n active yield surfaces, $n \leq N$. It is defined by $J_{act} := \{\alpha \in 1, 2, ...N | f_\alpha(\boldsymbol{\sigma}, \boldsymbol{\zeta}) = 0\}$. All yield surfaces obey the loading/unloading conditions [17],

$$f_\alpha \leq 0, \quad d\lambda_\alpha \geq 0, \quad f_\alpha d\lambda_\alpha = 0 \, , \tag{7}$$

along with the consistency requirement

$$d\lambda_\alpha df_\alpha = 0 \, . \tag{8}$$

As for ξ, an evolution law of the Arrhenius type reflects the thermally activated nature of the chemical reaction [3][36],

$$\dot{\xi} = \tilde{A}(\xi) \exp\left(-\frac{E_a}{RT}\right), \tag{9}$$

where E_a is the activation energy and R is the universal constant for ideal gas. $\tilde{A}(\xi)$ stands for the chemical affinity which is the driving force of the chemical reaction. Consequently, it describes the kinetics (i.e., the reaction rate) of the hydration process. The assumption of $\tilde{A}$ depending only on ξ expresses a de-coupling hypothesis [36]; the possible additional dependence of $\tilde{A}$ on T will be discussed in Sect. 3.

According to [4], the flow strains obey

$$\dot{\boldsymbol{\varepsilon}}^f = \frac{1}{\eta_f} \mathbf{G} : \boldsymbol{\sigma} , \tag{10}$$

with η_f standing for the viscosity governing the kinetics of the aforementioned dislocation-like processes. It can be written as

$$\frac{1}{\eta_f} = cS \exp\left[-\frac{2U}{\bar{k}}\left(\frac{1}{T} - \frac{1}{\bar{T}}\right)\right] , \tag{11}$$

where $U/\bar{k} \approx 2700$ K reflects the thermally activated nature of long-term creep [3]. $\bar{T}$ is a reference temperature, $c = 1$ MPa^{-2}s^{-1} is a dimensional constant. The rate of the so-called micro-prestress force S [4] is linked to the rate of viscous slip γ via [28]

$$\dot{S} = -H\dot{\gamma} , \tag{12}$$

with H denoting a constant material parameter [28]. The evolution law for the viscous slip γ is given as [33]

$$\dot{\gamma} = cS^2 \exp\left[-\frac{U}{\bar{k}}\left(\frac{1}{T} - \frac{1}{\bar{T}}\right)\right] . \tag{13}$$

Like aging in the context of hydrating concrete, short-term creep is related to the accumulation of hydrates which are initially free of microstress. For the description of this process, an infinitesimal (incremental) formulation for short-term creep [22][12] is used. The evolution equation for the viscous strains $\boldsymbol{\varepsilon}^v$ contains the integral over stress increments $d\boldsymbol{\sigma}(t')$ which occurred until the current time instant t,

$$\dot{\boldsymbol{\varepsilon}}^v(t) = \frac{1}{\tau_w[\xi(t)]} \left[\int_{t'=0}^{t} J^v_\infty[\xi(t')] \mathbf{G} : d\boldsymbol{\sigma}(t') - \boldsymbol{\varepsilon}^v(t) \right] . \tag{14}$$

The expression for $\dot{\boldsymbol{\varepsilon}}^v(t)$ contains a Stieltjes integral, wherein $d\boldsymbol{\sigma}(t')$ is an infinitesimal increment of stress. The Stieltjes integral, however, is also defined for stress jumps, i.e., for finite values of stress increments. In Equation (14), $\mathbf{G} = E(\xi)\mathbf{C}(\xi)^{-1}$ stands for the normalized compliance tensor. τ_w denotes the characteristic time of the short-term creep process; J^v_∞ represents the final viscous compliance of short-term creep. Both τ_w and J^v_∞ depend on the degree of hydration ξ. According to (14) $\dot{\boldsymbol{\varepsilon}}^v$ is controlled by the difference between the Stieltjes integral containing the stress history (driving force of short-term creep) and the current state of the viscous strains $\boldsymbol{\varepsilon}^v$.

3 Experimental determination of material properties

Altogether six intrinsic material functions between ξ and mechanical or chemical properties are required for the described material model. These properties are:

- the normalized chemical affinity $\tilde{A}$,
- the compressive strength f_c,
- Young's modulus E,
- the chemical shrinkage strain ε^s,
- the characteristic time for short-term creep, τ_w, and
- the final viscous compliance J^v_∞.

In addition, the material constant H for the description of long-term creep needs to be evaluated. As an example, the determination of the intrinsic material function $\tilde{A}(\xi)$ will be described in the following. As for the determination of other material functions, the reader is referred to [36][15][28] (see Fig. 1).

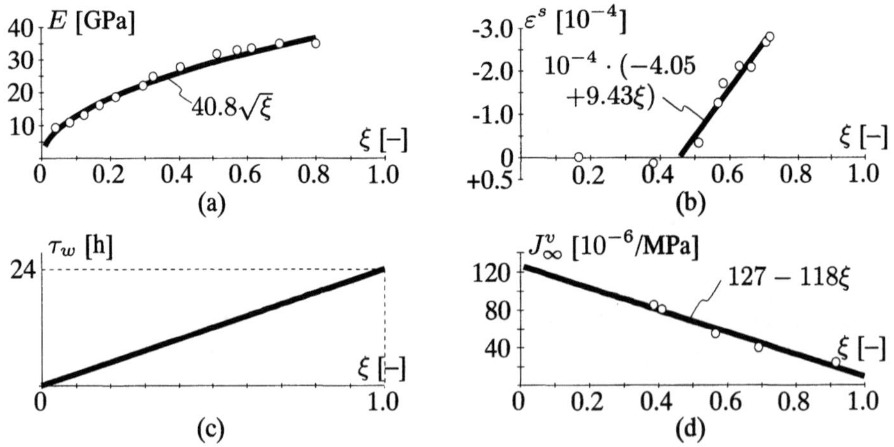

Fig. 1. Typical intrinsic material functions for shotcrete: (a) aging elasticity [13][11], (b) chemical shrinkage [15][16], (c) characteristic time for short-term creep [28], and (d) final viscous compliance [22]

3.1 Chemical kinetics – chemical affinity

The intrinsic character of $\tilde{A}(\xi)$ for early-age concrete was shown in [36]: from two independent tests (with different intervening physical properties and boundary conditions), namely, (i) an adiabatic test and (ii) a strength evolution test, the same material function $\tilde{A}(\xi)$ was derived.

In the following, $\tilde{A}(\xi)$ will be determined twice, employing results from two strength evolution tests performed at different temperatures (see [31] for experimental details). On the macroscopic level of a concrete sample, the hydration kinetics can be followed indirectly through the cross-effects between the hydration and the evolution of strength. The evolution of the compressive strength of shotcrete $f_c(t)$ is standardly determined according to the Austrian guideline for shotcrete [10], see Fig. 2b for the experiments described in [31]. As for the kind of chemoplastic coupling, reflected by $f_c(\xi)$, experimental

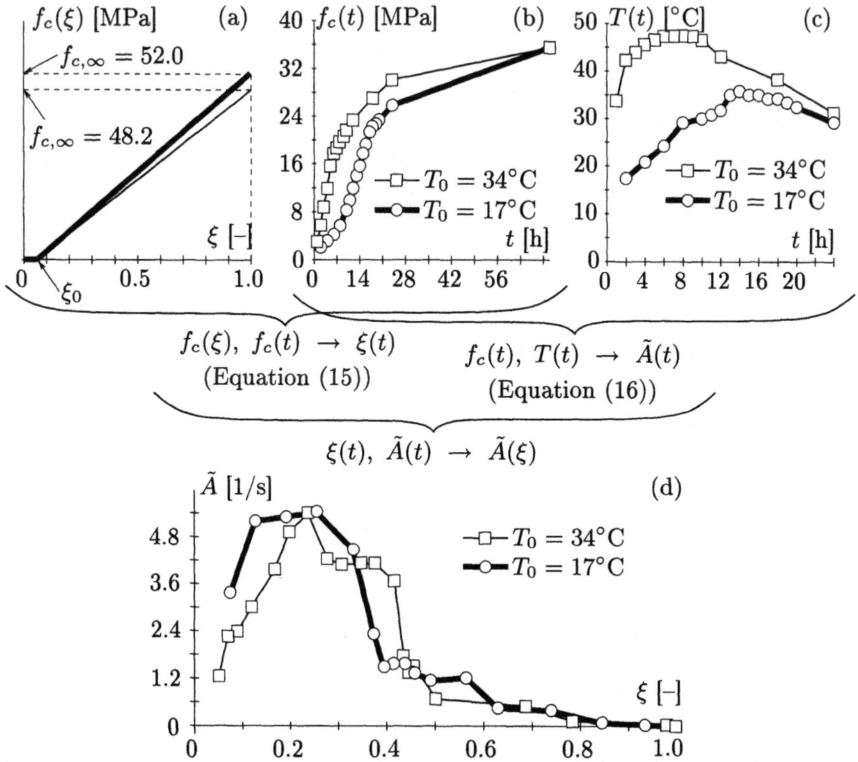

Fig. 2. Determination of chemical affinity from strength evolution tests documented in [31]: (a) intrinsic material function for the uniaxial compressive strength (linear chemical hardening); evolution of (b) compressive strength and (c) temperature according to [31] ($T_0 = 34°$: $f_{c,\infty} = 48.2$ MPa; $T_0 = 17°$: $f_{c,\infty} = 52.0$ MPa); (d) intrinsic material function for chemical affinity (ξ_0: percolation threshold; T_0: initial temperature)

results suggest a linear chemical hardening, as illustrated in Fig. 2a. In this figure, ξ_0 denotes the percolation threshold [1] beyond which the material can support a stress deviator. From this threshold on, the material strength is increasing linearly with the hydration degree. Combining $f_c(t)$ (Fig. 2b) and $f_c(\xi)$ (Fig. 2a), yields the evolution of the hydration degree, described by $\xi(t)$:

$$\xi(t) = \xi_0 + (1 - \xi_0)\frac{f_c(t)}{f_{c,\infty}}, \tag{15}$$

where $f_{c,\infty}$ is the strength at $\xi = 1$. Substituting (15) into (9), yields a parameter form of the kinetics function $\tilde{A}$ with t as the parameter:

$$\tilde{A}(t) = (1 - \xi_0)\frac{df_c(t)/dt}{f_{c,\infty}} \exp\left(\frac{E_a}{RT(t)}\right), \tag{16}$$

where $E_a/R \cong 4000$ K [6] and T is the (absolute) shotcrete temperature, see Fig. 2c for its evolution in time. Combining Equations (15) and (16), gives direct experimental access to the kinetics function $\tilde{A}(\xi)$ for shotcrete, as shown in Fig. 2d.

The similarity of the two functions $\tilde{A}(\xi)$ derived from the two strength evolution tests indicates that $\tilde{A}$ is independent of the temperature and that the thermal activation term $\exp(-E_a/RT)$ in (9) accounts properly for the effect of temperature on the reaction rate.

3.2 Microcracking – multi-surface plasticity

Microcracking under multi-axial stress states is considered in the framework of multi-surface plasticity [29]. The Drucker-Prager criterion is chosen to describe the behavior of shotcrete. This criterion is well suited for biaxial and uniaxial compressive stress states as observed in shotcrete tunnel shells. The Drucker-Prager loading surface is given by

$$f_{DP} = \sqrt{J_2} + \alpha I_1 - \zeta_{DP}(\chi_{DP}, \xi)/\beta = 0\ , \tag{17}$$

where I_1 denotes the first invariant of the stress tensor, J_2 stands for the second invariant of the stress deviator, and α and β are calibration parameters. Three Rankine surfaces are employed for the simulation of the tensile behavior of shotcrete [23][19], reading

$$f_{RA} = \sigma_A - \zeta_R(\chi_R, \xi) = 0, \text{ with } A = 1, 2, 3\ , \tag{18}$$

where σ_A represents the A-th principal stress ($\sigma_1 > \sigma_2 > \sigma_3$). Figure 3 shows the loading surfaces in the $I_1/\sqrt{3}$-$\sqrt{2J_2}$- stress space.

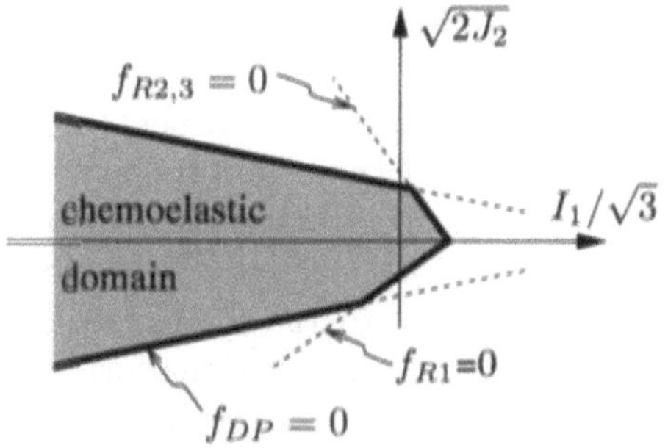

Fig. 3. Multi-surface plasticity model for shotcrete: Rankine – Drucker-Prager criterion in the $I_1/\sqrt{3}$-$\sqrt{2J_2}$- stress space

Calibration. The material parameters α and β can be determined by means of the uniaxial compressive strength f_c and the biaxial compressive strength f_b. Experimental results [6] suggest the assumption of isotropic chemical hardening. In this case, the ratio $\kappa = f_b/f_c \approx 1.16$ remains constant during hydration. Inserting the triplet of principal normal stresses $(-f_c, 0, 0)$ obtained from an uniaxial compressive test and the corresponding triplet $(-f_b, -f_b, 0)$ resulting from a biaxial compressive test into (17), α and β are obtained as

$$\alpha = \frac{\kappa - 1}{\sqrt{3}(2\kappa - 1)} \quad \text{and} \quad \beta = \frac{\sqrt{3}(2\kappa - 1)}{\kappa} \; . \tag{19}$$

Employing a quadratic plastic hardening law, ζ_{DP} can be expressed as

$$\zeta_{DP}(\chi_{DP}, \xi) = \begin{cases} \omega f_c(\xi) + (f_c(\xi) - \omega f_c(\xi)) \left[1 - \frac{(\chi_{DP} - \bar{\chi}_{DP})^2}{\bar{\chi}_{DP}^2}\right] \\ \qquad \text{...for } \chi_{DP} < \bar{\chi}_{DP}, \\ f_c(\xi) \quad \text{...for } \chi_{DP} \geq \bar{\chi}_{DP}, \end{cases} \tag{20}$$

where ω denotes the ratio of the elastic limit under uniaxial compressive loading, f_y, and the compressive strength f_c: $\omega = f_y/f_c \approx 0.25$. Similar to κ, ω remains constant during hydration. $\bar{\chi}_{DP}$ can be determined from a uniaxial compression test at complete hydration,

$$\chi_{DP,u} = \varepsilon^u_{c,\infty} - \frac{f_{c,\infty}}{E_\infty} \; , \tag{21}$$

where $\varepsilon^u_{c,\infty} \approx 0.0022$ is the total strain at peak stress. $f_{c,\infty}$ and E_∞ are the final values of the compressive strength and Young's modulus, respectively.

Concerning the Rankine criterion, perfect chemoplasticity (purely chemical hardening) is assumed, i.e.,

$$\zeta_R(\chi_R, \xi) = \zeta_R(\xi) = f_t(\xi) \; . \tag{22}$$

Thus, stress relaxation resulting from plastic softening is neglected. For the extension of the described material model to plastic softening, the reader is referred to [21,20].

4 Algorithmic formulation

The starting point for the algorithmic treatment of the material model is time integration of the state and the evolution equations in the context of the incremental-iterative solution strategy used in nonlinear FE analysis. Time integration within one time increment is performed on the basis of a backward Euler scheme leading to the discretized form of the state and evolution equations. The discretized equations form a set of equations for the unknown state variables at the end of the considered time increment. For the case of a thermochemoplastic material response, this set of equations becomes nonlinear. Hence, an iterative procedure, the so-called return map algorithm (RMA), is employed for the solution of these equations. Consistent linearisation of the RMA is a necessary prerequisite for a quadratic rate of convergence of the global Newton-Raphson iteration employed in nonlinear FE analysis.

Because of the decoupling hypothesis $\tilde{A} = \tilde{A}(\xi)$, Equations (9), (3), and (2) can be solved separately from the mechanical problem in a thermochemical FE analysis [14], delivering the the temporal evolutions and the spatial distributions of ξ and T as results (see, e.g., Chap. 6 and Chap. 10). These fields serve as input for the subsequent chemomechanical analysis, where the state variables $\boldsymbol{\varepsilon}, \boldsymbol{\varepsilon}^p, \boldsymbol{\varepsilon}^v, \boldsymbol{\varepsilon}^f, \boldsymbol{\chi}$, and γ are determined. In the following, this will be outlined in a more detailed manner.

4.1 Motivation: update of state variables

It is assumed that the state of the considered chemoplastic system at the end of the n-th time interval of an incremental-iterative FE analysis is known, i.e.,

$$\{\boldsymbol{\varepsilon}, \boldsymbol{\varepsilon}^p, \boldsymbol{\varepsilon}^v, \boldsymbol{\varepsilon}^f, \boldsymbol{\chi}, \gamma\}\big|_{t=t_n} = \{\boldsymbol{\varepsilon}_n, \boldsymbol{\varepsilon}^p_n, \boldsymbol{\varepsilon}^v_n, \boldsymbol{\varepsilon}^f_n, \boldsymbol{\chi}_n, \gamma_n\}. \tag{23}$$

Then, knowing the displacement increment in the next time step, $\Delta\mathbf{u}_{n+1}$, and, consequently, the respective increment of the total strains,

$$\Delta\boldsymbol{\varepsilon}_{n+1} = \nabla^s(\Delta\mathbf{u}_{n+1}), \tag{24}$$

the state of the system at $t = t_{n+1}$, $\{\boldsymbol{\varepsilon}_{n+1}, \boldsymbol{\varepsilon}^p_{n+1}, \boldsymbol{\varepsilon}^v_{n+1}, \boldsymbol{\varepsilon}^f_{n+1}, \boldsymbol{\chi}_{n+1}, \gamma_{n+1}\}$, can be determined. For this purpose, the state equations and the evolution equations of the underlying material model have to be discretized with respect to time.

4.2 Discretization of state and evolution equations

Discretized form of state equations. The finite counterpart of the infinitesimal (incremental) formulation for the stresses (4) for the $(n+1)$-st load increment reads as

$$\Delta\boldsymbol{\sigma}_{n+1} = \mathbf{C}(\xi_{n+1}) : \left[\Delta\boldsymbol{\varepsilon}_{n+1} - \Delta\boldsymbol{\varepsilon}^{p}_{n+1} - \mathbf{1}\Delta\varepsilon^{s}_{n+1}(\xi_{n+1}) \right.$$
$$\left. -\mathbf{1}\alpha_T \Delta T_{n+1} - \Delta\boldsymbol{\varepsilon}^{v}_{n+1} - \Delta\boldsymbol{\varepsilon}^{f}_{n+1}\right] . \tag{25}$$

Evaluating the state equation for the vector of the hardening forces at $t = t_{n+1}$ yields

$$\boldsymbol{\zeta}_{n+1} = \boldsymbol{\zeta}(\boldsymbol{\chi}_{n+1}, \xi_{n+1}) . \tag{26}$$

From integration of (12), the micro-prestress force at the end of the $(n+1)$-st time increment is obtained as

$$S_{n+1} = S_0(\xi_0) - H\gamma_{n+1} , \tag{27}$$

where S_0 denotes the initial value of the micro-prestress force. It is related to the time instant, at which the hydration process is passing the percolation threshold ξ_0 , i.e., $\xi = \xi_0$. This time instant is referred to as t_0. Assuming isothermal conditions up to t_0, with $T = T_0 = T(t_0)$, Equation (12) together with (13) can be integrated analytically, yielding

$$S_0 = S(t_0) = \frac{1}{Hc\,t_0 \exp\left[-\frac{U}{R}\left(\frac{1}{T_0} - \frac{1}{T}\right)\right]} . \tag{28}$$

Discretization of plastic evolution equations. The evolution equations of plasticity (6) as well as the loading/unloading conditions (7) are integrated by means of a backward Euler scheme for the interval $[t_n, t_{n+1}]$, resulting in

$$\Delta\boldsymbol{\varepsilon}^{p}_{n+1} = \sum_{\alpha \in J_{act}} \Delta\lambda_{\alpha,n+1} \frac{\partial f_{\alpha,n+1}}{\partial \boldsymbol{\sigma}}, \Delta\boldsymbol{\chi}_{n+1} = \sum_{\alpha \in J_{act}} \Delta\lambda_{\alpha,n+1} \frac{\partial f_{\alpha,n+1}}{\partial \boldsymbol{\zeta}}, \tag{29}$$

and

$$f_{\alpha,n+1} \le 0, \quad \Delta\lambda_{\alpha,n+1} \ge 0, \quad f_{\alpha,n+1}\Delta\lambda_{\alpha,n+1} = 0 . \tag{30}$$

Discretization of the evolution equations for creep. Use of a backward Euler scheme for the discretization of the evolution equation for the viscous creep strains (14) leads to

$$\frac{\Delta\boldsymbol{\varepsilon}^{v}_{n+1}}{\Delta t_{n+1}} = \frac{1}{\tau_w(\xi_{n+1})} \left[\sum_{k=1}^{n+1} [J^{v}_{\infty}(\xi_k)\mathbf{G} : \Delta\boldsymbol{\sigma}_k] - \boldsymbol{\varepsilon}^{v}_{n+1}\right] , \tag{31}$$

where $\Delta t_{n+1} = t_{n+1} - t_n$. In the same manner, the discretized form of the evolution law for flow creep (10) is obtained as

$$\frac{\Delta\boldsymbol{\varepsilon}^{f}_{n+1}}{\Delta t_{n+1}} = \frac{1}{\eta_f(S_{n+1})} \mathbf{G} : \boldsymbol{\sigma}_{n+1} , \tag{32}$$

where $1/\eta_f$ is obtained from (11) evaluated at $t = t_{n+1}$:

$$\frac{1}{\eta_f(S_{n+1})} = cS_{n+1} \exp\left[-\frac{2U}{k}\left(\frac{1}{T_{n+1}} - \frac{1}{\bar{T}}\right)\right] . \tag{33}$$

The discretized form of the evolution law for the viscous slip (13) reads

$$\frac{\Delta\gamma_{n+1}}{\Delta t_{n+1}} = cS^2_{n+1} \exp\left[-\frac{U}{R}\left(\frac{1}{T_{n+1}} - \frac{1}{\bar{T}}\right)\right] . \tag{34}$$

Reformulation of stress-strain law (25) according to [28]. In the discretized stress-strain law (25) the plastic strain increments $\Delta\boldsymbol{\varepsilon}^p_{n+1}$ and the increments of the viscous and flow strains, $\Delta\boldsymbol{\varepsilon}^v_{n+1}$ and $\Delta\varepsilon^f_{n+1}$, respectively, depend on the unknown stress increment $\Delta\boldsymbol{\sigma}_{n+1}$. On the right-hand side of the respective law for classical elastoplasticity, only $\Delta\boldsymbol{\varepsilon}^p_{n+1}$ depends on $\Delta\boldsymbol{\sigma}_{n+1}$. To obtain the same mode of dependence for the discretized stress-strain law (25) as in classical elastoplasticity, the Equations (31) and (32) are reformulated as

$$\Delta\boldsymbol{\varepsilon}^v_{n+1} = \frac{1}{1 + \frac{\tau_w(\xi_{n+1})}{\Delta t_{n+1}}}\left[\sum_{k=1}^{n+1}[J^v_\infty(\xi_k)\mathbf{G} : \Delta\boldsymbol{\sigma}_k] - \boldsymbol{\varepsilon}^v_n\right] ,$$
$$\Delta\boldsymbol{\varepsilon}^f_{n+1} = \frac{\Delta t_{n+1}}{\eta_f(S_{n+1})}\mathbf{G} : \boldsymbol{\sigma}_{n+1} , \tag{35}$$

where $\boldsymbol{\varepsilon}^v_{n+1}$ was replaced by $\boldsymbol{\varepsilon}^v_n + \Delta\boldsymbol{\varepsilon}^v_{n+1}$. Equations (35) are inserted into (25). Accounting for

$$\boldsymbol{\sigma}_{n+1} = \boldsymbol{\sigma}_n + \Delta\boldsymbol{\sigma}_{n+1} \tag{36}$$

and

$$\sum_{k=1}^{n+1}(\bullet)_k\Delta\boldsymbol{\sigma}_k = \sum_{k=1}^{n}(\bullet)_k\Delta\boldsymbol{\sigma}_k + (\bullet)_{n+1}\Delta\boldsymbol{\sigma}_{n+1} \tag{37}$$

yields

$$\Delta\boldsymbol{\sigma}_{n+1} = \mathbf{C}(\xi_{n+1}) : \Bigg[\Delta\boldsymbol{\varepsilon}_{n+1} - \Delta\boldsymbol{\varepsilon}^p_{n+1} - \mathbf{1}\Delta\varepsilon^s_{n+1}(\xi_{n+1}) - \mathbf{1}\alpha_T\Delta T_{n+1}$$
$$- \frac{1}{1 + \frac{\tau_w(\xi_{n+1})}{\Delta t_{n+1}}}\left[\sum_{k=1}^{n}[J^v_\infty(\xi_k)\mathbf{G} : \Delta\boldsymbol{\sigma}_k] + J^v_\infty(\xi_{n+1})\mathbf{G} : \Delta\boldsymbol{\sigma}_{n+1} - \boldsymbol{\varepsilon}^v_n\right]$$
$$- \frac{\Delta t_{n+1}}{\eta_f(S_{n+1})}\mathbf{G} : [\boldsymbol{\sigma}_n + \Delta\boldsymbol{\sigma}_{n+1}]\Bigg] . \tag{38}$$

Transferring $\Delta\boldsymbol{\sigma}_{n+1}$ to the left-hand side and using $\mathbf{C} : \mathbf{G} = \mathbf{C} : \mathbf{C}^{-1}E = \mathbf{I}E$ finally gives [28]:

$$\Delta\boldsymbol{\sigma}_{n+1} = \frac{1}{K_v}\mathbf{C}(\xi_{n+1}) : \Bigg[\Delta\boldsymbol{\varepsilon}_{n+1} - \Delta\boldsymbol{\varepsilon}^p_{n+1} - \mathbf{1}\Delta\varepsilon^s_{n+1}(\xi_{n+1}) - \mathbf{1}\alpha_T\Delta T_{n+1} - \frac{1}{1+\frac{\tau_w(\xi_{n+1})}{\Delta t_{n+1}}}\left[\sum_{k=1}^{n}[J^v_\infty(\xi_k)\mathbf{G} : \Delta\boldsymbol{\sigma}_k] - \boldsymbol{\varepsilon}^v_n\right] - \frac{\Delta t_{n+1}}{\eta_f(S_{n+1})}\mathbf{G} : \boldsymbol{\sigma}_n\Bigg] , \tag{39}$$

with the following expression for the coefficient K_v:

$$K_v = 1 + \frac{1}{1+\frac{\tau_w(\xi_{n+1})}{\Delta t_{n+1}}}E(\xi_{n+1})J^v_\infty(\xi_{n+1}) + \Delta t_{n+1}\frac{E(\xi_{n+1})}{\eta_f(S_{n+1})} . \tag{40}$$

Evaluation of microprestress force S_{n+1}. The microprestress force S_{n+1}, appearing in the definition of K_v (40) and in (39), is obtained from the known temperature at t_{n+1}, T_{n+1}, and the time increment Δt_{n+1}. Inserting (27) into the discretized form of the evolution law for viscous slip according to (34) leads to a quadratic equation for the increment of $\Delta\gamma_{n+1}$. The physically relevant solution ($\Delta\gamma_{n+1} > 0$) of this quadratic equation is given as [28]:

$$\Delta\gamma_{n+1} = \frac{1}{2}\left\{\frac{1 + 2c\Delta t_{n+1}HS_0\exp\left[-\frac{U}{R}\left(\frac{1}{T_{n+1}} - \frac{1}{T}\right)\right]}{c\Delta t_{n+1}H^2\exp\left[-\frac{U}{R}\left(\frac{1}{T_{n+1}} - \frac{1}{T}\right)\right]} - \sqrt{\Xi}\right\} - \gamma_n, \tag{41}$$

where Ξ is defined as

$$\Xi = \left[\frac{1 + 2c\Delta t_{n+1}HS_0\exp\left[-\frac{U}{R}\left(\frac{1}{T_{n+1}} - \frac{1}{T}\right)\right]}{c\Delta t_{n+1}H^2\exp\left[-\frac{U}{R}\left(\frac{1}{T_{n+1}} - \frac{1}{T}\right)\right]}\right]^2 - 4\frac{\gamma_n + c\Delta t_{n+1}S_0^2\exp\left[-\frac{U}{R}\left(\frac{1}{T_{n+1}} - \frac{1}{T}\right)\right]}{c\Delta t_{n+1}H^2\exp\left[-\frac{U}{R}\left(\frac{1}{T_{n+1}} - \frac{1}{T}\right)\right]} . \tag{42}$$

Knowing $\Delta\gamma_{n+1}$, S_{n+1} can be calculated according to (27).

4.3 Solution of discretized state and evolution equations – RMA

Consideration of creep by means of the scalar factor K_v led to a reduction of the number of equations to be considered during the solution of the discretized form of the state and evolution equations. The remaining equations for evaluation of the unknowns $\Delta\boldsymbol{\sigma}_{n+1}$ and $\boldsymbol{\zeta}_{n+1}$ (or $\Delta\boldsymbol{\varepsilon}^p_{n+1}$ and $\boldsymbol{\chi}_{n+1}$) are

- the stress-strain law (39), with $\Delta\boldsymbol{\varepsilon}^p_{n+1}$ according to $(29)_1$, and
- the state equation for the hardening forces (26), with $\Delta\boldsymbol{\chi}_{n+1}$ according to $(29)_2$,

where the solution for $\Delta\boldsymbol{\sigma}_{n+1}$ and $\boldsymbol{\zeta}_{n+1}$ (or $\Delta\boldsymbol{\varepsilon}^p_{n+1}$ and $\boldsymbol{\chi}_{n+1}$) must satisfy the conditions (30). In general, these equations are nonlinear. However, if no plastic evolution occurs in the $(n+1)$-st load increment, $\Delta\boldsymbol{\sigma}_{n+1}$ can be computed explicitly.

Chemoelastic predictor – trial state. In order to find out whether or not plastic loading will occur, the so-called trial stress state [29] is introduced. It is characterized by the assumption of zero plastic evolution in the $(n+1)$-st load increment, i.e., $\Delta\boldsymbol{\varepsilon}^p_{n+1} = \mathbf{0}$. With this assumption, the trial stress increment follows from (39) as

$$\Delta\boldsymbol{\sigma}^{trial}_{n+1} = \frac{1}{K_v}\mathbf{C}(\xi_{n+1}) : \Bigg[\Delta\boldsymbol{\varepsilon}_{n+1} - \mathbf{1}\Delta\varepsilon^s_{n+1}(\xi_{n+1}) - \mathbf{1}\alpha_T\Delta T_{n+1} - \frac{1}{1+\frac{\tau_w(\xi_{n+1})}{\Delta t_{n+1}}}\left[\sum_{k=1}^{n}[J^v_\infty(\xi_k)\mathbf{G} : \Delta\boldsymbol{\sigma}_k] - \boldsymbol{\varepsilon}^v_n\right] - \frac{\Delta t_{n+1}}{\eta_f(S_{n+1})}\mathbf{G} : \boldsymbol{\sigma}_n\Bigg] \,. \tag{43}$$

In contrast to standard plasticity, in chemoplasticity the trial state is defined not only by a trial stress increment, $\Delta\boldsymbol{\sigma}^{trial}_{n+1}$, but also by a trial chemical hardening state [15],

$$\zeta^{trial}_{n+1} \neq \zeta_n \,. \tag{44}$$

ζ^{trial}_{n+1} is computed from the degree of hydration at t_{n+1}. Hence, $\zeta^{trial}_{n+1} = \zeta(\chi_n, \xi_{n+1})$.

If $f_\alpha(\boldsymbol{\sigma}_n + \Delta\boldsymbol{\sigma}^{trial}_{n+1}, \zeta^{trial}_{n+1}) \le 0$ for $\alpha = 1, 2, \ldots N$, then the assumption of zero plastic evolution in the $(n+1)$-st load increment is correct and the stress increment is given by $\Delta\boldsymbol{\sigma}_{n+1} = \Delta\boldsymbol{\sigma}^{trial}_{n+1}$.

Plastic corrector. If, however, $f^{trial}_\alpha(\boldsymbol{\sigma}_n + \Delta\boldsymbol{\sigma}^{trial}_{n+1}, \zeta^{trial}_{n+1}) > 0$ for $\alpha \in \{1, 2, \ldots N\}$, then plastic loading occurs. In this case, a correction of the trial state is required. It is referred to as plastic correction. During the plastic corrector step, the set of the nonlinear state and evolution equations ((29), (26), and (30)) is solved iteratively. Commonly, the Newton iteration scheme is employed for this purpose. When using Rankine-type failure surfaces, the performance of the Newton iteration scheme can be considerably improved if the plastic corrector is performed in the principal stress space [19]. In contrast to the standard form of the Rankine surface (see, e.g., [8])

$$f_R(I_1, J_2, \vartheta) = 2\sqrt{3}\sqrt{J_2}\cos\vartheta + I_1 - 3f_t = 0 \,, \tag{45}$$

with I_1 as the first invariant of the stress tensor, J_2 as the second invariant of the stress deviator, and ϑ as the Lode angle, the formulation of the Rankine criterion in the principal stress space becomes trivial. It is characterized by three Rankine surfaces, given as

$$f_{R1} = \sigma_1 - f_t = 0\ , \qquad f_{R2} = \sigma_2 - f_t = 0\ , \qquad f_{R3} = \sigma_3 - f_t = 0\ . \tag{46}$$

In order to perform the plastic corrector in the principal stress space, transformation of $\boldsymbol{\sigma}$ to the principal axes is required. The spectral decomposition of the unknown stress tensor $\boldsymbol{\sigma}_{n+1}$ has the form:

$$\boldsymbol{\sigma}_{n+1} = \sum_{A=1}^{3} \sigma_{A,n+1} \mathbf{n}_A \otimes \mathbf{n}_A\ , \tag{47}$$

where $\mathbf{n}_A$ indicates the directions of the yet unknown principal axes of $\boldsymbol{\sigma}_{n+1}$. This stress state is obtained from the trial stress (Equation (43)) and the increment of plastic strains, $\Delta\boldsymbol{\varepsilon}^p_{n+1}$, reading [19]

$$\boldsymbol{\sigma}_{n+1} = \frac{1}{K_v} \mathbf{C}(\xi_{n+1}) : \left[\bar{\boldsymbol{\varepsilon}}^{trial}_{n+1} - \Delta\boldsymbol{\varepsilon}^p_{n+1}\right]\ , \tag{48}$$

where

$$\bar{\boldsymbol{\varepsilon}}^{trial}_{n+1} = K_v \mathbf{C}^{-1}(\xi_{n+1}) : \left[\boldsymbol{\sigma}_n + \Delta\boldsymbol{\sigma}^{trial}_{n+1}\right]\ . \tag{49}$$

Since both the hydration-dependent stiffness and creep (short-term and long-term) are considered by means of the scalar coefficients E and K_v, respectively, Equation (48) represents an isotropic stress-strain law. Hence, the principal axes of $\boldsymbol{\sigma}_{n+1}$ and $[\bar{\boldsymbol{\varepsilon}}^{trial}_{n+1} - \Delta\boldsymbol{\varepsilon}^p_{n+1}]$ are coinciding. Moreover, the use of isotropic yield functions f_α gives

$$\Delta\boldsymbol{\varepsilon}^p_{n+1} = \sum_{\alpha \in J_{act}} \left[\sum_{A=1}^{3} \frac{\partial f_{\alpha,n+1}(\sigma_1, \sigma_2, \sigma_3)}{\partial \sigma_A} \mathbf{n}_A \otimes \mathbf{n}_A \right]\ . \tag{50}$$

Consequently, $\Delta\boldsymbol{\varepsilon}^p_{n+1}$ has the same principal axes as $\boldsymbol{\sigma}_{n+1}$. Since the principal directions of both $[\bar{\boldsymbol{\varepsilon}}^{trial}_{n+1} - \Delta\boldsymbol{\varepsilon}^p_{n+1}]$ and $\Delta\boldsymbol{\varepsilon}^p_{n+1}$ are coinciding with the principal directions of $\boldsymbol{\sigma}_{n+1}$, $\mathbf{n}_A$, it follows that $\mathbf{n}_A$ are also the principal directions of $\bar{\boldsymbol{\varepsilon}}^{trial}_{n+1}$. Hence, based on the employed

- isotropic stress-strain law, and
- the isotropic yield functions,

the principal directions of the final stress state $\boldsymbol{\sigma}_{n+1}$, $\mathbf{n}_A$, can be determined already from the trial strains $\bar{\boldsymbol{\varepsilon}}^{trial}_{n+1}$ at the beginning of the plastic correction step:

$$\bar{\boldsymbol{\varepsilon}}^{trial}_{n+1} = \sum_{A=1}^{3} \bar{\varepsilon}^{trial}_{A,n+1} \mathbf{n}_A \otimes \mathbf{n}_A\ . \tag{51}$$

Thereafter, the trial stress state obtained from $\boldsymbol{\sigma}_{n+1}^{trial} = \boldsymbol{\sigma}_n + \Delta\boldsymbol{\sigma}_{n+1}^{trial}$ is transferred to the principal directions, giving $\sigma_{1,n+1}^{trial}$, $\sigma_{2,n+1}^{trial}$, and $\sigma_{3,n+1}^{trial}$.

The stress update within the principal-stress-space formulation reads [19]

$$\begin{Bmatrix} \sigma_{1,n+1} \\ \sigma_{2,n+1} \\ \sigma_{3,n+1} \end{Bmatrix} = \begin{Bmatrix} \sigma_{1,n+1}^{trial} \\ \sigma_{2,n+1}^{trial} \\ \sigma_{3,n+1}^{trial} \end{Bmatrix} - \frac{1}{K_v}\mathbf{c}(\xi_{n+1}) \begin{Bmatrix} \Delta\varepsilon_{1,n+1}^{p} \\ \Delta\varepsilon_{2,n+1}^{p} \\ \Delta\varepsilon_{3,n+1}^{p} \end{Bmatrix}$$
$$= \begin{Bmatrix} \sigma_{1,n+1}^{trial} \\ \sigma_{2,n+1}^{trial} \\ \sigma_{3,n+1}^{trial} \end{Bmatrix} - \frac{1}{K_v}\mathbf{c}(\xi_{n+1}) \sum_{\alpha\in J_{act}} \Delta\lambda_{\alpha,n+1} \begin{Bmatrix} \partial f_{\alpha,n+1}(\sigma_1,\sigma_2,\sigma_3)/\partial\sigma_1 \\ \partial f_{\alpha,n+1}(\sigma_1,\sigma_2,\sigma_3)/\partial\sigma_2 \\ \partial f_{\alpha,n+1}(\sigma_1,\sigma_2,\sigma_3)/\partial\sigma_3 \end{Bmatrix} \quad (52)$$

where $\mathbf{c}$ is the material matrix referred to the principal stress space. In Equation (52), J_{act} denotes the set of active yield surfaces defined as $J_{act} := \{\alpha \in 1,2,...,N | f_{\alpha,n+1} = 0\}$. In multi-surface plasticity, however, the set of active yield functions specified on the basis of the trial state, $J_{act}^{trial} := \{\alpha \in 1,2,...,N | f_{\alpha,n+1}^{trial} > 0\}$, is not necessarily equal to the correct set. Hence, an iterative procedure, characterized by a sequence of plastic corrections, is required, see [19] for details.

Evaluation of $\boldsymbol{\sigma}$ and update of state variables. After the plastic correction in the principal stress space $\boldsymbol{\sigma}_{n+1}$ is computed from the principal stresses $\sigma_{A,n+1}$, $A = 1,2,3$. The increments of the viscous strains, $\Delta\boldsymbol{\varepsilon}_{n+1}^{v}$, and the flow strains, $\Delta\boldsymbol{\varepsilon}_{n+1}^{f}$, are computed according to Equations (35). The increment of the plastic strains, $\Delta\boldsymbol{\varepsilon}_{n+1}^{p}$, is obtained from

$$\Delta\boldsymbol{\varepsilon}_{n+1}^{p} = \Delta\boldsymbol{\varepsilon}_{n+1} - \mathbf{1}\Delta\varepsilon_{n+1}^{s} - \mathbf{1}\alpha_T\Delta T_{n+1} - \Delta\boldsymbol{\varepsilon}_{n+1}^{v} - \Delta\boldsymbol{\varepsilon}_{n+1}^{f}$$
$$-\mathbf{C}^{-1}(\xi_{n+1})\Delta\boldsymbol{\sigma}_{n+1}\ , \quad (53)$$

where $\Delta\boldsymbol{\sigma}_{n+1}$ is given by $\boldsymbol{\sigma}_{n+1} - \boldsymbol{\sigma}_n$.

Linearisation of time integration scheme – consistent tangent. The consistent tangent is the basic requirement for a quadratic rate of convergence of the global Newton-Raphson iteration in nonlinear FE analysis [29]. It is defined as:

$$\mathbf{C}_{T,n+1} = \frac{d\boldsymbol{\sigma}_{n+1}}{d\boldsymbol{\varepsilon}_{n+1}}\ . \quad (54)$$

For a detailed description of the determination of the consistent tangent in the principal stress space, the reader is referred to [23]. The extension of the algorithmic procedure outlined in [23] towards consideration of chemoplasticity and creep is described in [19].

5 Numerical example

To illustrate the performance of the material model for shotcrete, a creep test is analyzed numerically. At an age of 8 h, a shotcrete cube of 1 m × 1 m × 1 m is loaded uniaxially with 10 MPa . The duration of loading is 292 h. At an age of the shotcrete cube of 300 h, the specimen is unloaded. Figure 4 shows the strain histories in the direction of loading resulting from the individual phenomena considered in the material model. In comparison to

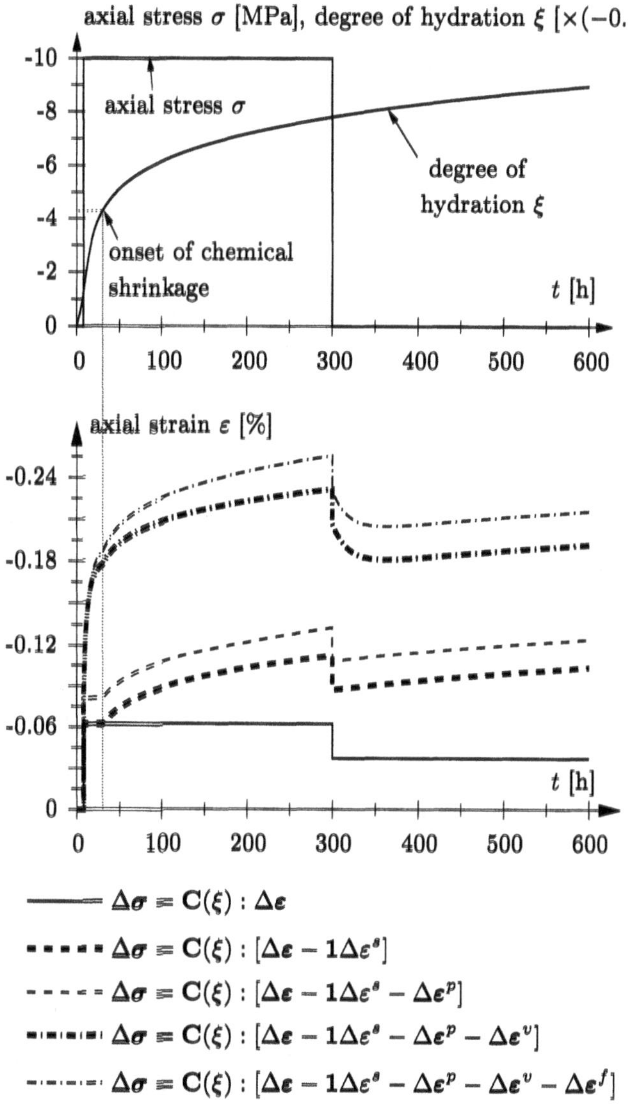

Fig. 4. Significance of thermochemomechanical material law for shotcrete: strain for uniaxial loading at different stages of constitutive modeling

pure chemoelasticity (solid line) consideration of chemical shrinkage (thick, dashed line) results in an increase of the deformation starting at a time instant of approximately 30 hours. This time instant marks the onset of chemical shrinkage characterized by $\varepsilon^s = 0$. According to the parameters given in Tab. 1, ε^s is equal to zero at $\xi = -a_s/b_s = 0.43$ (see also Fig. 1(b)). Microcracking (thin, dashed line) only occurs at the time instant of loading, at $t = 8$ h. Consideration of viscous creep (thick, dash-dotted line) results in an increase of deformation by a factor of almost two. In accordance with the final value of the characteristic time for viscous creep, $\tau_{w,\infty}$, of 24 hours (see Tab. 1), the main evolution of the viscous strain occurs during the first 50 hours after application of the load. The flow-creep strain, on the other hand, does not have an asymptotic behavior. It develops until the specimen is unloaded (thin, dash-dotted line).

Acknowledgements

The described research work has drawn great benefit from the close collaboration with Prof. Franz-Josef Ulm, Department of Civil and Environmental Engineering, Massachusetts Institute of Technology, Cambridge, USA, and the contributions of Dr. Jérôme Sercombe, Concrete Department, Commissariat á l'Energie Atomique, Saclay, France, who was involved in the SITU project in the years 1998 and 1999.

cement content	c [kg/m^3]	380
aggregate/cement-ratio	a/c [-]	4.79
water/cement-ratio	w/c [-]	0.60
compressive strength	$f_{c,\infty}$ [MPa]	39.6
$f_c(\xi) = f_{c,\infty}\frac{\xi-\xi_0}{1-\xi_0}$	ξ_0 [-]	0.01
biaxial compressive strength factor	f_b/f_c [-]	1.16
Young's modulus $E(\xi) = E_\infty\sqrt{\xi}$	E_∞ [MPa]	40800
Poisson's ratio	ν [-]	0.2
characteristic time $\tau_w(\xi) = \xi\tau_{w,\infty}$	$\tau_{w,\infty}$ [h]	24
viscous compliance $J^v_\infty = J^{v,0}_\infty(1-\xi)$	$J^{v,0}_\infty$ [1/MPa]	$127{\cdot}10^{-6}$
softening modulus for flow creep	H [MPa]	$1/7{\cdot}10^{6}$
reference temperature	$\bar{T}$ [°C]	20
chemical affinity	a_A [1/s]	7.313
$\tilde{A}(\xi) = a_A\frac{1-\exp(-b_A\xi)}{1+c_A\xi^{d_A}}$	b_A [-]	10.46
	c_A [-]	169.3
	d_A [-]	4.37
chemical dilation coefficient	a_s [-]	$-0.405{\cdot}10^{-3}$
$\varepsilon^s(\xi) = a_s + b_s\xi$	b_s [-]	$0.943{\cdot}10^{-3}$

Table 1. Mixture characteristics of shotcrete and material parameters of Lafarge "shotcrete cement"

References

[1] Acker P. (1988) Comportement mécanique du béton: apports de l'approche physico-chimique [Mechanical behaviour of concrete: a physico-chemical approach]. *Technical Report Res. Rep. LCPC 152, Laboratoires des Ponts et Chaussées*, Paris, France, in French

[2] Acker P., Ulm F.-J. (2001) Creep and shrinkage of concrete: physical origins and practical measurements. *Nuclear Engineering and Design*, **203**:148–158

[3] Bažant Z.P., (ed) (1988) Mathematical modeling of creep and shrinkage in concrete. Wiley, Chichester

[4] Z.P. Bažant, A.B. Hauggard, S. Baweja, and F.-J. Ulm (1997) Microprestress solidification theory for concrete creep, part I: Aging and drying effects. *Journal of Engineering Mechanics, ASCE*, **123**(11):1188-1194

[5] A. Boumiz, C. Vernet, and F. Cohen Tenoudij (1996) Mechanical properties of cement pastes and mortars at early age. *Advanced Cement Based Materials*, **3**:94–106

[6] J. Byfors (1980) Plain concrete at early ages. *Technical report, Swedish Cement and Concrete Research Institute*, Stockholm, Sweden

[7] P. Catharin (1978) Hydratationswärme und Festigkeitsentwicklung [Hydration heat and strength evolution]. *Technical Report 31, Forschungsinstitut des Vereins der österreichischen Zementfabrikanten*, Vienna, Austria, in German

[8] W.F. Chen (1982) Plasticity in reinforced concrete. McGraw-Hill, London, England

[9] O. Coussy (1995) Mechanics of porous continua. Wiley, Chichester

[10] Guideline (1997) Richtlinie für Spritzbeton [Guideline for shotcrete]. Österreichischer Betonverein, Vienna, Austria, in German

[11] Ch. Hellmich (1999) Shotcrete as part of the New Austrian Tunneling Method: from thermochemomechanical material modeling to structural analysis and safety assessment of tunnels. Ph.D. thesis, Vienna University of Technology, Vienna, Austria

[12] Ch. Hellmich, M. Lechner, R. Lackner, J. Macht, and H.A. Mang (2001) Creep in shotcrete tunnel shells. In S. Murakami and N. Ohno, editors, *Creep in Structures 2000 – Proceedings of the 5th IUTAM Symposium on Creep in Structures*, pages 217–229, Nagoya, Japan. Kluwer Academic Publishers, Dordrecht

[13] Ch. Hellmich, H.A. Mang, E. Schön, and R. Friedle (2003) Materialmodellierung von Spritzbeton - vom Experiment zum konstitutiven Gesetz [Material modeling of shotcrete - from the experiment to the constitutive law]. In Th. Varga, editor, *Proceedings of the conference held at the 1998 general assembly of the Austrian Society for Material Testing*, Vienna, Austria, in print, in German

[14] Ch. Hellmich, F.-J. Ulm, and H. A. Mang (1999) Consistent linearisation in finite element analysis of coupled chemo-thermal problems with exo- or endothermal reactions. *Computational Mechanics*, **24**(4):238–244

[15] Ch. Hellmich, F.-J. Ulm, and H. A. Mang (1999) Multisurface chemoplasticity I: Material model for shotcrete. *Journal of Engineering Mechanics (ASCE)*, **125**(6):692–701

[16] H.G. Huber (1991) Untersuchungen zum Verformungsverhalten von jungem Spritzbeton im Tunnelbau [Investigation of the deformation behavior of young shotcrete in tunneling]. M.Sc. thesis, University of Innsbruck, Innsbruck, Austria, in German

[17] W. Karush (1939) Minima of functions of several variables with inequalities as side constraints. M.Sc. thesis, Department of Mathematics, University of Chicago, Chicago, USA
[18] W.T. Koiter (1960) General theorems for elastic-plastic solids, volume I, Chapter IV, pages 167–218. North-Holland Publishing Company, Amsterdam
[19] R. Lackner, Ch. Hellmich, and H.A. Mang (2002) Constitutive modeling of cementitious materials in the framework of chemoplasticity. *International Journal for Numerical Methods in Engineering*, **53**(10):2357–2388
[20] R. Lackner and H.A. Mang (2003) Chemoplastic material model for the simulation of early-age cracking: from the constitutive law to numerical analyses of massive concrete structures. *Cement and Concrete Composites*, tentatively accepted for publication
[21] R. Lackner and H.A. Mang (2003) Cracking in shotcrete tunnel shells. *Engineering Fracture Mechanics*, **70**(7-8):1047–1068
[22] M. Lechner, Ch. Hellmich, and H.A. Mang (2001) Short-term creep of shotcrete - thermochemoplastic material modeling and nonlinear analysis of a laboratory test and of a NATM excavation by the finite element method. In P.A. Vermeer, S. Diebels, W. Ehlers, H.J. Herrmann, S. Luding, and E. Ramm, editors, *Continuous and discontinuous modeling of cohesive-frictional materials, Lecture Notes in Physics*, **568**:47–62, Springer, Berlin
[23] G. Meschke (1996) Consideration of aging of shotcrete in the context of a 3D viscoplastic material model. *International Journal for Numerical Methods in Engineering*, **39**:3123–3143
[24] S. Mindess, J.F. Young, and F.-J. Lawrence (1978) Creep and drying shrinkage of calcium silicate pastes. I: specimen preparation and mechanical properties. *Cement and Concrete Research*, **8**:591–600
[25] R. Rokahr and K.H. Lux (1987) Einfluß des rheologischen Verhaltens des Spritzbetons auf den Ausbauwiderstand [Influence of the rheological behavior of shotcrete on the lining resistance]. *Felsbau*, **5**:11–18, in German
[26] W. Ruetz (1966) Das Kriechen des Zementsteins im Beton und seine Beeinflussung durch gleichzeitiges Schwinden [Creep of cement in concrete as influenced by simultaneous shrinkage]. *Deutscher Ausschuß für Stahlbeton*, Heft 183, in German
[27] P. Schubert (1988) Beitrag zum rheologischen Verhalten von Spritzbeton [Contribution to the rheological behavior of shotcrete]. *Felsbau*, **6**:150–153, in German
[28] J. Sercombe, Ch. Hellmich, F.-J. Ulm, and H. A. Mang (2000) Modeling of early-age creep of shotcrete. I: model and model parameters. *Journal of Engineering Mechanics (ASCE)*, **126**(3):284–291
[29] J.C. Simo and T.J.R. Hughes (1998) Computational inelasticity. Springer, Berlin, Germany
[30] G. Swoboda, A. Moussa, and N. Hafez (1994) Two and three dimensional modeling of layered shotcrete lining. In P.K.K. Lee, L.G. Tham, and Y.K. Cheung, editors, *Proceedings of the International Conference on Computational Methods in Structural and Geotechnical Engineering*, pages 1077–1084, Hong Kong. China Translation and Printing Services, Hong Kong
[31] M. Testor (1995) Trockenspritzbeton mit neuen Bindemitteln - Temperatureinfluss, Staubeinfluss und Rückprallreduktion [Dry-mix shotcrete with new cements - influence of temperature and of dust, reduction of rebound]. M. Sc. thesis, University of Innsbruck, Innsbruck, Austria, in German

[32] P. Torrenti (1992) La résistance du béton au très jeune age [Strength of concrete at very early age]. *Bulletin liaison des Laboratoires des Ponts et Chaussées*, **179**:31–41, in French

[33] F-J. Ulm (1998) Couplages thermochémomécaniques dans les bétons : un premier bilan. [Thermochemomechanical couplings in concretes : a first review]. *Technical report, Laboratoires des Ponts et Chaussées*, Paris, France, in French

[34] F.-J. Ulm and P. Acker (1998) Le point sur le fluage et la recouvrance des bétons [Concrete creep and recovery: a review]. *Special issue of the "Bulletin des Laboratoires des Ponts et Chaussées"*, XX:73–82, in French

[35] F.-J. Ulm and O. Coussy (1995) Modeling of thermochemomechanical couplings of concrete at early ages. *Journal of Engineering Mechanics (ASCE)*, **121**(7):785–794

[36] F.-J. Ulm and O. Coussy (1996) Strength growth as chemo-plastic hardening in early age concrete. *Journal of Engineering Mechanics (ASCE)*, **122**(12):1123–1132

[37] F.-J. Ulm, O. Coussy, and Ch. Hellmich (1998) Chemoplasticity: a review of evidence. In R. de Borst, N. Bicanić, H. Mang, and G. Meschke, editors, *Computational Modeling of Concrete Structures, Proceedings of the EURO-C 1998 Conference*, pages 421–440, Bad Gastein, Austria. Balkema, Rotterdam.

[38] F.H. Wittmann (1982) Creep and shrinkage mechanisms, Chapter 6, pages 129–161. Wiley, Chichester

Chapter 6
Material model for soil and applications

Roman Lackner, Yvonne Spira, Christian Pichler, and Herbert A. Mang

Institute for Strength of Materials, Vienna University of Technology

Abstract. This chapter deals with the development of a realistic material model for the simulation of the mechanical behavior of granular soil. The material model is formulated in the framework of multi-surface viscoplasticity, involving a Mohr-Coulomb criterion for the simulation of compressive loading and a tension-cut-off for capturing the response of soil under tensile loading. Whereas the tension-cut-off is assumed to be ideally-plastic, friction harding is considered in the Mohr-Coulomb criterion. The commonly observed change in the dilational behavior of granular soil during plastic deformations is accounted for by a non-associative flow rule. After presentation of the algorithmic formulation of the proposed material model, its applicability for investigations of real-life structures is demonstrated by means of numerical analyses of two Austrian tunnels, focusing on (i) the effect of ground improvement by means of jet grouting and (ii) the safety of tunnels under fire load.

1 Multi-surface plasticity model for granular soil

During shearing of granular soil, particle movement does not occur exclusively in the direction of the shear force. At low to moderate confining pressures, particles push adjacent grains out of their way or move up and over them. Such pushing or climbing is termed interlocking. It generates an increase of the shear strength of the material and, hence, an increase of the angle of internal friction φ. Figure 1 shows the increase of the shear force resulting from the increase of shear strength by means of a schematic load-displacement curve of a simple-shear test.

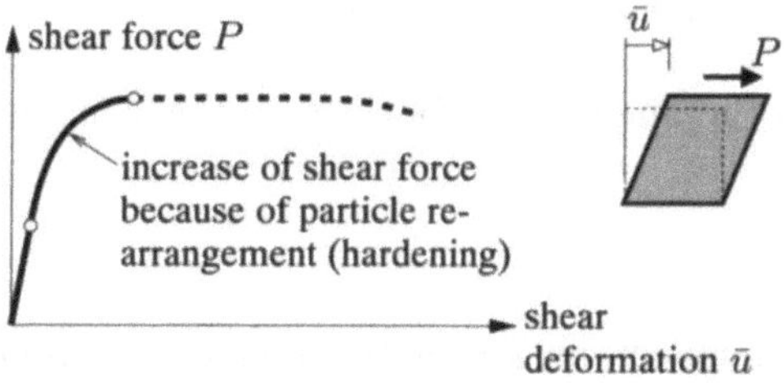

Fig. 1. Schematic plot of the evolution of the shear force obtained from a simple-shear test ($\bar{u}$: prescribed displacement)

1.1 Yield surfaces

In the proposed material model, the change of the friction angle φ is considered in the context of the Mohr-Coulomb yield criterion. This criterion is well-suited for the description of the inelastic behavior of granular materials under moderate compressive loading states. The loading function for this criterion is given by

$$f_{MC}(\boldsymbol{\sigma}, \kappa_{MC}, k_{MC}) = \frac{1}{2}(\sigma_1 - \sigma_3) + \kappa_{MC}\left[\frac{1}{2}(\sigma_1 + \sigma_3) - k_{MC}\right] , \tag{1}$$

where σ_1 and σ_3 represent the largest and the smallest principal stress. κ_{MC} and k_{MC} are material parameters. They are defined by

$$\kappa_{MC} = \sin\varphi \quad \text{and} \quad k_{MC} = \frac{c}{\tan\varphi} , \tag{2}$$

where c and φ denote the cohesion and the friction angle, respectively. For description of the material response under tensile loading, the tension-cut-off criterion is employed. It is characterized by the following loading function:

$$f_{TC}(\boldsymbol{\sigma}) = I_1 - f_t . \tag{3}$$

I_1 denotes the first invariant of the stress tensor; f_t represents the uniaxial tensile strength of the material which, in general, is negligible.

1.2 Plastic flow rule

Granular soils exhibit volume dilation when subjected to shear deformations. Use of an associative flow rule for the described Mohr-Coulomb criterion would result in an overestimation of this dilation. Consequently, a non-associative flow rule is adopted. The plastic potential g_{MC} is introduced. It is obtained from modification of the Mohr-Coloumb criterion (1) with respect to its volumetric part, reading

$$g_{MC} = \frac{1}{2}(\sigma_1 - \sigma_3) + \bar{\kappa}_{MC}\frac{1}{2}(\sigma_1 + \sigma_3) . \tag{4}$$

$\bar{\kappa}_{MC}$ is related to the angle of dilatancy ψ in the same way as κ_{MC} to φ in Eqn. (2) [2]

$$\bar{\kappa}_{MC} = \sin\psi . \tag{5}$$

For the tension-cut-off, an associative flow rule is used. I.e., $g_{TC} = f_{TC}$. Finally, the evolution equation for the plastic strain tensor is given by

$$\dot{\boldsymbol{\varepsilon}}^p = \dot{\gamma}_{MC}\frac{\partial g_{MC}}{\partial \boldsymbol{\sigma}} + \dot{\gamma}_{TC}\frac{\partial f_{TC}}{\partial \boldsymbol{\sigma}} , \tag{6}$$

where γ_{MC} and γ_{TC} are consistency parameters.

1.3 Hardening rule

Whereas ideally-plastic behavior is assumed for the tension-cut-off, isotropic hardening is considered in the context of the Mohr-Coulomb criterion. Two basic modes of isotropic hardening can be distinguished [29]: cohesion and friction hardening (Fig. 2). For modeling of cohesionless materials such as

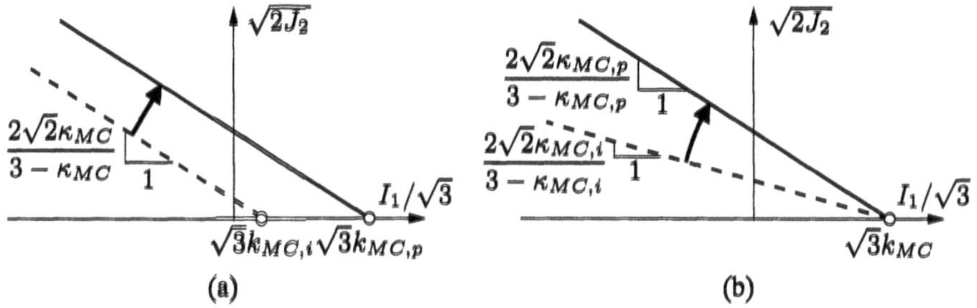

Fig. 2. Illustration of hardening in the framework of the Mohr-Coulomb criterion (compressive meridian): (a) cohesion and (b) friction hardening (subscripts i and p refer to the initial and the peak value, respectively)

granular soils, friction hardening is appropriate [19]. Accordingly, k_{MC} is set equal to

$$k_{MC} = \frac{c}{\tan \varphi_p} = \text{constant} , \tag{7}$$

where φ_p denotes the peak value of the friction angle φ. Friction hardening is controlled by the value of κ_{MC}. A quadratic increase of κ_{MC} from $\kappa_{MC,i} = \sin \varphi_i$ to $\kappa_{MC,p} = \sin \varphi_p$ is considered:

$$\kappa_{MC} = \kappa_{MC,i} + (\kappa_{MC,p} - \kappa_{MC,i}) \left[1 - \frac{(\chi_{MC} - \bar{\chi}_{MC})^2}{\bar{\chi}_{MC}^2}\right] \tag{8}$$

for $\chi_{MC} < \bar{\chi}_{MC}$, where $\bar{\chi}_{MC}$ is a calibration parameter. For $\chi_{MC} \geq \bar{\chi}_{MC}$, $\kappa_{MC} = \kappa_{MC,p} =$ constant. The evolution of κ_{MC} is controlled by means of the strain-like internal variable χ_{MC}. Since friction hardening is connected with deviatoric deformations, χ_{MC} is related to the deviatoric part of the plastic strain tensor, $\mathbf{e}^p$ [29,21,23]:

$$\dot{\chi}_{MC} = \sqrt{\frac{2}{3}\dot{\mathbf{e}}^{p,T}\dot{\mathbf{e}}^p} = \sqrt{\frac{2}{3}}\,\|\dot{\mathbf{e}}^p\| \,. \tag{9}$$

Consideration of $\dot{\mathbf{e}}^p = \dot{\gamma}_{MC}\partial g_{MC}/\partial \mathbf{s}$ in Eqn. (9), where $\mathbf{s}$ denotes the deviatoric stress tensor, gives

$$\dot{\chi}_{MC} = \dot{\gamma}_{MC}\sqrt{\frac{2}{3}}\,\|\frac{\partial g_{MC}}{\partial \mathbf{s}}\| \,. \tag{10}$$

1.4 Angle of dilatancy ψ

The change in dilatancy of granular soils during shear loading is accounted for by a variable angle of dilatancy ψ [29]. Under medium compressive loading, the angle of dilatancy ψ is assumed to increase during shear deformation from an initial value ψ_i to a peak value ψ_p. The respective values of $\bar{\kappa}_{MC}$ are obtained as

$$\bar{\kappa}_{MC,i} = \sin\psi_i \quad \text{and} \quad \bar{\kappa}_{MC,p} = \sin\psi_p \; . \tag{11}$$

The increase of $\bar{\kappa}_{MC}$ is considered in a similar manner as the increase of κ_{MC} according to Eqn. (8) [28], reading

$$\bar{\kappa}_{MC} = \bar{\kappa}_{MC,i} + (\bar{\kappa}_{MC,p} - \bar{\kappa}_{MC,i}) \left[1 - \frac{(\chi_{MC} - \bar{\chi}_{MC})^2}{\bar{\chi}_{MC}^2}\right] \tag{12}$$

for $\chi_{MC} < \bar{\chi}_{MC}$. For $\chi_{MC} \geq \bar{\chi}_{MC}$, $\bar{\kappa}_{MC} = \bar{\kappa}_{MC,p} =$ constant.

1.5 Algorithmic formulation – integration of evolution equations

For the integration of the evolution equations (6) and (10), an implicit integration scheme is applied. For the $(n+1)$-st load increment, with $t_n \leq t \leq t_{n+1}$, time integration yields the plastic strain tensor and the internal variable at t_{n+1}:

$$\boldsymbol{\varepsilon}^p_{n+1} = \boldsymbol{\varepsilon}^p_n + \Delta\gamma_{MC}\frac{\partial g_{MC}}{\partial \boldsymbol{\sigma}} + \Delta\gamma_{TC}\frac{\partial f_{TC}}{\partial \boldsymbol{\sigma}} \; , \tag{13}$$

$$\chi_{MC,n+1} = \chi_{MC,n} + \Delta\gamma_{MC}\sqrt{\frac{2}{3}}\,\|\frac{\partial g_{MC}}{\partial \mathbf{s}}\| \; . \tag{14}$$

Equations (13) and (14) represent a set of nonlinear equations for the unknowns $\boldsymbol{\varepsilon}^p_{n+1}$, $\chi_{MC,n+1}$, $\Delta\gamma_{MC}$, and $\Delta\gamma_{TC}$, collected in the vector $\mathbf{x}$, with $\mathbf{x} = \lfloor \boldsymbol{\varepsilon}^p_{n+1}, \chi_{MC,n+1}, \Delta\gamma_{MC}, \Delta\gamma_{TC} \rfloor^T$. The respective residual form of the Eqns. (13) and (14) reads

$$\mathbf{R}(\mathbf{x}) = \left\{ \begin{array}{c} -\boldsymbol{\varepsilon}^p_{n+1} + \boldsymbol{\varepsilon}^p_n + \Delta\gamma_{MC}\frac{\partial g_{MC}}{\partial \boldsymbol{\sigma}} + \Delta\gamma_{TC}\frac{\partial f_{TC}}{\partial \boldsymbol{\sigma}} \\ -\chi_{MC,n+1} + \chi_{MC,n} + \Delta\gamma_{MC}\sqrt{\frac{2}{3}}\,\|\frac{\partial g_{MC}}{\partial \mathbf{s}}\| \\ \hline f_{MC} \\ f_{TC} \end{array} \right\} = \mathbf{0} \; . \tag{15}$$

Equation (15) is based on the assumption of two active yield surfaces. Accordingly, the constraint conditions $f_{MC} = 0$ and $f_{TC} = 0$ are considered in the residual $\mathbf{R}$. For determination of the vector of unknowns $\mathbf{x}$ from Eqn. (15), the Newton scheme is employed, reading

$$\mathbf{R} + \frac{d\mathbf{R}}{d\mathbf{x}}\Delta\mathbf{x} = \mathbf{0} \; . \tag{16}$$

During the Newton iteration, history variables corresponding to the previous load increment and the prescribed strains at t_{n+1} remain constant. Hence, the differentials of these quantities are zero, i.e.,

$$d\boldsymbol{\varepsilon}_n^p = \mathbf{0}\ , \quad d\chi_{MC,n} = 0\ , \quad \text{and} \quad d\boldsymbol{\varepsilon}_{n+1} = \mathbf{0}\ . \tag{17}$$

Accounting for $d\boldsymbol{\sigma}_{n+1} = \mathbf{C}(d\boldsymbol{\varepsilon}_{n+1} - d\boldsymbol{\varepsilon}_{n+1}^p) = \mathbf{C}(-d\boldsymbol{\varepsilon}_{n+1}^p)$ and introducing $\bar{D} = -d\bar{\kappa}_{MC}/d\chi_{MC}$ and $D = -d\kappa_{MC}/d\chi_{MC}$, the matrix $d\mathbf{R}/d\mathbf{x}$ in Eqn. (16) becomes

$$\frac{d\mathbf{R}}{d\mathbf{x}} = \left[\begin{array}{c|cc} \mathbf{A}^{-1} & \boldsymbol{\nabla} g_{MC} & \boldsymbol{\nabla} f_{TC} \\ \hline \boldsymbol{\nabla} f_{MC}^T & 0 & 0 \\ \boldsymbol{\nabla} f_{TC}^T & 0 & 0 \end{array}\right] \left[\begin{array}{cc|cc} -\mathbf{C} & \mathbf{0} & \mathbf{0} & \mathbf{0} \\ \mathbf{0}^T & -D & 0 & 0 \\ \hline \mathbf{0}^T & 0 & 1 & 0 \\ \mathbf{0}^T & 0 & 0 & 1 \end{array}\right] \tag{18}$$

with

$$\mathbf{A}^{-1} = \begin{bmatrix} \mathbf{C}^{-1} & \Delta\gamma_{MC} \frac{\partial^2 g_{MC}}{\partial\boldsymbol{\sigma}\partial\bar{\kappa}_{MC}} \bar{D} D^{-1} \\ \mathbf{0}^T & D^{-1} + \Delta\gamma_{MC} \sqrt{\frac{2}{3}}\, \left\|\frac{\partial g_{MC}}{\partial \mathbf{s}}\right\|^{-1} \frac{\partial^2 g_{MC}}{\partial \mathbf{s}\partial\bar{\kappa}_{MC}} \bar{D} D^{-1} \end{bmatrix}, \tag{19}$$

and

$$\boldsymbol{\nabla} g_{MC} = \left\{ \begin{array}{c} \frac{\partial g_{MC}}{\partial\boldsymbol{\sigma}} \\ \sqrt{\frac{2}{3}}\, \left\|\frac{\partial g_{MC}}{\partial \mathbf{s}}\right\| \end{array} \right\}, \boldsymbol{\nabla} f_{MC} = \left\{ \begin{array}{c} \frac{\partial f_{MC}}{\partial\boldsymbol{\sigma}} \\ \frac{\partial f_{MC}}{\partial\kappa_{MC}} \end{array} \right\}, \boldsymbol{\nabla} f_{TC} = \left\{ \begin{array}{c} \frac{\partial f_{TC}}{\partial\boldsymbol{\sigma}} \\ 0 \end{array} \right\}. \tag{20}$$

After time integration, i.e., after convergence of the Newton iteration, the stresses are given by $\boldsymbol{\sigma}_{n+1} = \mathbf{C}(\boldsymbol{\varepsilon}_{n+1} - \boldsymbol{\varepsilon}_{n+1}^p)$. For the evaluation of the consistent tangent $\mathbf{C}_T$, the time integration algorithm is linearized in closed form [22]. For this purpose, the differential form of the discrete (algorithmic) flow rules and the yield criteria is required:

$$d\mathbf{R}(\mathbf{x}_{n+1}) = \mathbf{0}\ . \tag{21}$$

Use of the differential form of the constitutive law, $d\boldsymbol{\sigma}_{n+1}=\mathbf{C}(d\boldsymbol{\varepsilon}_{n+1}-d\boldsymbol{\varepsilon}_{n+1}^p)$, leads to the following system of equations with the unknowns $d\boldsymbol{\sigma}_{n+1}$, $d\kappa_{MC,n+1}$, $d\Delta\gamma_{MC}$, and $d\Delta\gamma_{TC}$:

$$d\mathbf{R} = \left[\begin{array}{c|cc} \mathbf{A}^{-1} & \boldsymbol{\nabla} g_{MC} & \boldsymbol{\nabla} f_{TC} \\ \hline \boldsymbol{\nabla} f_{MC}^T & 0 & 0 \\ \boldsymbol{\nabla} f_{TC}^T & 0 & 0 \end{array}\right] \left\{ \begin{array}{c} d\boldsymbol{\sigma}_{n+1} \\ d\kappa_{MC,n+1} \\ \hline d\Delta\gamma_{MC} \\ d\Delta\gamma_{TC} \end{array} \right\} - \left\{ \begin{array}{c} d\boldsymbol{\varepsilon}_{n+1} \\ 0 \\ \hline 0 \\ 0 \end{array} \right\} = \mathbf{0}\ . \tag{22}$$

Solving Eqn. (22) for $\lfloor d\boldsymbol{\sigma}_{n+1}, d\kappa_{MC,n+1} \rfloor^T$ gives

$$\begin{aligned} \left\{ \begin{array}{c} d\boldsymbol{\sigma}_{n+1} \\ d\kappa_{MC,n+1} \end{array} \right\} = \Big[\mathbf{A} &- \mathrm{g}_{11}^{inv} \mathbf{A} \boldsymbol{\nabla} g_{MC} \boldsymbol{\nabla} f_{MC}^T \mathbf{A} \\ &- \mathrm{g}_{12}^{inv} \mathbf{A} \boldsymbol{\nabla} g_{MC} \boldsymbol{\nabla} f_{TC}^T \mathbf{A} \\ &- \mathrm{g}_{21}^{inv} \mathbf{A} \boldsymbol{\nabla} f_{TC} \boldsymbol{\nabla} f_{MC}^T \mathbf{A} \\ &- \mathrm{g}_{22}^{inv} \mathbf{A} \boldsymbol{\nabla} f_{TC} \boldsymbol{\nabla} f_{TC}^T \mathbf{A}\Big] \left\{ \begin{array}{c} d\boldsymbol{\varepsilon}_{n+1} \\ 0 \end{array} \right\}, \end{aligned} \tag{23}$$

where $\mathbf{g}^{inv} = \mathbf{g}^{-1}$ and

$$\mathbf{g} = \begin{bmatrix} \nabla f_{MC}^T \mathbf{A} \nabla g_{MC} & \nabla f_{MC}^T \mathbf{A} \nabla f_{TC} \\ \nabla f_{TC}^T \mathbf{A} \nabla g_{MC} & \nabla f_{TC}^T \mathbf{A} \nabla f_{TC} \end{bmatrix} . \tag{24}$$

The consistent tangent $\mathbf{C}_T$ is obtained by extracting $d\boldsymbol{\sigma}_{n+1} = \mathbf{C}_{T,n+1} d\boldsymbol{\varepsilon}_{n+1}$ from Eqn. (23). It is noteworthy that $\mathbf{C}_T$ is non-symmetric, resulting in a non-symmetric tangent matrix for finite element (FE) analysis. In case of an associative flow rule,. characterized by $g_{MC} = f_{MC}$, $\mathbf{C}_T$ becomes symmetric.

2 Extension of the material model for granular soil to viscoplasticity

The described multi-surface plasticity model is extended to viscoplasticity by means of the viscoplastic law of [4]. Hereby, the stresses $\boldsymbol{\sigma}$ are given as

$$\boldsymbol{\sigma} = \mathbf{C}[\boldsymbol{\varepsilon} - \boldsymbol{\varepsilon}^{vp} - \mathbf{1}\alpha_T(T - T_0)] , \tag{25}$$

where $\mathbf{C}$ is the stiffness tensor, α_T is the thermal dilation coefficient, and T_0 is the initial temperature. $\mathbf{1}$ represents the second-order volumetric unit tensor. $\boldsymbol{\varepsilon}^{vp}$ denotes the viscoplastic strain tensor. The evolution laws for the viscoplastic strains and for the hardening parameter read

$$\dot{\boldsymbol{\varepsilon}}^{vp} = \frac{1}{\tau}\mathbf{C}^{-1}(\boldsymbol{\sigma} - \boldsymbol{\sigma}^{\infty}) \quad \text{and} \quad \dot{\kappa}_{MC} = -\frac{1}{\tau}(\kappa_{MC} - \kappa_{MC}^{\infty}) , \tag{26}$$

where τ denotes the relaxation time and $\boldsymbol{\sigma}^{\infty}$ and κ_{MC}^{∞} correspond to the solution for infinitely slow loading, coinciding with the stress state and hardening parameter of rate-independent plasticity.

3 Application I: Ground improvement by means of jet grouting

When driving tunnels according to the NATM, the load-carrying capacity of the surrounding soil formation is activated by the deformations resulting from creep of shotcrete. In urban areas, however, these deformations must be minimized in order to avoid damage of buildings and infrastructure. In order to achieve this goal, the load-carrying capacity of the soil surrounding the tunnel may be improved by means of grouting techniques or soil freezing. Recently, horizontal jet grouting (HJG) was used for this purpose. As implied by the designation, HJG is performed horizontally at the tunnel face. It is characterized by the construction of a cone-shaped support ring made of improved ground in front of the tunnel face. Applications of HJG are reported in [10,8,16,9,30].

The underground station "Taborstraße" in Vienna, which is part of the planned extension of the Vienna underground system, is investigated in this section. In this stretch of the underground track, the vertical distance between the top of the tunnel and foundations of surface buildings is less than one tunnel diameter. Moreover, geologically, the area around the Taborstraße belongs to the alluvium of the river Danube. Hence, the soil mainly consists of granular materials such as sands and gravels. Apparently, driving a tunnel with a large cross-sectional area (station tube) close to existing foundations in the mentioned ground conditions is a very challenging task.

3.1 Situation at the construction site

Figure 3 and Fig. 4 show a cross section and the longitudinal section, respectively, of the underground station "Taborstraße". The geometric dimensions of the considered cross section are listed in Fig. 3. The design of the support ring is characterized by 37 jet-grouted columns of 14 m length each. The overlapping of successive support rings in the longitudinal direction is 4 m. As known from previous applications of HJG in Vienna [16], the diameter of each column was set equal to 60 cm. In general, single-rod jet grouting is employed in HJG.

During construction of the support ring, a specific jet-grouting scheme must be followed in order to avoid adjacent "fresh" columns. Figure 5 shows this scheme, assuming a construction time of 1.75 hours for each column.

After jet grouting, the tunnel is excavated according to the scheme depicted in Fig. 4. The time span between the end of jet grouting and the beginning of the excavation is of great importance. Large time spans result in a delay of construction work. Continuation of the excavation right after jet

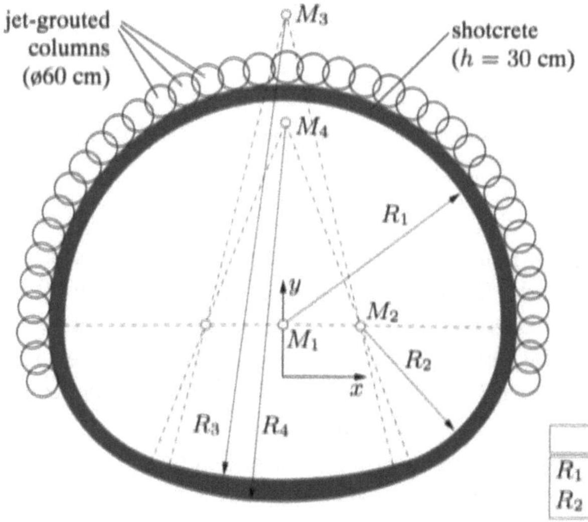

	x [m]	y [m]
M_1	0.00	1.11
M_2	1.60	1.10
M_3	0.00	7.40
M_4	0.00	5.22

radius [m]			
R_1	4.55	R_3	9.45
R_2	2.95	R_4	7.68

Fig. 3. Underground station "Taborstraße": cross section and geometric properties

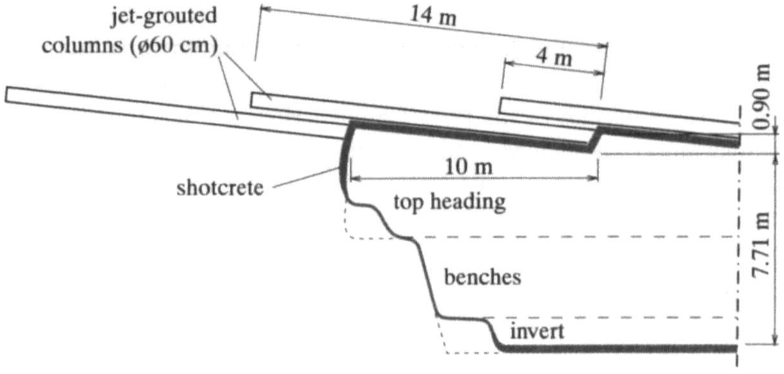

Fig. 4. Underground station "Taborstraße": longitudinal section

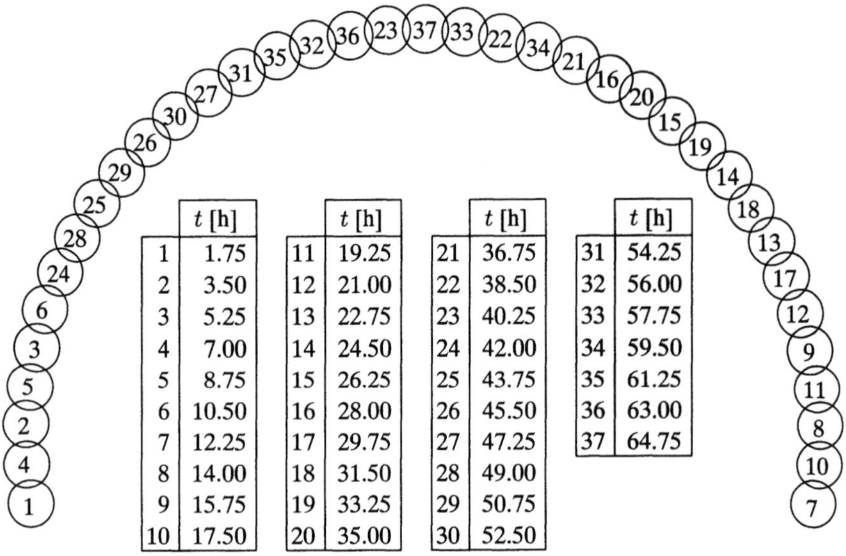

	t [h]		t [h]		t [h]		t [h]
1	1.75	11	19.25	21	36.75	31	54.25
2	3.50	12	21.00	22	38.50	32	56.00
3	5.25	13	22.75	23	40.25	33	57.75
4	7.00	14	24.50	24	42.00	34	59.50
5	8.75	15	26.25	25	43.75	35	61.25
6	10.50	16	28.00	26	45.50	36	63.00
7	12.25	17	29.75	27	47.25	37	64.75
8	14.00	18	31.50	28	49.00		
9	15.75	19	33.25	29	50.75		
10	17.50	20	35.00	30	52.50		

Fig. 5. Underground station "Taborstraße": jet-grouting scheme for construction of support ring consisting of 37 jet-grouted columns

grouting, on the other hand, results in loading of the young jet-grouted soil mass. The small strength and stiffness of the jet-grouted soil mass at the beginning of the hydration process (see [3]) could even cause larger settlements than the ones that would be obtained without jet grouting. In the present analysis, the time span between the end of jet grouting and the beginning of the excavation is assumed as 26 h (see Fig. 6).

The geological conditions at the construction site are characterized by fill material right beneath the ground surface, vast strata of sand and gravel, and a thin layer of silt, separating sand from gravel (Fig. 7). The groundwater level is located above the top of the station tube. During construction the

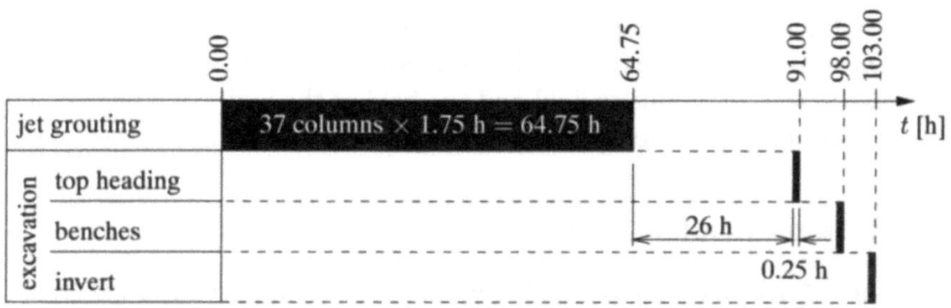

Fig. 6. Underground station "Taborstraße": time table for construction of jet-grouted support ring and excavation

groundwater level is lowered, by means of pumping, to at least 50 cm below the bottom of the tube.

For the numerical simulation of cementitious materials such as jet-grouted soil mass and shotcrete, the thermochemomechanical material model outlined in Chap. 5 is employed. Similar to the analyses of shotcrete shells based on of hybrid methods (see Chap. 10), the influence of mechanical quantities on chemical and thermal processes is negligible. Hence, the analysis is split into a thermochemical analysis followed by a chemomechanical analysis. The latter

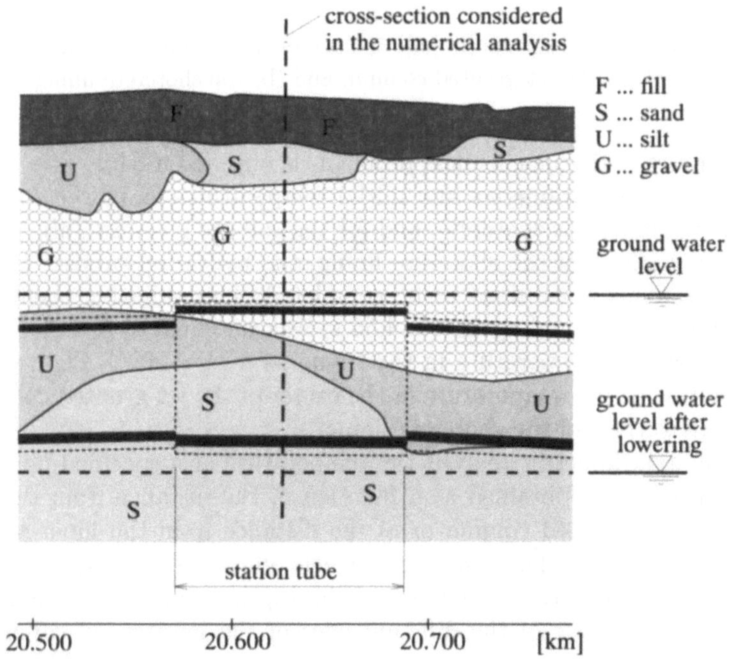

Fig. 7. Underground station "Taborstraße": longitudinal section showing geological conditions at the construction site

accounts for the influence of chemical processes on the mechanical deformations and properties of jet-grouted soil mass. Results of the thermochemical analysis are the temperature field and the field of the hydration degree. They serve as input for the chemomechanical analysis.

3.2 Thermochemical analysis

Both the jet-grouted column and the shotcrete lining are characterized by $R/L \ll 1$. R and L are the radius and the length of the jet-grouted column, respectively. For the shotcrete lining, R and L refer to the radius of the tunnel cross-section and the length of the tunnel, respectively. $R/L \ll 1$ allows to use axisymmetric models for the analysis of the jet-grouted column and the shotcrete lining (see Fig. 8). The finite element formulation for the solution

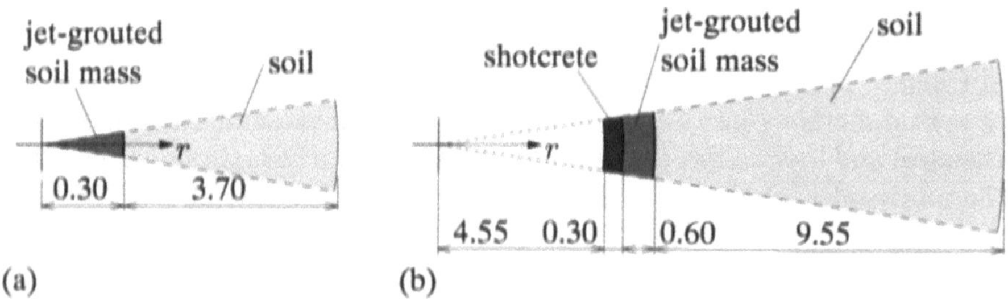

Fig. 8. Geometric dimensions (in [m]) of the numerical model used in thermochemical analysis of (a) the jet-grouted column, and (b) the shotcrete lining

of axisymmetric thermochemical problems is outlined in [15].

The material parameters and initial temperatures employed for the jet-grouted soil mass, for shotcrete, and for the soil are given in Tab. 1.

In the analysis of the shotcrete lining, the hydration process in the jet-grouted columns is assumed to be completed. The radiation coefficient for the shotcrete-air boundary at $r = 4.55$ m was set equal to $\alpha = 40.0$ kJ/(m^2 h K) [15] with a temperature in the tunnel of $T^\infty = 20$°C. Figure 9 shows the evolution of the temperature in the center of the jet-grouted column and the middle surface of the shotcrete lining.

¿From axisymmetric analysis the temperature at a specified point of the structural model is obtained as a function of the distance from the axis of the closest jet-grouted column or of the distance from the inner surface of the shotcrete lining. The temperature field illustrated in Fig. 10 refers to a time instant after the excavation of the top heading and the application of shotcrete. Because of the different ages of the jet-grouted columns (see Fig. 5) the temperature level varies from column to column. The shotcrete lining, which was installed right after the excavation of the top heading, has an initial temperature of 20°C (see Tab. 1).

Table 1. Thermochemical analyses: material parameters and initial temperatures

		shotcrete	jet-grouted soil mass	soil
activation term E_a/R [K]		4000	4000	–
heat capacity c [kJ/(kg K)]		1.0	1.4	1.1
density ρ [kg/m^3]		2428	2200	1800
thermal conductivity k [kJ/(m h K)]		12.6	4.0	3.0
cement content z [kg/m^3]		450*	150	–
latent heat $\ell_\xi = 600\,z$ [kJ/m^3]		270000	90000	–
chemical affinity $\tilde{A}$	a_A [1/s]	7.31	8.24	–
$\tilde{A} = a_A[1 - \exp(-b_A\xi)]/[1 + c_A\xi_A^{d}]$	b_A [–]	10.5	13.3	–
	c_A [–]	169.0	117.0	–
	d_A [–]	4.37	6.63	–
initial temperature T_0 [°C]		20	20	10

* ... because of rebound, the cement content was increased from 315 to 450 [15]

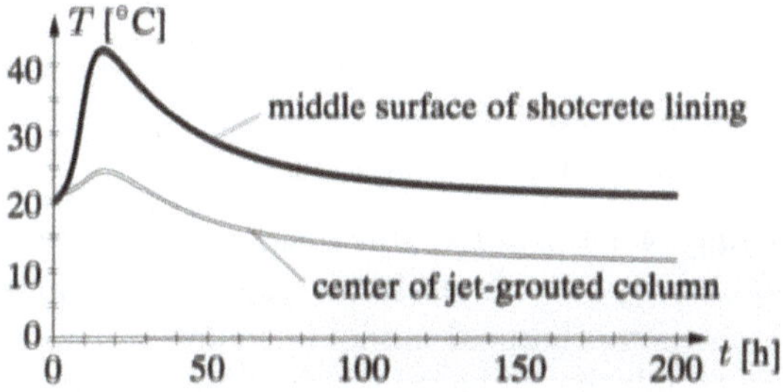

Fig. 9. Thermochemical analysis: evolution of temperature in the center of the jet-grouted column and the middle surface of the tunnel lining

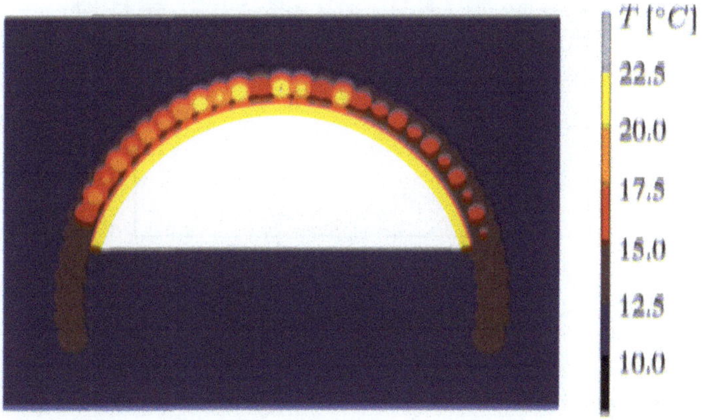

Fig. 10. Temperature distribution obtained from thermochemical analysis of jet-grouted column and shotcrete lining after excavation of the top heading and application of shotcrete

The obtained temperature profiles as well as the field of the degree of hydration (see Chap. 5) serve as input for the following chemomechanical analysis.

3.3 Chemomechanical analysis

In general, numerical simulations of the construction process of tunnels driven according to the NATM are very complicated. The two main reasons for this complexity are:

- the complexity of the material behavior of the soil and the shotcrete which, in the most general case, encompasses thermochemomechanical couplings for shotcrete as well as hydromechanical couplings for the soil;
- the necessity to model the construction process of a complex 3D structure, comprising numerous construction states.

The special situation at the considered underground station in Vienna allows to restrict the investigation to

- drained conditions as far as modeling of soil is concerned, i.e., hydromechanical couplings are not accounted for;
- 2D plane-strain conditions as a consequence of the cone-shaped support ring of improved soil; its stiffness is the reason of the dominance of the load transfer in the circumferential direction.

As regards modeling of jet-grouted soil mass and shotcrete, the main phenomena observed in early-age cementitious material are accounted for by the described thermochemomechanical model formulation (see Chap. 5).

The numerical simulation is based on the plane-strain model depicted in Fig. 11. Two layers of soil are considered. The material properties employed

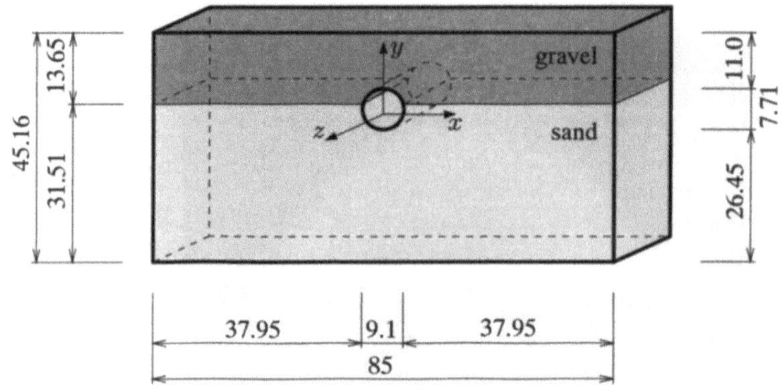

Fig. 11. Underground station "Taborstraße": geometric dimensions (in [m]) of the structural model

in the numerical analyses are listed in Tab. 2.

Table 2. Underground station "Taborstraße": material properties used in the analyses (f_b: biaxial compressive strength)

	gravel	sand	jet-grouted soil mass	shotcrete
density ρ [kg/m^3]	1700	1800	2200	2428
(final) Young's modulus E [N/mm^2]	80	40	2000	40800
Poisson's ratio ν [–]	0.32	0.35	0.2	0.2
thermal dilation coefficient α_T [–]	$1 \cdot 10^{-5}$	$1 \cdot 10^{-5}$	$1 \cdot 10^{-5}$	$1 \cdot 10^{-5}$
cohesion c [N/mm^2]	0	0.015	–	–
initial value of friction angle φ_i [–]	27.6°	23.4°	–	–
peak value of friction angle φ_p [–]	32.5°	27.5°	–	–
calibration parameter $\bar{\chi}_{MC}$ [–]	0.03	0.03	–	–
relaxation time τ [h]	0.015	0.015	–	–
(final) compr. strength $f_{c,\infty}$ [N/mm^2]	–	–	10.0	39.6
tensile strength f_t	–	–	$f_c/10$	$f_c/10$
f_b/f_c [–]	–	–	1.16	1.16
relaxation modulus H [N/mm^2]	–	–	$1/10 \cdot 10^6$	$1/7 \cdot 10^6$
shrinkage parameter a_s [–]	–	–	$-4.05 \cdot 10^{-4}$	$-4.05 \cdot 10^{-4}$
$\varepsilon^s = a_s + b_s\xi$ b_s [–]	–	–	$9.43 \cdot 10^{-4}$	$9.43 \cdot 10^{-4}$
asymptotic viscous compliance $J^v_\infty = a_v(1-\xi)$ a_v [μm/m/MPa]	–	–	149	445
char. time for short-term creep τ_w [h]	–	–	48	48
activation term U/R [K]	–	–	2700	2700
reference temperature $\bar{T}$ [°C]	–	–	20	20

Initial stress state. The primary stress state represents the mechanical condition in the soil prior to ground improvement and the excavation of the tunnel. It is caused by dead load. During application of dead load in the analysis the soil behaves elastically. The resulting stress state can be expressed in terms of the volume force density γ_{soil}, the depth measured from the soil surface, h, and the respective Poisson's ratio ν:

$$\sigma_y = -\gamma_{soil} h \ , \quad \sigma_x = \sigma_z = -\frac{\nu}{1-\nu}\gamma_{soil} h = -K_0 \gamma_{soil} h, \tag{27}$$

where K_0 is the ratio of the horizontal and the vertical component of the *in situ* stress, referred to as lateral pressure coefficient. According to [24], K_0 can be estimated from

$$K_0 = 1 - \sin\varphi_p \tag{28}$$

giving $K_0 = 0.46$ and 0.54 for gravel and sand, respectively (computed from φ_p given in Tab. 2). Based on the known values of K_0, the corresponding values of ν can be computed from Eqn. (27), giving $\nu = K_0/(1+K_0) = 0.32$ and 0.35 (these values are also listed in Tab. 2).

After application of the dead load of the soil, the displacements and strains are reset to zero. Hence, the analysis provides relative displacements as a consequence of construction works such as HJG and the excavation of the tunnel.

Consideration of jet grouting. In the numerical analyses, jet grouting is simulated by changing the material properties in the finite elements representing the jet-grouted column from *in situ* soil to young jet-grouted soil mass. According to [3], this change results in negligibly small strength and stiffness of the grouted material. Since the original microstructure of the soil is destroyed during jet grouting, disregard of strength and stiffness of the young jet-grouted soil mass is appropriate.

The increase of the hydrostatic pressure in the soil during jet grouting may be considered by adding a hydrostatic stress state Δp to the initial stress state after jet grouting. The latter results from dead load of the jet-grouted soil mass. In what is called compensation grouting, this hydrostatic pressure is increased by preventing backflow of grouting material through the bore hole. The increased hydrostatic pressure results in heaving of the surface which "compensates" settlements caused by the tunnel excavation [6,7]. At the tunnel site investigated in this section, however, compensation grouting will not be applied. Accordingly, Δp can be set equal to zero. Therefore, the numerical results reported in this section represent the upper limit of the settlements caused by jet grouting. Hence, these results represent the worst case.

Presentation of results. In order to assess the influence of HJG on the surface settlements during the construction of the station tube "Taborstraße", two analyses were performed. In the first analysis, HJG was considered. In the second analysis, only the shotcrete lining was assumed to support the tunnel: HJG was not considered in this analysis. Figure 12 shows the evolutions of the surface settlement at $x = 0$ obtained from both analyses. Modeling

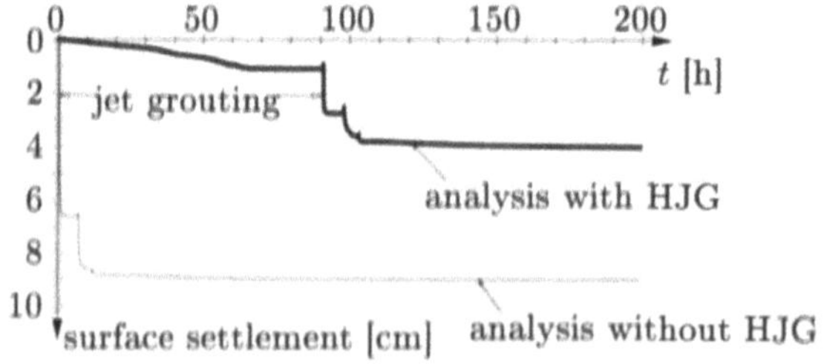

Fig. 12. Underground station "Taborstraße": evolutions of surface settlement at $x = 0$

of jet grouting by changing the material properties in the finite elements representing jet-grouted columns from soil to young jet-grouted soil mass results in a settlement of 10 mm prior to the excavation process. This result fits well into measurements performed during HJG at different tunnel sites ([10,1,16] reported on settlements between 7 and 10 mm observed in the course of jet grouting). The excavation of the tunnel results in an increase of the settlement from 10 mm to 4 cm. The jumps in the evolutions of the settlement depicted in Fig. 12 refer to different excavation steps, namely, excavation of the top heading, the benches, and the invert. The numerical analysis characterized by disregarding HJG resulted in settlements up to 9 cm (Fig. 12). Hence, based on the numerical results, application of HJG leads to a reduction of surface settlements by more than 50%. A similar decrease of settlements resulting from the application of HJG at the Bad Godesberg tunnel, Germany, was reported in [30]. Surface settlements at the beginning of the construction process reaching 6.8 cm signaled the necessity of ground improvement, which was done in the remaining part of the tunnel. The settlements measured in the part of the tunnel supported by a jet-grouted ring ranged from 1.6 to 4.5 cm, with a mean value of 3.4 cm [30]. Hence, the settlements were reduced by 50%.

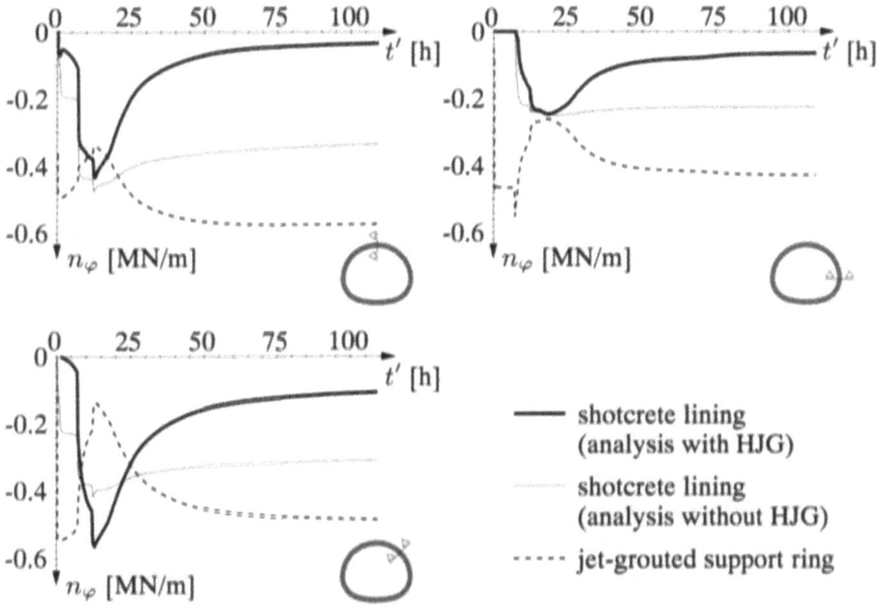

Fig. 13. Underground station "Taborstraße": evolution of n_φ in the shotcrete lining and the jet-grouted support ring ($t' = 0$ refers to the time instant at the start of the excavation)

Reduction of surface settlements is the main objective in tunneling in urban areas. Optimization of the structural performance of the shotcrete lining, on the other hand, is of minor importance. In general, the thickness of the shotcrete lining required to reduce the deformations such as to avoid damage of surface buildings leads to low stress levels and, hence, to a low level of loading of the lining. Figure 13 illustrates the loading of the lining at three different shell sections on the basis of the evolution of the axial force in the circumferential direction, n_φ, with

$$n_\varphi = \int_h \sigma_\varphi \, dr \ , \tag{29}$$

where h is the thickness of the shotcrete lining.

In the analysis without consideration of HJG, the excavation of the top heading results in compressive forces of 0.2 MN/m at the top of the tunnel. The analysis characterized by consideration of a jet-grouted support ring indicates a smaller loading of the lining. The main part of the load is carried by the support ring. It amounts to compressive forces up to 0.5 MN/m (see the evolution of n_φ in the jet-grouted support ring, which is also shown in Fig. 13). Continuation of the excavation is accompanied by the extension of the shotcrete lining to the benches and the invert, finally resulting in a closed shotcrete shell. The stiffness of the closed lining attracts forces, resulting in an increase of compressive loading. In the jet-grouted support ring, on the other hand, unloading is observed during the excavation of the benches and the invert. Consideration of creep of shotcrete results in relaxation of the compressive force in the lining. In addition to relaxation, the different creep properties of shotcrete and of the jet-grouted soil mass (see Tab. 2) lead to a redistribution of loading: unloading in the lining is accompanied by an increase of compressive loading in the jet-grouted support ring.

The reduction of the loading in the shotcrete lining because of creep is illustrated by the distribution of n_φ in the lining 13 and 130 h, respectively, after the start of the excavation (Fig. 14).

Consideration of HJG resulted in an increase of compressive loading at the shoulders and a reduction of loading in the top (see Fig. 14a). This situation is caused by bending (see, e.g., [1]), taken up by additional axial forces in the shotcrete – jet-grouted soil mass compound structure. Depending on the type of bending, the compressive force in the shotcrete lining is decreasing or increasing. For the analysis characterized by disregarding HJG, bending has no effect on the distribution of n_φ. 130 h after the start of the excavation, creep of shotcrete resulted in a redistribution of loading from the shotcrete shell to the jet-grouted support ring. The loading at the shoulders is still larger than the loading at the top of the shotcrete shell (Fig. 14b).

The loading of the surrounding soil is illustrated by the distribution of the internal variable of the Mohr-Coulomb criterion, χ_{MC} (Fig. 15). Application of the primary stress state in the soil according to Eqn. (27) does not result in plastic loading in the soil. The Mohr-Coulomb criterion becomes active during redistribution of stresses in the soil in consequence of the excavation process. For the (compressive) primary stress state $|\sigma_y| > |\sigma_x| = |\sigma_z|$ (see

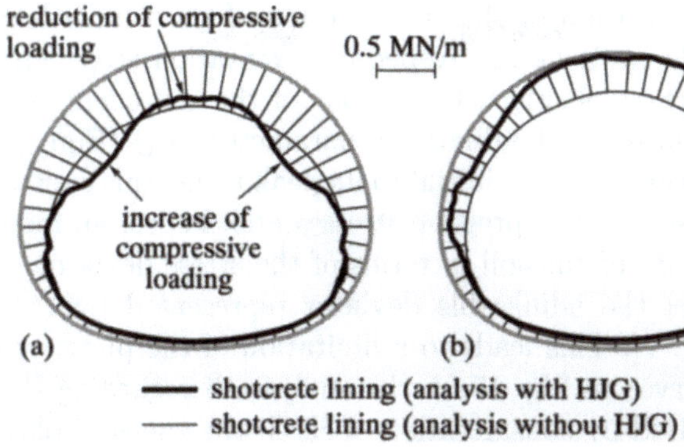

Fig. 14. Underground station "Taborstraße": axial force n_φ in the circumferential direction in the shotcrete lining (a) 13 h and (b) 130 h after the start of the excavation

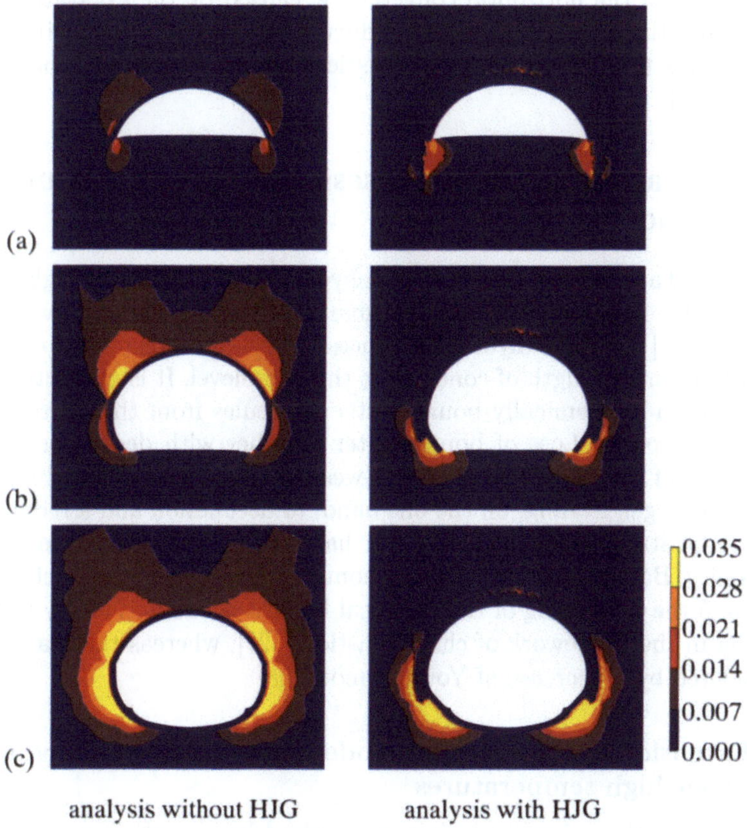

Fig. 15. Underground station "Taborstraße": distribution of χ_{MC} after excavation of (a) the top heading, (b) the benches, and (c) the invert (5-fold magnification of displacements)

Eqn. (27)), the excavation of the cross section of the tunnel leads to large stresses in the side walls at the respective excavation step. These are the regions where the first plastic deformations of the soil occur. According to the employed evolution of the dilatancy parameter $\bar{\kappa}_{MC}$ (Eqn. (12)), which increases quadratically from its initial to its peak value, this results in volume dilation. Dilation evokes compressive stresses and, hence, an increase of the load-bearing capacity of the soil in terms of the stress deviator (the greater $-I_1/\sqrt{3}$, the larger the admissible deviator represented by $\sqrt{2J_2}$, see the failure lines in Fig. 2). This leads to a limitation of the plastic zone around the tunnel as observed in Fig. 15 for the analysis disregarding HJG. For the analysis characterized by consideration of HJG, the zones of plastic loading of the soil are concentrated around the footings of the jet-grouted support ring. At the top and the shoulders of the tunnel, the jet-grouted support ring leads to an increase of the confinement in the soil. This has the same effect as plastic dilation, namely, an increase of the load-bearing capacity of the soil. The increase of the horizontal convergence, caused by the excavation of the benches and the invert, results in a reduction of the confinement in the soil. This explains the narrow bands of plastic material response at the benches of the tunnel.

4 Application II: Safety assessment of tunnels under fire load

In the second application the effect of dehydration of concrete at high temperatures on the overall structural performance of tunnels will be investigated. According to [17], the dehydration process results in a degradation of both the stiffness and strength of concrete at the microlevel. It is characterized by microdiffusion of chemically-bound water molecules from the micropores to the capillary pores. Loss of bound water together with decomposition and dissociation of the hydration products weakens the chemical bond structure of the cement gel, leading, on the one hand, to decohesion and a reduction of the material strength and, on the other hand, to a reduction of the modulus of elasticity. Based on the thermochemomechanical material model outlined in Chap. 5 the weakening of the chemical bond is accounted for by means of softening in the framework of chemoplasticity [27], whereas thermal damage is considered by a decrease of Young's modulus.

4.1 Extension of the material model for concrete to dehydration under high temperatures

Whereas hydration of concrete is characterized by a continuous increase of the degree of hydration ξ, dehydration of concrete under high temperatures is described by a decrease of ξ. The time scale of *microdiffusion* of chemically-bound water molecules from the micropores to the capillary pores is significantly smaller than the time scale corresponding to *heat transport* into the structure. Accordingly, the microdiffusion of chemically-bound water can be regarded as instantaneous, allowing to relate the degree of hydration directly to the temperature [27], giving $\xi = \xi(T)$ instead of $\xi = \xi(T, t)$. Hence, there is no need for specification of a kinetics law for the description of the dehydration process. The dehydration function $\xi(T)$ is determined from experimental data. In general, Young's modulus and the compressive strength are given as functions of the temperature T (see, e.g., [20]), specifying $E = E(T)$ and $f_c = f_c(T)$ (see Fig. 16). Based on linear chemomechanical coupling between the degree of hydration and Young's modulus (see Fig. 17a) [27], the dehydration function $\xi(T)$ can be extracted from $E(T)$ and $E(\xi)$ (see Fig. 17b). Knowledge of this function allows to determine the intrinsic material function describing thermal decohesion of concrete $f_c(\xi)$ from $f_c(T)$ depicted in Fig. 16.

In contrast to the hydration process, characterized by the formation of hydrates in a state free of microstress (see Chap. 5), hydrates are loaded when they are decomposed under high temperatures. This fact is reflected by the use of a total stress-strain law (as opposed to incremental stress-strain law, see Chap. 5) in the form:

$$\boldsymbol{\sigma}_{n+1} = \mathbf{C}_{n+1} : [\boldsymbol{\varepsilon}_{n+1} - \boldsymbol{\varepsilon}^p_{n+1} - \mathbf{1}\alpha_T T_{n+1}] \tag{30}$$

or equivalently

$$\begin{aligned}\boldsymbol{\sigma}_{n+1} &= \mathbf{C}_{n+1} : [\boldsymbol{\varepsilon}_n - \boldsymbol{\varepsilon}^p_n - \mathbf{1}\alpha_T T_n] \\ &\quad + \mathbf{C}_{n+1} : [\Delta\boldsymbol{\varepsilon}_{n+1} - \Delta\boldsymbol{\varepsilon}^p_{n+1} - \mathbf{1}\alpha_T \Delta T_{n+1}] \\ &= (E_{n+1}/E_n) \cdot \underbrace{\mathbf{C}_n : [\boldsymbol{\varepsilon}_n - \boldsymbol{\varepsilon}^p_n - \mathbf{1}\alpha_T T_n]}_{\boldsymbol{\sigma}_n} \\ &\quad + \mathbf{C}_{n+1} : [\Delta\boldsymbol{\varepsilon}_{n+1} - \Delta\boldsymbol{\varepsilon}^p_{n+1} - \mathbf{1}\alpha_T \Delta T_{n+1}] \, .\end{aligned} \tag{31}$$

Based on linear chemomechanical coupling of E and ξ, E_{n+1}/E_n can be replaced in Eqn. (31) by ξ_{n+1}/ξ_n, finally giving the stress update proposed in [27]. Figure 18 illustrates the decomposition of hydrates for the $(n+1)$-st time (load) increment, given by $\Delta m_{n+1} < 0$. The release of microstress in the decomposed hydrate results in a reduction of the applied macrostress σ_n by the factor E_{n+1}/E_n (or ξ_{n+1}/ξ_n), see Eqn. (31).

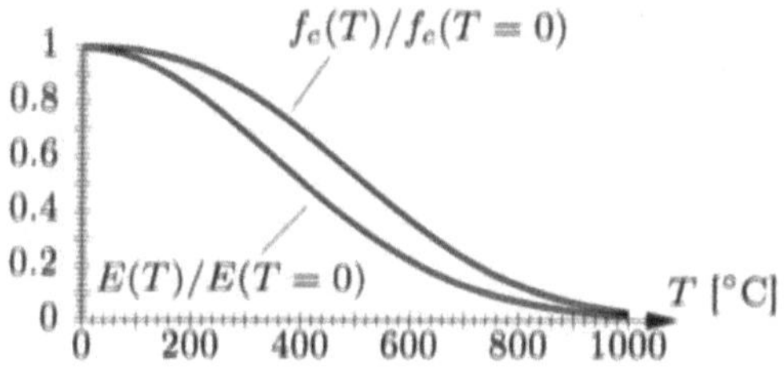

Fig. 16. Dependence of Young's modulus E and of the compressive strength f_c on the temperature T

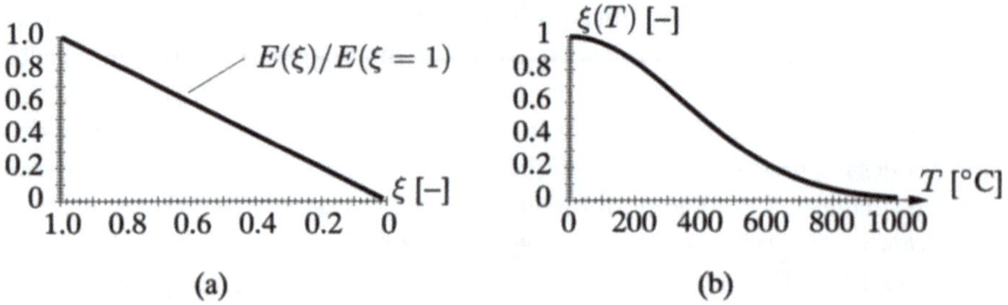

Fig. 17. (a) Intrinsic material function $E(\xi)$ and (b) dehydration function $\xi(T)$ employed in the analysis

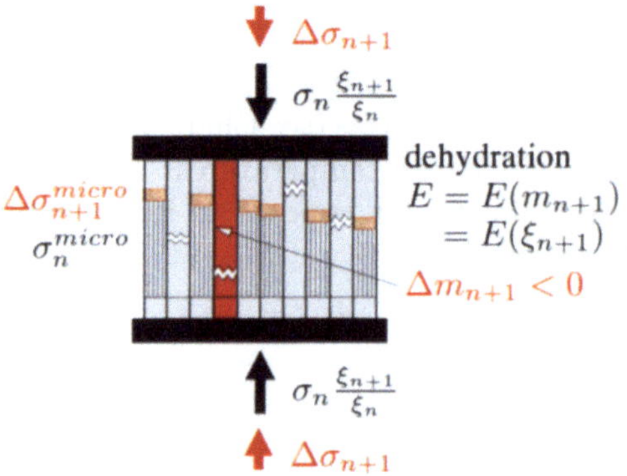

Fig. 18. 1D model of thought, illustrating the decomposition of hydrates during dehydration (σ^{micro}: microstress in the hydrates)

4.2 Application

Figure 19 shows the geometric dimensions of the considered tunnel cross-section. In the numerical simulation, the excavation of the tunnel is considered by means of four excavation steps. At first, the top heading I, top heading II, and a temporary invert are excavated, constituting a closed shotcrete shell. Later on, the temporary invert is removed and the final invert is installed. Finally, a (cast-in-situ) concrete lining of 40 cm thickness is installed. Figure 20 contains the time table employed in the numerical analysis. When the shotcrete lining is closed either by the temporary or the final invert, a time span of approximately 20 h between the excavation and the application of shotcrete is considered. Disregard of this time lag which follows from the sequential nature of the process of excavation and shotcreting would prevent the stress release in the ground immediately after the excavation, resulting in an underestimation of the deformations and an overestimation of the loading of the lining.

Geological conditions and structural model. Figure 21 illustrates the

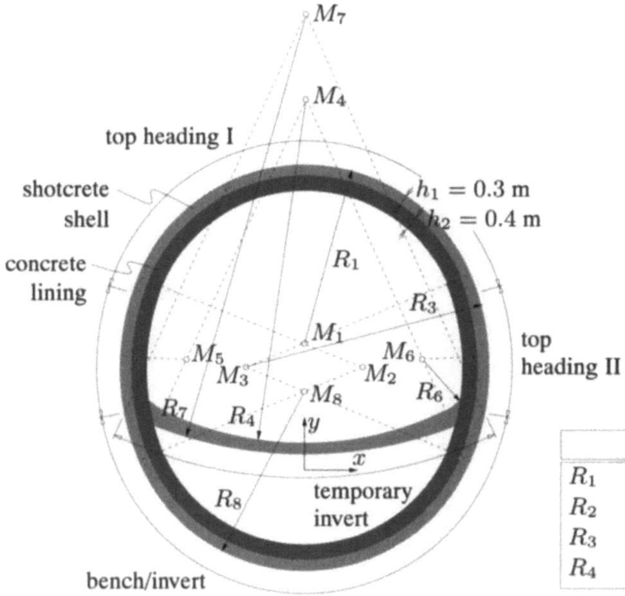

	x [m]	y [m]
M_1	0.00	3.40
M_2	1.51	2.62
M_3	-1.51	2.62
M_4	0.00	9.30
M_5	-2.76	2.87
M_6	2.76	2.87
M_7	0.00	11.50
M_8	0.00	1.73

	radius [m]		
R_1	4.38	R_5	1.50
R_2	6.08	R_6	1.50
R_3	6.08	R_7	11.00
R_4	8.50	R_8	4.33

Fig. 19. Cross section of considered single-track tube

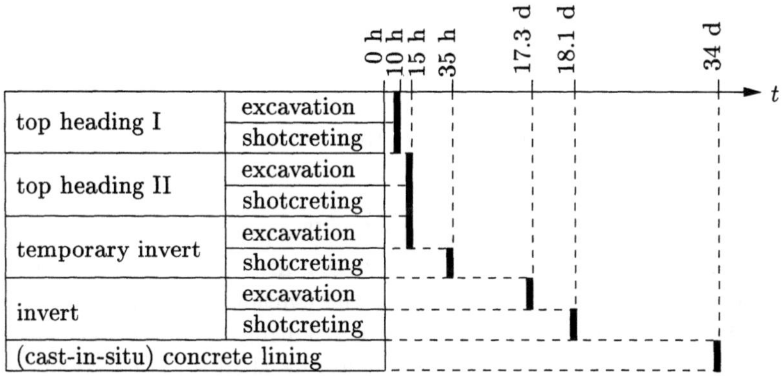

Fig. 20. Time table of construction

geological conditions at and in the vicinity of the considered tunnel cross-section. For the simulation of (a) the excavation process, (b) the installation of the shotcrete shell and the concrete lining, and (c) the fire accident, a 2D model is employed. It consists of the tunnel cross-section and the surrounding soil formations. The boundary on top of the structural model coincides with the free surface of the considered cross section (see Fig. 21). The symmetry of (a) the cross section at all stages of construction, (b) the soil stratification, and (c) the fire load allows to consider only one half of the cross section. The analysis was performed by means of the finite element method (FEM).

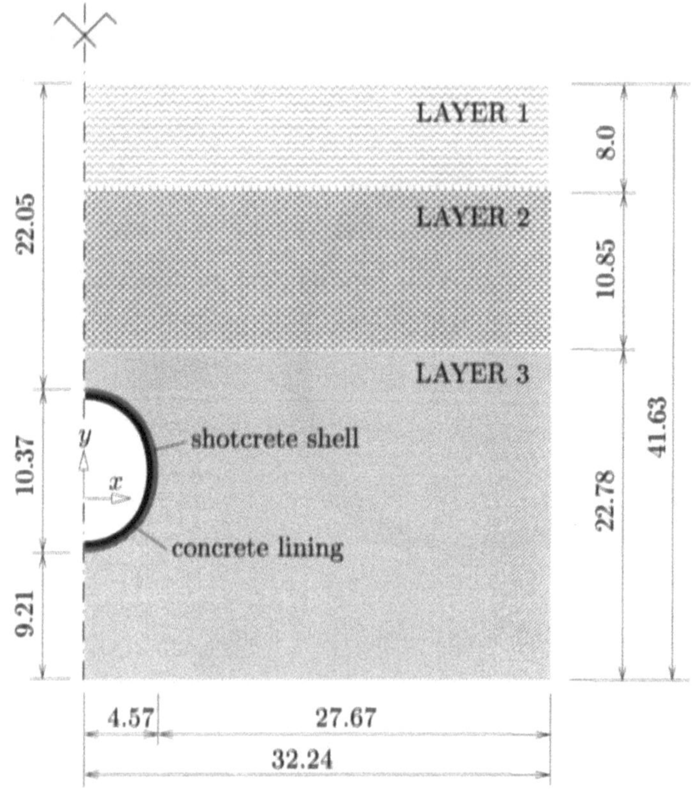

Fig. 21. Geometric dimensions (in [m]) of the structural model

Initial state for fire-accident analysis. The proposed mode of determination of the loading of tunnel linings under fire load is characterized by taking into account the entire history of the considered cross section of the tunnel. The deformation and stress state resulting from the excavation of the tunnel and the installation of the shotcrete shell and the concrete lining are obtained from a preceding analysis. Hereby, the construction scheme according to Fig. 20 was considered. The employed material parameters for the soil are listed in Tab. 3. The material parameters for shotcrete are taken from Chap. 5. For the concrete lining, concrete with $E = 30000$ N/mm^2 and $f_{c,\infty} = 22.5$ N/mm^2 (B30 according to [5]) was specified in the structural design.

Figure 22 shows the evolution of the circumferential force n_φ (see Eqn. (29)) in the shotcrete shell and the concrete lining at the top of the tunnel. The evolution of n_φ is characterized by a load transfer from the shotcrete shell towards the concrete lining. This can be explained by the rather high creep compliance of shotcrete (see [18]). Since the loading of the concrete lining is continuously increasing, a rather large simulation time was chosen

Table 3. Material properties for the ground

	LAYER 1	LAYER 2	LAYER 3
density ρ [kg/m^3]	1800	1800	2000
(final) Young's modulus E [N/mm^2]	60	60	30
Poisson's ratio ν [–]	0.33	0.33	0.40
thermal dilation coefficient α_T [–]	$1 \cdot 10^{-5}$	$1 \cdot 10^{-5}$	$1 \cdot 10^{-5}$
cohesion c [N/mm^2]	0	0	0.04
friction angle φ [–]	30°	30°	20°
relaxation time τ [h]	0.015	0.015	1.0*

* ... obtained from back analysis

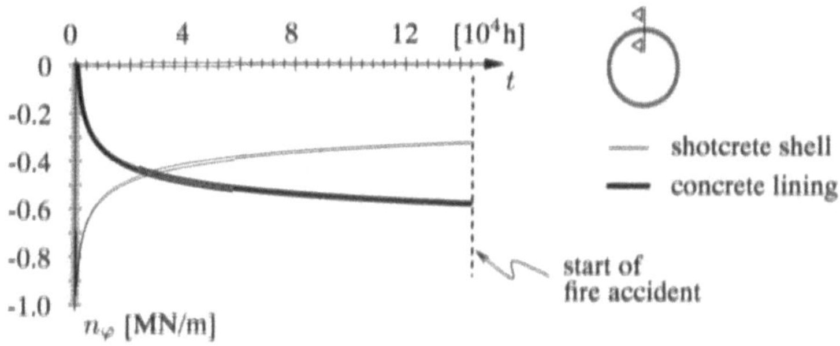

Fig. 22. Evolution of n_φ

for determination of the initial state for the fire-accident analysis. At $t > 15$ y, the load transfer from the shotcrete shell to the concrete lining is almost complete. Accordingly, the fire accident was assumed to occur at $t > 15$ y.

The distribution of n_φ at the time instant just before the fire accident, is shown in Fig. 23a. In the concrete lining, only compressive forces are observed. The maximum load occurs at the shoulders. In this area of the cross section, bending of the compound shotcrete–concrete structure results in an increase of the compressive loading in the concrete lining and in unloading in the shotcrete shell. Accordingly, the distribution of the circumferential force in the shotcrete shell is characterized by reduced compressive loading at the shoulders of the cross section.

Figure 23b shows the distribution of the axial force in the longitudinal direction, n_z with

$$n_z = \int_h \sigma_z dr \, , \qquad (32)$$

at the time instant just before the fire accident. The increase of loading of the concrete lining in the circumferential direction results in compressive forces in the longitudinal direction. Unloading of the shotcrete shell in the circumferential direction yields tensile loading in the longitudinal direction.

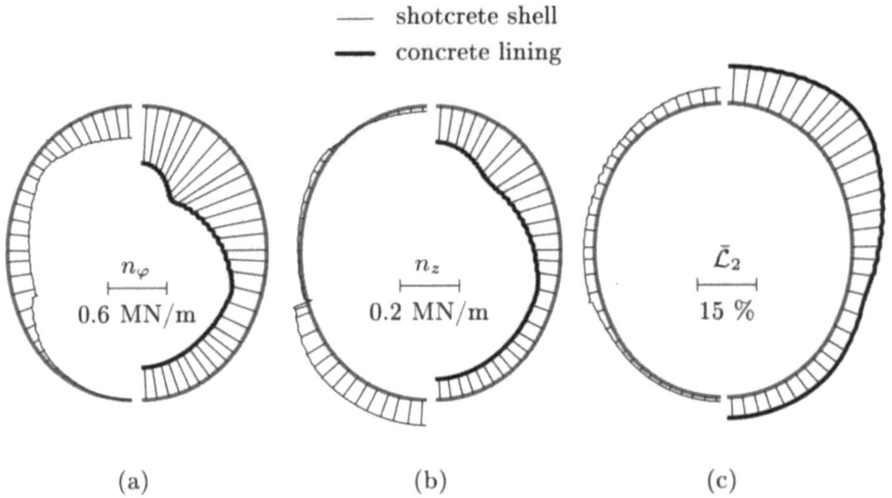

Fig. 23. Distribution of (a) n_φ, (b) n_z, and (c) $\bar{\mathcal{L}}_2$ at the time instant just before the fire accident (for the definition of $\bar{\mathcal{L}}_2$, see Chap. 5)

In order to assess the safety of the structural components, the level of loading $\mathcal{L}_2$ is determined for the shotcrete shell and the concrete lining (see Chap. 10 for the definition of $\mathcal{L}_2$). Figure 23c shows the spatial distribution of the level of loading in the shotcrete shell and the concrete lining averaged over the respective thickness, $\bar{\mathcal{L}}_2$ with $\bar{\mathcal{L}}_2 = 1/h \int_h \mathcal{L}_2 dr$, at the time instant just before the fire accident. Because of the rather small values of the axial forces, the maximum value for the level of loading is small, namely below 15%.

Numerical analysis of the fire accident. A well-defined temperature history was employed for the simulation of a fire accident. The employed temperature history is depicted in Fig. 24. The small influence of the mechanical

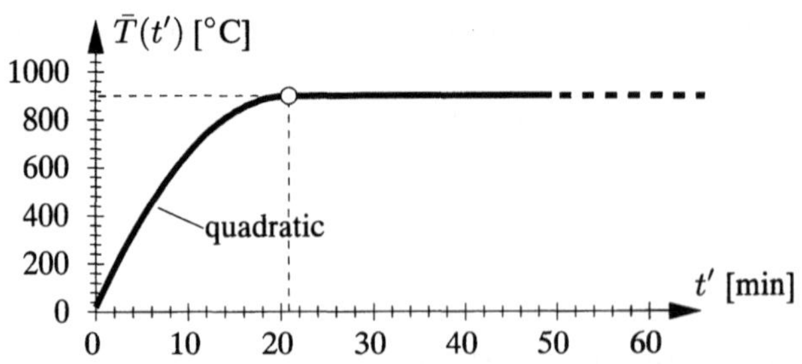

Fig. 24. Temperature history $\bar{T}$ for the simulation of the fire accident (curve representing the fire of stacked car tires; $t' = 0$ refers to the time instant at the start of the fire accident)

behavior on thermal and chemical processes (decoupling hypothesis) allows to divide the analysis of the fire accident into two parts:

- First, the thermochemical problem is solved, accounting for the change of latent heat related to the dehydration of concrete / shotcrete. This part of the analysis provides the history of the temperature field $T(\mathbf{x}, t')$.
- This field serves as input for the second part of the analysis, concerned with the solution of the chemomechanical problem.

Solution of the thermochemical problem. The thermochemical problem is described by a field equation derived from the first law of thermodynamics [26]:

$$C\dot{T} - \ell_\xi \dot{\xi} = -\mathrm{div}\,\mathbf{q} \ , \tag{33}$$

where C denotes the volume heat capacity, ℓ_ξ represents the latent heat of the (de)hydration process, with $\ell_\xi > 0$. $\mathbf{q}$ is the heat flow vector. It is related to the temperature via Fourier's law of heat conduction:

$$\mathbf{q} = -k\,\mathrm{grad}\,T \ , \tag{34}$$

where k denotes the thermal conductivity. Accounting for the direct dependence of the degree of hydration on the temperature, $\xi = \xi(T)$ (see Subsection 4.1), $\dot{\xi}$ in Eqn. (33) can be replaced by $d\xi/dT \cdot \dot{T}$, finally giving

$$\left(C - \ell_\xi \frac{d\xi}{dT} \right) \dot{T} = -\mathrm{div}\,\mathbf{q} \ . \tag{35}$$

Since $d\xi/dT < 0$ in the case of dehydration and $\ell_\xi > 0$, the effect of the change of latent heat can be interpreted as an increase of the volume heat capacity. Hence, the heat propagation into the lining is delayed by the dehydration of concrete / shotcrete. The material properties employed in the thermochemical analysis are shown in Tab. 4.

Table 4. Material parameters and initial temperatures used in thermochemical analysis

	soil	shotcrete	concrete
volume heat capacity C [kJ/(m^3 K)]	2000	2400	2400
thermal conductivity k [kJ/(m h K)]	3.0	12.0	12.0
latent heat ℓ_ξ [kJ/m^3]	–	190000	190000
radiation coefficient α [kJ/(m^2 h K)]	–	–	72
initial temperature T_0 [°C]	10	10	10

For the thermochemical analysis, the same 2D model was employed as the one used for determination of the initial state. Accordingly, no temperature flow was considered in the direction of the tunnel. For fire accidents in tunnels, which are characterized by a fire load over several meters, this is a reasonable assumption. The fire load is assumed to affect mainly the top of the tunnel. In this part of the cross section, the temperature history depicted in Fig. 24 is prescribed at the inner surface of the concrete lining (see Fig. 25). At the

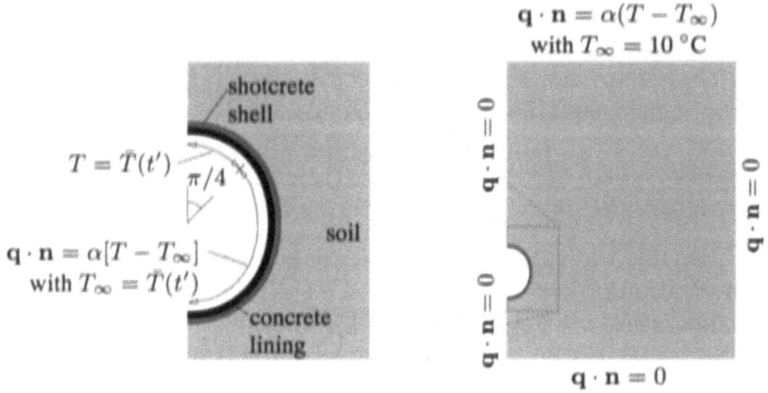

Fig. 25. Numerical model and boundary conditions for the solution of the thermochemical problem ($t' = 0$ refers to the time instant at the start of the fire accident)

benches and the invert of the cross section, the fire load is assumed to be smaller. Accordingly, temperature radiation is considered in this area of the cross section (see [25] for a similar type of fire load).

The prescribed-temperature boundary condition, which is characterized by a large increase during the first 10 min and a constant temperature of 900 °C after 20.8 min, entails a pronounced temperature gradient in the radial direction in the first few centimeters of the concrete lining (see Fig. 26). This gradient results in a heat flow into the concrete / shotcrete compound structure. The state of dehydration is governed by the thermal dehydration function given in Fig. 17b.

The sector of the cross section exposed to temperature radiation is characterized by much smaller temperature gradients and, hence, by a delayed propagation of the temperature and the dehydration front into the lining (Fig. 27). After a fire exposure of $t' = 20$ h, the temperature at the inner surface amounts to 640 °C. The region of the model where the thermal boundary condition changes from prescribed temperature to radiation is characterized by a thermal gradient and, hence, a heat flow in the circumferential direction (see Fig. 28).

Solution of the chemomechanical problem. The thermochemical analysis provides the temperature history at each integration point of the numerical model. The individual temperature histories are stored. They serve

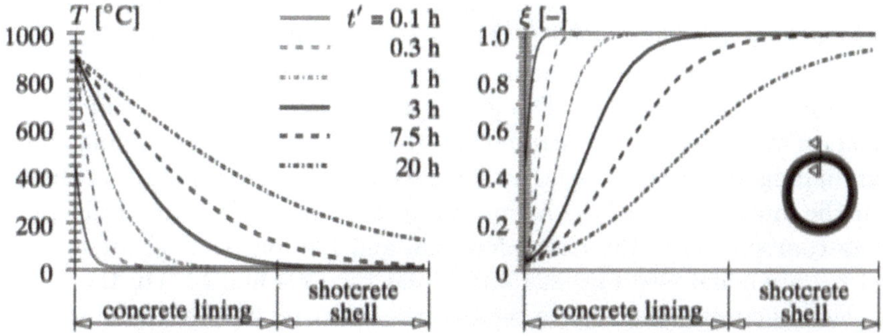

Fig. 26. Temperature profiles and distribution of the degree of hydration at various time instants in a section located at the top of the tunnel (prescribed-temperature boundary condition)

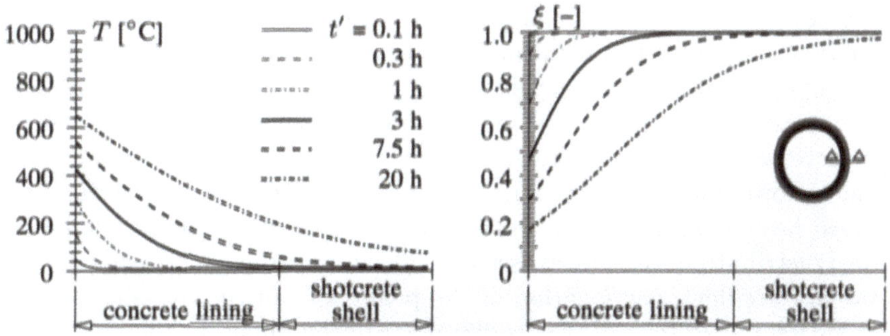

Fig. 27. Temperature profiles and distribution of the degree of hydration at various time instants in a section located at the bench of the tunnel (radiation boundary condition) ($t' = 0$ refers to the time instant at the start of the fire accident)

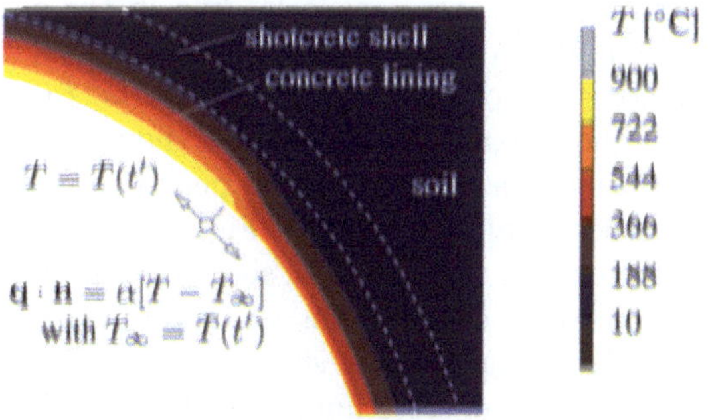

Fig. 28. Temperature field $T(\mathbf{x}, t')$ after 11 h of fire exposure ($t' = 0$ refers to the time instant at the start of the fire accident)

as input for the subsequent chemomechanical analysis, resulting in temperature strains in consequence of thermomechanical coupling. Because of the plane-strain condition assumed in the analysis, the strains in the longitudinal direction are equal to zero. The generation of strains in the circumferential direction of the tunnel is restrained by the colder parts of the lining and the surrounding soil. Accordingly, a biaxial compressive stress state is obtained near the inner surface of the lining. At the beginning of the fire accident, the stress components in the circumferential and the longitudinal direction, σ_φ and σ_z, are equal (see Fig. 29a and b and Fig. 30a and b). On the basis of an elastic material response, a temperature increase of only 70 °C would be sufficient for an increase of the biaxial loading up to the range of the biaxial compressive strength $f_{b,\infty}$, obtained as

$$\frac{E}{1-\nu}\,\alpha_T\,\Delta T = \frac{30000}{1-0.2}\,1\cdot 10^{-5}\cdot 70 = f_{b,\infty} = 1.16\cdot f_{c,\infty} = 26\ \mathrm{N/mm^2}\,. \quad (36)$$

Obviously, however, the temperature increase near the inner surface of the lining is accompanied by plastic material response. The respective stress state represents a point on the Drucker-Prager yield surface of the material model for concrete. In addition to the generation of thermal strains, the temperature increase leads to dehydration of concrete treated in the framework of chemical softening (thermochemical coupling). This leads to a reduction of the load level at regions close to the inner surface of the lining, giving bell-shaped distributions of stresses (see Fig. 29a and b). With increasing duration of the fire accident, the location of the peak value of the bell-shaped stress distribution moves towards the colder parts of the lining. In the longitudinal direction, the assumed plane-strain condition results in peak stresses in the range of the uniaxial and the biaxial compressive strength, respectively (see Fig. 29b). Since the generation of strains in the circumferential direction is restrained by the colder parts of the lining and the surrounding soil, these parts of the structure experience tensile loading in the circumferential direction (see Fig. 29a). When the tensile stress reaches the tensile strength, cracking occurs and the compliance of the colder parts of the lining increases significantly. Hence, the restraint in the plane of the cross section of the tunnel is reduced, resulting in a decrease of the peak value of the compressive stresses in the circumferential direction (see Fig. 29a). Because of the restriction of the tensile force n_φ in the shotcrete shell (see the thin line in Fig. 29d for $t' \geq 0.75$ h), the increase of n_φ in the concrete lining is reduced after σ_φ has reached the tensile strength in the shotcrete shell. The evolution of the compressive force n_z in the concrete lining, on the other hand, is characterized by a continuous increase until a peak value is reached at approximately $t' = 7.5$ h. Because of the plane-strain condition, the loading of the shotcrete shell in the longitudinal direction is not affected by the loading of the concrete lining. The tensile force in the shell at the beginning of the fire accident is induced via Poisson's effect by tensile loading in the circumferential direction. In the course of the fire accident, thermal dilation outbalances the initial

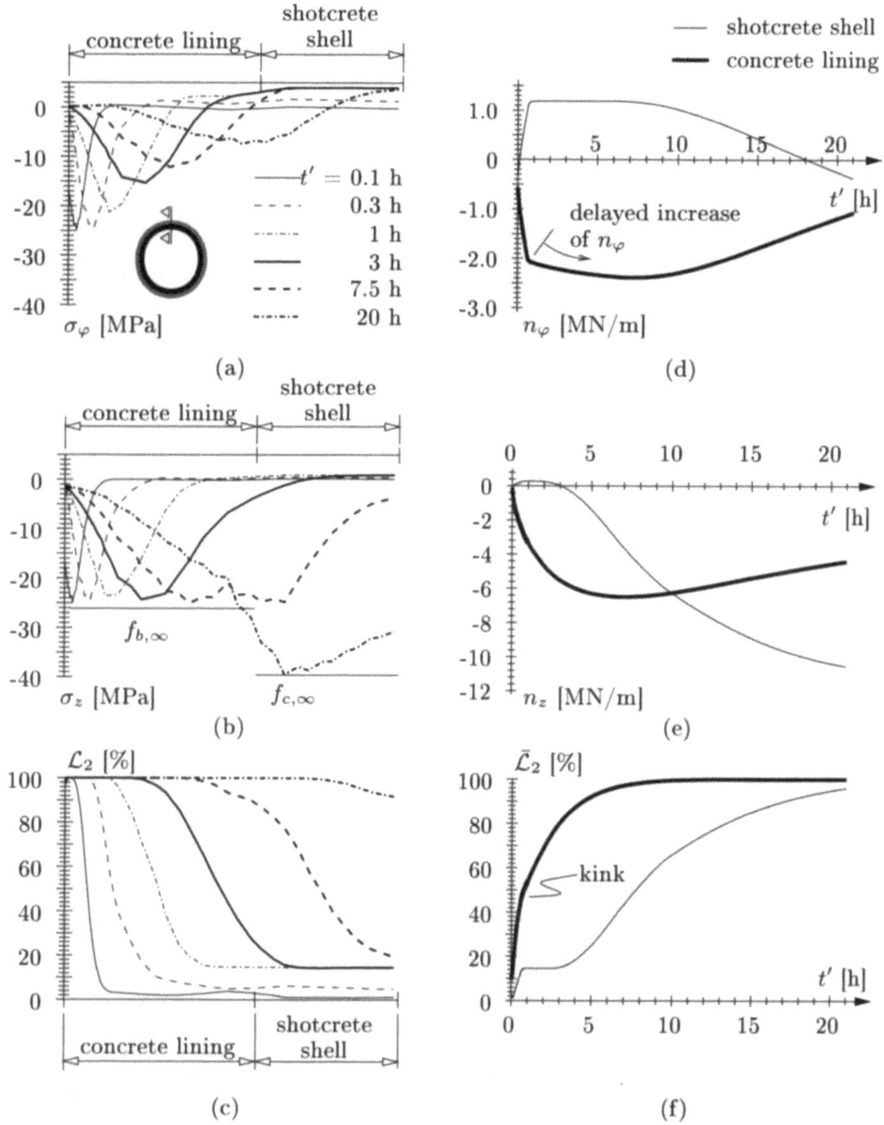

Fig. 29. Distribution of (a) σ_φ, (b) σ_z, and (c) $\mathcal{L}_2$ at different time instants and evolution of (d) n_φ, (e) n_z, and (f) $\bar{\mathcal{L}}_2$ of the concrete lining and the shotcrete shell in a section located at the top of the tunnel ($t' = 0$ refers to the time instant at the start of the fire accident; for the definition of $\mathcal{L}_2$ and $\bar{\mathcal{L}}_2$, see Chap. 10)

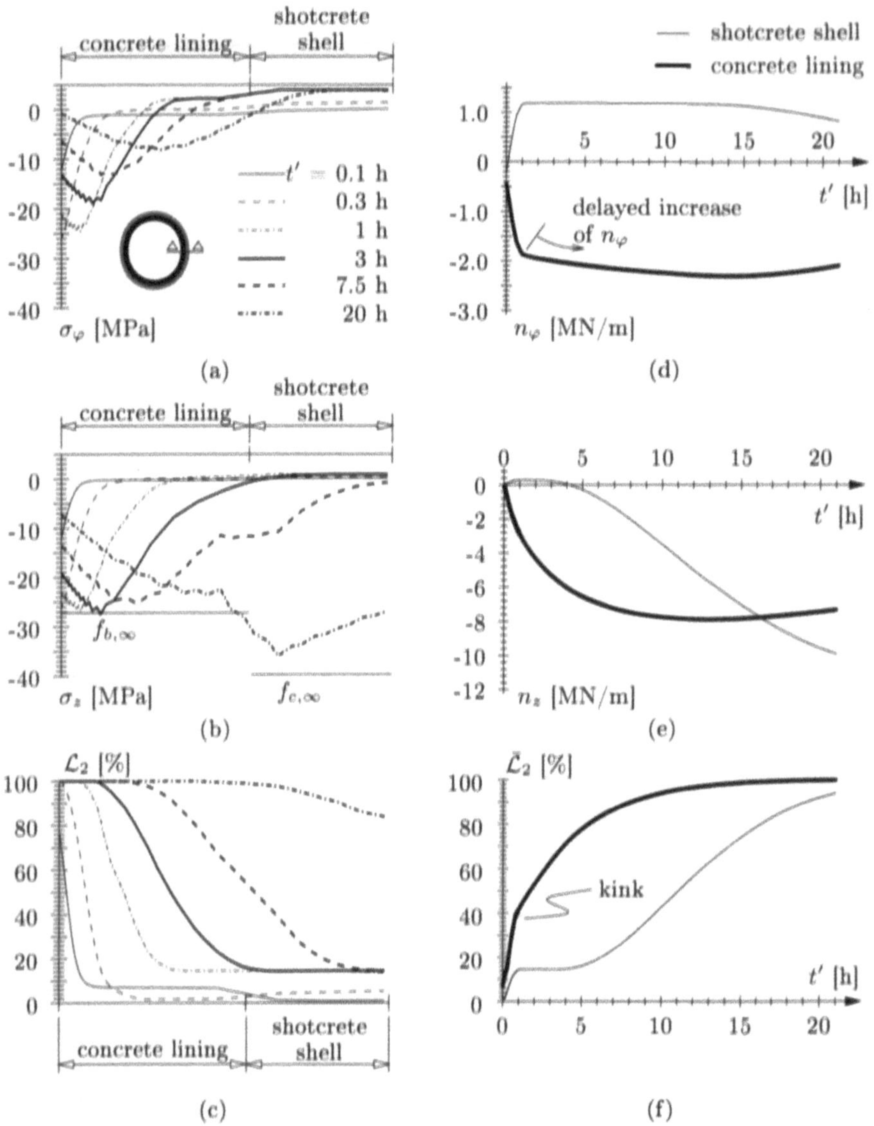

Fig. 30. Distribution of (a) σ_φ, (b) σ_z, and (c) $\mathcal{L}_2$ at different time instants and evolution of (d) n_φ, (e) n_z, and (f) $\bar{\mathcal{L}}_2$ of the concrete lining and the shotcrete shell in a section located at the bench of the tunnel ($t' = 0$ refers to the time instant at the start of the fire accident; for the definition of $\mathcal{L}_2$ and $\bar{\mathcal{L}}_2$, see Chap. 10)

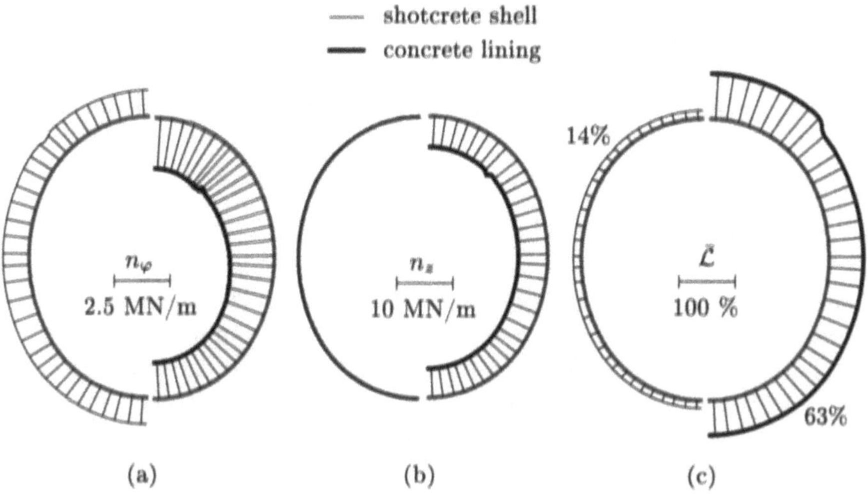

Fig. 31. Distribution of (a) n_φ, (b) n_z, and (c) $\bar{\mathcal{L}}_2$ at $t' = 3$ h ($t' = 0$ refers to the time instant at the start of the fire accident)

tensile stresses, finally resulting in large compressive loading of the shotcrete shell (see the thin line in Fig. 29e). The stress distributions in sections of the lining characterized by the radiation boundary condition (Fig. 30a and b) are similar to those obtained on the basis of the prescribed-temperature boundary condition (Fig. 29a and b). However, because of the smaller temperature gradient in the vicinity of the inner surface of the concrete lining, the propagation of the temperature and of the dehydration front is delayed.

Figure 31a and b show the distributions of n_φ and n_z for $t' = 3$ h. For each one of the two parts of the cross section governed by one of the two considered thermal boundary conditions, an almost uniform distribution is observed. Slight variations of n_φ stem from the loading of the lining prior to the fire accident (see Fig. 23a).

In the course of the fire accident, the level of loading $\mathcal{L}_2$ is rapidly increasing. The biaxial stress state close to the the exposed surface of the tunnel lining results in a level of loading $\mathcal{L}_2$ of 100% (see Fig. 29c and Fig. 30c). Moreover, the region characterized by $\mathcal{L}_2 = 100\%$ is ranging from the inner surface of the lining to the position of the peak of the bell-shaped stress distribution shown in Fig. 29b and Fig. 30b. Since the position of the peak value moves towards the colder parts of the lining, the portion of the lining with $\mathcal{L}_2 = 100\%$ is increasing. Accordingly, the average value of the level of loading increases, reaching $\bar{\mathcal{L}}_2 = 80\%$ at $t' = 3$ h in sections of the concrete lining governed by the prescribed-temperature boundary condition. In sections where radiation is accounted for, $\bar{\mathcal{L}}_2 = 63\%$ (see Fig. 31c). The kink in the evolution of $\bar{\mathcal{L}}_2$ in Fig. 29f and Fig. 30f refers to the time instant at which cracking of the shotcrete shell begins. The reduced increase of the compressive loading in the concrete lining for $t' > 0.75$ h is reflected by the reduced increase of $\bar{\mathcal{L}}_2$ after the kink.

The quantity $\bar{\mathcal{L}}_2$ provides insight into the safety of the tunnel and, hence, allows assessing the safety of fire-fighting units approaching the location of the fire accident.

5 Concluding remarks

Consideration of the soil in the numerical analyses presented in this chapter allowed to determine the state of loading of the surrounding soil, which in turn permitted to quantify the surface settlements and the stability of the tunnel. In the analysis of the underground station "Taborstraße", uncertainties arising from material parameters of the soil layers were accounted for by performing two analyses (one with and one without jet-grouted support) and relating the results to each other. In the analysis of the tunnel subjected to fire load, measurement data of the deformation of the lining corresponding to the construction process served as input for determination of uncertain material parameters of the soil via back analysis. The latter, however, was based on a trial-and-error scheme and, hence, required great computational effort. Automatization of back analysis of material parameters together with a reduction of computational effort is the goal of *parameter identification based on soft computing* (for an outline of the theoretical basis and applications to NATM tunneling, the reader is referred to Chap. 9).

References

[1] Beiche H., Erichsen C., Kargerer W., Schilcher E., Wooge M. (1992) Praktische Lösungen bei der vorauseilenden Schirminjektion unter Einsatz von Hochdruckinjektionen (HDI) [Practical solutions for the antecedent canopy injection using jet grouting]. Taschenbuch für den Tunnelbau, **16**, Glückauf GmbH, Essen, in German

[2] Borja R.I., Regueiro R.A., Lai T.Y. (2000) FE modeling of strain localization in soft rock. *Journal of Geotechnical and Geoenvironmental Engineering*, **126**(4):1–9

[3] Ch. Brandstätter, R. Lackner, and H.A. Mang (2002) Bestimmung von Materialeigenschaften zementgebundener Werkstoffe mittels Temperaturmessungen: Anwendung im Rahmen des Düsenstrahlverfahrens [Evaluation of properties of cementitious materials by means of temperature measurements: Application to jet grouting]. *Bauingenieur*, **77**(2):51–57, in German

[4] G. Duvaut and J.L. Lions (1972) Les inequations en mechanique et en physique [The inequations in mechanics and physics]. Dunod, Paris, in French

[5] European Institute for Standardization (1992) Design of Concrete Structures

[6] E. Falk (1997) Underground works in urban environment. In *Proceedings of the 14th International Conference on Soil Mechanics and Foundation Engineering*, **3**:1401–1406, Hamburg, Germany. Balkema, Rotterdam

[7] E. Falk (2001) Compensation grouting – Stand der Anwendung im Tunnelbau [Compensation grouting – State of the art in tunneling]. In H. Brandl, editor, *Proceedings of the 3rd Austrian Symposium on Geotechniques*, pages 167–180, Vienna, Austria, in German

[8] D. Fitl (1997) Der Tunnel Frankfurter Kreuz [The Frankfurter Kreuz tunnel]. *Felsbau*, **15**(5):347–350, in German

[9] G. Gioda and L. Locatelli (1999) Back analysis of the measurements performed during the excavation of a shallow tunnel in sand. *International Journal for Numerical and Analytical Methods in Geomechanics*, **23**:1407–1425

[10] G. Guatteri, P. Mosiici, A. Koshima, and V. Doro-Altan (1994) Application of jet grouting to tunnel portals and top headings in NATM tunneling: Brazilian experience. In A.L. Bell, editor, *Grouting in the ground*, pages 455–471, London, England. Thomas Telford, London

[11] Ch. Hellmich, H.A. Mang, and F.-J. Ulm (1999) Hybrid method for quantification of stress states in shotcrete tunnel shells: combination of 3D *in-situ* displacement measurements and thermochemoplastic material law. In W. Wunderlich, editor, *CD-ROM Proceedings of the 1st European Conference on Computational Mechanics*, Munich, Germany, 1999. Also published in *Computers and Structures*, **79**:2103–2115, 2001

[12] R. Lackner, Ch. Hellmich, and H.A. Mang (2002) Constitutive modeling of cementitious materials in the framework of chemoplasticity. *International Journal for Numerical Methods in Engineering*, **53**(10):2357–2388

[13] R. Lackner, J. Macht, Ch. Hellmich, and H.A. Mang (2002) Hybrid analyses of segmented shotcrete tunnel linings in squeezing rock. *Journal of Geotechnical and Geoenvironmental Engineering (ASCE)*, **128**(4):298–308

[14] R. Lackner and H.A. Mang (2003) Cracking in shotcrete tunnel shells. *Engineering Fracture Mechanics*, **70**(7-8):1047–1068

[15] R. Lackner and H.A. Mang (2002) Modeling of early-age fracture of shotcrete: application to tunneling. In B.L. Karihaloo, editor, *Proceedings of the IUTAM Symposium on Analytical and Computational Fracture Mechanics of Non-homogeneous Materials*, pages 197–210, Cardiff, Wales. Kluwer Academic Publishers, Dordrecht

[16] H. Liebsch and C. Haberland (1998) Soft ground tunneling in urban areas by NATM. *Felsbau*, **16**(2):78–90

[17] W.-M. Lin, T.D. Lin, and L.J. Powers-Couche (1996) Microstructure of fire-damaged concrete. *ACI Materials Journal*, **93**(3):199–205

[18] J. Macht, R. Lackner, Ch. Hellmich, and H.A. Mang (2001) Shotcrete creep properties – review of material tests and structural analyses of tunnels. In F.-J Ulm, Z.P. Bažant, and F.H. Wittmann, editors, *Proceedings of the 6th International Conference on Creep, Shrinkage and Durability Mechanics of Concrete and other Quasi-Brittle Materials*, pages 285–300, Cambridge, USA. Elsevier Science Ltd., Amsterdam

[19] T. Marcher and P.A. Vermeer (2001) Macromodelling of softening in non-cohesive soils. In P.A. Vermeer, S. Diebels, W. Ehlers, H.J. Herrmann, S. Luding, and E. Ramm, editors, *Continuous and discontinuous modeling of cohesive-frictional materials, Lecture notes in Physics*, **568**:89–108. Springer, Berlin

[20] L.T. Phan and N.J. Carino (1998) Review of mechanical properties of HSC at elevated temperatures. *Journal of Materials in Civil Engineering, ASCE*, **10**(1):58–64

[21] R.A. Regueiro and R.I. Borja (1999) A finite element model of localized deformation in frictional materials taking a strong discontinuity approach. *Finite Elements in Analysis and Design*, **33**:283–315
[22] J.C. Simo and T.J.R. Hughes (1998) Computational Inelasticity. Springer, Berlin
[23] D. Sterpi (1999) An analysis of geotechnical problems involving strain softening effects. *International Journal for Numerical and Analytical Methods in Geomechanics*, **23**:1427–1454, 1999
[24] K. Terzaghi, R.B. Peck, and G. Mesri (1996) Soil mechanics in engineering practice. John Wiley and Sons, New York
[25] F.-J. Ulm, P. Acker, and M. Lévy (1999) The "Chunnel" fire. II: Analysis of concrete damage. *Journal of Engineering Mechanics (ASCE)*, **125**(3):283–289
[26] F.-J. Ulm and O. Coussy (1995) Modeling of thermochemomechanical couplings of concrete at early ages. *Journal of Engineering Mechanics (ASCE)*, **121**(7):785–794
[27] F.-J. Ulm, O. Coussy, and Z.P. Bažant (1999) The "Chunnel" fire. I: Chemoplastic softening in rapidly heated concrete. *Journal of Engineering Mechanics, ASCE*, **125**(3):272–282
[28] I. Vardoulakis, M. Goldscheider, and G. Gudehus (1978) Formation of shear bands in sand bodies as a bifurcation problem. *International Journal for Numerical and Analytical Methods in Geomechanics*, **2**:99–128
[29] P.A. Vermeer and R. de Borst (1984) Non-associated plasticity for soils, concrete and rock. *HERON*, **29**(3):3–64
[30] W. Wittke, B. Pierau, and C. Erichsen (2000) Der Einsatz von Hochdruckinjektionen zur Baugrundverbesserung und für den Tunnelbau im Lockergestein [Application of jet grouting in ground improvement and tunneling in granular material]. In G. Riedmüller, W. Schubert, and S. Semprich, editors, *Düsenstrahlverfahren: Möglichkeiten und Grenzen der Anwendung*, Christian Veder Kolloquium, pages 155–183, Graz, Austria. Gruppe Geotechnik Graz, Graz, in German
[31] D.M. Wood (1990) Soil Behavior and Critical State Soil Mechanics. Cambridge University Press, Cambridge

Chapter 7
Damage mechanics of jointed rock in tunnelling

Qiang Yang[1], Gunter Swoboda[2], Dean Zhao[2,3], and Franz Laabmayr[4]

[1] Department of Hydraulic Engineering, Tsinghua University
[2] Institute for Structural Analysis and Strength of Materials, University of Innsbruck
[3] Department of Civil Engineering, Lanzhou Railway Institute
[4] Laabmayr and Partner Consulting Engineers

Abstract. In this chapter, an anisotropic damage model is established in strain space to describe the behavior of jointed rock masses under compression-dominated stress fields. The research work focuses on rate-independent and small-deformation behavior during isothermal processes. It is emphasized that the damage variables should be defined microstructurally rather than phenomenologically for jointed rock masses, and a second-order "fabric tensor" is chosen as the damage variable. Starting from it, a one-parameter damage-dependent elasticity tensor is deduced based on tensorial algebra and thermodynamic requirements; An equivalent state is developed to exclude the macroscopic stress/strain explicitly from the relevant constitutive equations. Finally, some numerical results are worked out to illustrate the mechanical behavior of this model. In order to eliminate the complexity and arbitrariness in the formulation of phenomenological anisotropic model, systematic micromechanical analysis has been done on damage elasticity, damage evolution laws and micromechanics of different types of microdefects.

1 Introduction

There exist high geo-stresses in natural rocks, especially with high overburden or under complicated geological conditions. During tunnel excavation processes, huge stress release loading is applied to the surrounding rocks and in turn changes the internal structure of the rocks or *damaging*. One of the behaviors caused by the damaging is the observed time-dependent deformation on tunnel faces. Without effective support measures, the interaction between the stress release loading and damaging may finally lead to tunnel collapse. In view of the complicated stress states and high stress level, and full range deformation processes from strain hardening to softening with time effects in surrounding rocks, no doubt, mechanical properties of rocks are revealed most sufficiently in tunnelling of all geotechnical engineering.

Except for very unusual cases, natural rock masses are always cracked, jointed and faulted mediums. The discontinuity system in rock masses as shown in Fig. 1 plays a central role in the damaging processes and is a key factor to the stability of tunnels. From the point of view of engineering geology, the influence of the crack size which is much smaller than the tunnel size is negligible, see Fig. 1a, and the rock mass with such cracks can be described very well by classical theories of elasticity or plasticity. On the other hand, large discontinuities, e.g. the fault in Fig. 1d, should be considered one by one and usually are simulated by joint or contact elements in the finite element method. The regularly distributed joint sets of the same order of the tunnel size, as shown in Fig. 1b and c, are of the special importance to tunnelling and is the research focus of this chapter.

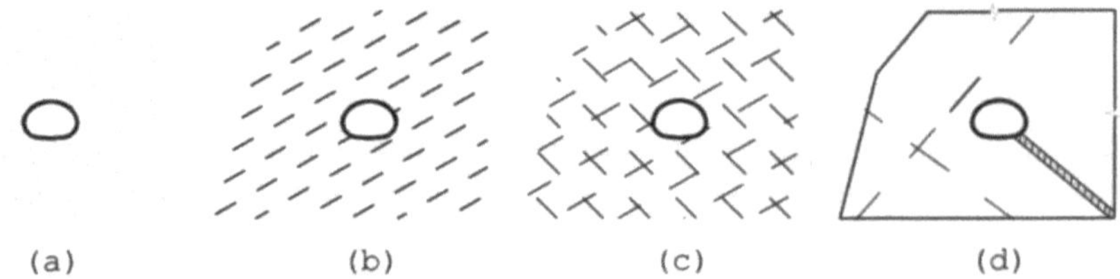

Fig. 1. Tunnels in the rock mass with different discontinuities

In general, there are two approaches to take the joint network into account in numerical methods, discontinuous and continuous approach. One typical discontinuous method is the distinct element method (DEM) proposed by Cundall [9]. In spite of its clear physical background, the requirement of definite joint network and the time-consuming feature make the discontinuous approach less feasible for large-scale calculations, especially for three dimensional calculation. Up to now, the continuous approach with the help of FEM is still the only feasible way to deal with large-scale calculations.

The well-established classic elasto-plastic theory is a certain formalism which treat the concerned solids as "black boxs" and it is difficult to incorporate the internal structure of materials into the constitutive equations. Continuum damage mechanics (CDM) is no doubt an appealing framework for jointed rock since the damage variables are indicators of the internal structures of materials.

1.1 A brief review of continuum damage mechanics

Jointed rock masses have complex mechanical behaviour, such as stress-induced anisotropy, hysteresis, dilatancy, irreversible strain and strongly path-dependent stress-strain relations, which is generally associated with the existence of a great deal of micro- and macro-cracks and their propagation. Continuum damage mechanics (CDM), which employs some continuum variables to describe the micro-defects, has been an appealing framework for modeling rock-like materials, see e.g. Ju [16], Dragon and Mroz [12], Dragon et al. [11] Kawamoto et al. [19], Zhang [48] and Stumvoll and Swoboda [37].

In most of these models, the damage variables are defined through a phenomenological way originating from the classical Kachanov's [17] damage variable. Although they are often interpreted as the net area reduction caused by the distributed microscopic cracks and cavities and associated with a net stress (see, e.g. Lemaitre [22]), as shown in Fig. 2, these damage variable are essentially the measures of the damage's mechanical effects rather than the microstructure. Consider a simple one-dimensional case. The explanation of net area reduction leads to the concept of net stress $\overline{\sigma} = \frac{\sigma}{1-\omega}$ where ω is the damage variable and σ is the macroscopical stress. The introduction of the *hypothesis of strain equivalence* by Lemaitre [22] further leads to $E = E_0(1-\omega)$ where E and E_0 are the apparent and intrinsic Young's modulus, respectively. On the other hand, its equivalent form $\omega = 1 - \frac{E}{E_0}$ demonstrates the true physical meaning of the damage variable. Ju [16] developed the idea to the fullest by taking the elasticity tensor directly as the damage variable.

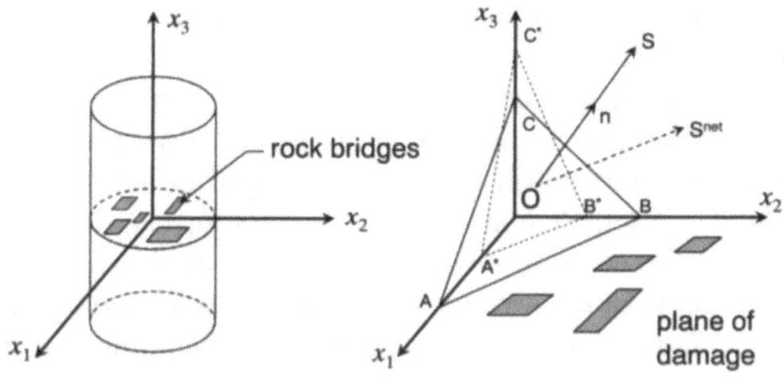

Fig. 2. Schematic one and three dimension net area reduction

Some problems arise in extending the phenomenological damage definition to jointed rock masses : (**1**) rock stiffness is significantly dependent on the sign of stress/strain. Thus, the phenomenological damage variable also becomes stress/strain-dependent, which violates the independent principle of state variables. (**2**) The phenomenological damage measure cannot take into account the geological data of jointed rock masses, which is measurable and provided important parameters for engineering designs. (**3**) The phenomenological damage definition furnishes a simple damage-elasticity relation (e.g., $E = E_0(1-\omega)$) at the cost of damage evolution laws. Since the phenomenological damage definition lacks a definite microscopical meaning, the corresponding damage evolution law has been the weakest and most arbitrary aspect. (**4**) The triaxial generalization of the net stress leads to a nonsymmetric net stress tensor, which must further be symmetrized in some arbitrary ways in order to obtain a practical formulation, see, e.g. Murakami [26].

In general, these problems stems from the fact that the essence of damage is the distributed micro-defects, and its the mechanical behaviours presents in many aspects and generally there do not exist one-to-one correspondence between the behaviour and the damage. The relation between the phenomenological damage definition and the microstructure is not definite and unique but correlative, and the deviation sometimes becomes significant. In this chapter, a second-order fabric tensor (see e.g. Oda [28] and Cowin [8], which is a geometric measure of material microstructure, is chosen as the damage variable, see (1).

1.2 Introduction to this research

The main task of this research is to incorporate the microstructural parameter of jointed rock, i.e., the second-order fabric tensor, to its constitutive equations. Within the framework of continuum damage mechanics, a second-order damage tensor leads to anisotropic damage models. The geometric damage measurement and anisotropy pose two challenges to this research, and there are no existing frameworks which can accommodate the starting points[1]. Based on such a damage tensor, the damage-dependent elasticity tensor cannot be deduced based on the hypothesis of strain equivalence. Therefore, the overall micromechanical analysis is necessary otherwise the direct phenomenological formulation will be too complicated or arbitrary.

In this chapter, Sects. 3–5 are devoted to micromechanical analysis of microcracked solids. Based on these micromechanical analysis, a phenomenological damage model with the second damage tensor is formulated in Sects. 6-8. Finally, some material and structural examples, including the underground theater of Linz, are presented in Sect. 9.

Some micromechanical attempts have been made to establish the damage elasticity relation based on the fabric tensor, see e.g., Oda [29] and Lemaitre [22]. However, their deduced analytic expressions involve second- and fourth-rank fabric tensors and basically only furnish a linear approximation with respect to the principal values of the damage tensor. It will further complicate the expression to take into account crack interaction. In Sect. 3, we develop an explicit expression of second-order fabric-tensor dependent effective compliance tensor of cracked solids [47] based on the work of Budiansky and O'Connell [6] and properties of orientation distribution functions (ODF), within the framework of non-interaction approximation.

From a phenomenological point of view, Cowin [7] developed the most general fabric tensor-dependent elasticity tensor by tensorial algebra, but

[1] Kawamoto et al. [19] extended the classical damage theory to model jointed rock with the damage tensor based on statistical geological data. However, their efforts failed at the point where the geometrical damage tensor is not compatible with the classical damage theory.

its expression involves too many unknown parameters. In Sect. 7, a one-parameter damage-dependent elasticity tensor [40] is deduced by tensorial algebra and thermodynamic requirements, see (78) and (83), which furnishes a fourth-order approximation.

For rock-like materials, most micromechanical analyses treat the damage microscopically as the distributed microcracks and often resort to the microcrack model of Budiansky and O'Connell [6]. However, this model is mainly valid in a tension-stress dominated field which is quite different with the stress state in the jointed rock masses. In Sect. 4, a general method with the so called energy-based equivalent inclusion size is proposed to evaluate the mechanical effects of various defects, see also Yang et al. [46].

In rock-like materials, the damaging process is often coupled with plastic flow. Certain special loading processes, e.g. hydraulic fracturing, have significant contribution to damage propagation. To describe such complicated mechanisms, only conjugate-force-based damage evolution laws can keep a unitary and compact form. As summarized by Chow and Lu [7], many damage evolution laws of second damage tensors, e.g, from Chaboche, Lee, Murakami-Ohno, Sidoroff-Cordebois, etc., can be covered by linear irreversible thermodynamics, $\dot{\boldsymbol{\Omega}} = \mathbf{J} : \mathbf{Y}$ where $\mathbf{Y}$ is the thermodynamic force conjugate to the damage tensor and $\mathbf{J}$ is a fourth-order damage characteristic tensor, no matter how complicated the internal dissipative mechanism is.

The conjugate force $\mathbf{Y}$ is uniquely determined by the corresponding free energy function. In Sect. 8, a general free energy function is developed, which includes damage, plastic and hydraulic fracturing internal variables. The characteristic tensor $\mathbf{J}$ has lacked of a definite definition due to the phenomenological attribute of the damage tensor. In Sect. 5, the micromechanical basis of nonlinear irreversible thermodynamics are disscussed in the framework Rice's [32][33] *normality structure.* Following strictly the definition of the fabric tensor, an analytic expression of $\mathbf{J}$ in (75) is developed by Swoboda and Yang [41], and the further refinement of the expression has been give by Yang et al. [44].

It is arguable to apply conjugate-force-based damage evolution laws to rock-like materials. As discussed in Sect. 5, these laws are suitable to a class of materials in which the influence of the macroscopic stress/strain on each defect appears only through the conjugate forces. In jointed rock masses, the macroscopic stress/strain changes the contact manner (close or open) of crack surfaces which in turn changes the crack propagation rule quantitatively and qualitatively, so it is difficult to exclude the macroscopic stress/strain explicitly from damage evolution laws. To overcome the difficulty and make a practical formulation, some authors simply assume that the drive force behind the damage propagation is the maximum tensile net stress. Obviously, it is a very coarse assumption. Damage generally propagates not along the direction of its conjugate force; the current existing damage also has influence on the direction (mainly through **J**).[2] Furthermore, the conjugate force is much better than the stress to take into account other mechanisms contributing to damage propagation. In the following phenomenological damage model, the conjugate-force-based damage evolution law is used and the influence of macroscopic stress/strain is reflected implicitly by introducing *an equivalent state* in Sect. 6.

2 Definition of damage

Classics continuum mechanics is a certain local approach, so is the continuum damage mechanics. In this sense, strictly speaking, CDM can only deal with the case of Fig. 1a, and certain nonlocal approach should be used for the cases of Fig. 1b and c, see, e.g. Bazant [4]. Two types of damage definitions are considered in this chapter, micromechanical and engineering definitions. The difference between the two definitions:

2.1 Micromechanical definition

The defects in rock-like materials consist of microcavities and microcracks. It is reasonable to assume that the governing defects in rock-like materials are microcracks. Consider a material sample with the size V^0 weakened by n microcracks. If each crack can be idealized as a penny crack, it can be defined uniquely by its radius and normal vector by neglecting the minor influence of its location. The damage tensor can be defined as

$$\boldsymbol{\Omega} = \frac{1}{V}\sum_{\alpha=1}^{n} r_\alpha^3 \mathbf{n}^\alpha \mathbf{n}^\alpha. \tag{1}$$

where r_α and $\mathbf{n}^\alpha$ are the radius and normal vector of the α-th crack. Essentially, (1) is the discrete expression of Oda's *fabric tensor* [28]. In order to further establish the damage evolution laws, the definition can be elaborated into an equivalent multi-level definition as follows.

The microstructure of the sample can be characterized by one set of basic internal variables

$$\boldsymbol{\xi} = \{\xi_1, \xi_2 \ldots, \xi_n\}, \qquad \xi_\alpha = \{\mathbf{n}^\alpha, r_\alpha\} \tag{2}$$

namely that one crack is represented by one internal variable ξ_α. Evidently, it consists of too many variables. A condensed damage definition [32] is expressed as

[2] The governing principle is that this direction will make the system dissipate maximum energy for a certain damage growth.

$$\boldsymbol{\zeta} = \{\zeta_1, \zeta_2 \ldots, \zeta_m\},$$
$$\zeta_\mu = \zeta_\mu(\xi_1, \xi_2 \ldots, \xi_n; V) \qquad (\mu = 1, 2, \ldots, m \ll n) \tag{3}$$

where V indicates averaging over the volume. For crack distribution, the condensation is usually done by regrouping microcracks into m sets of roughly parallel cracks. Each group is represented by $\zeta_\mu = \{\mathbf{n}^\mu, \omega_\mu\}$, which is called a damage vector. If $m = 1$, the damage vector can be defined as

$$\mathbf{n} = \frac{1}{n}\sum_{\alpha=1}^{n} \mathbf{n}^\alpha, \qquad \omega = \frac{1}{V}\sum_{\alpha=1}^{n} r_\alpha^3. \tag{4}$$

For multiple groups of cracks, the expression is the same, but the summation is only for the corresponding group of cracks. Therefore, the damage tensor defined in (1) can be redefined as the condensation of the damage vectors $\boldsymbol{\zeta}$

$$\boldsymbol{\Omega} = \sum_{\mu=1}^{m} \omega_\mu \mathbf{n}^\mu \mathbf{n}^\mu. \tag{5}$$

If the cracks in one group are exactly parallel, the damage definition (5) is equivalent to (1). Krajcinovic [20] endowed ω_μ with the implication of net-area reduction by assuming that, for one cross-section as shown in Fig. 3, all the microcracks within the influence band $(-l < \eta < l)$ will contribute to the damage variable ω of the cross-section with a weight function $w(\eta)$, and then a result similar to (4)b is obtained.

2.2 Engineering definition

Damage must be defined by a continuous homogeneous variable as an internal variable in continuum thermodynamics theory. In geology, each set of average distributed joints is usually described by three parameters (see Fig. 4), i.e., ω, the separation factor of one set of cracks, β, the angle between horizontal and the line of maximum dip of the joint plane and α, the angle between north and the horizontal projection of the line of maximum dip of the joint plane. To integrate these information, a second-order damage tensor is introduced as follows:

$$\Omega_{ij} = \omega n_i n_j, \tag{6}$$

where $\mathbf{n} = [l, m, n]^T$ is the direction vector of damage tensor $\boldsymbol{\Omega}$ of the crack set. To position the joint plane in a rectangular coordinate system, an additional angle between x axis and north, γ is introduced (see Fig. 4). Thus, the spherical coordinates can be written as follows:

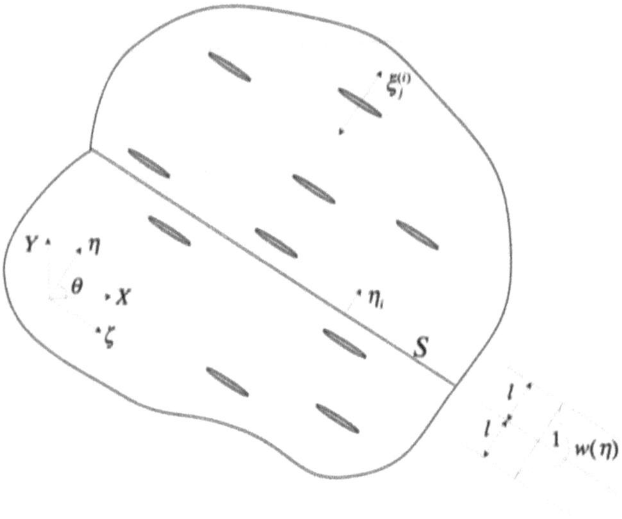

Fig. 3. A system of parallel microcracks

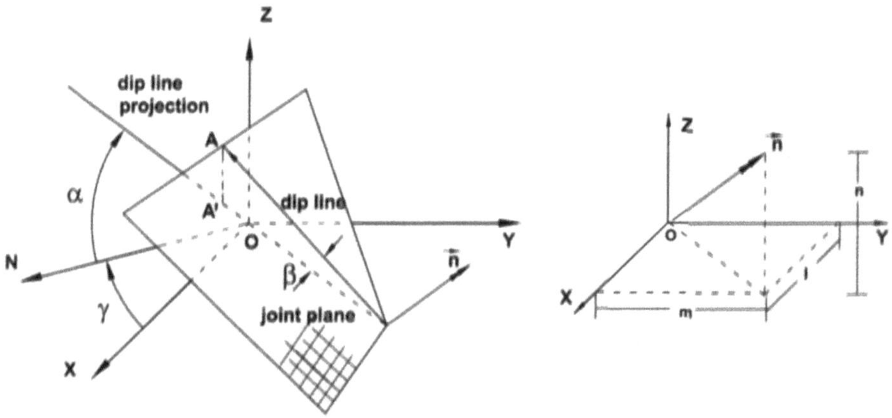

Fig. 4. Description of spatial joint plane and spherical coordinate system

$$\begin{cases} l = \sin\beta\cos[\pi - (\alpha+\gamma)] \\ m = \sin\beta\sin[\pi - (\alpha+\gamma)] \\ n = \cos\beta \end{cases} . \tag{7}$$

For the damage state of multiple sets of joints, the global damage tensor can be obtained by employing the principle of energy equivalence, i.e., the energy dissipated by the global damage tensor is equivalent to the sum of the energy dissipated by the damage tensor related to each set of joins. Refer to the damage elasticity described in the latter section of this chapter; the explicit formulation of calculating the global damage tensor is given in the following:

$$\boldsymbol{\Omega}_g = \mathbf{I} - \left[\sum_{n=1}^{N} (\mathbf{I} - \boldsymbol{\Omega}_n)^{-1} - (N-1)\mathbf{I} \right]^{-1} \tag{8}$$

3 Micromechanical analysis of damage elasticity

Except for the phenomenological approach for the damage elasticity $\mathbf{D}(\boldsymbol{\Omega})$ discussed in Sect. 7, the relation can also be established by micromechanics. One of the advantages is that micromechanics-based damage elasticity has no unknown coefficients. Here, the elastic compliance tensor $\mathbf{C} = \mathbf{D}^{-1}$. The following decomposition is usually assumed,

$$\mathbf{C}(\boldsymbol{\Omega}) = \mathbf{C}^0 + \Delta\mathbf{C}(\boldsymbol{\Omega}), \tag{9}$$

where $\mathbf{C}^0 = (\mathbf{D}^0)^{-1}$. With the work of Budiansky and O'Connell [6], the compliance disturbance due to the presence of a typical penny-shaped crack $\{\mathbf{n}, r\}$ in volume V is

$$\hat{\mathbf{C}} = \frac{8}{3}\frac{1-\nu}{2-\nu}\frac{1}{2\mu}\frac{r^3}{V}(\mathbf{T}' - 2\nu\mathbf{N}), \tag{10}$$

where μ, ν are shear modulus and Poisson's ratio of the matrix, respectively, and

$$\mathbf{T}' = T'_{ijkl} = n_i n_k \delta_{jl} + n_i n_l \delta_{jk} + n_j n_k \delta_{il} + n_j n_l \delta_{ik}, \qquad \mathbf{N} = \mathbf{nnnn}. \tag{11}$$

Within the framework of non-interaction approximation, the compliance increment $\Delta\mathbf{C}$ for a solid of size V^0 weakened by n cracks can be calculated by

$$\Delta\mathbf{C} = \sum_{\alpha=1}^{n} \hat{\mathbf{C}}_\alpha. \tag{12}$$

3.1 Discrete and continuous fabric tensor

Similar to (1), the fourth-order fabric tensor in discrete form is introduced as

$$\boldsymbol{\Omega} = \frac{1}{V}\sum_{\alpha=1}^{n} r_\alpha^3 \mathbf{n}^\alpha\mathbf{n}^\alpha\mathbf{n}^\alpha\mathbf{n}^\alpha \qquad \text{or} \qquad \Omega_{ijkl} = \frac{1}{V}\sum_{\alpha=1}^{n} r_\alpha^3 n_i^\alpha n_j^\alpha n_k^\alpha n_l^\alpha. \tag{13}$$

and the zero-th order fabric tensor

$$\omega = \frac{1}{V}\sum_{\alpha=1}^{n} r_\alpha^3, \qquad \omega = \mathrm{tr}\boldsymbol{\Omega} = \Omega_{kk}. \tag{14}$$

Equations (13)e9a can be considered as the discrete form of the fabric tensors. Here, its continuous form is established. The crack density distribution function $\rho(\mathbf{n})$ varies in all direction and is an orientation distribution function (ODF). Any possible direction $\mathbf{n}$ can be represented by one point on the surface of a unit sphere, $\mathbf{n} = \{\sin\theta\cos\phi, \sin\theta\sin\phi, \cos\theta\}$. An infinitesimal area on the unit sphere about the end point of $\mathbf{n}$ is denoted $d\mathbf{n} = \sin\theta\, d\phi\, d\theta$ which represents a bundle of directions about $\mathbf{n}$. The crack number in the infinitesimal area is $\frac{\rho(\mathbf{n})}{4\pi}d\mathbf{n}$. It is required that

$$\frac{1}{4\pi}\oint \rho(\mathbf{n})d\mathbf{n} = \rho_0 = \frac{n}{V}, \tag{15}$$

where ρ_0 is the crack volumetric density in the sample and the integral is defined by

$$\oint (\cdot)\, d\mathbf{n} = \int_0^{2\pi}\int_0^{\pi} (\cdot)\sin\theta\, d\phi\, d\theta. \tag{16}$$

Obviously, $\rho(\mathbf{n}) = \rho_0$ for a uniform crack distribution. The crack radius distribution in a material point can be described by another ODF $r(\mathbf{n})$. If set $\hat{\rho}(\mathbf{n}) = \rho(\mathbf{n})r^3(\mathbf{n})$, the continuous form of the fabric-tensors could be

$$\omega = \frac{1}{4\pi}\oint \hat{\rho}(\mathbf{n})d\mathbf{n}, \tag{17a}$$

$$\Omega_{ij} = \boldsymbol{\Omega} = \frac{1}{4\pi}\oint \hat{\rho}(\mathbf{n})\mathbf{n}\mathbf{n}d\mathbf{n}, \tag{17b}$$

$$\Omega_{ijkl} = \boldsymbol{\Omega} = \frac{1}{4\pi}\oint \hat{\rho}(\mathbf{n})\mathbf{n}\mathbf{n}\mathbf{n}\mathbf{n}d\mathbf{n}, \tag{17c}$$

$$\dots$$

On the other hand, the combined ODF $\hat{\rho}(\mathbf{n})$ can be approximated by

$$\hat{\rho}(\mathbf{n}) \approx \omega, \tag{18a}$$

$$\hat{\rho}(\mathbf{n}) \approx \frac{3}{2}(5n_in_j\Omega_{ij} - \omega), \tag{18b}$$

$$\hat{\rho}(\mathbf{n}) \approx \frac{15}{8}(21n_in_jn_kn_l\Omega_{ijkl} - 14n_in_j\Omega_{ij} + \omega), \tag{18c}$$

$$\dots\dots \tag{18d}$$

see, e.g. Yang et al. [45].

3.2 Compliance increment

In view of (10), $\hat{\mathbf{C}}$ can be written as

$$\hat{\mathbf{C}} = \frac{\gamma}{V} r^3 \mathbf{M}, \qquad \gamma = \frac{8}{3}\frac{1-\nu}{2-\nu}\frac{1}{2\mu}, \qquad \mathbf{M} = \mathbf{T}' + p\mathbf{N}, \qquad p = -2\nu. \quad (19)$$

Using (12), (1) and (13), one obtains,

$$\begin{aligned} \Delta\mathbf{C} &= \sum_{\alpha=1}^{m} \hat{\mathbf{C}}_\alpha = \frac{\gamma}{V}\sum_{\alpha=1}^{m} r_\alpha^3 \mathbf{M}_\alpha = \frac{\gamma}{V}\sum_{\alpha=1}^{m} r_\alpha^3 \mathbf{T}'_\alpha + \frac{p\gamma}{V}\sum_{\alpha=1}^{m} r_\alpha^3 \mathbf{N}_\alpha \\ &= \gamma(\Omega_{ik}\delta_{jl} + \Omega_{il}\delta_{jk} + \Omega_{jk}\delta_{il} + \Omega_{jl}\delta_{ik}) + p\gamma\Omega_{ijkl}. \end{aligned} \quad (20)$$

This is a well established expression, see e.g., Lemaitre [22]. However, the expression involves the fourth-order fabric tensor. Yang et al. [45] has solved the crux by substituting (18b) into (17c),

$$\begin{aligned} \Omega_{ijkl} &= \frac{1}{4\pi}\oint \hat{\rho}(\mathbf{n}) n_i n_j n_k n_l d\mathbf{n} \approx \frac{3}{8\pi}\oint (5 n_m n_n \Omega_{mn} - \omega) n_i n_j n_k n_l d\mathbf{n} \\ &= -\frac{1}{35}\omega(\delta_{ij}\delta_{kl} + \delta_{ik}\delta_{jl} + \delta_{il}\delta_{jk}) + \frac{1}{7}(\Omega_{ij}\delta_{kl} + \Omega_{kl}\delta_{ij}) \\ &+ \frac{1}{7}(\Omega_{ik}\delta_{jl} + \Omega_{il}\delta_{jk} + \Omega_{jk}\delta_{il} + \Omega_{jl}\delta_{ik}). \end{aligned} \quad (21)$$

Then, substituting (21) into (20) leads to the explicit expression of $\Delta\mathbf{C}(\Omega_{ij})$ is obtained

$$\begin{aligned} \Delta\mathbf{C}(\boldsymbol{\Omega}) \approx& \frac{2\gamma\nu}{35}\omega(\delta_{ij}\delta_{kl} + \delta_{ik}\delta_{jl} + \delta_{il}\delta_{jk}) - \frac{2\gamma\nu}{7}(\Omega_{ij}\delta_{kl} + \Omega_{kl}\delta_{ij}) \\ &+ \frac{\gamma}{7}(7 - 2\nu)(\Omega_{ik}\delta_{jl} + \Omega_{il}\delta_{jk} + \Omega_{jk}\delta_{il} + \Omega_{jl}\delta_{ik}). \end{aligned} \quad (22)$$

The simplest linear function of $\mathbf{C}(\boldsymbol{\Omega})$ have been proposed by Sayers and Kachanov [36] and Prat and Bazant [31]

$$\Delta\mathbf{C}(\boldsymbol{\Omega}) = \eta_1(\Omega_{ij}\delta_{kl} + \Omega_{kl}\delta_{ij}) + \eta_2(\Omega_{ik}\delta_{jl} + \Omega_{jl}\delta_{ik}). \quad (23)$$

where the parameters η_1 and η_2 are calculated by fitting the overall effective elastic and shear moduli E_{eff} and G_{eff} of the solid filled with uniformly distributed cracks. The moduli E_{eff} and G_{eff} is solved by self-consistent scheme or differential scheme. Evidently, the phenomenological tensorial expression of Sayers and Kachanov [36] or Prat and Bazant [31] has missed the isotropic term and is not complete by comparing (23) with (22).

4 Micromechanics of defects

There are many types of preexisting defects in rocks, e.g., cracks, cavities, inhomogeneities or inclusions. Even for cracks, the behaviors of cracks under tension and compression are quite different. In this section, a general method with the so called energy-based equivalent inclusion size is proposed to evaluate the mechanical effects of various defects.

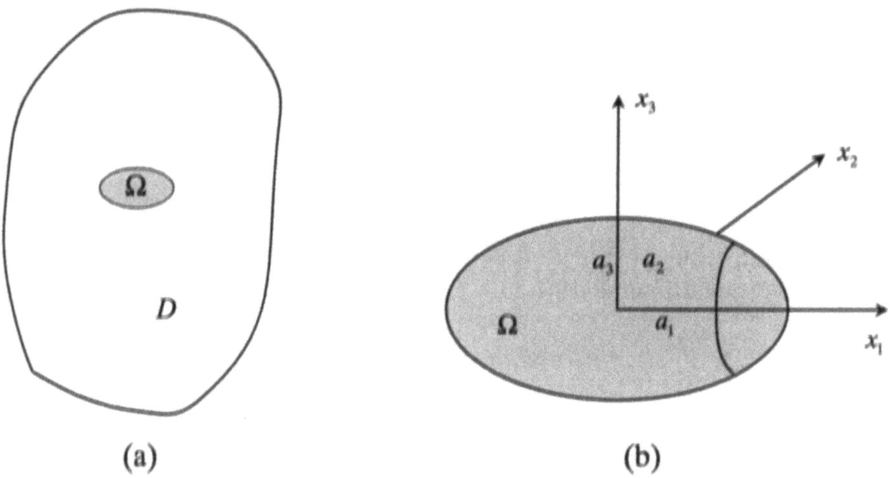

Fig. 5. (a) Inclusion or inhomogeneity Ω; (b) An ellipsoidal inclusion with principal half axes a_1, a_2 and a_3

$$U^* = \frac{1}{2}\int_{\Omega} \boldsymbol{\sigma} : \boldsymbol{\varepsilon}^* dD = \frac{V}{2}\boldsymbol{\sigma} : \boldsymbol{\varepsilon}^* \tag{24}$$

since the eigenstrain $\boldsymbol{\varepsilon}^*$ is uniform in domain Ω, where $\boldsymbol{\sigma}$ is the far-field stress and V is the size of the spheroidal inhomogeneity

$$V = \frac{4\pi}{3}a_1 a_2 a_3 = \frac{4\pi}{3}a^3\frac{a_3}{a} = \frac{4\pi}{3}a^3\xi, \qquad \xi = \frac{a_3}{a}. \tag{25}$$

If the inhomogeneity is in a representative unit cell with volume V^0, the compliance disturbance due to the inhomogeneity is

$$\hat{\mathbf{C}} = \frac{1}{V^0}\frac{\partial^2 U^*}{\partial \boldsymbol{\sigma}^2}. \tag{26}$$

Since $V \to 0$ due to $\xi \to 0$, for a macroscopically sensible penny-shaped inhomogeneity, the eigenstrain $\boldsymbol{\varepsilon}^*$ should be singular, otherwise its energy disturbance U^* is trivial and the inhomogeneity is macroscopically insensible. The asymptotic expressions of the singular eigenstrain ε^* have been obtained by Yang et al. [46], and the singular factors η contained in the singular $\boldsymbol{\varepsilon}^*$ have been identified. If moving the singular factor η from ε^* to V, the equivalent size is $V^{eq} = V\eta$ and the equivalent eigenstrain is $\varepsilon^*_{eq} = \frac{\boldsymbol{\varepsilon}^*}{\eta}$. Both the equivalent size V^{eq} and eigenstrain $\boldsymbol{\varepsilon}^*_{eq}$ are finite quantities, and the energy disturbance can be rewritten as $U^* = \frac{V^{eq}}{2}\boldsymbol{\sigma}^0 : \boldsymbol{\varepsilon}^*_{eq}$, which is the so-called *energy-based inclusion method* [46]. For a weak penny-shaped inhomogeneity, its singular factor η is

$$\eta = \frac{1}{\Gamma}, \qquad \Gamma = \frac{E^*}{E^0} < 1 \quad \Rightarrow \quad V^{eq} = \frac{4\pi}{3}a^3\xi_{eq}, \quad \xi_{eq} = \frac{\xi}{\Gamma} \tag{27}$$

where E^* and E^0 are the Young's modulus of the inhomogeneity and matrix, respectively. As shown in Fig. 6, the equivalent spheroid is much thicker than the real inhomogeneity.

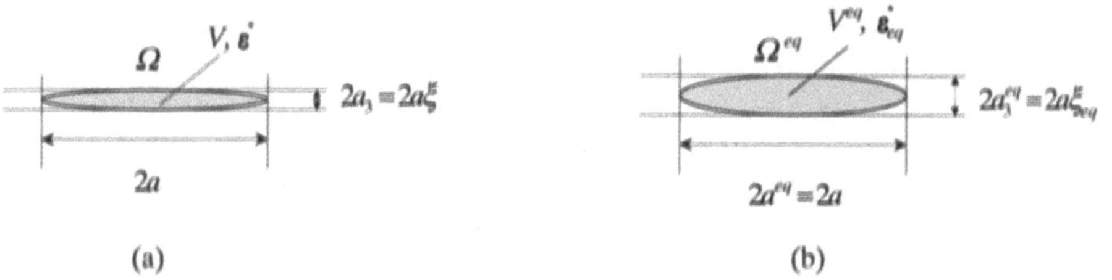

Fig. 6. (a) The Eshelby's equivalent inclusion Ω of the penny-shaped inhomogeneity; (b) The energy-based equivalent inclusion Ω^{eq} of the same inhomogeneity

As a special case of penny-shaped inhomogeneities, the singular factor of a crack is just $1/\xi$, so the corresponding equivalent inclusion is just a sphere, see Fig. 7.

Fig. 7. (a) The Eshelby's equivalent inclusion of a crack; (b) The spherical energy-based equivalent inclusion of the same crack

Naturally, $\eta = 1$ for the inhomogeneities of finite size, and then $V^{eq} = V$. Generally speaking, the Eshelby's and energy-based equivalent eigenstrain can be expressed as

$$\boldsymbol{\varepsilon}^* = \eta \boldsymbol{\varepsilon}^*_{eq}, \qquad \boldsymbol{\varepsilon}^*_{eq} = \boldsymbol{\Pi} : \boldsymbol{\varepsilon}^0. \tag{28}$$

Some quantities related to different inhomogeneities are listed in Tab. 1, where $V^{sp} = \frac{4\pi}{3} a^3$ is a reference size of a sphere with the radius a; the fourth-order tensor $\mathbf{S}$ is the limit of the Eshelby tensor for $\xi \to 0$, and $\mathbf{S}_t$ and $\mathbf{S}_n$ are shear and normal components of $\mathbf{S}$, respectively; $\mathbf{1}$ and $\mathbf{I}$ are identity tensors of rank two and four, respectively.

Table 1. Some characteristic quantities of different inhomogeneities

Inhomogeneities	$\frac{V}{V^{sp}}$	$\frac{V^{eq}}{V^{sp}}$	η	$\boldsymbol{\Pi}$
Penny-shaped inhomogeneity for $\Gamma \to 0$, $\xi_{eq} = \frac{\xi}{\Gamma} \ll 1$	ξ	ξ_{eq}	$\frac{1}{\Gamma}$	$\mathbf{S}$
Penny-shaped inhomogeneity for $\Gamma \to \infty$, $\xi_{eq} = \xi\Gamma \ll 1$	ξ	ξ_{eq}	Γ	$\mathbf{S} - \mathbf{I}$
Penny-shaped crack	ξ	1	$\frac{1}{\xi}$	$c_1 \mathbf{S}_n + c_2 \mathbf{S}_t$
Spherical cavity	1	1	1	$\frac{3(1-\nu^0)}{7-5\nu^0}\left[\frac{1-\nu^0}{2(1-2\nu^0)}\mathbf{1}\mathbf{1} + 5\mathbf{I}\right]$
Rigid sphere	1	1	1	$-\frac{3(1-\nu^0)}{2(4-5\nu^0)}\left(\frac{1-5\nu^0}{1+\nu^0}\mathbf{1}\mathbf{1} + 5\mathbf{I}\right)$

As shown in Tab. 1, all the $\boldsymbol{\Pi}$ tensors of the different inhomogeneities are of the same order of magnitude. Note that $U^* = \frac{1}{2} V^{eq} \boldsymbol{\sigma}^0 : \boldsymbol{\varepsilon}^{eq} = \frac{1}{2} V^{eq} \boldsymbol{\sigma}^0 : \boldsymbol{\Pi} : \boldsymbol{\varepsilon}^0$. Therefore, the size V^{eq} of the energy-based equivalent inclusion Ω^{eq} becomes the characteristic geometric measurement of the energy disturbance caused by the presence of the inhomogeneity. Naturally, the dimensionless damage variable can be defined as (assuming only one single inhomogeneity)

$$\omega = \frac{V^{eq}}{V^0} \tag{29}$$

where V^0 is the volume of a representative unit cell in the microstructure. If the inhomogeneity is a crack, the damage variable is identical to that of Budiansky and O'Connell [6] defined in (14) just scaled by $\frac{4\pi}{3}$. For an inhomogeneity of finite size, $\eta = 1$, then

$$\omega = \frac{V^{eq}}{V^0} = \frac{V}{V^0} \tag{30}$$

which is just Gurson's damage variable [14], the void volume fraction of a porous solid. We call V^{eq} the generic damage measurement because it is applicable to all types of inhomogeneities: cavities, weak and strong inhomogeneities, and rigid inhomogeneities of infinitesimal or finite sizes. It is also convenient to depict the damaging processes of inhomogeneities by V^{eq}. Consider a penny-shaped inhomogeneity with the relative modulus Γ_0 and radius a_0 (state 0), as shown in Fig. 8. The damaging process would be: a stiffness degrading of the inhomogeneity, state 0 $\rightarrow$ 2; b debonding between the inhomogeneity and matrix, state 2 $\rightarrow$ 3; and c cracking, state 3 $\rightarrow$ 5. All the state variations, mechanically or geometrically, can be illustrated unitarily by a series of spheroids or spheres of the energy-based equivalent inclusions, as shown in Fig. 8. The irreversible energy dissipation implies that the current spheroid or sphere is always expanding during the damaging process.

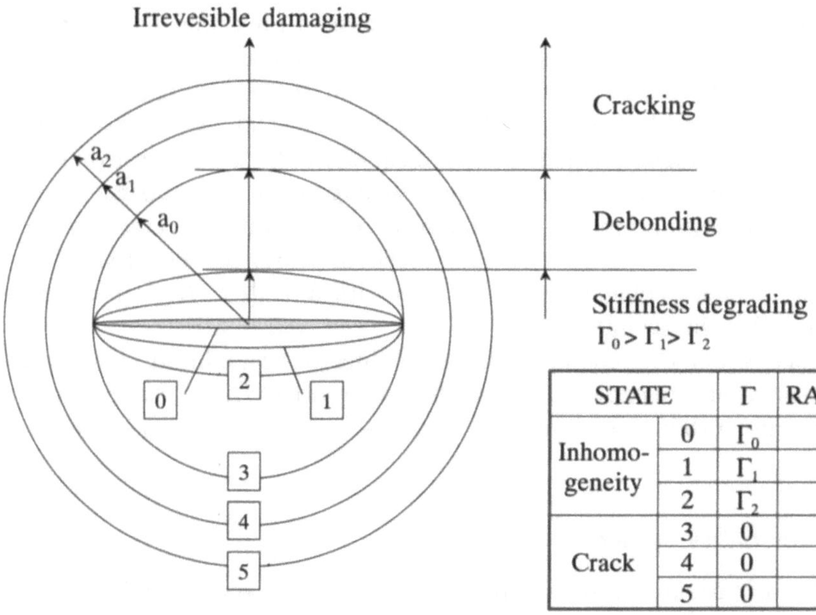

STATE		Γ	RADIUS
Inhomo-geneity	0	Γ_0	a_0
	1	Γ_1	a_0
	2	Γ_2	a_0
Crack	3	0	a_0
	4	0	a_1
	5	0	a_2

Fig. 8. The evolution of the energy-based equivalent inclusion Ω^{eq} during the damaging process of a penny-shaped inhomogeneity: stiffness degrading, debonding and cracking

5 Micromechanics basis of anisotropic damage evolution laws

Damage evolution laws, especially for anisotropic damaging behaviors, have been the most elusive parts of continuum damage mechanics, owing to their complex tensorial and high-degree nonlinear properties. It is usually assumed that there exists a scalar damage dissipation potential Q in phenomenological damage models, and then the damage evolution laws are derived from it by normality condition,

$$\dot{\boldsymbol{\Omega}} = \frac{\partial Q}{\partial \mathbf{Y}} \tag{31}$$

where $\boldsymbol{\Omega}$ denotes a damage variable and is considered as a second-order tensor here without a loss of generality, and $\mathbf{Y}$ is the generalized thermodynamic force conjugate to $\boldsymbol{\Omega}$. If further assuming that Q is a quadratic function in the conjugate force $\mathbf{Y}$, e.g., $Q = \frac{1}{2}\mathbf{Y} : \mathbf{J} : \mathbf{Y}$, the phenomenological equation or linear irreversible thermodynamics appears from (31),

$$\dot{\boldsymbol{\Omega}} = \mathbf{J} : \mathbf{Y} \tag{32}$$

where $\mathbf{J}$ is termed damage characteristic tensor of rank four. Chow and Lu [7] have shown that many classical damage evolution laws can be covered by (32), e.g., Chaboche, Lee, Murakami and Ohno and Sidoro and Cordebois, etc, see also Swoboda and Yang [40][41] and Yang et al. [44]. However, it is evident that the linear phenomenological equation (32) is not consistent with microscopic damaging mechanism, e.g. cracking where the propagating rate of cracks possesses a high-degree nonlinear relation with the cracking drive force, stress intensity factors or energy release rates, see e.g., (63). Thus, nonlinear phenomenological equation (32) with $\mathbf{J} = \mathbf{J}(\mathbf{Y}, \boldsymbol{\Omega}, \cdots)$ are more appealing than linear ones.

In this section, the general conditions leading to the nonlinear phenomenological equations are explored within the framework of Rice's [32,33] normality structure, a thermodynamic theory of irreversible processes with internal variables.

5.1 Normality structure

Consider a material sample of size V. Introduce the specific free energy ϕ and its Legendre transform ψ with respect to strain

$$\phi = \phi(\boldsymbol{\varepsilon}, \vartheta, H), \qquad \psi = \psi(\boldsymbol{\sigma}, \vartheta, H) = \boldsymbol{\varepsilon} : \frac{\partial \phi}{\partial \boldsymbol{\varepsilon}} - \phi, \tag{33}$$

where ϑ denotes temperature; $\boldsymbol{\varepsilon}$ denotes any strain tensor; $\boldsymbol{\sigma}$ denotes the symmetric conjugate stress; H denotes symbolically the current pattern of microstructural rearrangement of constituent elements of the materials. Consider two neighboring patterns of microstructural rearrangement denoted by H, $H + dH$. It is assumed that a set of incremental internal variables $d\xi_1, d\xi_2 \ldots, d\xi_n$ characterize the specific local rearrangements, which are represented collectively by dH, at sites throughout the material sample. The $d\xi$'s and dH are related by

$$\frac{1}{V} f_\alpha d\xi_\alpha = -d^p\phi = d^p\psi, \tag{34}$$

where $d^p\phi = \phi(\boldsymbol{\varepsilon}, \vartheta, H + dH) - \phi(\boldsymbol{\varepsilon}, \vartheta, H)$. In this section, Einstein summation convention is adopted. However, if an index range is listed like α in (36), the index is considered as a free index without the summation convention. Equation (34) also defines the thermodynamic forces $f_1, f_2, \ldots, f_n$ (collectively $\mathbf{f}$) conjugate to the variables. The corresponding set of total internal variables, $\boldsymbol{\xi} = \{\xi_1, \xi_2, \ldots, \xi_n\}$, generally are not state variables in the sense that thermodynamic state functions are not direct functions of $\boldsymbol{\xi}$, but instead depend on the path history of $\boldsymbol{\xi}$. The second law of thermodynamics further requires

$$\frac{1}{V} f_\alpha \dot{\xi}_\alpha \geq 0. \tag{35}$$

In the normality structure, a key assumption is that the kinetic rate laws of the internal variables take the form

$$\dot{\xi}_\alpha = \dot{\xi}_\alpha(f_\alpha, \vartheta, H), \qquad (\alpha = 1, 2, \ldots, n). \tag{36}$$

Therefore, the kinetic rate laws can be related to a flow potential Q and be recast as

$$\dot{\xi}_\alpha = V\frac{\partial Q}{\partial f_\alpha}, \qquad Q = Q(\mathbf{f}, \vartheta, H) = \frac{1}{V}\int_0^{\mathbf{f}} \dot{\xi}_\alpha(\mathbf{f}, \vartheta, H) df_\alpha, \tag{37}$$

where the integration is carried out at fixed ϑ and H, and defines a direct function Q of $\mathbf{f}$ since each term in the integrand is a total differential.

The set $\boldsymbol{\xi}$ generally contains numerous elements. One set of much reduced internal variables $\boldsymbol{\zeta}$ can be introduced as the average measurements of $\boldsymbol{\xi}$

$$\boldsymbol{\zeta} = \{\zeta_1, \zeta_2, \ldots, \zeta_m\}, \qquad \zeta_\mu = \zeta_\mu(\xi_1, \xi_2, \ldots, \xi_n; V)$$
$$(\mu = 1, 2, \ldots, m \ll n) \tag{38}$$

where V indicates averaging over the volume. The thermodynamic forces acting on the averaging variables $\boldsymbol{\zeta}$ are $g_1, g_2, \ldots, g_m$ (collectively $\mathbf{g}$). The equivalence that the averaging variables $\boldsymbol{\zeta}$ can describe the thermodynamic system characterized by $\boldsymbol{\xi}$, is achieved by requiring the equality for all $\delta\boldsymbol{\xi}$

$$g_\mu \delta\zeta_\mu = \frac{1}{V} f_\alpha \delta\xi_\alpha. \tag{39}$$

Substituting (38) into (39) leads to

$$f_\alpha = V g_\mu \frac{\partial \zeta_\mu(\boldsymbol{\xi}, V)}{\partial \xi_\alpha} = f_\alpha(\mathbf{g}, \vartheta, H) \quad \Rightarrow$$
$$\frac{\partial \zeta_\mu(\boldsymbol{\xi}, V)}{\partial \xi_\alpha} = \frac{1}{V}\frac{\partial f_\alpha(\mathbf{g}, \vartheta, H)}{\partial g_\mu}. \tag{40}$$

Therefore,

$$\dot{\zeta}_\mu = \frac{\partial \zeta_\mu}{\partial \xi_\alpha}\dot{\xi}_\alpha = \frac{1}{V}\dot{\xi}_\alpha \frac{\partial f_\alpha(\mathbf{g},\, \vartheta,\, H)}{\partial g_\mu} = \frac{\partial Q(\mathbf{g},\, \vartheta,\, H)}{\partial g_\mu}, \tag{41}$$

where

$$Q(\mathbf{g},\, \vartheta,\, H) = \frac{1}{V}\int_0^{\mathbf{f}(\mathbf{g},\, \vartheta,\, H)} \dot{\xi}_\alpha(\mathbf{f},\, \vartheta,\, H) df_\alpha = Q(\mathbf{f},\, \vartheta,\, H). \tag{42}$$

5.2 Homogeneous dissipation and potential functions

Let's define the dissipation functions at both microscopic and macroscopic levels,

$$\Phi(\mathbf{f},\, \vartheta,\, H) = \frac{1}{V} f_\alpha \dot{\xi}_\alpha, \qquad \Phi(\mathbf{g},\, \vartheta,\, H) = g_\mu \dot{\zeta}_\mu. \tag{43}$$

In fact, Φ/ϑ is just the entropy production rate. The introduction of Rice's [32,33] kinetic rate laws implies that the dissipation function $\Phi(\mathbf{f},\, \vartheta,\, H)$ is well-defined. On the other hand, (39) is just the requirement that the microscopic dissipation function should be equal to the macroscopic one, i.e., $\Phi(\mathbf{f},\, \vartheta,\, H) = \Phi(\mathbf{g},\, \vartheta,\, H)$. Due to (37) and (41), the dissipation and flow potential functions are related by

$$\Phi(\mathbf{f},\, \vartheta,\, H) = f_\alpha \frac{\partial Q(\mathbf{f},\, \vartheta,\, H)}{\partial f_\alpha}, \qquad \Phi(\mathbf{g},\, \vartheta,\, H) = g_\mu \frac{\partial Q(\mathbf{g},\, \vartheta,\, H)}{\partial g_\mu}. \tag{44}$$

Rice's [32][33] kinetic rate laws of the internal variables, (36), with each rate being stress dependent only via its conjugate thermodynamic force, are corner stones of the normality structure. However, they should be thought of only as an approximation and not as a physical law, as remarked by Rice [33]. This work focuses on a special type of kinetic rate laws that each rate $\dot{\xi}_\alpha$ is a homogeneous function of degree q in its conjugate force f_α,

$$\frac{\partial \dot{\xi}_\alpha(f_\alpha,\, \vartheta,\, H)}{\partial f_\alpha} f_\alpha = q\dot{\xi}_\alpha(f_\alpha,\, \vartheta,\, H) \qquad (\alpha = 1,\, 2,\, \cdots,\, n), \tag{45}$$

where Euler's Theorem on homogeneous functions is adopted as a definition. It is emphasized that all kinetic rate laws possess the same homogeneous degree q. With (45) and integration by parts, the flow potential Q defined in (37) can be recast as

$$\begin{aligned} Q(\mathbf{f},\, \vartheta,\, H) &= \frac{1}{V}\int_0^{\mathbf{f}} \dot{\xi}_\alpha df_\alpha = \frac{1}{V}\dot{\xi}_\alpha f_\alpha - \frac{1}{V}\int_0^{\mathbf{f}} q\dot{\xi}_\alpha df_\alpha \\ &= \Phi(\mathbf{f},\, \vartheta,\, H) - qQ(\mathbf{f},\, \vartheta,\, H) \end{aligned} \tag{46}$$

which leads to

$$\Phi(\mathbf{f},\,\vartheta,\,H) = (q+1)Q(\mathbf{f},\,\vartheta,\,H) \qquad \Rightarrow \\ \Phi(\mathbf{g},\,\vartheta,\,H) = (q+1)Q(\mathbf{g},\,\vartheta,\,H), \tag{47}$$

due to (42). This equation along with (44) indicate that both $\Phi(\mathbf{f},\,\vartheta,\,H)$, $Q(\mathbf{f},\,\vartheta,\,H)$ and $\Phi(\mathbf{g},\,\vartheta,\,H), Q(\mathbf{g},\,\vartheta,\,H)$ are homogeneous functions of degree $q+1$ in $\mathbf{f}$ and $\mathbf{g}$, respectively,

$$\frac{\partial\Phi}{\partial f_\alpha} f_\alpha = \frac{\partial\Phi}{\partial g_\mu} g_\mu = (q+1)\Phi, \qquad \frac{\partial Q}{\partial f_\alpha} f_\alpha = \frac{\partial Q}{\partial g_\mu} g_\mu = (q+1)Q. \tag{48}$$

Differentiating (44)(b) by g_μ, it follows that

$$\frac{\partial\Phi(\mathbf{g},\,\vartheta,\,H)}{\partial g_\mu} = \frac{\partial Q(\mathbf{g},\,\vartheta,\,H)}{\partial g_\mu} + g_\kappa \frac{\partial Q^2(\mathbf{g},\,\vartheta,\,H)}{\partial g_\kappa \partial g_\mu} \tag{49}$$

with the summation convention for κ $(\kappa = 1, 2, \ldots, m)$. Using (41) and (47)(b), one obtains

$$\dot{\zeta}_\mu = J_{\mu\kappa} g_\kappa, \qquad J_{\mu\kappa} = \frac{1}{q}\frac{\partial^2 Q(\mathbf{g},\,\vartheta,\,H)}{\partial g_\mu \partial g_\kappa} = J_{\kappa\mu} \tag{50}$$

which are exactly nonlinear phenomenological equations and Onsager reciprocal relations. The phenomenological equations can be written in matrix form

$$\dot{\zeta} = Jg, \qquad \dot{\zeta} = \{\dot{\zeta}_1,\, \dot{\zeta}_2,\, \cdots,\, \dot{\zeta}_m\}^T, \qquad g = \{g_1,\, g_2,\, \cdots,\, g_m\}^T, \tag{51}$$

where the nonlinear phenomenological coefficient matrix J is a $m \times m$ square matrix, and its element at μth row and κth column is $J_{\mu\kappa}$. Note that the Hessian matrix of the flow potential Q in g is denoted by $H(Q,\,g)$ and defined as

$$H(Q,\,g) = \begin{bmatrix} \dfrac{\partial^2 Q}{\partial g_1^2} & \dfrac{\partial^2 Q}{\partial g_1 \partial g_2} & \cdots & \dfrac{\partial^2 Q}{\partial g_1 \partial g_m} \\ \dfrac{\partial^2 Q}{\partial g_2 \partial g_1} & \dfrac{\partial^2 Q}{\partial g_2^2} & \cdots & \dfrac{\partial^2 Q}{\partial g_2 \partial g_m} \\ \vdots & \vdots & \ddots & \vdots \\ \dfrac{\partial^2 Q}{\partial g_m \partial g_1} & \dfrac{\partial^2 Q}{\partial g_m \partial g_2} & \cdots & \dfrac{\partial^2 Q}{\partial g_m^2} \end{bmatrix}. \tag{52}$$

Evidently, the matrix J is identical to the Hessian matrix scaled by $1/q$, i.e.,

$$J = \frac{1}{q} H(Q,\,g) \qquad \text{or} \qquad J = \frac{1}{q(q+1)} H(\Phi,\,g) \tag{53}$$

due to (47). Since Hessian matrixes are always symmetric, the Onsager reciprocal relations are incorporated implicitly. Similarly, (45) directly leads to the phenomenological equations at the microscopic level,

$$\dot{\xi}_\alpha = J_\alpha f_\alpha, \qquad J_\alpha = \frac{1}{q}\frac{\partial \dot{\xi}_\alpha}{\partial f_\alpha} = \frac{1}{q}\frac{\partial^2 Q}{\partial f_\alpha^2}, \qquad (\alpha = 1, 2, \cdots, n) \tag{54}$$

or in matrix form

$$\dot{\xi} = \tilde{J} f, \qquad \dot{\zeta} = \{\dot{\xi}_1, \dot{\xi}_2, \cdots, \dot{\xi}_n\}^T, \qquad f = \{f_1, f_2, \cdots, f_n\}^T, \tag{55}$$

where the nonlinear phenomenological coefficient matrix $\tilde{J}$ is a $n \times n$ square matrix and associated with the Hessian matrix of Q or Φ by

$$\tilde{J} = \frac{1}{q} H(Q, f) = \frac{1}{q(q+1)} H(\Phi, f). \tag{56}$$

Note that all the three square matrixes are diagonal matrixes since the off-diagonal elements of $H(Q, f)$ are $\frac{\partial^2 Q}{\partial f_\alpha \partial f_\beta} = \frac{\partial \dot{\xi}_\beta}{\partial f_\alpha} = 0$ for $\alpha \neq \beta$. The αth diagonal element of $\tilde{J}$ is just J_α.

Let's discuss the restriction of the second law of thermodynamics on the nonlinear phenomenological coefficient matrixes, see (35). In view of (51) and (55), it is required

$$\Phi = g^T J g = f^T \tilde{J} f \geq 0 \tag{57}$$

for any g or f. Thus, J and $\tilde{J}$ should be positive semidefinite, and then $H(Q, f)$, $H(Q, g)$, $H(\Phi, f)$ and $H(\Phi, g)$ should be also positive semidefinite. Obviously, If anyone of the six matrixes is positive semidefinite, the other ones are all positive semidefinite. Note that $\tilde{J}$ is a diagonal matrix, and it is positive semidefinite if

$$J_\alpha = \frac{1}{q}\frac{\partial^2 Q}{\partial f_\alpha^2} = \frac{1}{q}\frac{\partial \dot{\xi}_\alpha}{\partial f_\alpha} \geq 0, \qquad (\alpha = 1, 2, \cdots, n) \tag{58}$$

which require that $\dot{\xi}_\alpha$ is a monotonic increasing function of the conjugate force f_α. The requirement can be recast, due to (45),

$$J_\alpha = \frac{1}{q}\frac{\partial \dot{\xi}_\alpha}{\partial f_\alpha} = \frac{\dot{\xi}_\alpha}{f_\alpha} \geq 0 \qquad (\alpha = 1, 2, \cdots, n) \tag{59}$$

which is equivalent to, in the sense of non-negativeness,

$$f_\alpha \dot{\xi}_\alpha \geq 0 \qquad (\alpha = 1, 2, \cdots, n), \tag{60}$$

which can be considered as the conditions of positive semi-definiteness of the six matrixes. As compared with (35), it is evident that the homogeneous conditions (45) require that the intrinsic dissipation inequality hold for each internal variable or *locally*. Note that, if the Hessian matrix of a scalar function is positive semidefinite, the function is convex. Therefore, the convexity of the flow potential Q or dissipation function Φ is required by the second law of thermodynamics.

5.3 Concluding remarks

Rice's kinetic rate laws of local internal variables, with each rate being stress dependent only via its conjugate thermodynamic force, are corner stones of the normality structure. It is revealed in this section that nonlinear phenomenological equations and Onsager reciprocal relations emerge naturally from the normality structure if each rate is a homogeneous function of degree q in its conjugate force. Furthermore, the nonlinear phenomenological coefficient matrix is identical to the Hessian matrix of the flow potential function in conjugate forces only scaled by q.

It is evident that the framework is suitable for microcrack-dominated damage. Indeed, microcracks attracted, and still attract, most interest due to its relevance to the structural reliability and failure, as remarked by Krajcinovic [1]. Rice [34] has applied the normality structure to a material sample containing some distribution of Griffith cracks. Let the locus of all crack fronts be denoted by L and let da be a function of position along L describing the amount of local advance of the cracks, and hence constituting the structural rearrangements. It is assumed that the surfaces of cracking have continuously turning tangent planes, without abrupt forking or branching. Therefore, (34) becomes

$$Vd^p\psi = -Vd^p\phi = f_\alpha d\xi_\alpha \rightarrow \int_L [F da] dL \tag{61}$$

where F denotes the thermodynamic crack extension force per unit length along L. In this case, the kinetic rate law (36) and homogeneous condition (45) can be recast as,

$$\dot{a} = \dot{a}(F, \vartheta, H), \qquad \frac{\partial \dot{a}}{\partial F} F = q\dot{a} \tag{62}$$

at any local crack front. Macroscopic damage evolution corresponds to microscopic subcritical microcrack growth which is widely covered by the power-law, $\dot{a} \propto K^n$, where K is the stress intensity factor at the crack front. For example, $n = 13$ for the nickel-based superalloy Nimonic 80A at a temperature of 650°C [10]. The fatigue crack growth can also be described by similar power laws if taking cycle number N as the generalized time, e.g., the simple Paris equation $da/dN \propto (\Delta K)^n$ where the exponent n can take values as high as 15 to 50 in ceramics. As pointed by Rice [33], at any local crack front, $F = G - 2\gamma$, where G is Irwin's energy release rate and γ is the surface free energy. Note that $\gamma \ll G$ except for an "ideal" brittle cracking and $G \propto K^2$. Thus, the power law can be recast as

$$\dot{a} \propto G^q \qquad \text{or} \qquad \dot{a} = hG^q \approx hF^q, \tag{63}$$

where h is a scaling constant and $q = n/2$, which is exactly in the spirit of (62). Furthermore, (60) implies that the following inequality

$$F\dot{a} = (G - 2\gamma)\dot{a} \geq 0 \tag{64}$$

should hold pointwise along the crack front, which is just Rice's restriction on quasi-static growth of Griffith cracks [34]. Therefore, it is justified that, at least for microcrack-dominated damage, the damage evolution laws can be covered by nonlinear phenomenological equations.

6 Development of an equivalent state

In this section, an anisotropic damage model is formulated in strain space. The constitutive equations of a damage model mainly consists of two parts: damage elasticity and damage evolution law. The damage elasticity is the damage-dependent elasticity tensor $\mathbf{D}$ which generally takes the form $\mathbf{D} = \mathbf{D}(\boldsymbol{\Omega}, \boldsymbol{\varepsilon})$ due to the influence of strain $\boldsymbol{\varepsilon}$. In order to avoid a complicated formulation, the real state of a solid is mapped into an equivalent state on which the constitutive equations with a simple form like $\mathbf{D} = \mathbf{D}(\boldsymbol{\Omega})$ and $\dot{\boldsymbol{\Omega}} = \mathbf{J} : \mathbf{Y}$ hold true with respect to any strain state.

In rock-like materials, the frictional sliding on crack surfaces plays a central role in either inelastic deformation or cracking. Sliding causes dilation by opening the crack at asperities and by inducing local tensile cracking at some angle to the crack. Thus, the shear stress on the crack becomes the driving force of inelastic deformation and the materials become pressure-sensitive. In terms of macroscopic constitutive equations, sliding causes deviation from the normality structure, as pointed out by Rudnicki and Rice [35].

Frictional crack models have been investigated by many authors, see e.g. Nemat-Nasser et al. [27], Kachanov [18], etc. For local tensile cracking, as shown in Fig. 9a, the stress intensity factor of Mode I at crack tips can be computed with very good accuracy by considering an equivalent crack of length $2l$, subjected to a pair of collinear concentrated forces, τ, as well as the applied overall stress (see e.g. Nemat-Nasser et al. [27]), as shown in Fig. 9b. In this representation, τ denotes the resultant force transmitted across the preexisting crack.

Macroscopically, the states a and c are characterized by two sets of state variables and functions, respectively, see Tab. 2. In this chapter, the state (c) is termed the *equivalent state*; $\tilde{\boldsymbol{\Omega}}$ is termed the *effective damage tensor*, which corresponds to the equivalent cracks. Obviously, the fictitious cracks in the equivalent state are always in open state. The equivalent state is subjected

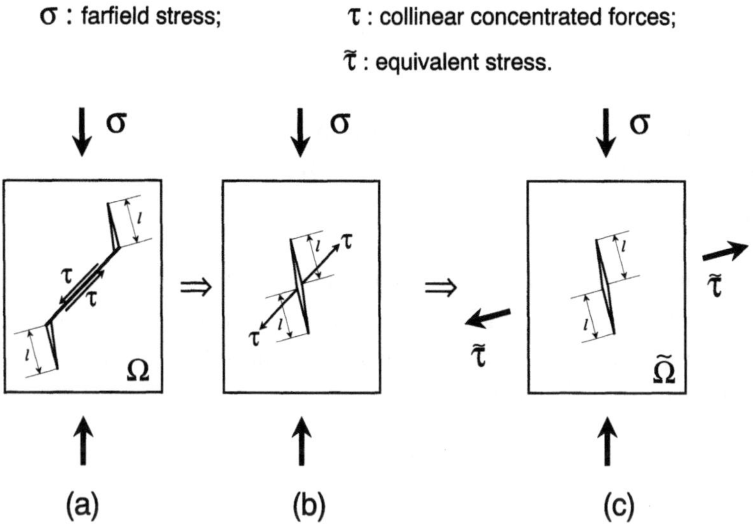

Fig. 9. Schematic frictional crack equivalence

Table 2. Macroscopic characterization of real and equivalent states

	Real state (a)	Equivalent state(c)	Relations
Strain	$\boldsymbol{\varepsilon}$	$\tilde{\boldsymbol{\varepsilon}}$	$\tilde{\boldsymbol{\varepsilon}} = \boldsymbol{\varepsilon}$
Damage tensor	$\boldsymbol{\Omega}$	$\tilde{\boldsymbol{\Omega}}$	$\tilde{\boldsymbol{\Omega}} = \mathbf{P}^{+} : \boldsymbol{\Omega}$
Stress	$\boldsymbol{\sigma}$	$\tilde{\boldsymbol{\sigma}}$	$\tilde{\boldsymbol{\sigma}} = \boldsymbol{\sigma} + \tilde{\boldsymbol{\tau}}$
Elasticity tensor	$\mathbf{D}$	$\tilde{\mathbf{D}}$	$\tilde{\mathbf{D}} = \mathbf{D}(\tilde{\boldsymbol{\Omega}})$
Free energy	ϕ	$\tilde{\phi}$	$\phi = \tilde{\phi} - g\tilde{\boldsymbol{\Omega}} : \boldsymbol{\varepsilon}$
Conjugate force	$\mathbf{Y}$	$\tilde{\mathbf{Y}}$	$\mathbf{Y} = \mathbf{P}^{+} : (\tilde{\mathbf{Y}} + g\boldsymbol{\varepsilon})$

to the effective stress $\tilde{\boldsymbol{\sigma}} = \boldsymbol{\sigma} + \tilde{\boldsymbol{\tau}}$ where $\tilde{\boldsymbol{\tau}}$ is the macroscopic representation of τ, see Fig. 9c. The plasticity effects are omitted in the formulation.

Let's consider the relations between the two states. It is postulated that the strain of the equivalent state under the effective stress is equal to the strain of the real state under the applied stress, namely, $\boldsymbol{\varepsilon} = \tilde{\boldsymbol{\varepsilon}}$. For the equivalent state, the stiffness and damage variable possess the one-to-one correspondence, i.e.,$\tilde{\mathbf{D}} = \mathbf{D}(\tilde{\boldsymbol{\Omega}})$ since $\tilde{\boldsymbol{\Omega}}$ corresponds to the open equivalent cracks. Therefore, the elasticity relation of the equivalent state, $\tilde{\boldsymbol{\sigma}} = \tilde{\mathbf{D}} : \tilde{\boldsymbol{\varepsilon}}$, leads to that of the real state

$$\boldsymbol{\sigma} = \mathbf{D}(\tilde{\boldsymbol{\Omega}}) : \boldsymbol{\varepsilon} - \tilde{\boldsymbol{\tau}}. \tag{65}$$

The free energy function and the conjugate force of the equivalent state can be defined as

$$\tilde{\phi} = \frac{1}{2}\boldsymbol{\varepsilon} : \tilde{\mathbf{D}} : \boldsymbol{\varepsilon}, \qquad \tilde{\mathbf{Y}} = -\frac{\partial \tilde{\phi}}{\partial \tilde{\boldsymbol{\Omega}}} = \frac{1}{2}\boldsymbol{\varepsilon} : \frac{\partial \mathbf{D}(\tilde{\boldsymbol{\Omega}})}{\partial \tilde{\boldsymbol{\Omega}}} : \boldsymbol{\varepsilon}. \tag{66}$$

Since τ acts on the equivalent crack as shown in Fig. 9b, it is reasonable to assume that its equivalent stress $\tilde{\boldsymbol{\tau}}$ is *parallel* to $\tilde{\boldsymbol{\Omega}}$, i.e.,

$$\tilde{\boldsymbol{\tau}} = g\tilde{\boldsymbol{\Omega}}, \tag{67}$$

where g is assumed to be a non-negative material constant. Note that, in the real state, the relation $\boldsymbol{\sigma} = \frac{\partial \phi}{\partial \boldsymbol{\varepsilon}}$ also leads to a stress-strain relation which should be consistent with (65). Based on this point and (67), the free energy function of the real state should be

$$\phi = \frac{1}{2}\boldsymbol{\varepsilon} : \mathbf{D}(\tilde{\boldsymbol{\Omega}}) : \boldsymbol{\varepsilon} - g\tilde{\boldsymbol{\Omega}} : \boldsymbol{\varepsilon} = \tilde{\phi} - g\tilde{\boldsymbol{\Omega}} : \boldsymbol{\varepsilon}. \tag{68}$$

6.1 Effective damage tensor

In principle, the effective damage tensor $\tilde{\boldsymbol{\Omega}}$ should be computed directly using (1) with respect to the equivalent cracks, but it is difficult to determine the equivalent cracks. Based on the fact that the equivalent cracks are roughly tensile-strain-orientated, the effective damage tensor is defined as

$$\tilde{\boldsymbol{\Omega}} = \mathbf{P}^+ : \boldsymbol{\Omega} \qquad \text{where} \quad P^+_{ijkl} = Q^+_{ik}Q^+_{jl}, \quad \mathbf{Q}^+ = \sum_{\nu=1}^{3} H(\varepsilon_\nu)\mathbf{p}^\nu\mathbf{p}^\nu, \tag{69}$$

where $\mathbf{p}^\nu$ and ε_ν are the ν-th principal normal vector and value, respectively, of strain $\boldsymbol{\varepsilon}$. Evidently, the definition leads to $\tilde{\Omega}_{ij} = \tilde{\Omega}_{ji} \Longrightarrow \tilde{\sigma}_{ij} = \tilde{\sigma}_{ji}$. The tensor $\mathbf{P}^+$ was termed *positive projection tensor* by Ju [16] and Dragon et al. [11].

Originally, $H(\cdot)$ was just the Heaviside function $\hat{H}(\cdot)$ and then the principal tensile strain can be determined as

$$\boldsymbol{\varepsilon}^+ = \sum_{\nu=1}^{3} \hat{H}(\varepsilon_\nu)\varepsilon_\nu\mathbf{p}^\nu\mathbf{p}^\nu = \mathbf{P}^+ : \boldsymbol{\varepsilon}. \tag{70}$$

In some cases for rock-like materials, the Heaviside function $\hat{H}$ will overestimate the difference between compression and tension. For example, in jointed rock, the joints are not perfectly contact but with some filling materials or roughness. Here H is defined as

$$H(\varepsilon_\nu) = (1-h)\hat{H}(\varepsilon_\nu) + h = \begin{cases} 1 & \text{if} \quad \varepsilon_\nu > 0 \\ h & \text{otherwise} \end{cases}, \tag{71}$$

where h $(0 \le h < 1)$ is a material constant to reflect the properties of crack contact. In view of (68) and (69), the conjugate force of the real state is

$$\mathbf{Y} = -\frac{\partial \phi}{\partial \boldsymbol{\Omega}} = \mathbf{P}^{+} : (\tilde{\mathbf{Y}} + g\boldsymbol{\varepsilon}) \tag{72}$$

which is tensile-strain-orientated.

6.2 Damage evolution law

For the equivalent state, the conjugate-force-based damage evolution law holds true for any stress state

$$d\tilde{\boldsymbol{\Omega}} = d\tilde{\lambda}\tilde{\mathbf{J}} : \tilde{\mathbf{Y}}, \tag{73}$$

where $\tilde{\mathbf{J}} = \mathbf{J}(\tilde{\boldsymbol{\Omega}})$. The function $\mathbf{J}(\boldsymbol{\Omega})$ is deduced by Swoboda and Yang [41]:

$$\mathbf{J}(\boldsymbol{\Omega}) = \sum_{\nu=1}^{3} \omega_\nu^2 \left(4\mathbf{T}_\nu + \frac{9}{4}\mathbf{N}_\nu \right) \tag{74}$$

$$\mathbf{T}_\nu = T_{ijkl}^\nu \equiv \frac{1}{4}(n_i^\nu n_k^\nu \delta_{jl} + n_i^\nu n_l^\nu \delta_{jk} + n_j^\nu n_k^\nu \delta_{il} + n_j^\nu n_l^\nu \delta_{ik}) - n_i^\nu n_j^\nu n_k^\nu n_l^\nu$$

$$\mathbf{N}_\nu = N_{ijkl}^\nu \equiv n_i^\nu n_j^\nu n_k^\nu n_l^\nu \qquad \text{(no summation for } \nu\text{)},$$

where $\mathbf{n}^\nu$ and ω_ν $(\nu = 1, 2, 3)$ are the principal directions and values, respectively, of damage tensor $\boldsymbol{\Omega}$. The associated damage surface is

$$\tilde{\mathcal{F}} = \tilde{\mathcal{G}} - \tilde{\mathcal{R}}, \qquad \tilde{\mathcal{G}}^2 = \frac{1}{2}\tilde{\mathbf{Y}} : \tilde{\mathbf{J}} : \tilde{\mathbf{Y}} \qquad \tilde{\mathcal{R}} = \max\{\mathcal{R}_0, \max \tilde{\mathcal{G}}\}, \tag{75}$$

where $\mathcal{R}_0$ is the damage threshold of virgin materials. The damage multiplier is proposed as

$$d\tilde{\lambda} = \beta \frac{d\xi}{\mathcal{G}}, \qquad d\xi = | de_{ij} | = \sqrt{de_{ij} de_{ij}}, \qquad e_{ij} = \varepsilon_{ij} - \frac{1}{3}\delta_{ij}\varepsilon_{kk}, \tag{76}$$

where $\beta > 0$ is a material constant.

6.3 Discussion on the equivalent state

The postulate $\boldsymbol{\varepsilon} = \tilde{\boldsymbol{\varepsilon}}$ is just Lemaitre's [22] hypothesis of strain equivalence that *the strain associated with a damaged state under the applied stress is equivalent to the strain associated with its undamaged state under the effective stress* if replacing the undamaged state with the equivalent state. One may argue that the undamaged state and its associated net stress can be used to furnish a much simpler formulation than can the equivalent state. Note that the net stress is not suitable to describe the anisotropic damaging, as discussed in the first section. It is also a very coarse concept for anisotropic damage induced elasticity. For example, the simplest symmetric net stress may be defined as $\dfrac{\overline{\sigma}_{ij}+\overline{\sigma}_{ji}}{2}$ which still leads to an unsymmetric elasticity tensor. An elasticity tensor is symmetric if $D_{ijkl}=D_{jikl}=D_{ijlk}=D_{klij}$.

The preceding formulation on the relations between the real state and the equivalent state is based on the three hypotheses: $\tilde{\varepsilon}=\varepsilon$, $\boldsymbol{\tau}=g\tilde{\boldsymbol{\Omega}}$, $\tilde{\boldsymbol{\Omega}}=\mathbf{P}^{+}:\boldsymbol{\Omega}$ which all are approximate descriptions of their physical realities. From a purely mathematical viewpoint, the introduction of the two parameters

$\varepsilon - g\tilde{\boldsymbol{\Omega}}$ to fit the

introduced the term

-generated residual

atancy, see Fig. 10.

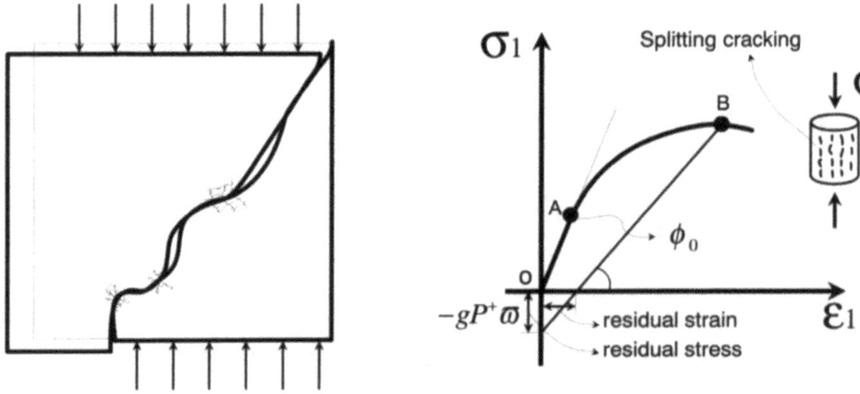

(a) (b)

Fig. 10. (a) Schematic dilatancy mechanism; (b) Typical stress-strain with residual stress and strain

It should be pointed out that, the main purpose and advantage to develop the conjugate-force-based damage evolution law and the equivalent state, is not for a simple brittle solid but for complicated internal dissipative mechanisms and mechanical behaviour. The formulation can be easily extended to describe the process coupled with plasticity just by choosing an appropriate free energy function, e.g. $\tilde{\phi}=\frac{1}{2}(\varepsilon-\varepsilon^{p}):\tilde{\mathbf{D}}:\frac{1}{2}(\varepsilon-\varepsilon^{p})$ where ε^{p} is the plastic strain. Then the influence of plasticity on damaging is reflected implicitly through the corresponding conjugate force $\tilde{\mathbf{Y}}$. The hydraulic fracturing mechanism can also be accommodated in the framework by this way, see Swoboda et al. [38,39] and Yang [43]. As usual, the damage's influence on plastic flow can be reflected implicitly thorough another effective stress base on another equivalent state (undamaged material). In this framework, such an effective stress is $\tilde{\boldsymbol{\sigma}}=\mathbf{D}^{0}:(\varepsilon-\varepsilon^{p})$ in line with Lemaitre's [22] hypothesis of strain equivalence, where $\mathbf{D}^0$ is the isotropic elasticity tensor of the virgin material.

7 Damage elasticity

An explicit expression of the damage-dependent elasticity tensor $\mathbf{D}(\boldsymbol{\Omega})$ is developed in the section. In general, an elasticity tensor is subject to the general principles of continuum mechanics (see e.g. Malvern [24]): (1) *symmetry condition* requires $D_{ijkl} = D_{jikl} = D_{ijlk} = D_{klij}$; (2) *material symmetry condition* requires that $\mathbf{D}(\boldsymbol{\Omega})$ be an isotropic tensor function; (3) *positive definite condition* requires the elastic potential function $W = \frac{1}{2}\varepsilon_{ij}D_{ijkl}\varepsilon_{kl}$ is positive-definite as a function of strain ε_{ij}. The isotropic tensor function $\mathbf{D}(\boldsymbol{\Omega})$, which satisfies the symmetric condition, takes the form:

$$
\begin{aligned}
D_{ijkl} &= A_1\delta_{ij}\delta_{kl} + A_2(\delta_{ik}\delta_{jl} + \delta_{il}\delta_{jk}) \\
&\quad + A_3(\Omega_{ij}\delta_{kl} + \Omega_{kl}\delta_{ij}) + A_4(\Omega_{ik}\delta_{jl} + \Omega_{il}\delta_{jk} + \Omega_{jk}\delta_{il} + \Omega_{jl}\delta_{ik}) \\
&\quad + A_5\Omega_{ij}\Omega_{kl} + A_6(\Omega_{ik}\Omega_{jl} + \Omega_{il}\Omega_{jk}) \\
&\quad + A_7(\Theta_{ij}\delta_{kl} + \Theta_{kl}\delta_{ij}) + A_8(\Theta_{ik}\delta_{jl} + \Theta_{il}\delta_{jk} + \Theta_{jk}\delta_{il} + \Theta_{jl}\delta_{ik}) \\
&\quad + A_9(\Theta_{ij}\Omega_{kl} + \Omega_{ij}\Theta_{kl}) + A_{10}(\Theta_{ik}\Omega_{jl} + \Omega_{ik}\Theta_{jl} + \Theta_{il}\Omega_{jk} + \Omega_{il}\Theta_{jk}) \\
&\quad + A_{11}\Theta_{ij}\Theta_{kl} + A_{12}(\Theta_{ik}\Theta_{jl} + \Theta_{il}\Theta_{jk}),
\end{aligned}
\tag{77}
$$

where $\Theta_{ij} = \Omega_{im}\Omega_{mj}$; A_1, A_2, $\cdots$, A_{12} are functions of invariants of Ω_{ij}. According to Cayley-Hamilton's theorem, the third- and higher-order terms can be expressed as linear combinations of linear and quadratic terms. Thus, terms like $\Omega_{im}\Theta_{mj}$, $\Theta_{im}\Theta_{mj}$ disappear in (78), see e.g. Bazant [3]. Obviously, Murakami [25], Stumvoll and Swoboda [37], and Cowin's [8] expressions of the damage elasticity are all special cases of (78).

It is a very complicated problem to impose the positive definite condition on (78). It is natural to consider such a strong condition to require all terms to be non-negative in any case. Obviously, the strong condition leads to $A_3 = A_4 = A_7 = A_8 = \cdots = 0$, which causes significant loss of generality. Here, the crux is solved by introducing an intermediate second-order tensor Φ_{ij} to define the elasticity tensor

$$
\begin{aligned}
D_{ijkl} &= \lambda\Phi_{ij}\Phi_{kl} + \mu(\Phi_{ik}\Phi_{jl} + \Phi_{il}\Phi_{jk}), \\
\Phi_{ij} &= C_1\delta_{ij} + C_2\Omega_{ij} + C_3\Omega_{im}\Omega_{mj},
\end{aligned}
\tag{78}
$$

where C_1, C_2 and C_3 are functions of the invariants of Ω_{ij}; λ and μ are two non-negative constant. Its expansion is

$$\begin{aligned}D_{ijkl} &= C_1C_1[\lambda\delta_{ij}\delta_{kl} + \mu(\delta_{ik}\delta_{jl} + \delta_{il}\delta_{jk})] \\ &+ C_1C_2[\lambda\Omega_{ij}\delta_{kl} + \lambda\Omega_{kl}\delta_{ij} + \mu(\Omega_{ik}\delta_{jl} + \Omega_{il}\delta_{jk} + \Omega_{jk}\delta_{il} + \Omega_{jl}\delta_{ik})] \\ &+ C_2C_2[\lambda\Omega_{ij}\Omega_{kl} + \mu(\Omega_{ik}\Omega_{jl} + \Omega_{il}\Omega_{jk})] \\ &+ C_1C_3[\lambda\Theta_{ij}\delta_{kl} + \lambda\Theta_{kl}\delta_{ij} + \mu(\Theta_{ik}\delta_{jl} + \Theta_{il}\delta_{jk} + \Theta_{jk}\delta_{il} + \Theta_{jl}\delta_{ik})] \\ &+ C_2C_3[\lambda\Theta_{ij}\Omega_{kl} + \lambda\Omega_{ij}\Theta_{kl} + \mu(\Theta_{ik}\Omega_{jl} + \Omega_{ik}\Theta_{jl} + \Theta_{il}\Omega_{jk} + \Omega_{il}\Theta_{jk})] \\ &+ C_3C_3[\lambda\Theta_{ij}\Theta_{kl} + \mu(\Theta_{ik}\Theta_{jl} + \Theta_{il}\Theta_{jk})]\end{aligned} \tag{79}$$

which is a complete tensorial polynomial as compared with (78). Then the positive definite condition requires

$$W = \frac{1}{2}\varepsilon_{ij}D_{ijkl}\varepsilon_{kl} = \frac{\lambda}{2}(\Phi_{ij}\varepsilon_{ij})^2 + \mu\varepsilon_{im}\Phi_{mj}\Phi_{in}\varepsilon_{nj} \geq 0. \tag{80}$$

Note that $\varepsilon_{im}\Phi_{mj}\Phi_{in}\varepsilon_{nj} \geq 0$ if Φ_{ij} is a positive definite tensor. Therefore, a fourth-order positive definite problem reduces to a second-order one.

Although not necessarily, the normalized damage variable is convenient to use. Here it is postulated that

$$\mathbf{D} = \begin{cases} \mathbf{D}^0 & \text{if} \quad \boldsymbol{\Omega} = \mathbf{0} \\ 0 & \text{if} \quad \boldsymbol{\Omega} = \mathbf{I}, \end{cases} \tag{81}$$

where $\mathbf{I} = \delta_{ij}$ and $\mathbf{D}^0$ is the isotropic elasticity tensor of the virgin material. Based on this condition, λ and μ in (78) are just Lamé's constants of virgin material and C_1, C_2, C_3 should satisfy

$$C_1 = 1 \quad \text{if} \quad \boldsymbol{\Omega} = \mathbf{0}; \qquad C_1 + C_2 + C_3 = 0 \quad \text{if} \quad \boldsymbol{\Omega} = \mathbf{I}. \tag{82}$$

If further assuming that C_1, C_2, C_3 are material constants and noting the positive definite requirement on Φ_{ij}, we have

$$C_1 = 1, \qquad C_2 = -k, \qquad C_3 = -(1-k), \qquad 0 \leq k \leq 1 \tag{83}$$

so there is only one independent material constant k. The elasticity tensor defined in (78) and (83) represents the orthotropic symmetry of elasticity. The so-called "hypothesis of complementary energy equivalence", see e.g. Lee et al. [21] and Zhang [48], leads to $\mathbf{D} = (\mathbf{I} - \boldsymbol{\Omega}) \cdot \mathbf{D}^0 \cdot (\mathbf{I} - \boldsymbol{\Omega})$, which is just a special case of (78) for $k = 1$.

8 Free energy function

The specific free energy ψ, taken as the thermodynamic potential, plays a central role in describing thermodynamic processes. It determines the stress-strain relation at the current state and the conjugate forces, which are the drive forces of state changes. Confined in isothermal and small deformation processes, it is assumed that there exists a free energy function $\psi = \psi(\boldsymbol{\varepsilon}, \boldsymbol{\varepsilon}^p, \mathbf{q}, \boldsymbol{\Omega}, \mathbf{p})$, which is the function of strain tensor $\boldsymbol{\varepsilon}$, plastic strain tensor $\boldsymbol{\varepsilon}^p$, one set of suitable plastic internal variables $\mathbf{q}$, damage tensor $\boldsymbol{\Omega}$, seepage pressure $\mathbf{p}$. The free energy can be uncoupled into elastic and plastic parts,

$$\psi(\boldsymbol{\varepsilon}, \boldsymbol{\varepsilon}^p, \mathbf{q}, \boldsymbol{\Omega}, \mathbf{p}) = \psi_e(\boldsymbol{\varepsilon}^e, \boldsymbol{\Omega}, \mathbf{p}) + \psi_p(\mathbf{q}) \tag{84}$$

where $\boldsymbol{\varepsilon}^e = \boldsymbol{\varepsilon} - \boldsymbol{\varepsilon}^p$ is the elastic strain tensor; $\psi_e(\boldsymbol{\varepsilon}^e, \boldsymbol{\Omega}, \mathbf{p})$ is the elastic strain energy; $\psi_p(\mathbf{q})$ is the free energy due to plastic hardening. As pointed out in Sect. 6, the real state and the equivalent state possess the same elastic strain energy. Based on this assumption, $\psi_e(\boldsymbol{\varepsilon}^e, \boldsymbol{\Omega}, \mathbf{p})$ can be computed according to the equivalent state and then uncoupled as

$$\psi_e(\boldsymbol{\varepsilon}^e, \boldsymbol{\Omega}, \mathbf{p}) = \psi_{ed}(\boldsymbol{\varepsilon}^e, \tilde{\boldsymbol{\Omega}}, \mathbf{p}) + \psi_{dp}(\tilde{\boldsymbol{\tau}}, \tilde{\boldsymbol{\Omega}}, \boldsymbol{\varepsilon}), \tag{85}$$

where ψ_{ed} is the free energy contributed by stress $\boldsymbol{\sigma}$ and water pressure $\mathbf{p}$ on the equivalent state; ψ_{dp} is the free energy contributed by the fictitious equivalent stress $\tilde{\boldsymbol{\tau}}$ and can be termed the *damage residual* term.

8.1 Damage elasticity and seepage

In this part, the explicit expression of $\psi_{ed}(\boldsymbol{\varepsilon}^e, \tilde{\boldsymbol{\Omega}}, \mathbf{p})$ is to be established. If water flow goes through a rock, the binary mixture of rock and water is a two-phase material. The free energy ψ is only associated with the rock skeleton, but the work increment $dw = \sigma_{ij} d\varepsilon_{ij}$ is done by external forces upon the total system of the two-phase material. Here, the water is assumed to be incompressible, so the work is only absorbed by the rock skeleton. Note that $\psi_{ed}(\boldsymbol{\varepsilon}^e, \tilde{\boldsymbol{\Omega}}, \mathbf{p})$ is a state function and path-independent. For an admissible isothermal process without internal dissipation, the first law of thermodynamics gives

$$\Delta\psi_{ed} = \psi_{ed}(\boldsymbol{\varepsilon}^e, \tilde{\boldsymbol{\Omega}}, \mathbf{p}) - \psi_{ed}(\mathbf{0}, \tilde{\boldsymbol{\Omega}}, \mathbf{0}) = \Delta w, \tag{86}$$

where Δw is the work done on the system by external forces during the process. If we further assume $\psi_{ed}(\mathbf{0}, \tilde{\boldsymbol{\Omega}}, \mathbf{0}) = 0$, then

$$\psi_{ed}(\boldsymbol{\varepsilon}^e, \tilde{\boldsymbol{\Omega}}, \mathbf{p}) = \Delta w. \tag{87}$$

The external forces include Cauchy stress σ_{ij} and water pressure $\mathbf{p} = \{p_1, p_2\}$, where p_1 is the pore water pressure and p_2 the fracture water pressure, as shown in Fig. 11b. Here, a "smeared" fracture system is assumed, namely that p_1 and p_2 are located at each spatial point. Such continuum is exactly the same as Barenblatt's *et al.* [2] "dual-porosity" media, see also Bai *et al.* [1].

Unlike an isotropic poroelastic medium, the deformation of anisotropic poroelastic medium is a controversial issue. Oda [30] assumed that the total

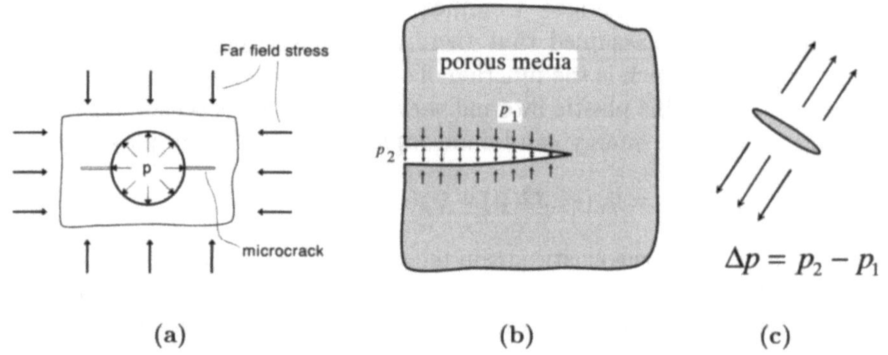

Fig. 11. (a) Hydraulic fracturing; (b) Water pressures near a fracture around porous media; (c) Acting mechanism of the water pressure difference Δp

stress and Terzaghi effective stress deform the rock matrix and cracks, respectively. With a modification, namely that Biot's effective stress deforms the rock matrix, a general anisotropic poroelastic relation is proposed as

$$\sigma_{ij} = \tilde{D}_{ijkl}\varepsilon^e_{kl} + p_1\Psi_{ij}, \quad \textit{where} \quad \Psi_{ij} = \delta_{ij} - \frac{1-2\nu}{E}(1+\alpha)\tilde{D}_{ijkk} \tag{88}$$

where α is Biot's poroelastic constant. If $\tilde{\boldsymbol{\Omega}} = \mathbf{0}$, (88) degenerates into the isotropic poroelastic relation, see e.g., Visser and Mier [42]. If $\alpha = 0$, (88) is simply Oda's expression [30]. If $\alpha = -1$, (88) represents the Terzaghi effective stress principle. Thus, the deformation work σ_{ij} and p_1 is obtained by

$$\Delta w_1 = \int_0^{\varepsilon^e_{ij}} \sigma_{ij} d\varepsilon^e_{ij} = \frac{1}{2}\varepsilon^e_{ij}\tilde{D}_{ijkl}\varepsilon^e_{kl} + p_1\Psi_{ij}\varepsilon^e_{ij}. \tag{89}$$

Although the fracture water pressure p_2 may make a negligible contribution to the poroelastic deformation, it is the dominant driving force for crack propagation. Consider the hydraulic fracturing in the wellbore as shown in Fig. 11a. The fracture water pressure can be divided into two parts $p_2 = p_1 + \Delta p$. The first part is assumed to have been taken into account by (88). Consider a spatial point representing the local microstructure. It is assumed that that Δp is fully enclosed in the cracks, as shown in Fig. 11c for a typical crack. Thus, Δp only disturbs the stress and strain around each crack but makes no contribution to the macroscopic or farfield stress and strain, which is consistent with (88). For a penny crack $\xi_\alpha = \{\mathbf{n}^\alpha, r_\alpha\}$ in Fig. 11c, the work done by Δp is

$$\hat{w}_\alpha = \eta r^3_\alpha \Delta p^2, \qquad \eta = \frac{8}{3\tilde{E}}(1-\tilde{\nu}^2), \tag{90}$$

where $\tilde{E}$ and $\tilde{\nu}$ are the effective Young's modulus and Poisson's ratio, respectively, of the damaged rock. See Budiansky and O'Connel [6] and Eqns (48) and (25) therein. As the total work in the unit volume, Δw_2 can be obtained as

$$\Delta w_2 = \frac{1}{V^0}\sum_{\alpha=1}^{n} \hat{w}_\alpha = \eta \Delta p^2 \left(\frac{1}{V^0}\sum_{\alpha=1}^{n} r_\alpha^3 \right) = \eta \Delta p^2 \omega \tag{91}$$

by using (4)(b). For multiple sets of parallel microcracks characterized by a damage tensor Ω_{ij} in (5), the expression can be easily generalized as

$$\Delta w_2 = \eta \Delta p^2 \sqrt{\Omega_{ij}\Omega_{ij}}. \tag{92}$$

By using (87), (89) and (92), the free energy function is

$$\psi_{ed}(\boldsymbol{\varepsilon}^e, \tilde{\boldsymbol{\Omega}}, \mathbf{p}) = \Delta w_1 + \Delta w_2 = \frac{1}{2}\varepsilon_{ij}^e \tilde{D}_{ijkl}\varepsilon_{kl}^e + p_1 \Psi_{ij}\varepsilon_{ij}^e + \eta \Delta p^2 \sqrt{\tilde{\Omega}_{ij}\tilde{\Omega}_{ij}}. \tag{93}$$

8.2 Damage residual term

As shown in Fig. 9b, the equivalent forces τ simply make the crack open. Thus, the work done by τ or its equivalent stress $\tilde{\boldsymbol{\tau}}$ during a deformation process without damage propagation is all converted into the strain energy, i.e., $\psi_{dp}(\tilde{\boldsymbol{\tau}}, \tilde{\boldsymbol{\Omega}}, \boldsymbol{\varepsilon})$. Equation (67) indicates that $\tilde{\boldsymbol{\tau}}$ is independent of stress or strain. Thus, ψ_{dp} can be calculated as

$$\psi_{dp}(\tilde{\boldsymbol{\tau}}, \tilde{\boldsymbol{\Omega}}, \boldsymbol{\varepsilon}) = -\tilde{\boldsymbol{\tau}} : \boldsymbol{\varepsilon} = -g\tilde{\Omega}_{ij}\varepsilon_{ij}, \tag{94}$$

where the negative sign indicates that $\tilde{\boldsymbol{\tau}}$ is not an external stress but an equivalent internal stress. ψ_{dp} is called the damage residual term because a *residual* elastic strain energy can be expected as $\psi_{dp} = -g\tilde{\Omega}_{kl}\varepsilon_{ij}^p$, after unloading, i.e., $\varepsilon_{ij}^e = 0$. Microscopically, the residual ψ_{dp} is caused by a residual τ produced by the residual shear deformation on the main crack surfaces due to the roughness. Frictional sliding eventually causes dilatancy and splitting cracking, as shown in Fig. 10.

The free energy function is obtained by substituting (93) and (94) into (85) and (84),

$$\psi(\boldsymbol{\varepsilon}, \boldsymbol{\varepsilon}^p, \mathbf{q}, \boldsymbol{\Omega}, \mathbf{p}) = \frac{1}{2}\varepsilon_{ij}^e \tilde{D}_{ijkl}\varepsilon_{kl}^e + p_1 \Psi_{ij}\varepsilon_{ij}^e + \eta \Delta p^2 \sqrt{\tilde{\Omega}_{ij}\tilde{\Omega}_{ij}} - g\tilde{\Omega}_{kl}\varepsilon_{ij} + \psi_p(\mathbf{q}). \tag{95}$$

Note that the water pressure $\mathbf{p}$ is also an external variable for the rock skeleton like strain ε_{ij}. The classical form of the Clausius-Duheim inequality $\sigma_{ij}\dot{\varepsilon}_{ij} - \dot{\psi} \geq 0$ should be modified as

$$\sigma_{ij}\dot{\varepsilon}_{ij} + \mathbf{P} \cdot \dot{\mathbf{p}} - \dot{\psi} \geq 0 \tag{96}$$

which means the rock skeleton is subjected to additional work done by the water pressure $\mathbf{p}$ through its conjugate force $\mathbf{P}$. Substituting (95) into (96) yields

9 Numerical results

Several numerical examples from simple specimens to practical structural analyses, have been worked out to illustrate the behaviour of the damage model.

9.1 Damage-dependent elasticity tensor

The behaviour of the damage-dependent elasticity tensor $\mathbf{D}(\tilde{\Omega})$ defined in (78) and (69) is illustrated by one set of uniaxial tests, as shown in Fig. 12. The parallel microcracks in the specimen can be characterized by a damage vector (n_i, ω), where $\omega = 0.5$ and the normal vector n_i is determined by α, the angle between loading direction and the damage vector. The damage vector corresponds to a damage tensor $\Omega_{ij} = \omega n_i n_j$. Then, (78) is graphed into curves illustrating the relation between the apparent Young's modulus along the loading direction and α with different k and h. In test (a), we change k but keep $h = 1.0$ which implies that $\tilde{\Omega}_{ij} = \Omega_{ij}$. In test (b), we change h but keep $k = 1.0$, and the results indicate that the Heaviside function ($h = 0.0$) overestimates the compressive effects for rock-like materials. The relation between apparent Young's modulus along the loading direction and α is also confirmed by the experiment done by Kawamoto et al. [19], as shown in Fig. 13.

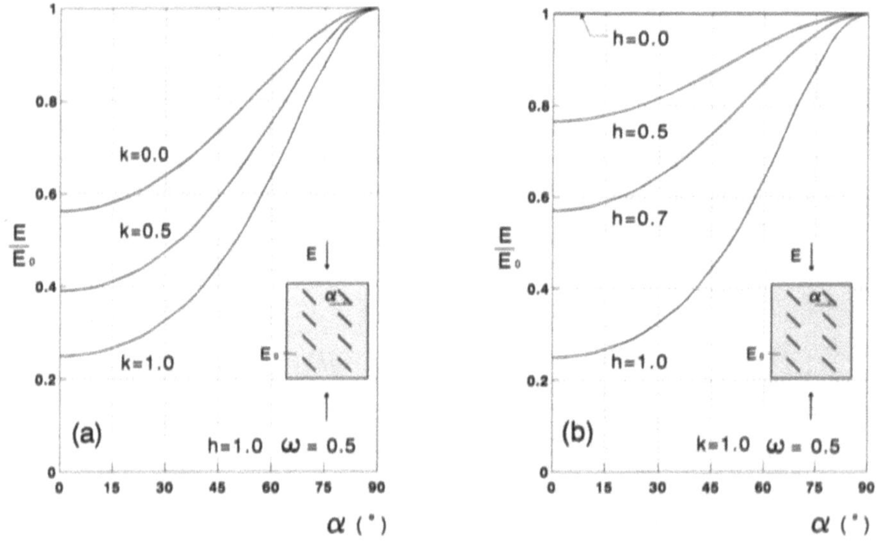

Fig. 12. Normalized apparent Young's modulus *versus* the crack direction with the variations (a) k and (b) h

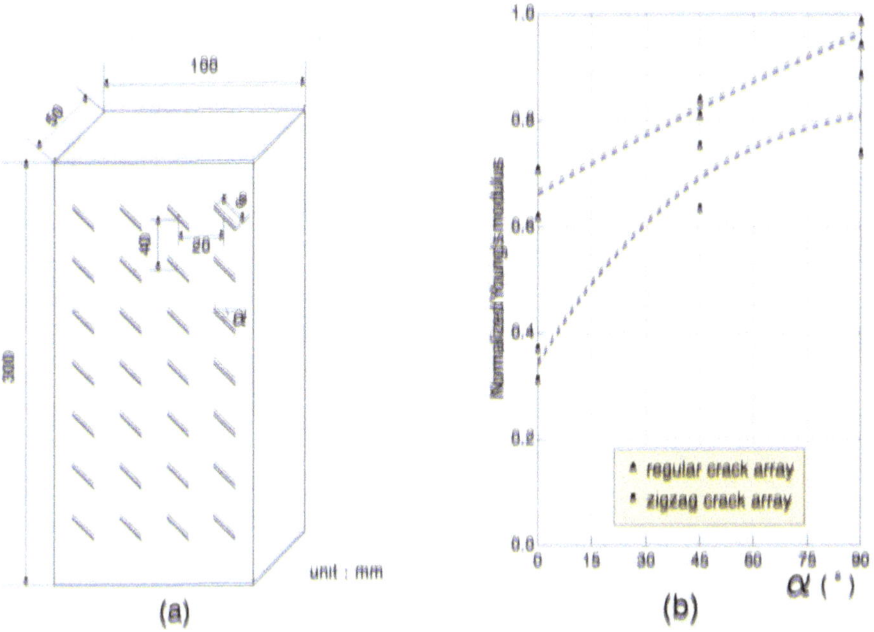

Fig. 13. (a) Experimental specimen; (b) apparent Young's modulus with changing crack direction

9.2 Uniaxial and biaxial loading tests

A uniaxial loading process is illustrated in Fig. 14. The material is initially virgin and the parameters are listed in Tab. 3. As a proportional loading process, $\dot{\boldsymbol{\Omega}}, \boldsymbol{\Omega}, \mathbf{Y}, \boldsymbol{\sigma}, \boldsymbol{\varepsilon}$ are all coaxial. Their process curves *versus* the axial strain are depicted in Fig. 14. The damage propagates mainly along the lateral direction in which the tensile strain is developed, as suggested by the conjugate force components Y_{11} and Y_{22}. The test reflects the typical "splitting" mechanism in rock-like materials.

Table 3. Material parameters for uniaxial and biaxial loading tests

E	50000.0	MPa
ν	0.25	–
h	0.1	–
k	1.0	–
$\mathcal{G}_0$	0.3	MPa
β	700000.0	–
g	50.0	MPa

The biaxial loading test with the same parameters is shown in Fig. 15. It indicates that the model can describe the main deformation characters of rock-like materials with few parameters. Using a conventional orthotropic elasticity model it is difficult to reflect such significant dilatancy.

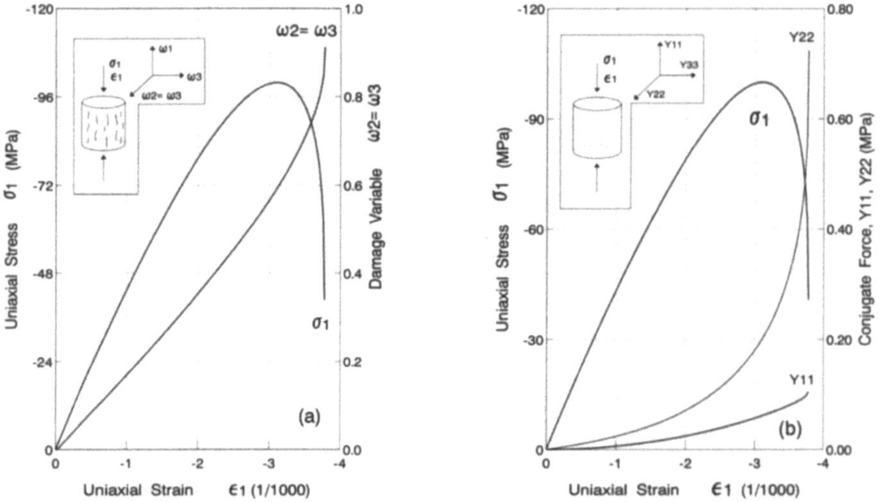

Fig. 14. (a) axial stress and damage variable; (b) axial stress and conjugate forces under increasing axial strain

9.3 Parameter identification

Quite obviously, the parameter identification can only be a trial-and-error procedure. By fitting the experimental curve of quartzite performed by Bieniawski [5], one set of parameters is obtained, as shown in Fig. 16. In order to identify the parameters efficiently, some parameter sensitivity tests have been done by Yang [43].

9.4 Design of an underground theater: analysis in jointed rock

For the design of an underground theater for 900 persons in the heart of Linz, the structural analysis aimed to consider the impact of the joint system's orientation. From intensive geological investigations conducted in the tunnel system constructed during World War II and with the help of borings performed to a depth of 55 m, the geological situation including the joint formations, as shown in Fig. 18, was well known. The main part of the theater and the garage were to be situated in a rock layer of gneiss. This layer is covered by marl and a thin layer of gravel. The quality of the gneiss layer at the site of the theater excavation is very high, but a strong impact of

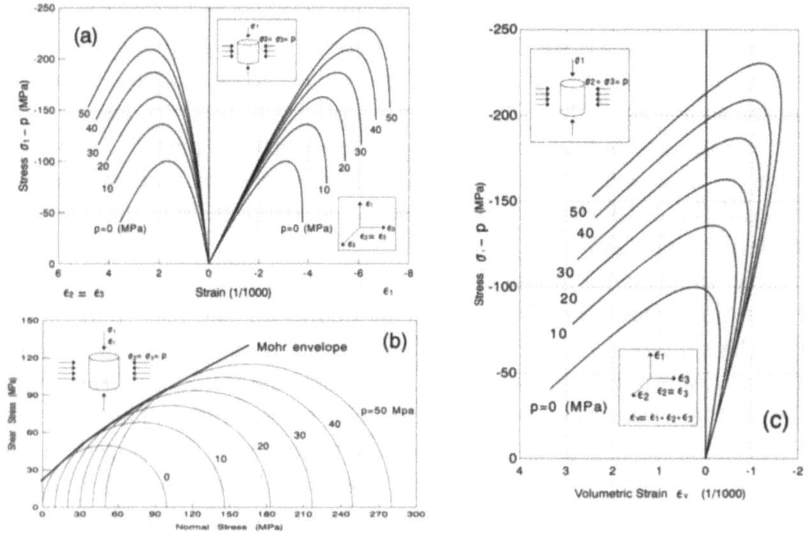

Fig. 15. Biaxial compressive tests of proposed damage model (a) Axial and lateral strain under increasing axial stress (b) Mohr envelope for strength failure (c) Relationship between volumetric strain and axial stress

E	84217.0	MPa
ν	0.1875	–
h	0.1	–
k	1.0	–
$\mathcal{G}_0$	77.0	MPa
β	41600.0	–
g	65.0	MPa

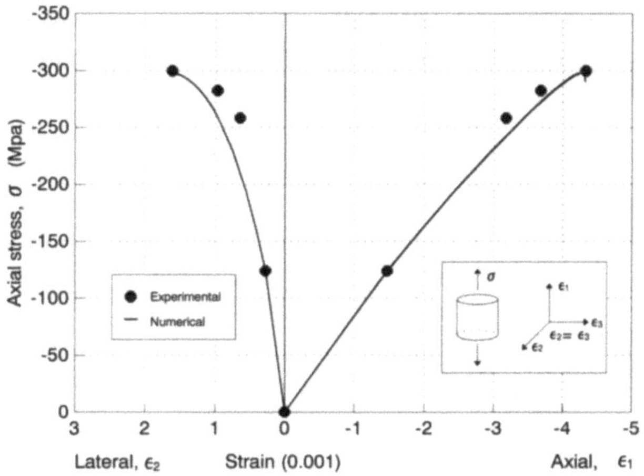

Fig. 16. Parameter identification for quartzite

the three main faults was expected. The layout of the complete underground structure including the stage, auditorium, entrance hall, garage system and the existing tunnel system is shown in Fig. 17. Due to the expansive layout of the entire structure a three-dimensional model was used to describe the impact of the underground excavation. The most critical part of the structure, namely a "rock bridge" with a span of 40 m above the auditorium, was meant to hold the old fortified wall of the castle above it. The models were based on damage mechanics.

The three-dimensional model shown in Fig. 19 with 15,900 elements and 190,000 equations was used. It includes modelling of the rock and of the support of the shotcrete lining. Excavation was simulated in four excavation steps. The initial stress field, which also includes the loading of the castle above the theater, was generated with the help of dead load. The resulting vertical stress is show in Fig. 20. The global damage tensor was based on the orientation of the joint system and of the model, as shown in Fig. 18. In order to interpret the design the plot of the damage tensor was of great importance. If the damage is zero, the joint system is closed. Any value above zero shows the amount of opening in the joint system. Once this value is known, the layout of the bolt system can be designed.

Figure 21 shows the damage in y direction. The bridge was seen to act as a beam in y direction. We found that the cracks open in horizontal direction y. The opening in vertical direction z was seen to be very good from a inside view of the model. The final normal forces in the shotcrete lining are presented in Fig. 23.

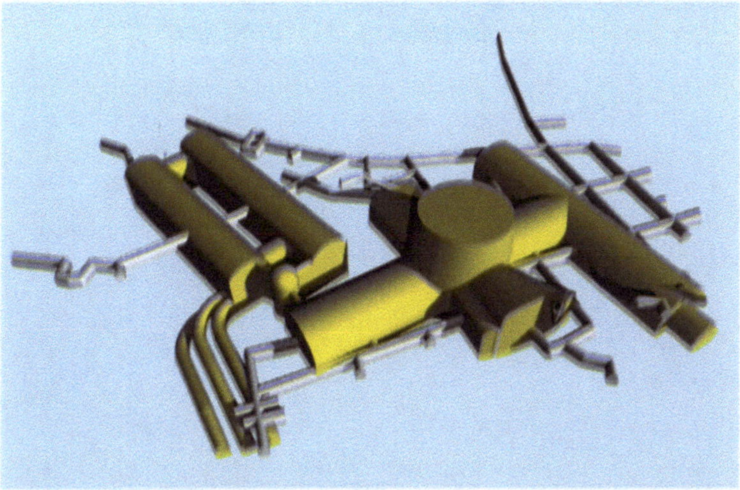

Fig. 17. Layout of the underground theater

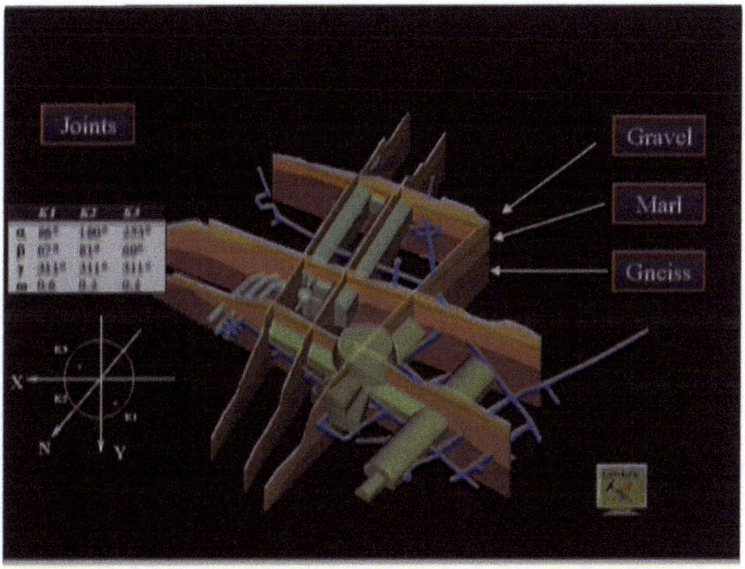

Fig. 18. Geological information, including three-set joint system

10 Conclusion

A microstructural damage variable, e.g. the second-order fabric tensor employed in this chapter, furnishes a much more essential characterization of distributed micro-defects than does a phenomenological damage variable, especially in jointed rock masses. Starting from it, a one-parameter damage-dependent elasticity tensor is developed based on tensorial algebra and ther-

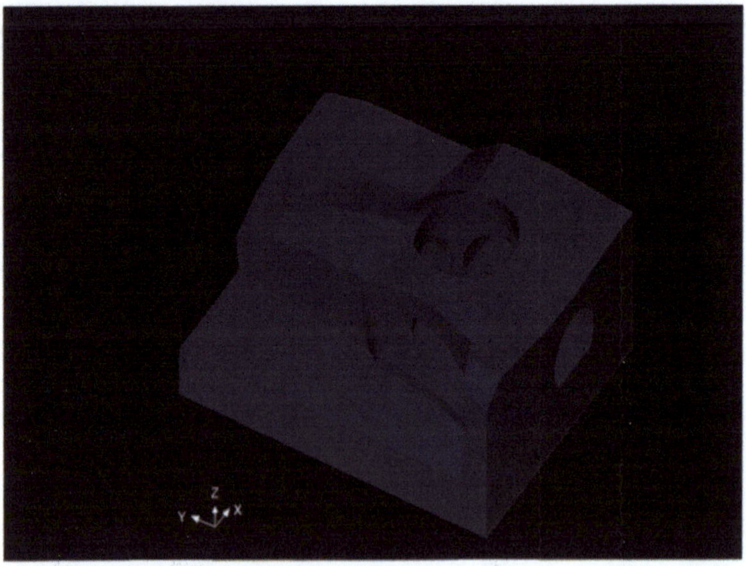

Fig. 19. Finite element model of the theater

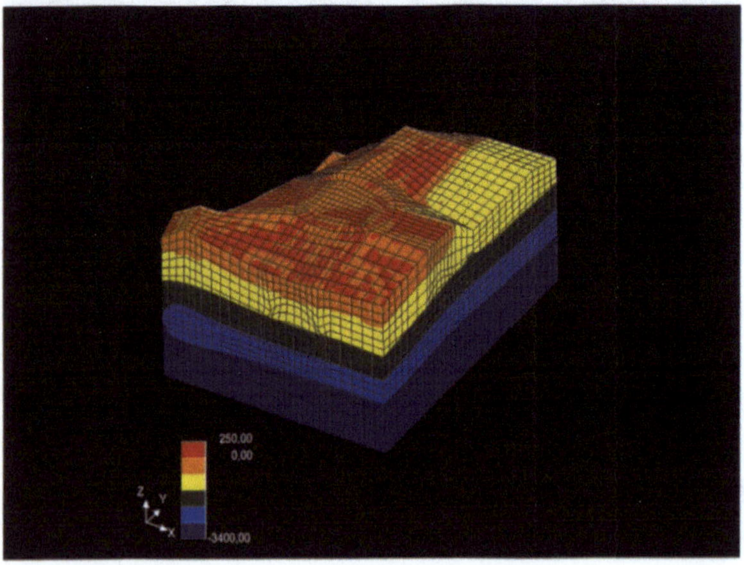

Fig. 20. Vertical initial stress

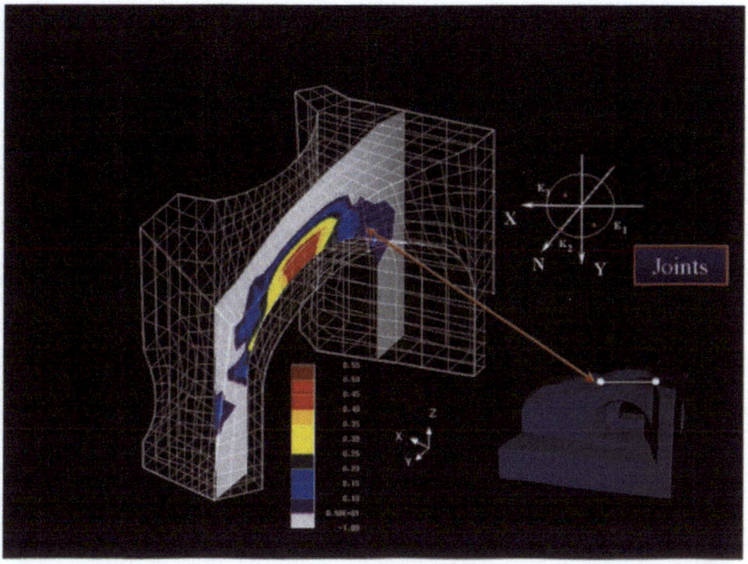

Fig. 21. Rock bridge, damage in y-direction

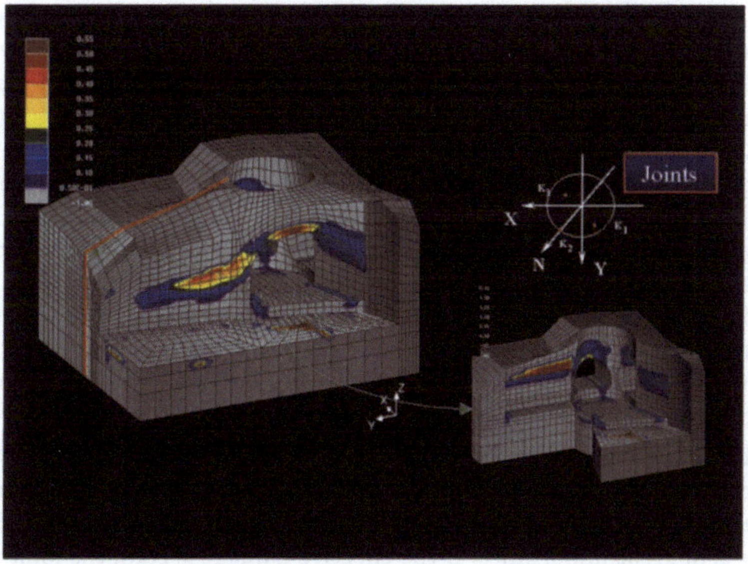

Fig. 22. Vertical damage in z-direction

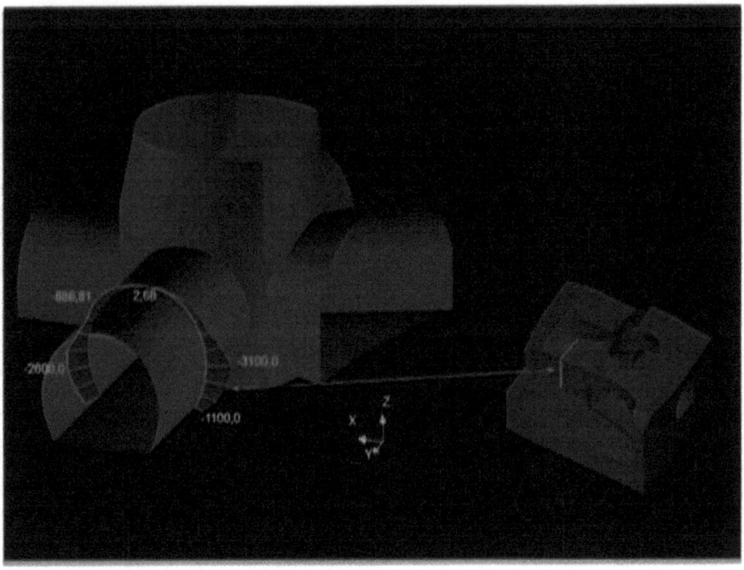

Fig. 23. Normal forces in shotcrete lining

modynamics requirements; a conjugate-force-based anisotropic damage evolution law is deduced within the framework of Rice's [32] "normality structure". Although the deduction is complicated and not straightforward as compared with the deduction based on the concept of net stress, it turns out a practical formulation with a rigorous basis. In order to extend the deduced constitutive equation to rock-like materials, the real state of rock-like materials is mapped into an equivalent state of the ideal materials. The equivalent state represents current open cracks in terms of microstructure and is subjected to the effective stress which can be used to simulate "splitting cracking", dilatancy and residual strain/stress. The effective stress and equivalent state have a parallel relation with the net stress and the associated undamaged state. Of the two, however, the latter can only furnish a very coarse description of anisotropic damage.

The numerical calculation of the underground theater of Linz shows that the proposed damage model can successfully take into account the orientation of the joint system. For interpretation of the design the plot of the damage tensor is of great importance. If the effective damage is zero, the joint system is closed. Any value above zero shows the amount of opening in the joint system. From this result the layout of the bolt system can be designed. Thus, the effective damage distribution is really an important indicator for tunnelling design.

There are many types of preexisting defects in rocks, e.g., cracks, cavities, inhomogeneities or inclusions. Even for cracks, the behaviors of cracks under tension and compression are quite different. It is shown in Sect. 4 that the so-called energy-based equivalent inclusion size is a common geometric damage measurement to various types of defects and can be used to unify the well-known Gurson [14] and Budiansky-O'Connel [6] damage definitions.

In order to take into account multiple dissipation mechanisms in a unitary framework, it is necessary to develop conjugate-force-based damage evolution laws and general free energy functions. It is revealed in Sect. 5 that nonlinear phenomenological equations and Onsager reciprocal relations emerge naturally from the so-called normality structures [32][33] if each rate of local internal variables is a homogeneous function in its conjugate force. Rice's restriction on quasi-static growth of Griffith cracks [34] can also be understood as the requirement of such kinetic rate laws, and macroscopic damage evolution laws are adequate to be covered by nonlinear phenomenological equations at least for microcrack-dominated damage.

Acknowledgements

The work reported here was also supported by the National Science Foundation of China with grant No. 59879005 and 59939190.

References

[1] Bai M., Ma Q., Roegiers J. C. (1994) Dual-Porosity Behaviour of Naturally Fractured Reservoirs. *Int. J. Numer. Anal. Meth. Geomech.* **18**, 359-376

[2] Barenblatt G. E., Zheltov I. P., Kochina I. N. (1960) Basic concepts in the theory of homogeneous liquids in fractured rocks. *J. Appl. Mech.* **24**, 1286-1303

[3] Bazant Z. P. (1983) Comment on orthotropic models for concrete and geomaterials. *J. Eng. Mech. ASCE* **109**, 849-865

[4] Bazant Z. P. (1987) Nonlocal damage theory. *J. Eng. Mech. ASCE* **113**, 1512-1533

[5] Bieniawski Z. T. (1968) Fracture dynamics of rock. *The International Journal of Fracture Mechanics*, **4**, 415-430

[6] Budiansky B., O'Connel R. J. (1976) Elastic moduli of a cracked solid. *Int. J. Solids Structures* **12**, 81-97

[7] Chow C. L., Lu T. J. (1989) On evolution laws of anisotropic damage. *Eng. Fracture Mech.* **34**, 679-701

[8] Cowin S. C. (1985) The relationship between the elasticity tensor and the fabric tensor. *Mech. of Mat.* **4**, 137-147

[9] Cundall P. A. (1971) A computer model for simulating prograssive, large scale movements in blocky rock system. *Proceedings of the Symposium of the International Society of Rock Mechanics*, Nancy, France. **1**, 11–18

[10] Delph T. J. (1999) A simple model for crack growth in creep resistant alloys. *Int. J. Fracture* **98**, 77-86

[11] Dragon A., Charlez Ph., Pham D., Shao J. F. (1993) A model of anisotropic damage by (micro)crack growth. in *Proceedings of the International Symposium on Assessment and Prevention of Failure Phenomena in Rock Engineering* (Edited by A. G. Pasamehmetoglu, *et al.*), pp. 71-78. Istanbul, Turkey.

[12] Dragon A. and Mroz Z. (1979) A continuum model for plastic-brittle behaviour of rock and concrete. *Int. J. Engng. Sci.* **17**, 121-137

[13] Eshelby J. D. (1957) The determination of the elastic field of an ellipsoidal inclusion and related problems. *Proc. Roy. Soc.* **A241**, 376-396

[14] Gurson A. L. (1977) Continuum theory of ductile rupture by void nucleation and growth, I. Yield criteria and flow rules for porous ductile media. *J. Eng. Mater. Technol.* **99**, 2-15

[15] Ju J.W., Lee X. (1991) Micromechanical Damage Models for Brittle Solids. *J. Eng. Mech. ASCE* **117**, 1495-1536

[16] Ju J.W. (1989) On energy-based coupled elastoplastic damage theories: constitutive modelling and computational aspects. *Int. J. Solids Structures* **25**, 803-833

[17] Kachanov L.M. (1958) Time of the rupture process under creep conditions. *IVZ Akad Nauk, S.S.R., Otd Tech Nauk,* **8**, 26-31

[18] Kachanov M. (1982) A microcrack model of rock inelasticity part II: propagation of microcracks. *Mech. Mater.* **1**, 29-41

[19] Kawamoto T., Ichikawa Y., Kyoya T. (1988) Deformation and fracturing behaviour of discontinuous rock mass and damage mechanics theory. *Int. J. Numer. Anal. Meth. Geomech.* **12**, 1-30

[20] Krajcinovic D. (1983) Continuum Damage Mechanics Revisited: Basic Concepts and Definitions. *J. Appl. Mech.* **52**, 829-834

[21] Lee H., Peng K., Wang J. (1985) An anisotropic damage criterion for deformation instability and its application to forming limit analysis of metal plates. *Eng. Fracture Mech.* **21**, 1031-1054

[22] Lemaitre J. (1990) *A Course on Damage Mechanics.* Second Edition, Springer.

[23] Linsbauer H. N., Ingraffea A. R., Rossmanith H. P., Wawrzynek P. A. (1989) Simulation of cracking in large arch dam. *J. Struct. Eng. ASCE* **115**, 1599-1630

[24] Malvern L. E. (1969) *Introduction to the mechanics of a continuous medium.* Prentice-Hall, Inc., Englewood Cliffs, N.J.

[25] Murakami S. (1983) Notion of continuum damage mechanics and its application to anisotropic creep damage theory. *J. Eng. Mater. Technol.* **105**, 99-105

[26] Murakami S. (1988) Mechanical Modeling of material damage. *J. Appl. Mech.* **55**, 280-286

[27] Nemat-Nasser S., Obata M. (1988) A microcrack model of dilatancy in brittle Materials. *J. Eng. Mech. ASCE* **55**, 24-35

[28] Oda M. (1983) A method for evaluating the effect of crack geometry on the mechanical behaviour of cracked rock masses. *Mech. Mater.* **2**, 163-171

[29] Oda M., Suzuki K., Maeshibu T. (1984) Elastic compliance for rock-like materials with random cracks. *Solids and Foundations,* **24**, 27-40

[30] Oda M. (1986) An Equivalent Continuum Model for coupled stress and fluid flow analysis in jointed rock masses. *Water Resour. Res.* **22**, 1845-1856

[31] Prat P. C., Bazant Z. P. (1997) Tangential stiffness of elastic materials with systems of growing or closing cracks. *J. Mech. Phys. Solids* **45**, 611-636

[32] Rice J. R. (1971) Inelastic constitutive relations for solids: an integral variable theory and its application to metal plasticity. *J. Mech. Phys. Solids* **19**, 433-455

[33] Rice J.R., (1975) Continuum mechanics and thermodynamics of plasticity in relation to microscale deformation mechanisms. In: Argon, A.S. (Ed.), Constitutive Equations in Plasticity. MIT Press, Cambridge, MA, pp. 23-79.

[34] Rice J.R. (1978) Thermodynamics of quasi-static growth of griffith cracks. *J. Mech. Phys. Solids* **26**, 61-78

[35] Rudnicki J. W., Rice J. R. (1975) Conditions for the localization of deformation in pressure-sensitive dilatancy materials. *J. Mech. Phys. Solids* **23**, 371-394

[36] Sayers C. M., Kachanov M. (1991) A simple technique for finding effective constants of cracked solids for arbitrary crack orientation statistics. *Int. J. Solids Structures* **27**, 671-680

[37] Stumvoll M., Swoboda G. (1993) Deformation behavior of ductile solids containing anisotropic damage. *J. Eng. Mech. ASCE* **119**, 1331-1352

[38] Swoboda G., Yang Q., Ito F. (1995) Damage Propagation model and its Application to Rock Engineering Problems. in *Proceedings of the Eighth International Congress on Rock Mechanics* (Edited by T. Fujii), pp. 159-164. Tokyo, Japan.

[39] Swoboda G., Yang Q. (1997) On energy-based anisotropic damage models. in *Proceedings of the 9th IACMAG Conference*, Wuhan, China.

[40] Swoboda G., Yang Q. (1999) An energy-based damage model of geomaterials — I. Formulation and numerical results. *Int. J. Solids Structures* **36**, 1719-1734

[41] Swoboda G., Yang Q. (1999) An energy-based damage model of geomaterials — II. Deduction of damage evolution laws. *Int. J. Solids Structures* **36**, 1735-1755

[42] Visser J. H. M., Mier J. G. M. (1992) Hydraulic fracturing of concrete and rock. in *Proceedings of the Second International Conference on Fracture and Damage of Concrete and Rock*(Edited by H. P. Rossmanith), pp. 435-444. Vienna, Austria, November 9-13, 1992.

[43] Yang Q. (1996) *Numerical Modelling for Discontinuous Geomaterials Considering Damage Propagation and Seepage.* Ph.D Dissertation, Faculty of Architecture and Civil Engineering, University of Innsbruck, Austria.

[44] Yang Q., Zhou W.Y., Swoboda G. (1999) Micromechanical identification of anisotropic damage evolution laws. *Int. J. Fracture* **98**, 55-76

[45] Yang Q., Li Zhongkui, Tham L. G. (2001) An explicit expression of second-order fabric-tensor dependent elastic compliance tensor. *Mechanics Research Communication*, **28**, 255-260

[46] Yang Q., Zhou W. Y., Swoboda G. (2001) Asymptotic Solutions of Penny-shaped Inhomogeneities in Global Eshelby's Tensor. *J. Appl. Mech.* **28**, 255-260

[47] Yang Q., Tham L. G., Swoboda G. (2002) Micromechanical basis of nonlinear phenomenological equations as damage evolution laws. *Mechanics Research Communication*, **29**, 131-136

[48] Zhang W. H. (1992) *Numerical Analysis of Continuum Damage Mechanics.* Ph.D Dissertation, School of Civil Engineering University of New South Wales, Australia.

Chapter 8
Parameter identification and its application in tunneling

Zhihai Xiang[1], Gunter Swoboda[2], and Zhangzhi Cen[1]

[1] Department of Engineering Mechanics, Tsinghua University
[2] Institute for Structural Analysis and Strength of Materials, University of Innsbruck

Abstract. This paper focuses on discussing the parameter identification techniques in geotechnical engineering. Firstly, the general formulation of the parameter identification process is presented. Secondly, the problem of identifiability is discussed and illustrated by an example of identifying the initial damage parameters of a damage model for jointed rocks. Then the algorithms of designing the optimal layout of displacement measurements are proposed, based on the analyses of the well-posednesses of the parameter identification processes with the Gauss-Newton method and the Complex method, respectively. The validities of these algorithms are proved by some academic and applied engineering examples. Finally, the advantages and drawbacks of the gradient-type methods and the direct-search methods are carefully compared.

1 Introduction

Mechanical properties of rocks and soils play an important role in geotechnical engineering. These properties obtained from experimental investigations may, however, fail to represent the behavior of *in-situ* rock and soil masses because of the complex and inhomogeneous geological conditions of the ground. The widely used solutions to this problem are to adopt a large factor of safety or to introduce certain hypotheses corresponding to other similar cases. The first solution is undoubtedly uneconomical, while the second one may cause some safety risks.

An alternative solution is proposed from Terzaghi's "observational construction method" [26], in which the expected behavior of the structures is compared with the field measurements during the construction period. Depending on the discrepancy, the geotechnical properties adopted in the design are reevaluated and the original design modified if necessary. This procedure is so flexible that the structures can be constructed economically and safely.

Over the past twenty or more years, more and more attention has been paid to developing methods with which to determine the properties of rocks and soils, such as mechanical parameters, initial stresses, permeability, etc., from field measurements. These methods are usually called *back-analysis* [10]. This paper discusses only some aspects involved in identifying geomechanical parameters and adopts displacement measurements, which are sensitive to many important parameters and can be accurately measured with ease.

The parameter identification process can be described as a mathematical programming problem [11], which follows some general formulations. Although these formulations are well developed, two questions concerning identifiability and reliability should be answered for a successful parameter identification process. The system sensitivity is usually the key to answering the first question. However, the second question is more complicated. This paper carefully tests the identifiabilities to the initial damage parameters of a damage model for jointed rock. Then the well-posednesses of the parameter identification processes, based on the Gauss-Newton method and the Complex method, are analyzed. Subsequently, the problem of determining the optimal layout of displacement measurements is thoroughly discussed and the advantages and drawbacks of these two optimization methods are compared.

2 The general formulation of parameter identification processes

Generally, the parameter identification process involves solving a certain kind of constrained nonlinear optimization problem:

$$\begin{aligned}&Min\ f(\mathbf{p})\\&subject\ to\\&\mathbf{K}(\mathbf{p})\,\mathbf{u}=\mathbf{F},\ \textit{with boundary conditions},\\&\mathbf{p}\in\mathbf{D}_p\end{aligned} \tag{1}$$

where f is the objective function; $\boldsymbol{p}$ is the parameter vector; $\mathbf{D}_p$ is the admissible set of the parameters; $\boldsymbol{u}$ is the displacement vector; $\mathbf{K}(\mathbf{p})\,\mathbf{u}=\mathbf{F}$is the system equation, which is presented in Finite Element (FE) formulation in this paper. For the sake of simplicity, time-dependent problems are not discussed here.

2.1 Objective functions

The general form of objective functions can be obtained from the extended Bayesian estimation (Yusuke et al., 1994) as

$$f(\mathbf{p})=\mathbf{R}^T\mathbf{C}_u^{-1}\mathbf{R}+\beta\left(\mathbf{p}-\mathbf{p}^0\right)^T\mathbf{C}_p^{-1}\left(\mathbf{p}-\mathbf{p}^0\right), \tag{2}$$

where $\mathbf{C}_u$ is the covariance matrix of measured displacements; $\mathbf{C}_p$ is the covariance matrix of parameters; $\mathbf{p}^0$ are pre-estimated parameters; β is a non-negative scalar parameter and $\boldsymbol{R}$ is the residual vector

$$\mathbf{R}(\mathbf{p})=\mathbf{u}_s^*-\mathbf{S}\mathbf{u}, \tag{3}$$

where $\mathbf{u}_s^*$ is the measured displacements and $\boldsymbol{S}$ is the selective matrix with respect to the locations of the measurements. If the prior information is unknown, i.e. $\mathbf{C}_p \to \infty$, the objective function in (2) degenerates to the maximum likelihood (ML) formulation, which is actually a kind of weighted least square (LS) formulation. Furthermore, (2) becomes the ordinary LS formulation, when all measurements are uncorrelated and have the same variance σ^2, i.e., $\mathbf{C}_u = \mathbf{I}\sigma^2$ (**I** is the identity matrix.).

Actually, the objective function in (2) can be regarded as an ML formulation with a kind of Tikhonov ([20] and [7]) regulator $\beta\left(\mathbf{p}-\mathbf{p}^0\right)^T \mathbf{C}_p^{-1}\left(\mathbf{p}-\mathbf{p}^0\right)$. This explains why the parameter identification process, based on extended Bayesian estimation, can ensure stable solutions. However, the Tikhonov regulator is only a mathematical trick. If β is larger, the identified parameters will be closer to $\mathbf{p}^0$, while the measurements will have more impact on the final solution if β is smaller. Therefore, it is necessary to compromise between stability and accuracy.

2.2 Optimization methods

Many kinds of optimization method have been used to solve the problem presented in (1). These methods are usually classified in two categories: the gradient-type methods and the direct-search methods [10].

Gradient-type methods. As a benefit of the gradient information, the gradient-type methods are characterized by fast and smooth convergence. Of these methods, the Gauss-Newton method is reported to be most effective, because the object functions from extended Bayesian estimation or other degenerated estimations are presented in nonlinear LS form [19].

The Gauss-Newton method involves a series of iterations. At iteration step k, the updated parameters are:

$$\mathbf{p}^k = \mathbf{p}^{k-1} + \Delta\mathbf{p}^{k-1}, \; k = 1,\ 2,\ \ldots, \tag{4}$$

where $\Delta\mathbf{p}^{k-1}$ is the update from step k-1 to step k.

The objective function f at step k can be expanded at $\mathbf{p}^{k-1}$ as

$$\begin{aligned} & f\left(\mathbf{p}^k\right) \approx q \equiv \\ & \left(\mathbf{p}^{k-1}\right) + \nabla_p f\left(\mathbf{p}^{k-1}\right) \Delta\mathbf{p}^{k-1} + \tfrac{1}{2}\left(\Delta\mathbf{p}^{k-1}\right)^T \nabla_p^2 f\left(\mathbf{p}^{k-1}\right) \Delta\mathbf{p}^{k-1}, \end{aligned} \tag{5}$$

where ∇_p is the gradient operator with respect to $\boldsymbol{p}$ and

$$\nabla_p f = 2\mathbf{J}^T\mathbf{C}_u^{-1}\mathbf{R} + 2\beta\,\mathbf{C}_p^{-1}\left(\mathbf{p}-\mathbf{p}^0\right) \tag{6}$$

$$\nabla_p^2 f = 2\mathbf{J}^T\mathbf{C}_u^{-1}\mathbf{J} + 2\mathbf{R}^T\mathbf{C}_u^{-1}\nabla_p^2\mathbf{R} \; + 2\beta\,\mathbf{C}_p^{-1} \tag{7}$$

$$\mathbf{J} = \nabla_p\mathbf{R} = -\mathbf{S}\frac{\partial\mathbf{u}}{\partial\mathbf{p}}. \tag{8}$$

The system sensitivity matrix $\frac{\partial \mathbf{u}}{\partial \mathbf{p}}$ can be derived from the system equation as (supposing $\boldsymbol{F}$ is independent of $\boldsymbol{p}$.)

$$\begin{aligned}&\frac{\partial \mathbf{K}}{\partial \mathbf{p}}\mathbf{u}+\mathbf{K}\frac{\partial \mathbf{u}}{\partial \mathbf{p}}=0 \Rightarrow \\ &\qquad\frac{\partial \mathbf{u}}{\partial \mathbf{p}}=-\mathbf{K}^{-1}\frac{\partial \mathbf{K}}{\partial \mathbf{p}}\mathbf{u}.\end{aligned} \tag{9}$$

For problems with many degrees of freedom, it could be difficult to calculate $\mathbf{K}^{-1}$ directly. However, this problem can be circumvented by introducing an intermediate matrix $\boldsymbol{L}$, which can be obtained from

$$\mathbf{KL}=\frac{\partial \mathbf{K}}{\partial \mathbf{p}}\mathbf{u} \tag{10}$$

Therefore, (8) can be rewritten as

$$\mathbf{J}=\mathbf{SL} \tag{11}$$

The update $\Delta \mathbf{p}^{k-1}$ should satisfy

$$\begin{aligned}&\nabla_{\Delta p} q=0\\ &\nabla_p f\left(\mathbf{p}^{k-1}\right)+\nabla_p^2 f\left(\mathbf{p}^{k-1}\right)\Delta \mathbf{p}^{k-1}=0\\ &\nabla_p^2 f\left(\mathbf{p}^{k-1}\right)\Delta \mathbf{p}^{k-1}=-\nabla_p f\left(\mathbf{p}^{k-1}\right)\end{aligned} \tag{12}$$

Substituting (6) and (7) into (12), yields:

$$\begin{aligned}&\left[\left(\mathbf{J}^{k-1}\right)^T \mathbf{C}_u^{-1}\mathbf{J}^{k-1}+\beta\,\mathbf{C}_p^{-1}\right]\,\Delta \mathbf{p}^{k-1}=\\ &\quad-\left[\left(\mathbf{J}^{k-1}\right)^T \mathbf{C}_u^{-1}\mathbf{R}^{k-1}+\beta\,\mathbf{C}_p^{-1}\left(\mathbf{p}^{k-1}-\mathbf{p}^0\right)\right]\end{aligned}, \tag{13}$$

where $\mathbf{R}^T\mathbf{C}_u^{-1}\nabla_p^2\mathbf{R}$ is omitted to avoid calculating the Hessian matrix of R.

Briefly put, the Gauss-Newton iteration process can be written as

$$\mathbf{p}^k=\mathbf{G}\left(\mathbf{p}^{k-1}\right),\ k=1,\ 2,\ \ldots, \tag{14}$$

where

$$\mathbf{G}(\mathbf{p})=\mathbf{p}-\left(\mathbf{J}^T\mathbf{C}_u^{-1}\mathbf{J}+\beta\,\mathbf{C}_p^{-1}\right)^{-1}\left[\mathbf{J}^T\mathbf{C}_u^{-1}\mathbf{R}+\beta\,\mathbf{C}_p^{-1}\left(\mathbf{p}-\mathbf{p}^0\right)\right]. \tag{15}$$

The iteration process stops when both of the following criteria of convergence are satisfied:

$$\frac{\left\|\mathbf{u}^k-\mathbf{u}^{k-1}\right\|}{\left\|\mathbf{u}^k\right\|}<10^{-3}, \tag{16}$$

$$Max\left\{\left|\frac{dp_i^k}{p_i^k}\right|\right\}<10^{-3},\ i=1,\ 2,\ \ldots,N_p, \tag{17}$$

where $\Delta p_i^k$is the increment of the i-th parameter; p_i^k is the value of the i-th parameter at step k, and N_p is the number of parameters to be identified.

The above parameter identification algorithm was coded into an FE program with the name FIBACK.

Direct-search methods. Although the gradient-type methods have the advantage of fast and smooth convergence, in many cases the analytical expression of the gradient cannot be determined, especially when the system equation is nonlinear. In these cases the finite difference method is a popular remedy. However, it is not easy to determine the optimal step size of this [7] and processing is quite time-consuming.

It is quite flexible to use direct-search methods to solve the above problem, because these methods can provide the optimal solution by means of only successive evaluations of the objective function. A typical and popular direct-search method is the Nelder-Mead variable polyhedron search method [18], which is an extension of Spendley's Simplex method [21]. In some cases, however, Nelder and Mead's method generates an irregular simplex, which may provide some fake local solutions. To overcome this drawback, Box proposed the complex method [2]. The parameter identification algorithm, based on the complex method, was coded into an FE program with the name CPL.

3 Damage parameter identification

Jointed rock is frequently encountered in geotechnical engineering. The construction strategies used for it are strongly influenced by the complex mechanical behavior, such as anisotropy, hysteresis, dilatancy, irreversible strain, etc. To facilitate numerical simulation, many damage models have been proposed to obtain the average response of the rock mass with joints [13]. However, these damage models can hardly be used for practical analysis without calibrating all parameters inside. This section aims to identify the initial damage parameters of a damage model proposed by Yang and Swoboda ([22] and [23])as shown in the previous chapter.

3.1 Algorithm

The detailed theories of this damage model are presented in previous papers in this book. In this model the parameters which need to be calibrated are: Yong's modulus E, Poisson's ratio ν, inclusion parameter h, material constant m and damage tensor $\boldsymbol{\Omega}$. To identify these parameters, the Gauss-Newton method is adopted in the following analysis.

Regardless of the prior information, (13) is rewritten as

$$\left(\mathbf{J}^{k-1}\right)^T \mathbf{J}^{k-1} \Delta\mathbf{p}^{k-1} = -\left(\mathbf{J}^{k-1}\right)^T \mathbf{R}^{k-1}, \tag{18}$$

where $\boldsymbol{J}$ is obtained by (11), which requires the evaluation of $\frac{\partial \mathbf{K}}{\partial \mathbf{p}}$. In FE formulation

$$\frac{\partial \mathbf{K}}{\partial \mathbf{p}} = \int \mathbf{B}^T \frac{\partial \mathbf{D}}{\partial \mathbf{p}} \mathbf{B} dV, \tag{19}$$

where $\boldsymbol{B}$ is the strain matrix, $\boldsymbol{D}$ is the damage elasticity [31]. Therefore, the aim is to calculate the derivations of $\boldsymbol{D}$ over different material parameters. The basic equations of Chap. 7 give

$$\frac{\partial D_{ijkl}}{\partial E} = \frac{\nu}{(1+\nu)(1-2\nu)}\Phi_{ij}\Phi_{kl} + \frac{1}{2(1+\nu)}\left(\Phi_{ik}\Phi_{jl} + \Phi_{il}\Phi_{jk}\right), \tag{20}$$

$$\frac{\partial D_{ijkl}}{\partial \nu} = \frac{E\left(1+2\nu^2\right)}{\left[(1+\nu)(1-2\nu)\right]^2}\Phi_{ij}\Phi_{kl} - \frac{E}{2(1+\nu)^2}\left(\Phi_{ik}\Phi_{jl} + \Phi_{il}\Phi_{jk}\right), \tag{21}$$

$$\begin{aligned}\frac{\partial D_{ijkl}}{\partial h} = {} & \frac{E\nu}{(1+\nu)(1-2\nu)}\left(\frac{\partial \Phi_{ij}}{\partial h}\Phi_{kl} + \Phi_{ij}\frac{\partial \Phi_{kl}}{\partial h}\right) + \\ & \frac{E}{2(1+\nu)}\left(\frac{\partial \Phi_{ik}}{\partial h}\Phi_{jl} + \Phi_{ik}\frac{\partial \Phi_{jl}}{\partial h} + \frac{\partial \Phi_{il}}{\partial h}\Phi_{jk} + \Phi_{il}\frac{\partial \Phi_{jk}}{\partial h}\right),\end{aligned} \tag{22}$$

$$\frac{\partial \Phi_{ij}}{\partial h} = -m\frac{\partial \hat{\Omega}_{ij}}{\partial h} - (1-m)\left(\frac{\partial \hat{\Omega}_{ik}}{\partial h}\hat{\Omega}_{kj} + \hat{\Omega}_{ik}\frac{\partial \hat{\Omega}_{kj}}{\partial h}\right), \tag{23}$$

$$\begin{aligned}\frac{\partial \hat{\Omega}_{ij}}{\partial h} &= \frac{\partial \hat{\mathrm{T}}_{ijkl}}{\partial h}\Omega_{kl} \\ &= \left(\frac{\partial \hat{Q}_{ia}}{\partial h}\hat{Q}_{jb} + \hat{Q}_{ia}\frac{\partial \hat{Q}_{jb}}{\partial h}\right) Q_{ak}Q_{bl}\Omega_{kl},\end{aligned} \tag{24}$$

$$\frac{\partial \hat{Q}_{ij}}{\partial h} = \sum_{k=1}^{3}\left[1 - \hat{H}\left(\varepsilon_k\right)\right] v_i^k v_j^k, \tag{25}$$

$$\begin{aligned}\frac{\partial D_{ijkl}}{\partial m} = {} & \frac{E\nu}{(1+\nu)(1-2\nu)}\left(\frac{\partial \Phi_{ij}}{\partial m}\Phi_{kl} + \Phi_{ij}\frac{\partial \Phi_{kl}}{\partial m}\right) + \\ & \frac{E}{2(1+\nu)}\left(\frac{\partial \Phi_{ik}}{\partial m}\Phi_{jl} + \Phi_{ik}\frac{\partial \Phi_{jl}}{\partial m} + \frac{\partial \Phi_{il}}{\partial m}\Phi_{jk} + \Phi_{il}\frac{\partial \Phi_{jk}}{\partial m}\right),\end{aligned} \tag{26}$$

$$\frac{\partial \Phi_{ij}}{\partial m} = -\hat{\Omega}_{ij} + \hat{\Omega}_{ik}\hat{\Omega}_{kj}. \tag{27}$$

Since the non-diagonal items of $\boldsymbol{\Omega}$ are dependent on each other, they cannot be identified simultaneously. They can, however, be obtained using an indirect approach. From (5) of Chap. 7, $\boldsymbol{\Omega}$ is rewritten as

$$\boldsymbol{\Omega} = \bar{\mathbf{v}}\bar{\mathbf{v}}^T, \tag{28}$$

where

$$\bar{\mathbf{v}} = \sqrt{\omega}\mathbf{n}. \tag{29}$$

Thus, all items of $\boldsymbol{\Omega}$ can be uniquely determined identifying $\bar{\mathbf{v}}$. The derivations of $\boldsymbol{D}$ over $\bar{\mathbf{v}}$ are:

$$\begin{aligned}\frac{\partial D_{ijkl}}{\partial \bar{v}_n} = & \frac{E\nu}{(1+\nu)(1-2\nu)}\left(\frac{\partial \Phi_{ij}}{\partial \bar{v}_n}\Phi_{kl} + \Phi_{ij}\frac{\partial \Phi_{kl}}{\partial \bar{v}_n}\right) + \\ & \frac{E}{2(1+\nu)}\left(\frac{\partial \Phi_{ik}}{\partial \bar{v}_n}\Phi_{jl} + \Phi_{ik}\frac{\partial \Phi_{jl}}{\partial \bar{v}_n} + \frac{\partial \Phi_{il}}{\partial \bar{v}_n}\Phi_{jk} + \Phi_{il}\frac{\partial \Phi_{jk}}{\partial \bar{v}_n}\right),\end{aligned} \quad (30)$$

$$\frac{\partial \Phi_{ij}}{\partial \bar{v}_n} = -m\frac{\partial \hat{\Omega}_{ij}}{\partial \bar{v}_n} - (1-m)\left(\frac{\partial \hat{\Omega}_{ik}}{\partial \bar{v}_n}\hat{\Omega}_{kj} + \hat{\Omega}_{ik}\frac{\partial \hat{\Omega}_{kj}}{\partial \bar{v}_n}\right), \quad (31)$$

$$\begin{aligned}\frac{\partial \hat{\Omega}_{ij}}{\partial \bar{v}_n} &= \hat{\mathrm{T}}_{ijkl}\frac{\partial \Omega_{kl}}{\partial \bar{v}_n} \\ &= \hat{\mathrm{T}}_{ijkl}\left(\delta_{kn}\bar{v}_l + \bar{v}_k\delta_{ln}\right)\end{aligned}. \quad (32)$$

3.2 Example

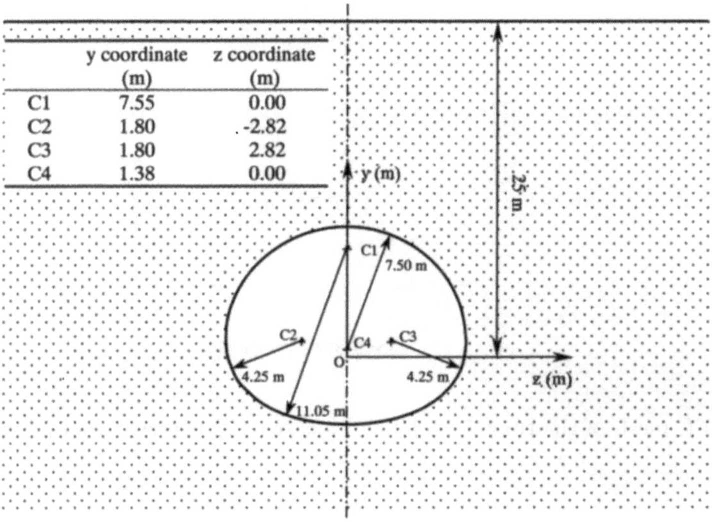

Fig. 1. Cross-section of the tunnel model

The identifiabilities of this algorithm for each initial damage parameter in this model are tested using the following tunnel excavation example. The cross-section of this tunnel is presented in Fig. 1. Figure 2 shows the corresponding FE mesh, in which only half of the tunnel is modeled for the sake of symmetry. Two steps are involved in simulation of the excavation procedure. In the first step the initial stresses and initial strains are obtained from the response to the dead load. In the second step the Young's moduli of the excavated elements are set to zero, and the reaction forces are applied to the surrounding elements. In both steps, additional iterations are added to determine the active damage tensor. The convergence criterion of these iterations is:

$$Max\left\{\left|\frac{{}^{k+1}U_i^{ob} - {}^{k}U_i^{ob}}{{}^{k+1}U_i^{ob}}\right|\right\} < 10^{-4}\,,\ i = 1, 2, \ldots, N^{ob}\,, \tag{33}$$

where ${}^{k}U_i^{ob}$ is the ith observed displacement at iteration step k, and N^{ob} is the number of observed displacements. Sometimes these iterations converge with difficulty if the initial values of the parameters are not properly specified. To avoid redundant iterations, the convergence tolerance is multiplied by a factor of 10 every four iteration steps.

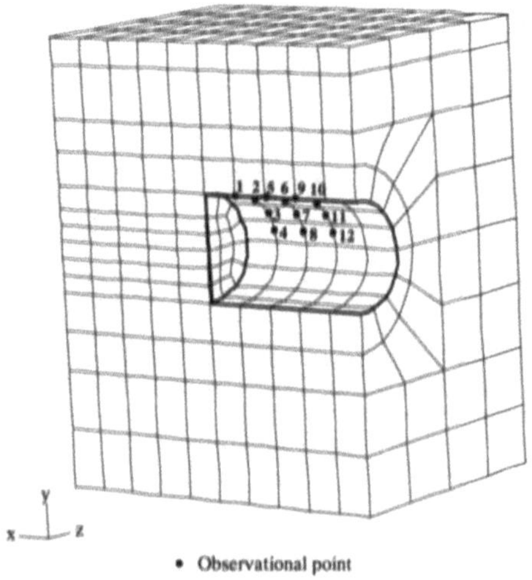

Fig. 2. FE mesh of the tunnel model and locations of observational points

Table 1. Input displacements of observational points

Observational point	1	2	3	4	5	6	7	8	9	10	11	12
x displacement												
y displacement	×	×	×		×	×	×		×	×	×	
z displacement			×	×			×	×			×	×

Table 2. Material properties of tunnel models

Model	1	2	3	4
E(Pa)	3.0×10^5	3.0×10^5	3.0×10^5	3.0×10^5
ν	0.3	0.3	0.3	0.3
h	0.1	0.2	0.3	0.8
m	0.5	0.5	0.5	0.5
ρ (kN / m^3)	3.0	3.0	3.0	3.0
Ω_{11}	0.021428571	0.021428571	0.021428571	0.021428571
Ω_{12}	0.064285714	0.064285714	0.064285714	0.064285714
Ω_{13}	0.042857143	0.042857143	0.042857143	0.042857143
Ω_{22}	0.192857143	0.192857143	0.192857143	0.192857143
Ω_{23}	0.128571429	0.128571429	0.128571429	0.128571429
Ω_{33}	0.085714286	0.085714286	0.085714286	0.085714286

Table 3. Parameter sets

Set	E	ν	H	m	Ω
1	×	×	×	×	
2	×	×			×
3	×	×		×	×
4	×	×	×		×

The displacements, calculated using the FE program FINAL [24], are used as the 'observed' input data. Twelve observational points are located at the crown of the tunnel (see Fig. 2 and Tab. 1). Four models are involved in this example. The only difference among the exact material parameters of each model is the inclusion parameter h (see Tab. 2). In model 4, an extremely large value of h is deliberately used to test its influences. Tab. 3 shows four parameter sets, which are tested for each model to evaluate the identifiabilities to various parameters. The identified results are given in Tab. 4, Fig. 3 and Fig. 4. From the results of Fig. 3 it is seen that:

(a) E, ν, h and m can always be successfully identified simultaneously.
(b) E, ν and $\mathbf{\Omega}$ can always be successfully identified simultaneously.
(c) the results from model 1 to model 4 reveal that it is easier to identify h, m and $\mathbf{\Omega}$ if the joint is open more. It is also observed that these parameters are very difficult to identify when h is very small. This is

because most of the regions are under pressure in this example. Therefore, if h is very small, the displacements are not so sensitive to m and $\boldsymbol{\Omega}$. This is confirmed by the test for set 2 of model 1, in which the identified $\boldsymbol{\Omega}$ is far from the exact value, but the average error of observed displacements is still very small (see Tab. 4).

(d) the convergence speed of E is much faster than that of other parameters. This is mainly because the displacements are more sensitive to E.

(e) Parameter m exceeds its lower bound in the first four iteration steps of model 4 set 3 (see Fig. 4a). However, thanks to the mechanism of keeping all parameters within their bounds, the iteration process converges to the exact parameters.

4 Optimal displacement layout

The results of geomechanical parameter identification are observed to be highly sensitive to the input displacements. Many researchers have tried to find a certain criterion to aid in designing the optimal displacement layout. Some popular views are reported as follows:

It is well known that successful parameter identification should use the measurement on which the parameters have great impact. From this point of view it is suggested to measure displacements with high system sensitivity [17]. However, this criterion can hardly help when there is more than one parameter to be identified [12].

Since large measurements are more reliable than small measurements, it is suggested that the displacements be measured with large magnitude [29].

The variances of the results from ordinary LS estimation are $\left(\mathbf{J}^T\mathbf{J}\right)^{-1}\sigma^2$, if the measuring errors of all displacements are uncorrelated and have the same variance σ^2 ([16], [6], [8]). Based on this theory, a natural requirement is to search for the measurements that give the minimum values of the items in $\left(\mathbf{J}^T\mathbf{J}\right)^{-1}$. Many criteria have been proposed [12] for this purpose. The most popular one is called 'D optimality' ([14], [16], [28], [25] and [8]), in which the optimal displacements are determined by searching for the maximum determinant of matrix $\mathbf{J}^T\mathbf{J}$.

There is no doubt that the number of displacements should be greater than or equal to the number of parameters. Otherwise, the unique results cannot be guaranteed. However, more measurements are not necessarily better. It seems that the quality of the identified parameters is determined by the quantity as well as the quality of measurements ([5] and [27])

So far, there is no recognized criterion for designing the optimal measurement layout although all the above views make a certain amount of sense. This section thoroughly discusses this problem by analyzing the well-posedness of the parameter identification processes using the Gauss-Newton method and the complex method.

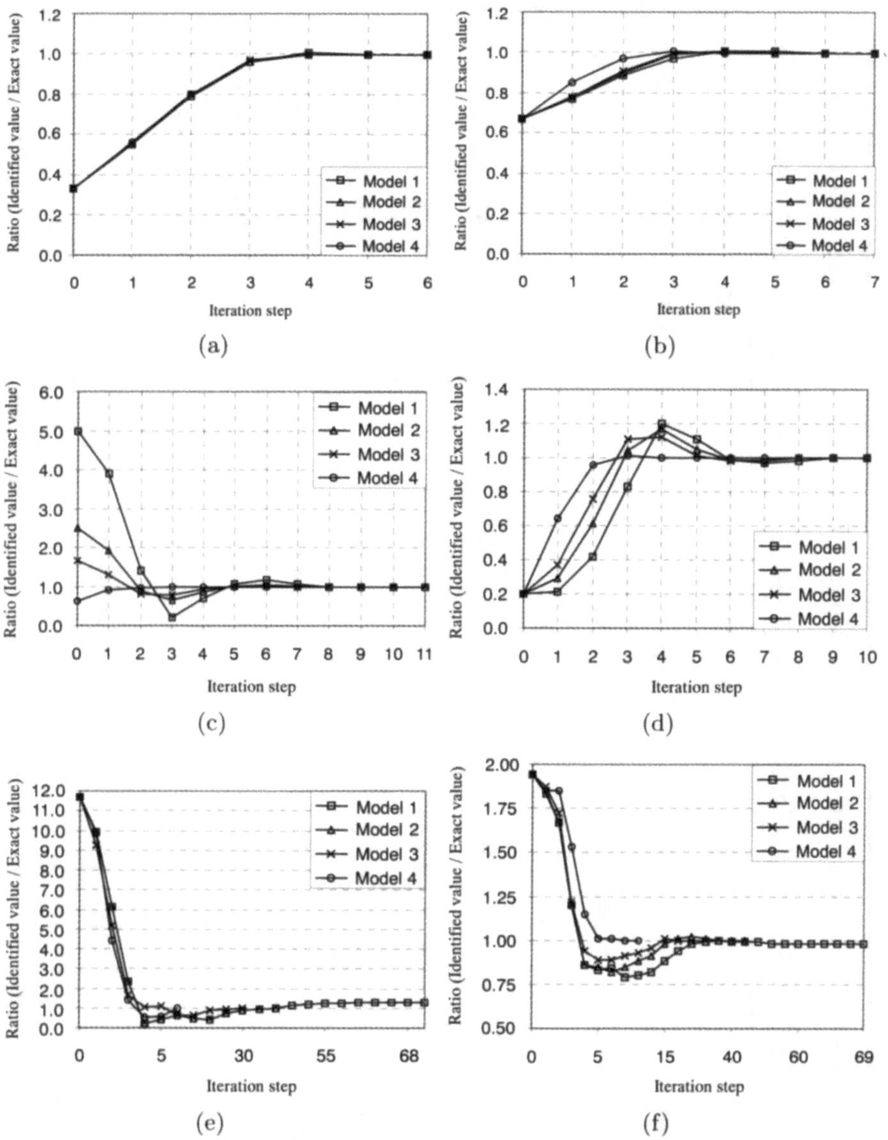

Fig. 3. Identification process for some typical parameters
(a) Young's Modulus (set 1) (b) Poisson's Ratio (set 1)
(c) h (set 1) (d) m (set 1)
(e) Ω_{11} (set 2) (f) Ω_{23} (set 2)

Table 4. Identified parameters' percentage errors

	Model 1 Error (%)				Model 2 Error (%)				Model 3 Error (%)				Model 4 Error (%)			
	Set 1	Set 2	Set 3	Set 4	Set 1	Set 2	Set 3	Set 4	Set 1	Set 2	Set 3	Set 4	Set 1	Set 2	Set 3	Set 4
E	0.00	-0.05	/	/	-0.01	-0.02	/	/	-0.01	-0.03	/	/	0.00	0.00	0.03	-0.01
ν	0.00	-0.06	/	/	0.00	-0.01	/	/	-0.02	-0.02	/	/	-0.01	0.02	-0.03	0.00
h	0.01			/	0.04			/	0.32			/	0.02			0.04
m	-0.12		/		-0.11		/		-0.29		/		-0.02		0.25	
Ω_{11}		29.98	/	/		-0.99	/	/		-1.94	/	/		0.45	-2.40	0.47
Ω_{12}		13.65	/	/		-0.27	/	/		-0.47	/	/		-0.05	-1.01	0.13
Ω_{13}		11.66	/	/		-0.71	/	/		-1.44	/	/		0.56	-1.48	0.33
Ω_{22}		-0.63	/	/		0.46	/	/		1.03	/	/		-0.55	0.39	-0.22
Ω_{23}		-2.37	/	/		0.02	/	/		0.04	/	/		0.06	-0.09	-0.01
Ω_{33}		-4.08	/	/		-0.43	/	/		-0.94	/	/		0.67	-0.56	0.20
Average error of observed displacements	0.00	0.02	/	/	0.00	0.01	/	/	0.00	0.02	/	/	0.00	0.02	0.01	0.01
Iteration steps	15	68	Fail	Fail	12	44	Fail	Fail	11	32	Fail	Fail	7	9	39	18

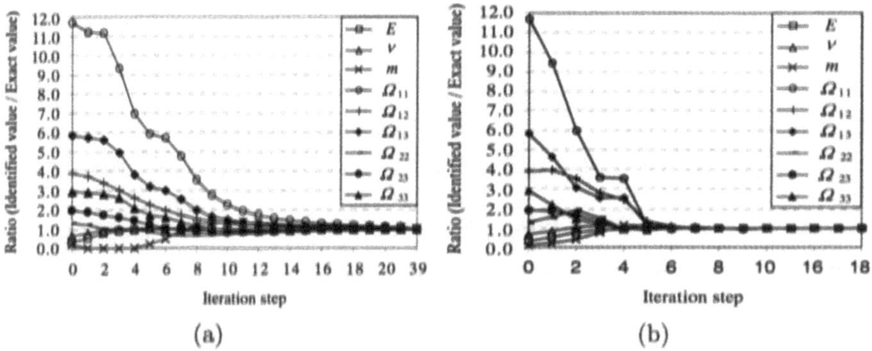

Fig. 4. Two identification processes for each parameter of model 4
(a) (set 3) (b) (set 4)

4.1 Parameter identification process using the Gauss-Newton method

Analysis of well-posedness. To discuss the well-posedness of the algorithm, Hadamard's original [7] should be referred to:

(a) For all admissible data, a solution exists.
(b) For all admissible data, the solution is unique.
(c) The solution continuously depends on the data.

Existence. Property (a) is violated when the numerical model fails to describe the mechanism of the physical problem or the measurements contain seriously polluted data. If a proper numerical model cannot be found for a physical problem, the mechanism of this problem is not clear and parameter identification is far from discussion. Practically, bad data must be filtered out before the parameter identification process. The problem of nonexistence is thus physically out of the question.

Mathematically, because $\mathbf{D}_p$ is a bounded, closed and convex set, if $\mathbf{G}(\mathbf{p})$ is a continuous function, then according to *Brouwer's fixed-point theorem* [9] at least one solution exists.

Uniqueness. Physically, satisfying property (b) requires that the measured data contain enough information to generate a unique solution.

Mathematically, supposing $\mathbf{p}_1$ and $\mathbf{p}_2$ are two arbitrary parameter points, $\mathbf{p}_1,\ \mathbf{p}_2 \in \mathbf{Q}_p$, then

$$\mathbf{G}(\mathbf{p}_1) - \mathbf{G}(\mathbf{p}_2) = \frac{\partial \mathbf{G}(\zeta)}{\partial \mathbf{p}} (\mathbf{p}_1 - \mathbf{p}_2), \tag{34}$$

where $\zeta = \mathbf{p}_1 + \eta(\mathbf{p}_1 - \mathbf{p}_2)$, $0 < \eta < 1$ and

$$\begin{aligned}\bar{\boldsymbol{\Omega}}(\mathbf{p}) &\equiv \frac{\partial \mathbf{G}(\mathbf{p})}{\partial \mathbf{p}} \\ &= \left(\mathbf{J}^T\mathbf{C}_u^{-1}\mathbf{J}+\beta\,\mathbf{C}_p^{-1}\right)^{-1}\frac{\partial\left(\mathbf{J}^T\mathbf{C}_u^{-1}\mathbf{J}\right)}{\partial \mathbf{p}}\left(\mathbf{J}^T\mathbf{C}_u^{-1}\mathbf{J}+\beta\,\mathbf{C}_p^{-1}\right)^{-1} \\ &\quad \left[\mathbf{J}^T\mathbf{C}_u^{-1}\mathbf{R}+\beta\,\mathbf{C}_p^{-1}\left(\mathbf{p}-\mathbf{p}^0\right)\right]- \\ &\quad \left(\mathbf{J}^T\mathbf{C}_u^{-1}\mathbf{J}+\beta\,\mathbf{C}_p^{-1}\right)^{-1}\frac{\partial \mathbf{J}^T}{\partial \mathbf{p}}\mathbf{C}_u^{-1}\mathbf{R}. \end{aligned} \tag{35}$$

Then

$$\|\mathbf{G}(\mathbf{p}_1)-\mathbf{G}(\mathbf{p}_2)\|_\infty \le \|\bar{\boldsymbol{\Omega}}(\zeta)\|_\infty \|\mathbf{p}_1-\mathbf{p}_2\|_\infty \le L_\Omega \|\mathbf{p}_1-\mathbf{p}_2\|_\infty, \tag{36}$$

where $\|\cdot\|_\infty$ is the norm of infinity and

$$L_\Omega \equiv Max\, \|\bar{\boldsymbol{\Omega}}(\mathbf{p})\|_\infty . \tag{37}$$

Because $\mathbf{G}(\mathbf{p}) \in \mathbf{D}_p$, according to the *contraction mapping theorem* [4] if $L_\Omega < 1$ the convergence and uniqueness of this iteration process can be guaranteed.

Stability. The discussion of property (c) is directly related to the evaluation of the continuity of function $\boldsymbol{G}$. Because $\boldsymbol{G}$ contains the gradient information, its smoothness is thus reduced. This is probably the main reason why parameter identification problems are usually unstable. Practically, some kinds of regulation are introduced to compensate the loss of continuity [3].

The initial parameters $\mathbf{p}^0$ can be presented as

$$\mathbf{p}^0 = \mathbf{p}^* + \mathbf{I}_p\delta\mathbf{p}, \tag{38}$$

where $\mathbf{p}^*$ is the actual parameters; $\mathbf{I}_p$ is a diagonal matrix with +1 or −1 at the diagonal positions and $\delta\mathbf{p}$is positive biases.

The measurements with noises can be presented as

$$\mathbf{u}_s^* = \mathbf{S}\left(\mathbf{I}+\mathbf{I}_\varepsilon\varepsilon\right)\mathbf{u}^*, \tag{39}$$

where $\mathbf{u}^*$ are displacements corresponding to $\mathbf{p}^*$; $\mathbf{I}_\varepsilon$ is a diagonal matrix with +1 or −1 on the diagonal positions and ε is the positive relative error of the measurements.

To trace the propagation of the measuring error during the identification process, define

$$\mathbf{h}^k = \mathbf{p}^k - \mathbf{p}^*. \tag{40}$$

Then

$$\begin{aligned}\mathbf{h}^k - \mathbf{h}^{k-1} &= \mathbf{p}^k - \mathbf{p}^{k-1} \\ &= \bar{\boldsymbol{\Psi}}\left(\mathbf{p}^{k-1}, \varepsilon\right) \\ &= \bar{\boldsymbol{\Psi}}\left(\mathbf{p}^*, 0\right) + \frac{\partial \bar{\boldsymbol{\Psi}}\left(\mathbf{p}^{\xi_{k-1}}, \varepsilon^{\xi_{k-1}}\right)}{\partial \mathbf{p}} \mathbf{h}^{k-1} + \\ &\quad \frac{\partial \bar{\boldsymbol{\Psi}}\left(\mathbf{p}^{\xi_{k-1}}, \varepsilon^{\xi_{k-1}}\right)}{\partial \varepsilon} \varepsilon, \end{aligned} \tag{41}$$

where $\mathbf{p}^{\xi_{k-1}} = \mathbf{p}^* + \gamma\left(\mathbf{p}^{k-1} - \mathbf{p}^*\right)$, $0 < \gamma < 1$, $\varepsilon^{\xi_{k-1}} \in (0, \varepsilon)$, and

$$\bar{\boldsymbol{\Psi}}(\mathbf{p}, \varepsilon) \equiv -\left(\mathbf{J}^T \mathbf{C}_u^{-1} \mathbf{J} + \beta \mathbf{C}_p^{-1}\right)^{-1} \left[\mathbf{J}^T \mathbf{C}_u^{-1} \mathbf{R} + \beta \mathbf{C}_p^{-1}\left(\mathbf{p} - \mathbf{p}^0\right)\right] \tag{42}$$

$$\bar{\boldsymbol{\Psi}}\left(\mathbf{p}^*, 0\right) = \left[\left(\mathbf{J}^*\right)^T \mathbf{C}_u^{-1} \mathbf{J}^* + \beta \mathbf{C}_p^{-1}\right]^{-1} \beta \mathbf{C}_p^{-1} \mathbf{I}_p \delta \mathbf{p}, \tag{43}$$

$$\frac{\partial \bar{\boldsymbol{\Psi}}(\mathbf{p}, \varepsilon)}{\partial \mathbf{p}} = \bar{\boldsymbol{\Omega}}(\mathbf{p}, \varepsilon) - \mathbf{I}, \tag{44}$$

$$\bar{\mathbf{A}}(\mathbf{p}) \equiv \frac{\partial \bar{\boldsymbol{\Psi}}(\mathbf{p}, \varepsilon)}{\partial \varepsilon} = -\left(\mathbf{J}^T \mathbf{C}_u^{-1} \mathbf{J} + \beta \mathbf{C}_p^{-1}\right)^{-1} \mathbf{J}^T \mathbf{C}_u^{-1} \mathbf{S} \mathbf{I}_\varepsilon \mathbf{u}^*. \tag{45}$$

Therefore,

$$\begin{aligned}\mathbf{h}^k - \mathbf{h}^{k-1} &= \left[\left(\mathbf{J}^*\right)^T \mathbf{C}_u^{-1} \mathbf{J}^* + \beta \mathbf{C}_p^{-1}\right]^{-1} \beta \mathbf{C}_p^{-1} \mathbf{I}_p \delta \mathbf{p} \\ &\quad + \left[\bar{\boldsymbol{\Omega}}\left(\mathbf{p}^{\xi_{k-1}}, \varepsilon^{\xi_{k-1}}\right) - \mathbf{I}\right] \mathbf{h}^{k-1} + \bar{\mathbf{A}}\left(\mathbf{p}^{\xi_{k-1}}\right) \varepsilon \\ \mathbf{h}^k &= \left[\left(\mathbf{J}^*\right)^T \mathbf{C}_u^{-1} \mathbf{J}^* + \beta \mathbf{C}_p^{-1}\right]^{-1} \beta \mathbf{C}_p^{-1} \mathbf{I}_p \delta \mathbf{p} + \\ &\quad \bar{\boldsymbol{\Omega}}\left(\mathbf{p}^{\xi_{k-1}}, \varepsilon^{\xi_{k-1}}\right) \mathbf{h}^{k-1} + \bar{\mathbf{A}}\left(\mathbf{p}^{\xi_{k-1}}\right) \varepsilon. \end{aligned} \tag{46}$$

Define

$$L_p \equiv \left\| \left[\left(\mathbf{J}^*\right)^T \mathbf{C}_u^{-1} \mathbf{J}^* + \beta \mathbf{C}_p^{-1}\right]^{-1} \beta \mathbf{C}_p^{-1} \mathbf{I}_p \delta \mathbf{p} \right\|_\infty, \tag{47}$$

$$L_A \equiv Max \left\| \bar{\mathbf{A}}(\mathbf{p}) \right\|_\infty. \tag{48}$$

It reads:

$$\begin{aligned}\left\|\mathbf{h}^k\right\|_\infty &\le L_p + L_\Omega \left\|\mathbf{h}^{k-1}\right\|_\infty + L_A \varepsilon \\ &\le \left(1 + L_\Omega\right) L_p + L_\Omega^2 \left\|\mathbf{h}^{k-2}\right\|_\infty + \left(1 + L_\Omega\right) L_A \varepsilon \\ &\quad \cdots \\ &\le \left(1 + L_\Omega + \cdots L_\Omega^{k-1}\right) L_p + L_\Omega^k \left\|\delta \mathbf{p}\right\|_\infty + \\ &\quad \left(1 + L_\Omega + \cdots L_\Omega^{k-1}\right) L_A \varepsilon. \end{aligned} \tag{49}$$

If $L_\Omega < 1$

$$\lim_{k\to\infty} \left\|\mathbf{h}^k\right\|_\infty \le \frac{1}{1-L_\Omega} L_p + \frac{1}{1-L_\Omega} L_A \varepsilon \approx \left(L_p + L_A\varepsilon\right)\left(1+L_\Omega\right), \quad (50)$$

The iteration process converges to the solution with bias $(L_p + L_A\varepsilon)\,(1+L_\Omega)$. L_Ω represents the speed of the convergence.

Equation (35) gives

$$\bar{\mathbf{\Omega}}(\mathbf{p},\varepsilon) \approx \bar{\mathbf{\Omega}}(\mathbf{p}^*,0) + \frac{\partial\bar{\mathbf{\Omega}}(\mathbf{p}^*,0)}{\partial\mathbf{p}}(\mathbf{p}-\mathbf{p}^*) + \frac{\partial\bar{\mathbf{\Omega}}(\mathbf{p}^*,0)}{\partial\varepsilon}\varepsilon, \quad (51)$$

where higher-order terms are omitted.

$$\begin{aligned}\bar{\mathbf{B}}(\mathbf{p}^*) &\equiv \bar{\mathbf{\Omega}}(\mathbf{p}^*,0)\\ &= -\left[(\mathbf{J}^*)^T\mathbf{C}_u^{-1}\mathbf{J}^* + \beta\,\mathbf{C}_p^{-1}\right]^{-1}\frac{\partial\left[(\mathbf{J}^*)^T\mathbf{C}_u^{-1}\mathbf{J}^*\right]}{\partial\mathbf{p}}\\ &\quad\left[(\mathbf{J}^*)^T\mathbf{C}_u^{-1}\mathbf{J}^* + \beta\,\mathbf{C}_p^{-1}\right]^{-1}\beta\,\mathbf{C}_p^{-1}\mathbf{I}_p\delta\mathbf{p},\end{aligned} \quad (52)$$

$$\begin{aligned}\bar{\mathbf{C}}(\mathbf{p}^*) &\equiv \frac{\partial\bar{\mathbf{\Omega}}(\mathbf{p}^*,0)}{\partial\mathbf{p}}\\ &= 2\left[(\mathbf{J}^*)^T\mathbf{C}_u^{-1}\mathbf{J}^* + \beta\,\mathbf{C}_p^{-1}\right]^{-1}\frac{\partial\left[(\mathbf{J}^*)^T\mathbf{C}_u^{-1}\mathbf{J}^*\right]}{\partial\mathbf{p}}\\ &\quad\left[(\mathbf{J}^*)^T\mathbf{C}_u^{-1}\mathbf{J}^* + \beta\,\mathbf{C}_p^{-1}\right]^{-1}\frac{\partial\left[(\mathbf{J}^*)^T\mathbf{C}_u^{-1}\mathbf{J}^*\right]}{\partial\mathbf{p}}\\ &\quad\left[(\mathbf{J}^*)^T\mathbf{C}_u^{-1}\mathbf{J}^* + \beta\,\mathbf{C}_p^{-1}\right]^{-1}\beta\,\mathbf{C}_p^{-1}\mathbf{I}_p\delta\mathbf{p}\\ &\quad-\left[(\mathbf{J}^*)^T\mathbf{C}_u^{-1}\mathbf{J}^* + \beta\,\mathbf{C}_p^{-1}\right]^{-1}\frac{\partial^2\left[(\mathbf{J}^*)^T\mathbf{C}_u^{-1}\mathbf{J}^*\right]}{\partial\mathbf{p}^2}\\ &\quad\left[(\mathbf{J}^*)^T\mathbf{C}_u^{-1}\mathbf{J}^* + \beta\,\mathbf{C}_p^{-1}\right]^{-1}\beta\,\mathbf{C}_p^{-1}\mathbf{I}_p\delta\mathbf{p}\\ &\quad+\left[(\mathbf{J}^*)^T\mathbf{C}_u^{-1}\mathbf{J}^* + \beta\,\mathbf{C}_p^{-1}\right]^{-1}(\mathbf{J}^*)^T\mathbf{C}_u^{-1}\frac{\partial\mathbf{J}^*}{\partial\mathbf{p}},\end{aligned} \quad (53)$$

$$\begin{aligned}\bar{\mathbf{D}}(\mathbf{p}^*) &\equiv \frac{\partial\bar{\mathbf{\Omega}}(\mathbf{p}^*,0)}{\partial\varepsilon}\\ &= -\left[(\mathbf{J}^*)^T\mathbf{C}_u^{-1}\mathbf{J}^* + \beta\,\mathbf{C}_p^{-1}\right]^{-1}\end{aligned}$$

$$\left\{\frac{\partial (\mathbf{J}^*)^T}{\partial \mathbf{p}} - \frac{\partial \left[(\mathbf{J}^*)^T \mathbf{C}_u^{-1}\mathbf{J}^*\right]}{\partial \mathbf{p}}\right.$$
$$\left.\left[(\mathbf{J}^*)^T \mathbf{C}_u^{-1}\mathbf{J}^* + \beta \mathbf{C}_p^{-1}\right]^{-1} (\mathbf{J}^*)^T\right\}$$
$$\mathbf{C}_u^{-1}\mathbf{S}\mathbf{I}_\varepsilon \mathbf{u}^* \tag{54}$$

$$\frac{\partial \mathbf{J}}{\partial \mathbf{p}} = \mathbf{S}\mathbf{K}^{-1}\left(\frac{\partial^2 \mathbf{K}}{\partial \mathbf{p}^2} - 2\frac{\partial \mathbf{K}}{\partial \mathbf{p}}\mathbf{K}^{-1}\frac{\partial \mathbf{K}}{\partial \mathbf{p}}\right)\mathbf{u}. \tag{55}$$

In each iteration step, if

$$\mathbf{p} \in \mathbf{D}'_p \equiv \{\, \mathbf{p}' \; |\mathbf{p}' \in \mathbf{D}_p \cap \; \|\mathbf{p}' - \mathbf{p}^*\|_\infty \le \|\delta \mathbf{p}\|_\infty \,\} \tag{56}$$

then approximately

$$\begin{aligned} \|\bar{\boldsymbol{\Omega}}(\mathbf{p},\varepsilon)\|_\infty &\le \|\bar{\mathbf{B}}(\mathbf{p}^*)\|_\infty + \|\bar{\mathbf{C}}(\mathbf{p}^*)\|_\infty \|\mathbf{p} - \mathbf{p}^*\|_\infty + \|\bar{\mathbf{D}}(\mathbf{p}^*)\|_\infty \varepsilon \\ &\le \|\bar{\mathbf{B}}(\mathbf{p}^*)\|_\infty + \|\bar{\mathbf{C}}(\mathbf{p}^*)\|_\infty \|\delta \mathbf{p}\|_\infty + \|\bar{\mathbf{D}}(\mathbf{p}^*)\|_\infty \varepsilon \end{aligned} \tag{57}$$

$$L_\Omega \approx \|\bar{\mathbf{B}}(\mathbf{p}^*)\|_\infty + \|\bar{\mathbf{C}}(\mathbf{p}^*)\|_\infty \|\delta \mathbf{p}\|_\infty + \|\bar{\mathbf{D}}(\mathbf{p}^*)\|_\infty \varepsilon. \tag{58}$$

Equation (45) gives

$$\bar{\mathbf{A}}(\mathbf{p}) \approx \bar{\mathbf{A}}(\mathbf{p}^*) + \frac{\partial \bar{\mathbf{A}}(\mathbf{p}^*)}{\partial \mathbf{p}}(\mathbf{p} - \mathbf{p}^*), \tag{59}$$

where

$$\frac{\partial \bar{\mathbf{A}}(\mathbf{p}^*)}{\partial \mathbf{p}} = \bar{\mathbf{D}}(\mathbf{p}^*). \tag{60}$$

If condition (56) is satisfied, then approximately

$$\begin{aligned} \|\bar{\mathbf{A}}(\mathbf{p})\|_\infty &\le \|\bar{\mathbf{A}}(\mathbf{p}^*)\|_\infty + \|\bar{\mathbf{D}}(\mathbf{p}^*)\|_\infty \|\mathbf{p} - \mathbf{p}^*\|_\infty \\ &\le \|\bar{\mathbf{A}}(\mathbf{p}^*)\|_\infty + \|\bar{\mathbf{D}}(\mathbf{p}^*)\|_\infty \|\delta \mathbf{p}\|_\infty, \end{aligned} \tag{61}$$

$$L_A \approx \|\bar{\mathbf{A}}(\mathbf{p}^*)\|_\infty + \|\bar{\mathbf{D}}(\mathbf{p}^*)\|_\infty \|\delta \mathbf{p}\|_\infty. \tag{62}$$

Design of measurement layout. From the above analyses it is concluded that the unique and stable solution can be guaranteed by the measured displacements, which cause $L_\Omega < 1$ and minimize $(L_p + L_A\varepsilon)\,(1 + L_\Omega)$. If these measured displacements can characterize the whole displacement field, other displacements calculated with the identified parameters can approximately match the actual ones. This implies that the optimal measurements should satisfy two conditions:

a) Necessary condition: the optimal measurements should represent the whole displacement field.
b) Sufficient condition: the optimal measurements should cause $L_\Omega < 1$ and minimize $(L_p + L_A\varepsilon)\,(1 + L_\Omega)$.

As stated above, extended Bayesian estimation is actually a kind of Tikhonov regulation. The results are dependent to some extent on the prior estimation of the parameters. Therefore, this approach should be avoided if the validity of the prior information cannot be ensured and stable solutions can be obtained by ordinary LS formulation or weighted LS formulation.

The design of the measurement layout, based on ordinary LS formulation or weighted LS formulation and the Gauss-Newton method, can be stated as:

For a given system, if ε and $\mathbf{C}_u^{-1}$ are known, for a guess of $\mathbf{p}^0$ and a given $\delta\mathbf{p}$

(a) if a matrix $\boldsymbol{S}$ can be found that causes $L_\Omega < 1$, then the identification process will converge.
(b) of all feasible $\boldsymbol{S}$, it is better to find the one that represents the whole displacement field and minimizes $L_A\,(1 + L_\Omega)\,\varepsilon$ (since $\beta = 0$, $L_p = 0$).

If a proper $\boldsymbol{S}$ is not found to obtain stable solutions, the extended Bayesian estimation method should be used. Since we have high confidence for $\mathbf{p}^0$, we assume $\delta\mathbf{p} \approx 0$. Then the design of the measurement layout, based on the extended Bayesian estimation and the Gauss-Newton method, can be stated as:

For a given system, if $\varepsilon, \beta, \mathbf{C}_u^{-1}$ and $\mathbf{C}_p^{-1}$ are known, for a confident prior estimation $\mathbf{p}^0$

(a) if a matrix $\boldsymbol{S}$ can be found that causes $L_\Omega < 1$, then the identification process will converge.
(b) of all feasible $\boldsymbol{S}$, it is better to find the one that represents the whole displacement field and minimizes $L_A\,(1 + L_\Omega)\,\varepsilon$ (since $\delta\mathbf{p} \approx 0$, $L_p \approx 0$).

Discussion:

(a) The measurement set that can represent the whole displacement field should contain the displacements sampled from large strain regions.

(b) Because $\mathbf{p}^*$ is unknown, it is impossible to calculate L_A and L_Ω. However, from (38) $\mathbf{p}^*$ can be guessed as

$$\mathbf{p}_g^* = \mathbf{p}^0 - \mathbf{I}_p \delta \mathbf{p}. \tag{63}$$

Then we can use L'_A and L'_Ω instead:

$$L'_A = \left\| \bar{\mathbf{A}} \left(\mathbf{p}_g^*\right) \right\|_\infty + \left\|\bar{\mathbf{D}} \left(\mathbf{p}_g^*\right) \right\|_\infty \left\|\delta \mathbf{p}\right\|_\infty, \tag{64}$$

$$L'_\Omega = \left\|\bar{\mathbf{C}} \left(\mathbf{p}_g^*\right) \right\|_\infty \left\|\delta \mathbf{p}\right\|_\infty + \left\|\bar{\mathbf{D}} \left(\mathbf{p}_g^*\right) \right\|_\infty \varepsilon. \tag{65}$$

Practically, if there are N_p parameters, we should search for all feasible $\boldsymbol{S}$ in $2^{N_p} + 1$ different cases (a special case is $\delta \mathbf{p} = 0$.). Of these feasible $\boldsymbol{S}$, we have more confidence for the ones that are the common feasible sets for all cases. Furthermore, of these common $\boldsymbol{S}$, the one that represents the whole displacement field and minimizes$L_A \left(1 + L_\Omega\right) \varepsilon$ (corresponding to $\mathbf{p}^0$) may produce better results.

(c) Supposing i measurements are selected from the candidate set U_c, which contains j measurements, there will be $C_j^i = \frac{j\,!}{i!\,(j-i)\,!}$ different measurement sets. Usually, it is impossible to check all measurement sets, because the combination C_j^i is a huge number. Although the algorithms proposed for 'D Optimality' ([16], [28] and [12]) can give some clues for obtaining a reasonable measurement set, they cannot guarantee that the global optimal solution will be found. Moreover, because the numerical model can only approximately describe the physical problem, (39) is merely an approximation. Therefore, we can expect to find only a reasonable measurement set. For this purpose, some simple algorithms may work.

(d) From the physical point of view, higher sensitivity means the error in the measurement has less influence on the identified parameters. From the mathematical point of view, smaller values of L_Ω and L_A can be expected if the elements in $\left(\mathbf{J}^T \mathbf{C}_u^{-1} \mathbf{J} + \beta\, \mathbf{C}_p^{-1}\right)^{-1}$ are smaller (see (51) to (62)). This coincides with the requirement to minimize the variances of parameters. Therefore, it is better to arrange $\boldsymbol{S}$ in such a way that the diagonal items in $\mathbf{J}^T \mathbf{C}_u^{-1} \mathbf{J} + \beta\, \mathbf{C}_p^{-1}$ are high sensitivities.

(e) Equation (51) to (62) reveal that smaller displacements can probably lead to smallerL_Ω and L_A. However, we have more confidence for the measured displacements in large magnitude. Therefore, a practical algorithm should specify the minimum magnitude of the displacements ($u_{\min}$) and consider small displacements together with large displacements.

(f) Equation (51) to (62) reveal that more measurements may lead to larger L_Ω and L_A although they can give a better representation for the whole displacement field. Therefore, we should choose the proper number of measurements within a limited range.

(g) The pair of symmetric and anti-symmetric measurements contains redundant data.

(h) The initial value of β can be set in such a way that two parts of the objective function in (2) are equally weighted. This is also the original reason to introduce βinto the Bayesian estimation [30].
(i) If condition (56) cannot be satisfied, the estimations of L_Ω ((58)) and L_A ((62)) may be smaller than the actual ones. However, if $\boldsymbol{p}$ is not far from $\mathbf{D}'_p$, (58) and (62) are still satisfactory estimations of L_Ω and L_A.

Algorithms. From the above discussions, algorithms for designing the optimal measurement layout are described as follows:

Ordinary LS or weighted LS approach

(a) Specify U_c, $\mathbf{C}_u^{-1}$, $\mathbf{p}^0$, $\delta\mathbf{p}$, ε, $u_{\min}$, the minimum and maximum number of measurements ($N_{\min}$ and $N_{\max}$).
(b) For each $\mathbf{p}_g^*$, from U_c, exclude the displacements whose absolute values are less than $u_{\min}$ and one of the displacements which belong to the symmetric or anti-symmetric pairs. Denote the number of the rest of the displacements in U_c as j. Sort these displacements in ascending order with respect to their absolute values. Denote this sorted vector as $\mathbf{U}_c^1$. Sort the remaining displacements in U_c in such an order that the diagonal items in $\mathbf{J}^T\mathbf{C}_u^{-1}\mathbf{J}$ are high sensitivities. Denote this sorted vector as $\mathbf{U}_c^2$. Denote the number of measurements as i. Set $i = N_{\min}$.
(c) If $C_j^i \geq 1 \times 10^6$ (this value is dependent on the speed of the computer), go to step (d). Otherwise, check all possible measurement sets in $\mathbf{U}_c^1$ and record all feasible $\boldsymbol{S}$, which causes $L_\Omega < 1$. Go to step (e).
(d) Construct the vector $\mathbf{U}_c^{JM} = \mathbf{U}_c^J \cup \mathbf{U}_c^M$, where $\mathbf{U}_c^J$ contains the first i displacements in $\mathbf{U}_c^2$ and $\mathbf{U}_c^M$ contains the first and the last i displacements in $\mathbf{U}_c^1$. Denote the number of displacements in $\mathbf{U}_c^{JM}$ as N_{JM}.
   I. Remove the displacement from the tail of $\mathbf{U}_c^{JM}$ ($N_{JM} = N_{JM} - 1$) until $C_{N_{JM}}^i < 1 \times 10^6$.
   II. Check all possible measurement sets in $\mathbf{U}_c^{JM}$ and record all feasible $\boldsymbol{S}$, which causes $L_\Omega < 1$.
(e) If $i < N_{\max}$, then $i = i + 1$, go to (c). Otherwise, go to (f).
(f) If all$\mathbf{p}_g^*$ guesses are checked, go to (g). Otherwise, calculate the next guess and go to (b).
(g) Find out all common $\boldsymbol{S}$ from the recorded feasible ones for each $\mathbf{p}_g^*$. The reasonable $\boldsymbol{S}$ is the one that can represent the whole displacement field and has the minimum $L_A\,(1 + L_\Omega)\,\varepsilon$ corresponding to the case of $\delta\mathbf{p} = 0$. If there is no common $\boldsymbol{S}$, the extended Bayesian estimation method should be used.

Extended Bayesian estimation approach

(a) Specify U_c, β, $\mathbf{C}_u^{-1}$, $\mathbf{C}_p^{-1}$, $\mathbf{p}^0$, ε, $u_{\min}$, the minimum and maximum number of measurements ($N_{\min}$ and $N_{\max}$).

(b) From U_c, exclude the displacements whose absolute values are less than $u_{\min}$ and one of the displacements which belong to the symmetric or anti-symmetric pairs. Denote the number of the remaining displacements in U_c as j. Sort these displacements in ascending order with respect to their absolute values. Denote this sorted vector as $\mathbf{U}_c^1$. Sort the remaining displacements in U_c in such an order that the diagonal items in $\mathbf{J}^T\mathbf{C}_u^{-1}\mathbf{J}$ are high sensitivities. Denote this sorted vector as $\mathbf{U}_c^2$. Denote the number of measurements as i. Set $i = N_{\min}$.
(c) If $C_j^i \geq 1 \times 10^6$ (this value is dependent on the speed of the computer), go to step (d). Otherwise, check all possible measurement sets in $\mathbf{U}_c^1$ and record all feasible $\boldsymbol{S}$, which causes $L_\Omega < 1$ and minimizes $L_A\,(1 + L_\Omega)\,\varepsilon$. Go to step (e).
(d) Construct the vector $\mathbf{U}_c^{JM} = \mathbf{U}_c^J \cup \mathbf{U}_c^M$, where $\mathbf{U}_c^J$ contains the first i displacements in $\mathbf{U}_c^2$ and $\mathbf{U}_c^M$ contains the first and the last i displacements in $\mathbf{U}_c^1$. Denote the number of displacements in $\mathbf{U}_c^{JM}$ as N_{JM}.

    I. Remove the displacements from the tail of $\mathbf{U}_c^{JM}$ $(N_{JM} = N_{JM} - 1)$ until $C_{N_{JM}}^i < 1 \times 10^6$.
    II. Check all possible measurement sets in $\mathbf{U}_c^{JM}$ and record all feasible $\boldsymbol{S}$, which causes $L_\Omega < 1$ and minimizes $L_A\,(1 + L_\Omega)\,\varepsilon$.

(e) If $i < N_{\max}$, then $i = i + 1$, go to (c). Otherwise, go to (f).
(f) The reasonable $\boldsymbol{S}$ is the one that can represent the whole displacement field and has the minimum $L_A\,(1 + L_\Omega)\,\varepsilon$.

All algorithms above were coded into an FE program with the name BAME.

4.2 Parameter identification process based on the complex method

Analysis of well-posedness

Existence. It is noticed that the objective functions from extended Bayesian estimation and other degenerated estimations are always nonnegative and the complex method generates a series of bounded descending objective values. Therefore, a solution can be obtained in a closed set $\mathbf{Q}_p$, if the search scheme of the complex method is flexible enough.

Uniqueness. The uniqueness of the solution from the complex method cannot be guaranteed, because it cannot observe that $\nabla_p^2 f$ is always positive defined on $\mathbf{Q}_p$. Practically, the best solution is selected from the results of many trials.

Stability. Supposing the iteration converges at step k, it should satisfy:

$$\frac{\partial f(\mathbf{p}^k,\,\varepsilon)}{\partial \mathbf{p}} = 0$$

$$\frac{\partial f(\mathbf{p}^*,\,0)}{\partial \mathbf{p}} + \frac{\partial^2 f(\mathbf{p}^*,\,0)}{\partial \mathbf{p}^2}\left(\mathbf{p}^k - \mathbf{p}^*\right) + \frac{\partial^2 f(\mathbf{p}^*,\,0)}{\partial \mathbf{p}\partial \varepsilon}\varepsilon \approx 0. \tag{66}$$

Substituting (3), (6), (7), and (38) to (40) into Equation (66), yields:

$$\beta \mathbf{C}_p^{-1}\left(\mathbf{p}^* - \mathbf{p}^0\right) + \left[\left(\mathbf{J}^*\right)^T \mathbf{C}_u^{-1}\mathbf{J}^* + \beta\,\mathbf{C}_p^{-1}\right]\mathbf{h}^k + \left(\mathbf{J}^*\right)^T \mathbf{C}_u^{-1}\mathbf{S}\mathbf{I}_\varepsilon \mathbf{u}^*\varepsilon \approx 0. \tag{67}$$

According to (45) and (47), we now read:

$$\mathbf{h}^k \approx \bar{\mathbf{A}}\left(\mathbf{p}^*\right)\varepsilon + \left[\left(\mathbf{J}^*\right)^T \mathbf{C}_u^{-1}\mathbf{J}^* + \beta\,\mathbf{C}_p^{-1}\right]^{-1}\beta\,\mathbf{C}_p^{-1}\delta\mathbf{p} \tag{68}$$

$$\left\|\mathbf{h}^k\right\|_\infty \le L_p + L_A\varepsilon. \tag{69}$$

Equation (69) reveals that the measuring errors do not accumulate during the iteration process of the complex method.

Design of measurement layout. Since the uniqueness of the solution provided by the complex method cannot be guaranteed, the maximum bias of the identified parameters is the main guidance to determine the optimal layout of measurements. Moreover, considering that large Conditional Number (CN) of $\left(\mathbf{J}^*\right)^T \mathbf{C}_u^{-1}\mathbf{J}^* + \beta\,\mathbf{C}_p^{-1}$ may cause some numerical difficulties, it is better to specify an upper threshold CN^{up}. If $L_\Omega < 1$ and $L_A\left(1 + L_\Omega\right)\varepsilon$ are replaced with $CN < CN^{up}$ and $L_A\varepsilon$ respectively, the algorithms proposed in section 4.1.2 can be implemented to determine the reasonable measurement layout for the complex method. These algorithms were coded into an FE program with the name COME.

4.3 Examples

Brick model. In this example parameters of a two-layered 10 × 10 × 10m brick, which is fixed at the bottom and subjected to two uniform surface pressures (see Fig. 5), are to be identified. Point 1 is the reference point. The candidate measurement set U_c contains the relative displacements (calculated by FINAL) of all points to point 1. $u_{\min}$is set to 1 mm. $\mathbf{C}_u^{-1}$ is an identity matrix, because the measurements are uncorrelated and have the same significance. As Tab. 5 shows, four cases are tested here. When not otherwise stated, the following results are obtained from ordinary LS estimation.

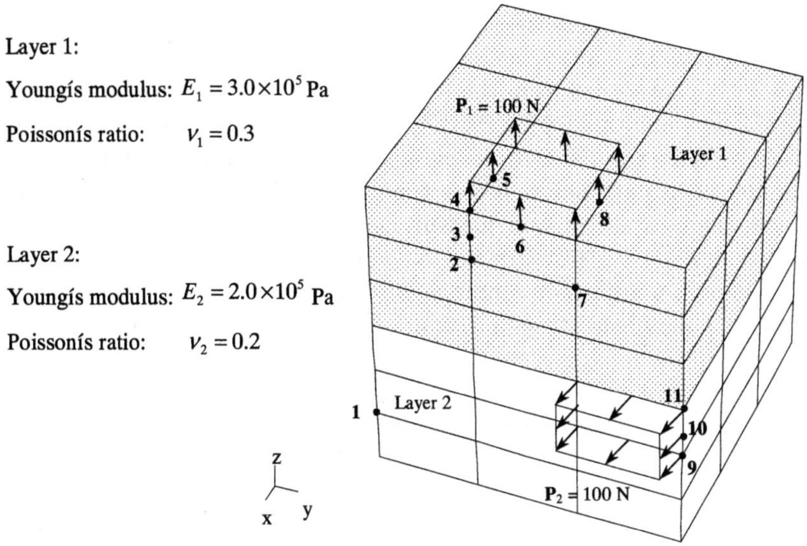

Fig. 5. A brick model

Table 5. Four testing cases for brick model

Case	Parameters p	Initial parameters $\mathbf{p}^0$ (E is in Pa)	Parameter biases $\delta\mathbf{p}$ (E is in Pa.)	ε (%)
1	(E_1, E_2)	$(3.0\times10^5, 2.0\times10^5)$	$(0,0)$	1
2	(E_1, E_2)	$(2.5\times10^5, 2.5\times10^5)$	$(0.8\times10^5, 0.8\times10^5)$	1
3	(E_1, E_2, ν_1, ν_2)	$(3.0\times10^5, 2.0\times10^5, 0.3, 0.2)$	$(0,0,0,0)$	10^{-4}
4	(E_1, E_2, ν_1, ν_2)	$(3.0\times10^5, 2.0\times10^5, 0.3, 0.2)$	$(0,0,0,0)$	0.1

Table 6. Good measurement sets for brick model

Case	Program / Parameter	$N_{\min}$	$N_{\max}$	Measurements (Node/Direction)
1	BAME / (E_1, E_2)	2	6	6/z 9/x
2	BAME / (E_1, E_2)	2	6	2/z 5/z 6/z 8/z 10/x
4	BAME/(Lamé)	4	12	3/z 7/z 8/z 11/x
	BAME/(Lamé) (Extended Bayesian)	4	12	8/z 9/x 10/x 11/x
	COME/(E_1, E_2, ν_1, ν_2)	4	12	5/z 6/z 9/x 11/x

Case 1 is used to check the validity of error estimation. The parameters are identified using the Gauss-Newton method. The best measurements are listed in Tab. 6. It is observed from the results in Tab. 7 that the maximum biases of parameters identified by FIBACK agree very well with those estimated by BAME. With the 1% error in measurements, the maximum relative error of the identified parameters is less than 2%. This is a very stable solution. However, using a set of bad measurements in Tab. 8, the parameters are identified as $E_1 = 1.9169 \times 10^5$ Pa, $E_2 = 4.7349 \times 10^5$ Pa. This confirms that $L_\Omega < 1$ is only a sufficient condition of convergence, and the measurements with $L_\Omega > 1$ may lead to some risk of obtaining unstable solutions.

Table 7. Results of case 1 for brick model

Measurements with errors		Maximum bias by BAME $L_A(1+L_\Omega)\varepsilon$	Identified E_1(Pa) by FIBACK	Identified E_2(Pa) by FIBACK	Maximum bias by FIBACK
6 z	9 x				
-ε	-ε	3060	♣3.0303×10^5	2.0202×10^5	3030
-ε	+ε	6041	♣3.0593×10^5	1.9744×10^5	5930
+ε	-ε	6041	♣2.9430×10^5	2.0261×10^5	5700
+ε	+ε	3060	♣2.9703×10^5	1.9802×10^5	2970

♣ where the maximum bias is located

Table 8. Bad measurement sets by BAME for brick model

	Node / Direction / Error				L_Ω
Case 1	2/z/+ε	7/z/-ε			1.54
Case 3	2/z/+ε	4/z/-ε	7/z/-ε	11/x/-ε	12182.93

Case 2 is a typical situation encountered in practice, where only some rough estimation of parameters is available. Using the good measurements (with -ε measuring relative errors) listed in Tab. 6, the parameters are identified as $E_1 = 3.0302 \times 10^5$ Pa and $E_2 = 2.0202 \times 10^5$ Pa using the Gauss-Newton method. This is a successful identification.

In the above two cases, the good measurements are located at the points that are directly subjected to the loading. These are regions with high sensitivities. This confirms that system sensitivity plays a very important role in the design of measurement layout.

Although the measurement error in case 3 is very small, the Gauss-Newton identification process cannot converge with the measurements listed in Tab. 8. This is a warning that the layout of the measurements should not be arbitrarily determined, even when the displacements are located at highly sensitive regions with very accurate measurements. It is further noticed that from all the sets with four to twelve measurements, $L_{\Omega} < 1$ is observed only in the sets with six, seven and eight measurements. This confirms the conclusion that there is no clear rule for the optimal number of measurements.

Table 9. The results of case 4 for brick model (LS approach)

Measurements with errors				Max. bias by BAME	Identified parameters by FIBACK ($\times 10^5$)				Max. bias by FIBACK
3 z	11 x	7 z	8 z		λ_1	λ_2	μ_1	μ_2	
-ε	-ε	-ε	-ε	173	♣1.7341	0.5557	1.1550	0.8341	330
-ε	-ε	-ε	+ε	7167	♣1.6676	0.5626	1.1556	0.8349	6320
-ε	-ε	+ε	-ε	3335	1.7351	♣0.5264	1.1642	0.8315	2920
-ε	-ε	+ε	+ε	7025	♣1.6681	0.5328	1.1648	0.8324	6270
-ε	+ε	-ε	-ε	810	♣1.7244	♣0.5620	1.1564	0.8316	640
-ε	+ε	-ε	+ε	8364	♣1.6583	0.5689	1.1570	0.8324	7250
-ε	+ε	+ε	-ε	2596	1.7253	♣0.5325	1.1657	0.8290	2310
-ε	+ε	+ε	+ε	8199	♣1.6587	0.5390	.1662	0.8298	7210
+ε	-ε	-ε	-ε	8199	♣1.8052	0.5721	1.1421	0.8367	7440
+ε	-ε	-ε	+ε	2596	1.7363	♣0.5793	1.1426	0.8376	2370
+ε	-ε	+ε	-ε	8364	♣1.8069	0.5418	1.1510	0.8342	7610
+ε	-ε	+ε	+ε	810	1.7375	♣0.5486	1.1515	0.8350	700
+ε	+ε	-ε	-ε	7025	♣1.7952	0.5786	1.1434	0.8342	6440
+ε	+ε	-ε	+ε	3335	1.7267	♣0.5859	1.1439	0.8350	3030
+ε	+ε	+ε	-ε	7168	♣1.7968	0.5481	1.1524	0.8316	6600
+ε	+ε	+ε	+ε	173	♣1.7279	0.5550	1.1529	0.8325	310

♣ where the maximum bias is located

For Case 4, no feasible ***S*** can be found uisng BAME. It seems that the identification process becomes very unstable if Young's Modulus and Poisson's Ratio are to be identified simultaneously. One possible reason is that some numerical difficulties may arise from the great disparity between the scales of these two parameters. From the relations among Young's modulus, Poisson's ratio and Lamé parameters, i.e., $\lambda = \frac{E\nu}{(1+\nu)(1-2\nu)}$ and $\mu = \frac{E}{2(1+\nu)}$, it is observed that λ and μ are approximately in the same scale. Therefore, the identification process would be numerically more stable if we tried to identify Lamé parameters λ and μ instead of E and ν. This idea is confirmed by the converged identification by means of FIBACK with reasonable measurements indicated by BAME (see Tab. 6). The identification results are given in Tab. 9. However, compared with the actual Lamé parameters: $\lambda_1 = 1.7308 \times 10^5$Pa, $\mu_1 = 1.1538 \times 10^5$Pa, $\lambda_2 = 0.5556 \times 10^5$Pa and $\mu_2 = 0.8333 \times 10^5$Pa, the maximum error of identified parameters is about 5.5%, which is much greater than the measuring error 1. This implies that aside from the numerical errors, there are other reasons for this unstable solution. One possible reason is that the displacement is insensitive to Poisson's Ratio. It is noticed that small perturbations ΔE and $\Delta\nu$ will cause the following biases of Lamé parameters:

$$\Delta\lambda \approx \frac{\nu^*}{(1+\nu^*)(1-2\nu^*)}\Delta E + \frac{E^*\left[1+2(\nu^*)^2\right]}{\left[(1+\nu^*)(1-2\nu^*)\right]^2}\Delta\nu, \tag{70}$$

$$\Delta\mu \approx \frac{1}{2(1+\nu^*)}\Delta E - \frac{E^*}{2(1+\nu^*)^2}\Delta\nu. \tag{71}$$

Because the influences of $\Delta\nu$ cannot be ignored, $\Delta\lambda$ and $\Delta\mu$ are usually quite significant.

Since the identification process with ordinary LS estimation is unstable for Case 4, the extended Bayesian estimation method is adopted. Set$\beta = 1 \times 10^{-10}$ and C_p^{-1}as diagonal matrix with the diagonal items:$\frac{1}{(0.01\lambda_1)^2}$, $\frac{1}{(0.01\lambda_2)^2}$, $\frac{1}{(0.01\mu_1)^2}$ and $\frac{1}{(0.01\mu_2)^2}$, the reasonable measurements obtained with BAME are listed in Tab. 6. The identified results are given in Tab. 10, in which the maximum relative error of the identified parameters is 0.001. This is a very stable solution.

Finally, the complex method is used to identify Young's Moduli and Poisson's Ratios for Case 4. At this time, the best measurements can be easily obtained using COME with $CN^{up} = 10^7$ (see Tab. 6). Setting $E \in \left[1 \times 10^5, 1 \times 10^6\right]$ Pa and $\nu \in [0.01, 0.49]$, all parameters can be successfully identified with CPL from these measurements. Table 11 shows the identified results from one trial. The maximum relative error of the identified parameters is about 3.3. Compared with the 1 error in measurements, this deviation is acceptable.

Munich subway tunnel. Application of these algorithms to the Munich subway tunnel (line 5/9) [1] is demonstrated here. As Fig. 6 shows, the chosen critical cross-section is located on Impler Street, where two tunnels join

Table 10. Results of case 4 for brick model (extended Bayesian approach)

Measurements with errors				Max. bias by BAME	Identified parameters by FIBACK ($\times 10^5$)				Max. bias by FIBACK
10 x	9 x	11 x	8 z		λ_1	λ_2	μ_1	μ_2	
-ε	-ε	-ε	-ε	77	1.7309	0.5556	1.1544	♣0.8341	80
-ε	-ε	-ε	+ε	78	1.7306	0.5555	1.1534	♣0.8341	80
-ε	-ε	+ε	-ε	44	1.7308	0.5556	1.1540	♣0.8338	50
-ε	-ε	+ε	+ε	77	1.7306	0.5556	♣1.1531	0.8338	70
-ε	+ε	-ε	-ε	67	1.7309	0.5556	♣1.1545	0.8336	70
-ε	+ε	-ε	+ε	29	1.7307	0.5555	1.1536	♣0.8336	30
-ε	+ε	+ε	-ε	32	1.7309	0.5556	♣1.1542	0.8333	40
-ε	+ε	+ε	+ε	60	1.7306	0.5556	♣1.1532	0.8333	60
+ε	-ε	-ε	-ε	60	1.7309	0.5556	♣1.1545	0.8334	70
+ε	-ε	-ε	+ε	32	1.7307	0.5555	♣1.1535	0.8334	30
+ε	-ε	+ε	-ε	29	1.7309	0.5556	♣1.1541	♣0.8330	30
+ε	-ε	+ε	+ε	67	1.7306	0.5556	♣1.1532	0.8331	60
+ε	+ε	-ε	-ε	77	1.7309	0.5555	♣1.1546	0.8329	80
+ε	+ε	-ε	+ε	44	1.7307	0.5555	1.1537	♣0.8329	40
+ε	+ε	+ε	-ε	78	1.7309	0.5556	1.1543	♣0.8325	80
+ε	+ε	+ε	+ε	77	1.7307	0.5555	1.1533	♣0.8326	70

♣ where the maximum bias is located

to form one large tunnel 13.0 m wide and 8.4 m high. During construction, TRIVEC Sliding Micrometers (SM) [15] were installed to measure the displacements at three different positions on cross-section A-A. This measurement system can give the full three-dimensional displacement vector along the measurement line at intervals of 1 meter. The relative measurement error is about 0.001. Laboratory tests were carried out for the different soil layers around the tunnel (see Fig. 7 and Tab. 12).

During excavation the tunnel was subjected to two kinds of loading. One was the traffic loading on the street, which was assumed as 10 kPa. The other was the pressure caused by the initial stresses due to dead load of each layer, which was calculated as $P_y = \sum \rho\, h$, $P_x = k_0\, P_y$. P_y and P_x are the pressures in y and x direction, respectively; ρ is the density of the soil; h is the depth from the surface, and $k_0 = 0.8$, as obtained from a laboratory test.

Table 11. Results of Case 4 for brick model (complex method)

Measurements with system errors				Max. bias using COME	Parameters identified using CPL				Max. bias using CPL
6z	9x	11x	5z		E_1 (10^5Pa)	E_2 (10^5Pa)	ν_1	ν_2	
-ε	-ε	-ε	-ε	300	♣3.0030	2.0020	0.3000	0.2000	301
-ε	-ε	-ε	+ε	154	♣3.0015	2.0010	0.2912	0.1999	149
-ε	-ε	+ε	-ε	661	2.9975	♣2.0066	0.2989	0.2053	665
-ε	-ε	+ε	+ε	563	2.9959	♣2.0057	0.2900	0.2052	567
-ε	+ε	-ε	-ε	969	♣3.0097	1.9944	0.3017	0.1965	971
-ε	+ε	-ε	+ε	823	♣3.0083	1.9934	0.2929	0.1963	825
-ε	+ε	+ε	-ε	422	♣3.0042	1.9990	0.3006	0.2017	423
-ε	+ε	+ε	+ε	276	♣3.0027	1.9980	0.2918	0.2016	270
+ε	-ε	-ε	-ε	276	♣2.9972	2.0020	0.3081	0.1198	282
+ε	-ε	-ε	+ε	422	♣2.9958	2.0010	0.2995	0.1983	421
+ε	-ε	+ε	-ε	823	♣2.9918	2.0066	0.3070	0.2037	822
+ε	-ε	+ε	+ε	969	♣2.9903	2.0057	0.2983	0.2036	969
+ε	+ε	-ε	-ε	563	3.0038	♣1.9944	0.3098	0.1949	558
+ε	+ε	-ε	+ε	661	3.0025	♣1.9934	0.3011	0.1948	658
+ε	+ε	+ε	-ε	154	♣2.9984	1.9990	0.3087	0.2001	158
+ε	+ε	+ε	+ε	300	♣2.9970	1.9980	0.3000	0.2000	300

♣where the maximum bias is located

Table 12. Soil parameters from laboratory tests

Material type	Density ρ (kN / m^3)	Young's Modulus E (kPa)	Poisson's ratio ν
Layer 1 Gravel	22	80	0.35
Layer 2 Silty clay	21	80	0.35
Layer 3 Sand	21	100	0.35
Layer 4 Silty clay	22	80	0.35

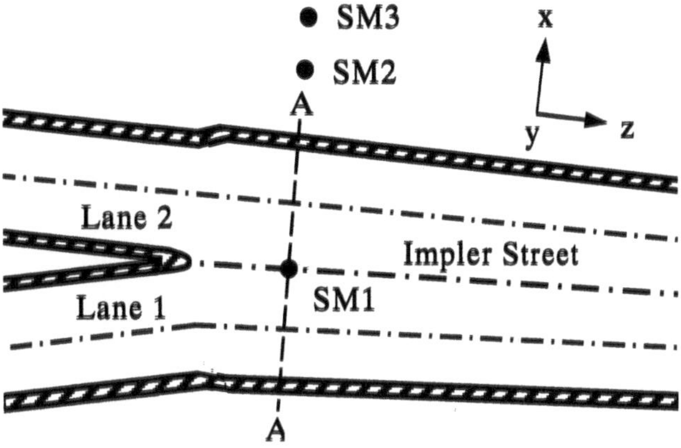

Fig. 6. General layout of Munich subway tunnel (line 5/9)

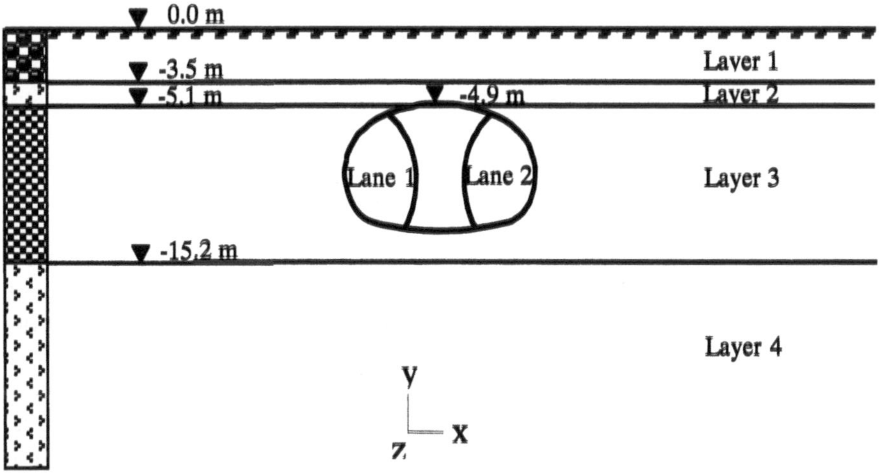

Fig. 7. Profile of cross-section A-A

One-step excavation is simulated using a plane strain FE model. This step involves full excavation of lane 2 and excavation of the crown of lane 1. Figure 8 shows the FE mesh, which is carefully generated to place the element nodes at the correct positions of the measuring points. The reference points SM1, SM2 and SM3 are denoted as point 1, point 2 and point 3, respectively. The measured displacements along SM1, SM2 and SM3 are plotted in Fig. 9. In the following identification processes only the measurements along SM1 in x and y direction are used. This is because: (1) the curve of the y displacements of SM3 (see Fig. 9e) has a big jump, which implies that SM3 was not properly mounted and most of the data are incorrect. (2) The original data of measurements along SM2 are missing. We obtained the y displacements

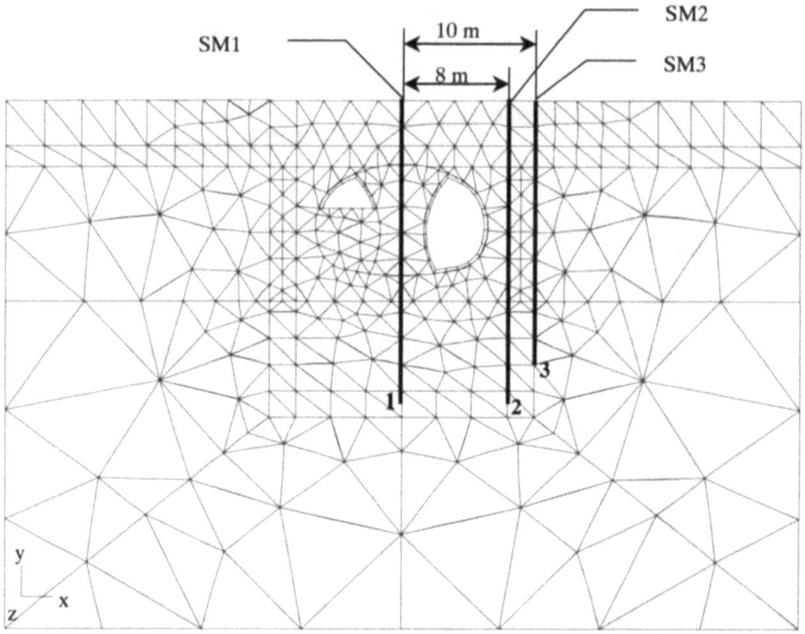

Fig. 8. FE mesh and the positions of sliding micrometers (SM)

by measuring from a drawing. It is further noticed that the y displacements below 14m suddenly approach zero (see Fig. 9c). Considering that the y displacements below 15m along SM1 and SM3 still have some reasonable nonzero values, we do not have much confidence in these measurements. Furthermore, the measured x displacements at 19m and above 4m along SM1 are discarded due to too much noise (see Fig. 9a). The following identification processes adopted that $\varepsilon = 0.001$, $\mathbf{C}_u^{-1} = \mathbf{I}$ and $u_{\min} = 1mm$.

Table 13. Good measurement sets (along SM1)

Program	Case	$N_{\min}$	$N_{\max}$	Measurements (Depth (m) / Direction)
BAME	1st trial LS	4	12	0/y 6/x 11/y 12/y 16/x
	2nd trial LS	4	12	1/y 6/x 10/y 12/y 16/x
	4th trial Bayesian	8	12	0/y 10/y 11/x 11/y 12/x 12/y 14/x 16/x
COME	5th trail LS	8	12	2/y 6/x 7/y 10/y 11/x 12/y 14/x 16/x
	6th trial Bayesian	8	12	0/y 6/x 7/x 10/y 11/y 12/y 14/x 16/x

Table 14. The parameters identified using the Gauss-Newton method

Case	Parameters	Layer 1	Layer 2	Layer 3	Layer 4	Iteration steps
LS (1st trial)	E(kPa)	42.145	56.525	97.792	47.739	6
	ν (fixed)	0.350	0.350	0.350	0.350	
	λ (kPa)	36.422	48.849	84.512	41.256	
	μ (kPa)	15.609	20.935	36.219	17.681	
LS (2nd trial)	E(kPa)	45.303	45.059	102.061	48.575	5
	ν (fixed)	0.350	0.350	0.350	0.350	
	λ (kPa)	39.151	38.940	88.201	41.978	
	μ (kPa)	16.779	16.689	37.803	17.991	
Extended Bayesian (4th trial)	E (kPa)	46.150	52.463	97.190	48.008	3
	ν	0.348	0.316	0.356	0.350	
	λ (kPa)	39.093	34.124	88.767	41.640	
	μ (kPa)	17.122	19.939	35.831	17.775	

The first trial is carried out only to identify Young's Modulus for each layer using ordinary LS estimation and the Gauss-Newton method. Poisson's Ratios are fixed as the values from laboratory tests. The inputs for BAME are set as $\delta\mathbf{p} = (16,\ 16,\ 20,\ 16)$ kPa and $\mathbf{p}^0 = (E_1,\ E_2,\ E_3,\ E_4) = (80,\ 80,\ 100,\ 80)$ kPa. The reasonable measurement set is given in Tab. 13, from which the material parameters are successfully identified (see Tab. 14). The comparison between the measured displacements and the calculated displacements is depicted in Fig. 9. BAME also points out one set of bad measurements, which is listed in Tab. 15. With these measurements, the identification process diverges. This warns us that the measurements should not be arbitrarily selected.

Table 15. One set of bad measurements (along SM1)

	Measurement (Depth (m) / Direction)					
Bad set	0/y	7/x	8/x	10/x	11/y	12/y

Because $\mathbf{p}^*$ is unknown, BAME can suggest only a 'probably good' measurement set. To verify the identified parameters above, the second trial is carried out with $\mathbf{p}^0 = (E_1,\ E_2,\ E_3,\ E_4) = (42.145,\ 56.525,\ 97.792,\ 47.739)$kPa and $\delta\mathbf{p} = 0$ kPa. The reasonable measurement set is listed in Tab. 13. The identified results (see Tab. 14 and Fig. 9) are similar to those from the first trial.

The third trial aims to identify the Lamé parameters of each layer using ordinary LS estimation and the Gauss-Newton method. According to the results of the second trial, the inputs for BAME are set as $\delta\mathbf{p} = 0$ kPa and

$$\mathbf{p}^0 = (\lambda_1, \lambda_2, \lambda_3, \lambda_4, \mu_1, \mu_2, \mu_3, \mu_4)$$
$$= (39.15, 38.94, 88.20, 41.97, 16.78, 16.69, 37.80, 17.99)\,\text{kPa}.$$

However, no common feasible measurement set is found. This implies the identification process will be very unstable.

The fourth trial is to identify all Lamé parameters with the extended Bayesian estimation. The rough prior information comes from the results of the first and second trials. According to this experience, the inputs for BAME are set as $\beta = 1.0 \times 10^{-9}$

$$\mathbf{p}^0 = (\lambda_1, \lambda_2, \lambda_3, \lambda_4, \mu_1, \mu_2, \mu_3, \mu_4)$$
$$= (37.79, 43.90, 86.36, 41.62, 16.19, 18.81, 37.01, 17.84)\,\text{kPa}$$

and $\mathbf{C}_p^{-1}$ is a diagonal matrix with the diagonal items: $\frac{1}{(0.036\lambda_1)^2}$, $\frac{1}{(0.113\lambda_2)^2}$, $\frac{1}{(0.021\lambda_3)^2}$, $\frac{1}{(0.009\lambda_4)^2}$, $\frac{1}{(0.036\mu_1)^2}$, $\frac{1}{(0.113\mu_2)^2}$, $\frac{1}{(0.021\mu_3)^2}$ and $\frac{1}{(0.009\mu_4)^2}$. The reasonable measurement set is listed in Tab. 13. The identified material parameters are given in Tab. 14. Comparison of measured and calculated displacements is also plotted in Fig. 9.

Ordinary LS estimation and the complex method are adopted in the fifth trial. According to the results of the second trial, the inputs for COME are set as $\delta\mathbf{p} = 0$, $CN^{up} = 10^9$ and

$$\mathbf{p}^0 = (E_1, E_2, E_3, E_4, \nu_1, \nu_2, \nu_3, \nu_4)$$
$$= (45.303, 45.059, 102.061, 48.575, 0.350, 0.350, 0.350, 0.350)$$

(Young's Modulus is in kPa). With the reasonable measurements listed in Tab. 13 and the admissible ranges given in Tab. 16, the identified parameters from ten trials are listed in Tab. 17.

Table 16. Admissible ranges for the complex method (Young's Modulus is in kPa)

	E_1	E_2	E_3	E_4	ν_1	ν_2	ν_3	ν_4
Upper bound	1×10^4	1×10^4	1×10^4	1×10^4	0.300	0.300	0.300	0.300
Lower bound	1×10^5	1×10^5	2×10^5	1×10^5	0.380	0.380	0.380	0.380

With the prior information in Tab. 17, it is possible to implement extended Bayesian estimation to identify all parameters. In the sixth trail the inputs

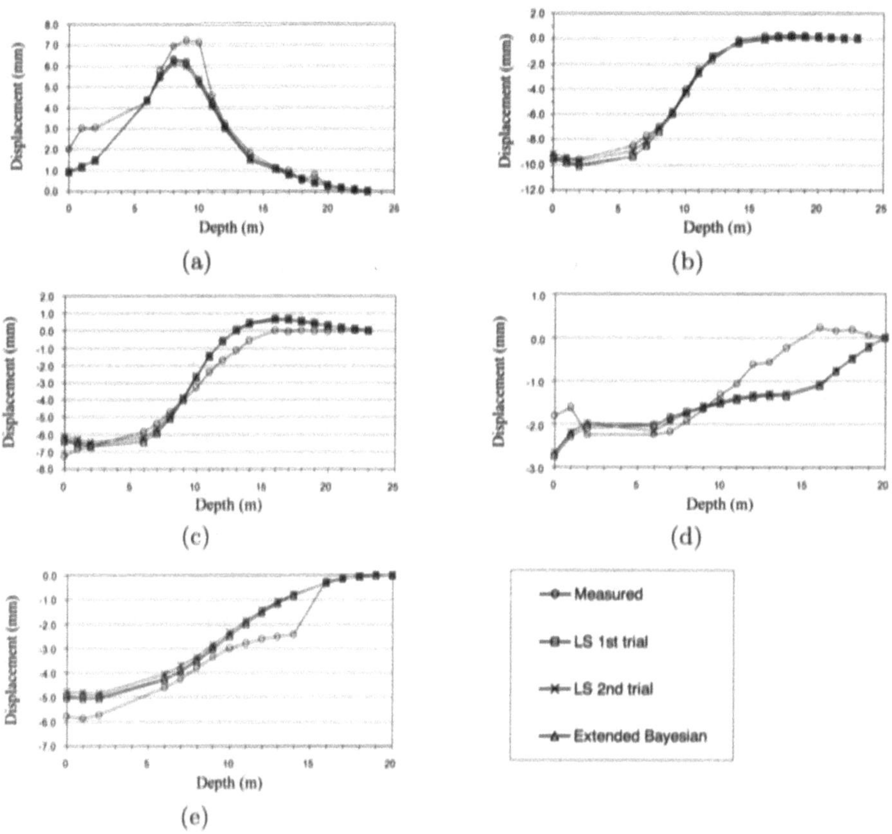

Fig. 9. Comparison of measured and calculated displacements
(a) x displacements along SM1 (b) y displacements along SM1
(c) y displacements along SM2 (d) x displacements along SM3
(e) y displacements along SM3

for COME are set as $\beta = 1.0 \times 10^{-10}$, $CN^{up} = 10^9$,

$$\begin{aligned}\mathbf{p}^0 &= (E_1, E_2, E_3, E_4, \nu_1, \nu_2, \nu_3, \nu_4)\\ &= (50.99, 35.52, 104.00, 53.96, 0.34, 0.34, 0.37, 0.34)\end{aligned}$$

(Young's Modulus is in kPa) and $\mathbf{C}_p^{-1}$ is a diagonal matrix with the diagonal items: $\frac{1}{(0.032E_1)^2}$, $\frac{1}{(0.010E_2)^2}$, $\frac{1}{(0.003E_3)^2}$, $\frac{1}{(0.028E_4)^2}$, $\frac{1}{(0.051\nu_1)^2}$, $\frac{1}{(0.044\nu_2)^2}$, $\frac{1}{(0.022\nu_3)^2}$ and $\frac{1}{(0.066\nu_4)^2}$. With the reasonable measurement set listed in Tab. 13 and the admissible ranges given in Tab. 16, the identified parameters from ten trials are listed in Tab. 18. Comparison of measured displacements and calculated displacements using the Gauss-Newton method and the complex method is plotted in Fig. 4.3. (The displacements from the complex method are calculated with the mean values of the identified parameters.)

Table 17. LS results from CPL (Young's Modulus is in kPa)

	E_1	E_2	E_3	E_4	ν_1	ν_2	ν_3	ν_4
1	50.585	35.532	104.470	52.088	0.314	0.371	0.380	0.314
2	53.777	34.966	103.680	52.244	0.371	0.342	0.379	0.300
3	47.981	36.092	104.450	55.310	0.327	0.314	.380	0.369
4	52.363	35.250	103.730	53.981	0.364	0.333	0.380	0.336
5	51.202	35.468	104.050	54.840	0.340	0.352	0.380	0.346
6	52.428	35.462	103.630	55.054	0.364	0.357	0.380	0.346
7	51.872	35.101	104.066	51.101	0.350	0.354	0.353	0.347
8	49.101	36.159	104.081	55.605	0.329	0.351	0.380	0.379
9	50.749	35.484	103.944	54.894	0.345	0.341	0.380	0.357
10	49.893	35.712	103.917	54.559	0.355	0.330	0.380	0.360
Mean	50.995	35.523	104.002	53.968	0.346	0.344	0.377	0.346
Mean square derivation	1.631	0.365	0.276	1.495	0.017	0.015	0.008	0.023
Mean square derivation /Mean	0.032	0.010	0.003	0.028	0.051	0.044	0.022	0.066

It is observed from Figs. 9 and 10 that there is not much difference among the calculated displacements of the first, second, fourth, fifth and sixth trail. Compared with the measured displacements, although only parts of the measurements are used, satisfactory results are observed not only at the points used for input but also at other measuring points. It is noticed that the calculated curve of y displacements along SM1 matches the measured curve very well. The calculated curve of x displacements along SM1 is quite similar to the measured curve, except for some points closed to the surface where the measuring noises are intolerable. The curve of the calculated y displacements along SM2 is also similar to the measured one. The calculated y displacements along SM3 are nearly the same at the points below 15 m. It seems that if SM3 were correctly mounted the calculated curve would be closer to the measured curve. Big differences are observed between the calculated and measured x displacement curves along SM3. This is probably because SM3 was improperly mounted.

Table 18. Extended Bayesian results from CPL (Young's Modulus is in kPa)

	E_1	E_2	E_3	E_4	ν_1	ν_2	ν_3	ν_4
1	48.587	35.471	101.677	42.638	0.308	0.345	0.380	0.300
2	50.972	35.314	101.114	44.230	0.327	0.331	0.359	0.345
3	45.496	35.907	101.920	44.149	0.302	0.303	0.380	0.349
4	51.489	35.535	100.705	44.082	0.335	0.335	0.360	0.341
5	47.163	35.696	101.771	44.381	0.300	0.342	0.378	0.342
6	47.130	35.604	101.816	44.478	0.300	0.341	0.375	0.352
7	47.428	35.567	102.141	42.568	0.300	0.308	0.378	0.301
8	49.261	35.423	101.593	42.681	0.306	0.363	0.380	0.300
9	49.545	35.715	100.532	44.333	0.337	0.350	0.380	0.337
10	50.487	35.468	100.228	45.378	0.344	0.357	0.380	0.349
Mean	48.756	35.570	101.350	43.892	0.316	0.337	0.375	0.332
Mean square derivation	1.840	0.161	0.625	0.894	0.017	0.018	0.008	0.021
Mean square derivation /Mean	0.038	0.005	0.006	0.020	0.053	0.055	0.021	0.063

It is also observed from Tab. 14, Tab. 17 and Tab. 18 that the identified results for Poisson's Ratio are not very stable. This is because the displacement is not very sensitive to these parameters. The stability of the identified Young's Moduli varies from layer to layer. Young's Modulus for layer three is most stable, because most of the tunnel is embedded in this layer. Young's Modulus for layer two is most unstable, because this layer is very thin and no displacement is sampled here.

Although the identified parameters are not exactly the same in different trials, we can conclude that these identifications are still successful, because the calculated curve of y displacements along SM1 is very close to the measured curve. In engineering, these displacements (settlements) are usually the critical factor in judging the stability of the tunnel.

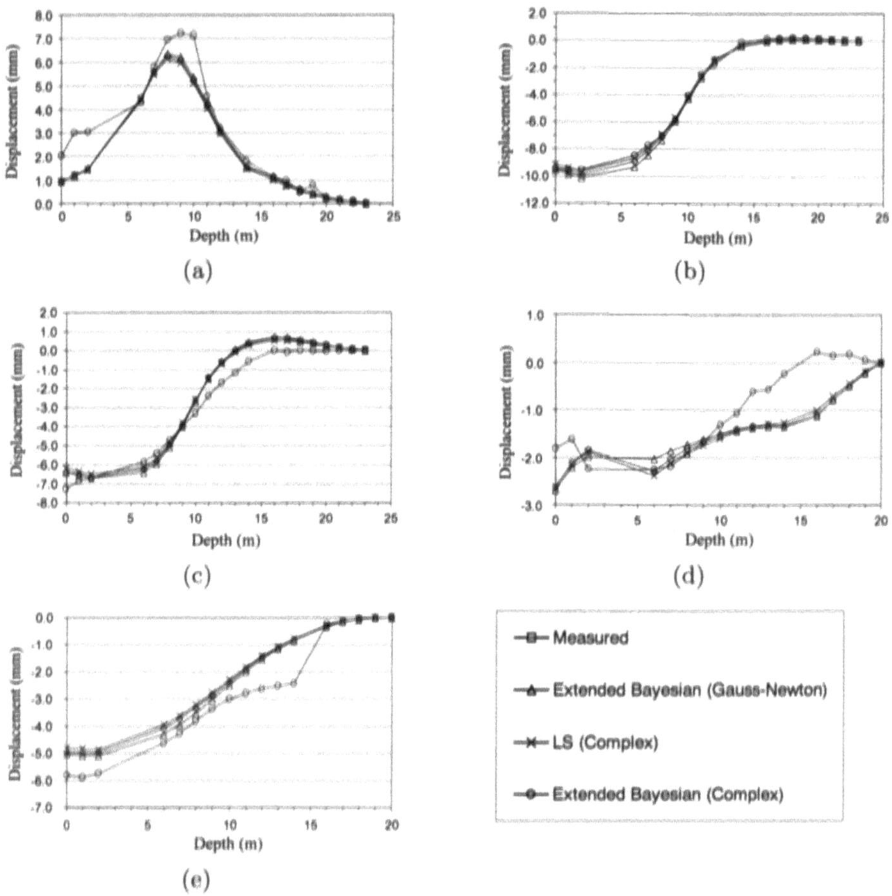

Fig. 10. Comparison of measured and calculated displacements using the Gauss-Newton method and the complex method
(a) x displacements along SM1 (b) y displacements along SM1
(c) y displacements along SM2 (d) x displacements along SM3
(e) y displacements along SM3

5 Comparisons between optimization methods

The Gauss-Newton method is a typical gradient-type method and the complex method is a typical direct-search method. From the theoretical analyses and numerical calculations in the last section, some properties of these two methods are compared in Tab. 19, in which the following points should be emphasized:

(a) No matter what method is adopted, the parameter with less impact on the measurements can be identified with difficulty.

(b) No matter what method is adopted, the initial estimation bias L_p (in extended Bayesian estimation) and the measuring bias $L_A \varepsilon$ cannot be avoided.
(c) Thus, can be called the basic bias. This bias is magnified by the factor $(1 + L_\Omega)$ during the iteration process of the Gauss-Newton method. There is no bias propagation and accumulation during the iteration process of the complex method. Therefore, if the global optimal solution can be found using the complex method, it must be more stable than the solution obtained with the Gauss-Newton method.
(d) With the guidance of the gradient information, the convergence speed of the Gauss-Newton method is much faster than that of the complex method. In our experience, for a same benchmark, the computing time for the Gauss-Newton method is about one-twentieth that of the complex method.
(e) In some cases, when the system sensitivity cannot be calculated analytically, the complex method is a simple and flexible solution.

Table 19. Comparisons between the Gauss-Newton method and the complex method

	Gauss-Newton method	complex method
Existence	The solution exists if $G(p)$ is continuous.	The solution exists if the search scheme is flexible enough.
Uniqueness	The solution is unique if $L_\Omega < 1$.	The uniqueness of the solution cannot be guaranteed.
Stability	The maximum bias of the identified parameters is about $(L_p + L_A \varepsilon)(1 + L_\Omega)$	The maximum bias of the identified parameters is about $L_p + L_A \varepsilon$
Complexity	Complex	Simple
Speed of convergence	Fast	Very slow
Sensitivity v.s. convergence	The iteration is easier to converge if measurements have higher sensitivities.	The convergence of the algorithm is irrelevant to system sensitivity.

Since both gradient-type methods and direct-search methods have their own advantages and limits, the proper method should be chosen to meet the specific situation. However, if the measurements are sensitive to all parameters and it is not difficult to calculate the gradients, the Gauss-Newton method is strongly recommended.

6 Conclusions

The techniques of back-analysis shine a bright light on obtaining various properties of the geotechnical structures. Of these techniques, the parameter identification strategies are specially emphasized in this paper.

Generally, the parameter identification process can be presented in a unified mathematical programming formulation. This formulation is used to easily formulate the algorithms for identifying the parameters of different kinds of models. However, the identifiability of these algorithms is strongly related to the system sensitivities of measurements. To achieve successful identification, the parameters should have great impact on the measurements. This conclusion is illustrated by identifying the initial damage parameters of a damage model for jointed rock.

The quality of the identified parameters is directly related to the reliability of the identification algorithm. To discuss this topic, it is necessary to analyze the well-posedness of the identification process. A consequent result of this analysis is the determination of the optimal layout of measurements, because the properties of the identification process are greatly influenced by the distribution of measurements. From the systematic analysis of the well-posednesses of the identification processes using the Gauss-Newton method and the complex method in this paper, it is concluded that:

(a) Every identification method has its own advantages and limits. The selection from gradient-type methods and direct-search methods is determined by the specific situation.
(b) For a specific identification method the quality of the identified parameters is determined by the synthetic influences of the measurement set. It is meaningless to discuss the property of a single measurement.
(c) Although the identification is usually facilitated by measurements with high system sensitivities, bad identification results can also be obtained from an improper combination of these measurements.
(d) There is no definite relation between the number of measurements and the optimal measurement layout.
(e) It is not always true that bigger measurements are better.

In a word, the identification method should be properly selected and measurement layout must be carefully designed.

References

[1] Amstad C., Kovari K., Grob H. (1984) The report on the displacement measuring of Munich subway tunnel line 5/9, Impler Street. Rock Engineering Section, Federal Institute of Technology, Zürich, (in German)

[2] Box M. J. (1965) A new method of constrained optimization and a comparison with other methods. *The Computer Journal*, **8**: 42-52.

[3] Cen Z. Z., Shen X. P. and Xu B. Y. (1994, 1995) A numerical method of material parameter identification in rock-engineering. *In: Computer Methods and Advances in Geomechanics (Morgantown)*, Siriwardane H J, Zaman, eds., Rotterdam, Balkema: 2229-2234
[4] Chicone C. (1999) Ordinary Differential Equations with Applications. Springer Verlag, New York
[5] Cividini A., Jurina L., Gioda G. (1981) Some aspects of 'characterization' problems in geomechnics. *International Journal of Rock Mechanics and Mining Sciences and Geomechanics Abstracts*, **18**: 487-503
[6] Cividini A., Maier G. and Nappi A. (1983) Parameter estimation of a static geotechnical model using a Bayes' approach. *International Journal of Rock Mechanics and Mining Sciences and Geomechanics Abstracts*, **20**: 215-226
[7] Engl H. W., Hanke M. and Neubauer A. (1996) Regularization of inverse problems. Netherlands: Kluwer Academic Publishers
[8] Fadale T. D., Nenarokomov A. V., Emery A. F. (1995) Two approaches to optimal sensor locations. *Journal of Heat Transfer*, **117**: 373-379
[9] Fulton W. (1995) Algebraic topology. Springer Verlag New York
[10] Gioda G. and Sakurai S. (1987) Back analysis procedures for the interpretation of field measurements in geomechanics. *International Journal for Numerical and Analytical Methods in Geomechanics*, **11**: 555-583
[11] Gioda G. (1990) Some applications of mathematical programming in geomechanics. *In: Numerical Methods and Constitutive Modelling in Geomechanics*, Desai, C. S., Gioda, G. eds., Springer Verlag New York, 319-350
[12] Haftka R. T., Scott E. P., Cruz J. R. (1998) Optimization and experiments: a survey. *Applied Mechanics Reviews*, **51**: 435-448
[13] Kawamoto T., Ichikawa Y., Kyoya T. (1988) Deformation and fracturing behaviour of discontinuous rock mass and damage mechanics theory. *International Journal for Numerical and Analytical Methods in Geomechanics*, **12**: 1-30
[14] Keeler S. E., Reilly P. M. (1992) The design of experiments when there are errors in all the variables. *The Canadian Journal of Chemical Engineering*, **70**: 774-779
[15] Köppel J., Amstad C., Kovari K. (1983) The measurement of displacement vectors with the TRIVEC borehole probe. *In: Field Measurements in Geomechanics*, Zurich, K. Kovari ed., Rotterdam: Balkema: 209-239
[16] Mitchell T. J. (1974) An algorithm for the construction of "D optimal" experimental designs. *Technometrics*, **16**: 203-220
[17] Murakami A., Hasegawa T., Sakaguchi H. (1991) Interpretation of ground performances based on back analysis results. *In: Computer Methods and Advances in Geomechanics*, Beer G., Booker J. R. and Carter J. P. eds., Rotterdam: Balkema: 1011-1015
[18] Nelder J. A., Mead R. (1964) A simplex method for function minimization. *The Computer Journal*, **7**: 308-313
[19] Nocedal J., Wright S. J. (1999) Numerical optimization, Springer Verlag New York
[20] Romanov V. G. (1987) Inverse Problems of Mathematical Physics, Utrecht, Vnuscience Press BV
[21] Spendley W., Hext G. R., Himsworth F R. (1962) Sequential application of simplex designs in optimization and evolutionary operation. *Technometrics*, **4**: 441-461

[22] Swoboda G., Yang Q. (1999) An energy-based damage model of geomaterials I. Formulation and numerical results. *International Journal of Solids and Structures*, **36**: 1719-1734

[23] Swoboda G., Yang Q. (1999) An energy-based damage model of geomaterials II. Deduction of damage evolution laws. *International Journal of Solids and Structures*, **36**: 1735-1755

[24] Swoboda G. (2001) The manual of FINAL 7.1. University of Innsbruck, Austria

[25] Taktak R., Beck J. V. and Scott E. P. (1993) Optimal experimental design for estimating thermal properties of composite materials. *International Journal of Heat and Mass Transfer*, **36**: 2977-2986

[26] Terzaghi K. and Peck R. B. (1948) Soil mechanics in engineering practice. John Wiley and Sons New York

[27] Venclik P. (1994) Development of an inverse back analysis code and its verification. *In: Numerical Methods in Geotechnical Engineering (Manchester).* Smith I. M. ed., Rotterdam: Balkema: 423-430

[28] Welch W. J. (1982) Branch-and-bound search for experimental designs based on D optimality and other criteria. *Technometrics*, **24**: 41-48

[29] Yang L. D. (1996) The theory and applications of back analysis in geotechnical problems. Beijing: Science Press, (in Chinese)

[30] Yusuke H., Lie W. T., Soumitra G. (1994) Inverse analysis of an embankment on soft clay by extended Bayesian method. *International Journal for Numerical and Analytical Methods in Geomechanics*, **18**: 709-734

[31] Zienkiewicz O. C. (1989) The finite element method. McGraw-Hill London

Chapter 9
Soft-computing-based parameter identification as the basis for prognoses of the structural behaviour of tunnels

Bernhard Pichler, Roman Lackner, Herbert A. Mang

Institute for Strength of Materials, Vienna University of Technology

Abstract. A parameter identification (PI) method for determination of unknown model parameters in tunnelling is presented. The PI method is based on measurement data provided by the construction site. Model parameters for finite element (FE) analyses are identified such that the results of these calculations meet the available measurements as well as possible. For the determination of the unknown parameter set, use of an artificial neural network (ANN) is proposed. The network is trained to approximate results of already performed FE simulations. A genetic algorithm (GA) uses the trained ANN to provide a prognosis for an optimal parameter set which, finally, must be assessed by an additional FE analysis. In contrast to other gradient-free methods requiring a large number of FE simulations, the proposed PI method renders back analysis of model parameters feasible even for large-scale models. Finally, the performance of this PI method as the basis for prognoses of the structural behaviour of a tunnel is demonstrated.

1 Introduction

In Chapt. 10 hybrid methods for the determination of the level of loading of shotcrete tunnel shells are presented. These methods require measured displacement histories at measurement points (MPs) of the shell. Obviously, such a history does not begin before the construction of the shell has reached the measurement cross-section (MCS) containing the considered MPs. Consequently, the aforementioned methods allow an assessment of the present situation but do not permit predictions of future situations. In contrast to such an assessment, genuine predictions of stress states in the shotcrete shell require consideration of the interaction between the shotcrete shell and the surrounding soil/rock formations. Whereas material properties of shotcrete are well known from laboratory tests, little is known about the material properties of the surrounding soil/rock. Lack of such parameters renders genuine predictions of stress states in shotcrete tunnel shells an almost impossible task.

In order to specify unknown soil/rock parameters, use of a gradient-free PI method resting on soft computing is proposed. Hereby, material parameters of soil/rock layers are related via numerical analyses to *in situ* measurements. Since both the shotcrete tunnel shell and the surrounding soil are considered in the numerical model, different types of measurements such as displacements at MPs, surface settlements, and extensometer measurements can be employed during PI. Unknown material parameters will be determined at specific cross sections of the tunnel, favourably at MCSs. During PI, the time span considered in the numerical analyses ranges from the time instant associated with the start of the excavation at the *considered* MCS to the day on which the analysis is performed. Based on the identified material parameters, future displacements and, thus, future stress states can be predicted. There are two different types of predictions (see Fig. 1):

(1) The first type of prediction is characterised by continuation of the numerical simulation for the *considered* MCS based on the identified parameters. This type of prediction provides information about future displacements and stress states at the *considered* (already installed) MCS.
(2) In case of moderately changing geological conditions, the soil/rock parameters determined at already installed MCSs may still be valid a stretch ahead of the tunnel face. Hence, based on identified parameters, numerical analyses of the tunnel excavation at future (not yet installed) MCSs are possible.

Both types of predictions would represent a significant progress in tunnel engineering. Beside the possibility of adaptation and optimisation of tunnel-driving parameters, they would allow to predict the safety of the tunnel and, hence, reduce the risks for the construction crew.

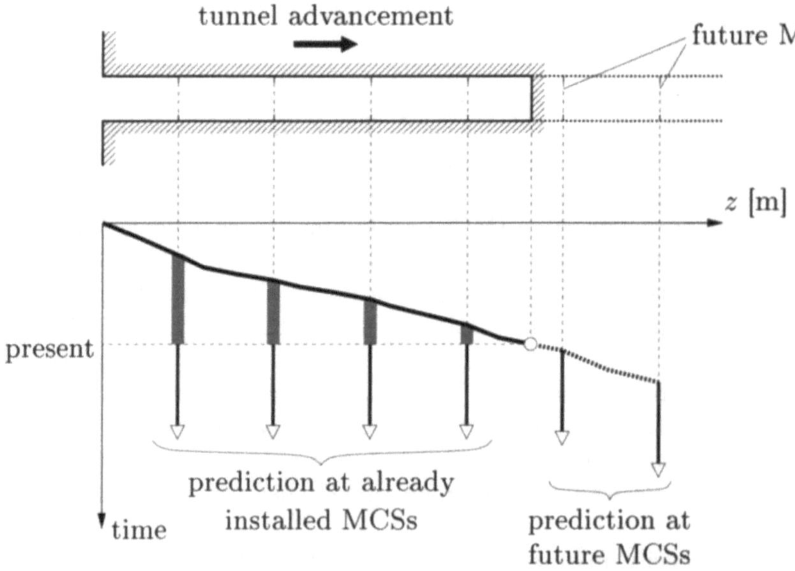

Fig. 1. Illustration of two types of prediction on the basis of soil/rock parameters obtained from PI, i.e., prediction either at already installed or at yet not installed MCSs

2 Back-analysis of soil/rock parameters in tunnelling

Already in the 1980s, the use of data from *in situ* measurements as input for back analyses in geotechnical engineering was proposed in [16] [4]. Such measurements are nowadays routinely performed during construction. Based on these data, values of unknown model parameters characterising mechanical properties of the soil/rock mass can be determined. Finally, values of unknown model parameters are obtained which are optimal in the sense that the best possible agreement between respective results and *in situ* measurements will be achieved, if these parameters are used as input for the numerical analysis. Hence, based on the identified model parameters, an appropriate numerical model provides a good numerical approximation of the monitored structural behaviour.

In order to solve a PI problem, an error function has to be defined. It represents the difference between selected values of n_m different *in situ* measurements and corresponding quantities obtained from a numerical analysis. The error function depends on the unknown parameters. It can be defined as

$$\mathcal{R}^h(\mathbf{p}) = \sqrt{\frac{1}{n_m}\sum_{c=1}^{n_m}\frac{1}{n_{o,c}}\sum_{r=1}^{n_{o,c}}\left(\frac{u^h_{c,r}(\mathbf{p}) - u_{c,r}}{\bar{u}_c}\right)^2} \tag{1}$$

with

$$\bar{u}_c = \frac{1}{n_{o,c}}\sum_{r=1}^{n_{o,c}}|u_{c,r}| \, , \tag{2}$$

where $u_{c,r}$ and $u^h_{c,r}$, $c = 1, 2, \ldots n_m$ and $r = 1, 2, \ldots n_{o,c}$, are the selected values of measurements and the corresponding numerical results, respectively. Obviously, the numerical results $u^h_{r,c}$ depend on the unknown model parameters collected in the vector $\mathbf{p}$. The definition of the error function given in (1) is characterised by relating the error of the r-th selected value of the c-th measurement curve to the mean value $\bar{u}_c$ of all selected measurements referred to the respective measurement curve [1]. Based on the error function (1), back analysis aims at determination of the absolute minimum of $\mathcal{R}^h$. This minimum provides the best agreement between *in situ* measurements and numerical results. Various techniques for solving such a minimisation problem have been proposed in the open literature. They can be divided into two groups:

(1) Gradient-free PI methods (e.g., GAs [5] or stochastic search techniques) are widely used to solve PI problems. They are characterised by searching for the optimal solution in the entire parameter space. Hereby, values of the error function, related to different parameter sets, are determined

[1] Summing up the values of measurements chosen for PI of all available measurement curves yields $n_o = \sum_{c=1}^{n_m} n_{o,c}$.

sequentially and compared. This strategy requires a large number of numerical analyses. Therefore, to render applications of gradient-free PI methods feasible, the computational effort for each analysis must be low.

(2) Gradient-based PI methods employ the gradient of $\mathcal{R}^h$, i.e., the derivative of $\mathcal{R}^h$ with respect to the parameter vector $\mathbf{p}$. In case of elastic material behaviour, this gradient can be computed analytically [10] [11] [17]. In most other cases, however, computation of the analytical expression of the gradient of $\mathcal{R}^h$ is a challenging task, requiring major modifications of the analysis tool (e.g., FE program) [9]. Moreover, gradient-based PI methods do not search in the entire parameter space for optimal solutions. They are characterised by a local search for a minimum of the error function, starting from an initial choice for the unknown parameters. Since the results strongly depend on these initial values of the unknown parameters, such algorithms may get trapped in a local minimum of the error function.

Numerical simulations in tunnelling are mostly performed by means of the finite element method (FEM). Unfortunately, the computational effort of such analyses is usually rather high. Moreover, realistic modelling requires consideration of different types of nonlinearities. They arise, e.g., from nonlinear boundary conditions such as the excavation process and/or nonlinear material behaviour accounting, e.g., for plastic deformations. Consequently, the error function $\mathcal{R}^h$ may have several local minima making back analysis a challenging task.

In order to render PI for large-scale models both feasible and reliable, an alternative mode of PI employing ANNs was presented in [7] [3] [12] [13] [14]. In [12], an ANN was employed to solve a PI problem in tunnelling according to the principles of the New Austrian Tunnelling Method (NATM). The ANN was trained to provide an approximation of FE analyses in tunnelling (Fig. 2). Accordingly, model parameters and respective numerical results from FE simulations were used for training the ANN. The trained ANN is able to reproduce the correlation between model parameters and results from FE analyses. Moreover, it provides a generalisation of this correlation. The ANN maps parameter sets that were not used as input for FE analyses onto interpolated values of numerical results. Based on the trained ANN, an estimate of an optimal parameter set can be computed. It corresponds to the minimum of the error function

$$\mathcal{R}^*(\mathbf{p}) = \sqrt{\frac{1}{n_m} \sum_{c=1}^{n_m} \frac{1}{n_{o,c}} \sum_{r=1}^{n_{o,c}} \left(\frac{u^*_{c,r}(\mathbf{p}) - u_{c,r}}{\bar{u}_c} \right)^2} , \tag{3}$$

where $u^*_{c,r}$, $c = 1, 2, \ldots n_m$ and $r = 1, 2, \ldots n_{o,c}$, are output quantities of the ANN. The quality of the estimated optimal parameters is assessed by the value of the error function $\mathcal{R}^h$, requiring an additional FE analysis.

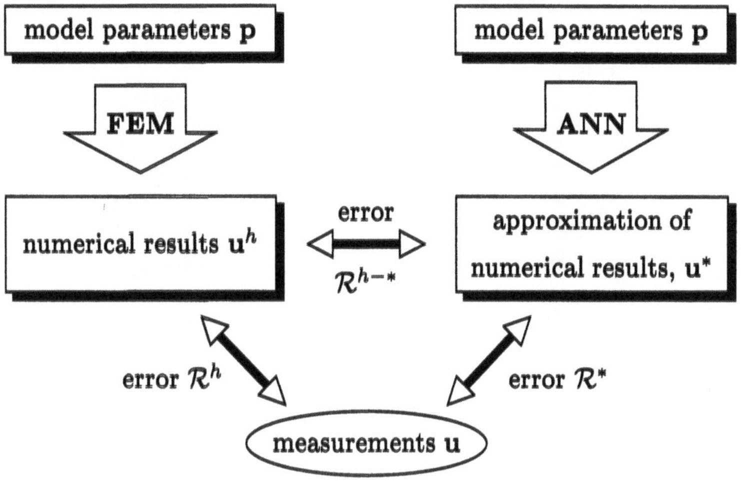

Fig. 2. Approximation of the FE analysis by an ANN. Definition of error functions.

The following section deals with the structure of the ANN, training of the network aiming at approximation of FE analyses, and determination of an optimal parameter set.

3 Soft-computing-based parameter identification method

3.1 Artificial neural network (ANN)

In the context of the proposed PI method, an error back-propagation ANN is used. This type of ANN was originally developed to model the governing processes of biological neural networks [15]. It consists of several layers of neurons and nodes (see Fig. 3). The input layer consists of n_i input nodes, the hidden layer of n_h neurons, and the output layer of n_o output neurons. Lines connecting two neurons or a node with a neuron are called *edges*. For the input layer and the hidden layer, an additional node containing the value 1 is considered (see Fig. 3). This modification accounts for the so-called *stimulus threshold* of biological neurons [15].

The ANN maps n_i-dimensional input vectors $\mathbf{p}$ onto n_o-dimensional output vectors $\mathbf{u}^*$. It can be trained to map *given* input vectors onto respective *given* output vectors referred to as *desired* output vectors. A pair consisting of such a *given* input vector $\mathbf{p}$ and the related *given* output vector $\mathbf{u}^h$ is called data set. Both vectors of a data set may contain values with different physical units and of different orders of magnitude. Hence, the input vectors of the different data sets, $\mathbf{p}$, are scaled to dimensionless values within the interval $[0.1, 0.9]$. The vectors containing the scaled and dimensionless input values are denoted as $\tilde{\mathbf{p}}$ (see Fig. 3).

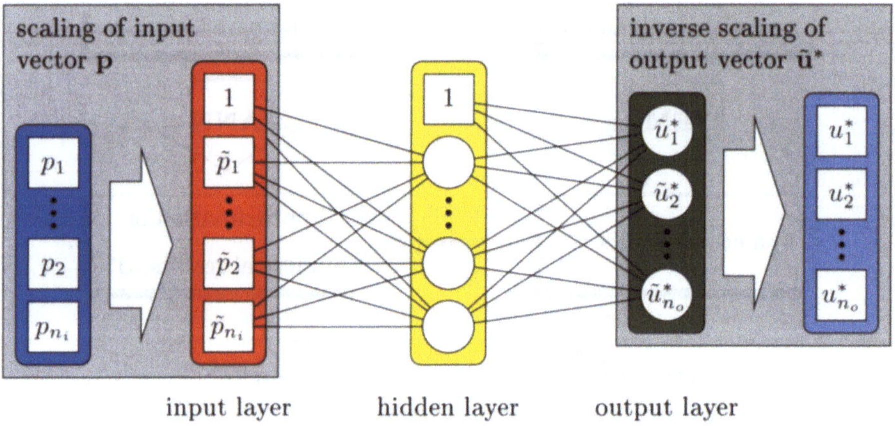

Fig. 3. Structure of an error back-propagating ANN with one hidden layer (circles: neurons; squares: nodes)

Traversing of such a vector $\tilde{\mathbf{p}}$ from the input layer of the ANN to its output layer is performed by means of a codified calculation scheme. The respective algorithm is called the *feedforward procedure*: dimensionless input values are fed into the ANN at the nodes of the input layer. If a dimensionless value $\tilde{p}_j$ moves from node j of the input layer (index i) along an edge to a neuron k of the hidden layer (index h), it will be multiplied by the *edge weight* w_{jk}^{ih}. This scalar value is also referred to as *network weight*. The network weights are parameters of the ANN. As weighted values from different edges enter a neuron k of the hidden layer, they are summed up. The sum σ_k^h is called the *excitation* of the respective neuron of the hidden layer:

$$\sigma_k^h = \sum_{j=0}^{n_i} \tilde{p}_j \, w_{jk}^{ih} \,, \tag{4}$$

with $\tilde{p}_0 := 1$ accounting for the additional node of the input layer containing the value 1 (see Fig. 3). A *sigmoidal activation function* is applied to this sum, yielding

$$s(\sigma_k^h) = \left[1 + \exp\left(-\sigma_k^h\right)\right]^{-1} \,, \tag{5}$$

where $s(\sigma_k^h)$ represents the output of neuron k of the hidden layer. The output of all neurons of the hidden layer are "shot" along the edges to the neurons of the output layer. Analogous to (4), the excitation of neuron r of the output layer (index o) is obtained as

$$\sigma_r^o = \sum_{k=0}^{n_h} s\left(\sigma_k^h\right) w_{kr}^{ho} \,, \tag{6}$$

with $s(\sigma_0^h) := 1$ accounting for the additional node of the hidden layer containing the value 1 (see Fig. 3). Application of the activation function (5) to σ_r^o yields the r-th dimensionless output value of the ANN: $\tilde{u}_r^* = s(\sigma_r^o)$. Considering (5), $\tilde{u}_r^* \in (0.0\,, 1.0)$.

In order to obtain correct physical units and correct orders of magnitude, output vectors $\tilde{\mathbf{u}}^*$ computed by means of the feedforward procedure are scaled. For this purpose, the operation used to scale the desired output vectors $\mathbf{u}^h$ to dimensionless values in the interval $[0.1\,, 0.9]$ is reversed. Output vectors of the ANN, characterised by the correct physical units and the correct order of magnitude, are denoted as $\mathbf{u}^*$ (see Fig. 3). They are functions of the respective input parameters $p_j\,,\ j = 1, 2, \ldots n_i$, and of the network weights. The number of network weights, n_w, for a certain network architecture is obtained from

$$n_w = (1 + n_i) \cdot n_h + (1 + n_h) \cdot n_o \ . \tag{7}$$

Minimisation of the differences between the obtained output vectors $\mathbf{u}^*$ and the desired output vectors $\mathbf{u}^h$ is referred to as ANN training.

3.2 Training of the ANN

An ANN can be trained to approximate a relationship between multi-dimensional vectors, which is described by n_{tot} given data sets. These sets are divided into two groups: (i) n_d training data sets and (ii) n_t test data sets, with $n_{tot} = n_d + n_t$. During training of the ANN, each input vector of the n_d training data sets, $\mathbf{p}_d\,,\ d = 1, 2 \ldots n_d$, is fed into the network and a feedforward procedure is performed. The so-obtained output vectors of the ANN, $\mathbf{u}^*(\mathbf{p}_d)$, are compared with the respective given output vectors $\mathbf{u}^h(\mathbf{p}_d)$ of the training data sets. The ANN approximation-error for the training data sets, $\mathcal{R}^{*-h}$, is evaluated as

$$\mathcal{R}^{*-h} = \sqrt{\frac{1}{n_d} \sum_{d=1}^{n_d} \frac{1}{n_m} \sum_{c=1}^{n_m} \frac{1}{n_{o,c}} \sum_{r=1}^{n_{o,c}} \left(\frac{u_{c,r}^*(\mathbf{p}_d) - u_{c,r}^h(\mathbf{p}_d)}{\bar{u}_c} \right)^2} \ . \tag{8}$$

Training of the ANN is equal to minimisation of $\mathcal{R}^{*-h}$ by adjusting the unknown network weights. The nonlinear minimisation problem $\mathcal{R}^{*-h}(w_{jk}^{ih}, w_{kr}^{ho}) \rightarrow$ minimum is solved by means of the gradient-descent method (GDM), referred to as *backpropagation algorithm* [15]. In an iterative manner, the network weights are updated continuously. The incremental network weights related to one GDM iteration step, Δw_{jk}^{ih} and Δw_{kr}^{ho}, are functions of $\partial \mathcal{R}^{*-h} / \partial w_{jk}^{ih}$ and $\partial \mathcal{R}^{*-h} / \partial w_{kr}^{ho}$, respectively, reading (see, e.g., [15])

$$\Delta w_{jk}^{ih} = \Delta w_{jk}^{ih} \left(\frac{\partial \mathcal{R}^{*-h}}{\partial w_{jk}^{ih}} \right) \quad \text{and} \quad \Delta w_{kr}^{ho} = \Delta w_{kr}^{ho} \left(\frac{\partial \mathcal{R}^{*-h}}{\partial w_{kr}^{ho}} \right) \ . \tag{9}$$

For the determination of the derivatives of $\mathcal{R}^{*-h}$ with respect to the network weights, the reader is referred to [13]. At the end of each GDM iteration step, all network weights are updated by Δw_{jk}^{ih} and Δw_{kr}^{ho}. In the next iteration step, a feedforward procedure of all n_d input vectors is performed on the basis of the updated network weights w_{jk}^{ih} and w_{kr}^{ho}. The so-obtained updated output vectors $\mathbf{u}^*(\mathbf{p}_d)$ allow to determine the new value of the error function $\mathcal{R}^{*-h}$ (see (8)).

The quality of the interpolations between the data sets used for the training of the ANN is assessed by the n_t test data sets. For each input vector of the test data sets, p_d, $d = 1, 2, \ldots n_t$, a feedforward procedure is performed. The obtained output vectors of the ANN, $\mathbf{u}^*(\mathbf{p}_d)$, are compared with the respective given output vectors $\mathbf{u}^h(\mathbf{p}_d)$ of the test data sets. Consequently, the value of the approximation-error $\mathcal{R}_{test}^{*-h}$ of the ANN, corresponding to the test data sets, can be computed as

$$\mathcal{R}_{test}^{*-h} = \sqrt{\frac{1}{n_t}\sum_{d=1}^{n_t}\frac{1}{n_m}\sum_{c=1}^{n_m}\frac{1}{n_{o,c}}\sum_{r=1}^{n_{o,c}}\left(\frac{u_{c,r}^*(\mathbf{p}_d) - u_{c,r}^h(\mathbf{p}_d)}{\bar{u}_c}\right)^2} . \tag{10}$$

In general, both $\mathcal{R}^{*-h}$ and $\mathcal{R}_{test}^{*-h}$ are decreasing at the beginning of the ANN training. Hence, both the approximation of the training data sets by the ANN and the ability of the ANN to interpolate between these sets are improved. At a certain stage of the training procedure, $\mathcal{R}_{test}^{*-h}$ attains a minimum. Continuation of training results in a further improvement of the approximation of the training data sets, i.e., $\mathcal{R}^{*-h}$ is decreasing. However, the value of the testing error $\mathcal{R}_{test}^{*-h}$ will increase. Hence, the ability of the network to interpolate between the training data sets is reduced. Consequently, ANN training will be terminated when $\mathcal{R}_{test}^{*-h}$ attains a minimum. The respective values of the error functions $\mathcal{R}_{test}^{*-h}$ and $\mathcal{R}^{*-h}$ represent measures of the success of the training procedure.

3.3 Training of the ANN in the context of PI in tunnelling

With respect to the proposed PI method, the ANN is trained to approximate results from FE analyses. In tunnelling such analyses are in generally characterised by a comparatively great computational effort. In order to keep the effort required for PI reasonably small, only few FE analyses are performed for determination of data sets prior to PI. Training and testing of the ANN on the basis of only a few data sets is controversial: $\mathcal{R}_{test}^{*-h}$ will attain a minimum at an early stage of the training algorithm and, hence, training of the ANN will be stopped. At this stage, however, the value of the error function $\mathcal{R}^{*-h}$, obtained on the basis of the training data sets, will still be rather large. Consequently, the approximation of the training data sets by the ANN will be poor, rendering interpolations between these data sets by the ANN useless.

In order to avoid such a situation, the precomputed data sets are *not* divided into training data sets and test data sets. All data sets available are used instead for training of the ANN. Hence, $n_d = n_{tot}$ and $n_t = 0$. Since no data sets are available for testing of the ANN, the criterion $\mathcal{R}^{*-h}_{test} \rightarrow$ minimum used for termination of network training is replaced by:

$$\mathcal{E}^{*-h}_{c,r,d} \leq \bar{\mathcal{E}}^{*-h} \quad \forall \quad \begin{cases} c &= 1,2,\ldots n_m \\ r &= 1,2,\ldots n_{o,c} \\ d &= 1,2,\ldots n_d \end{cases} \tag{11}$$

with

$$\mathcal{E}^{*-h}_{c,r,d} = \left| \frac{u^*_{c,r}(\mathbf{p}_d) - u^h_{c,r}(\mathbf{p}_d)}{\bar{u}_c} \right| , \tag{12}$$

where $\bar{\mathcal{E}}^{*-h}$ denotes a prespecified threshold value. If condition (11) is satisfied, the ANN will be able to reproduce, with a prespecified accuracy, each output value of each data set available.

In some cases, however, the training procedure fails to converge because of a small number of values $u^*_r(\mathbf{p}_d)$ violating condition (11). The reasons for this are:

(1) The GDM used for training of the ANN approaches a local minimum of $\mathcal{R}^{*-h}$ in the n_w-dimensional network-weight space.
(2) Values $u^*_{c,r}(\mathbf{p}_d)$ that already satisfy condition (11) result in contributions to the gradient of $\mathcal{R}^{*-h}$ contradictory to the contributions of values $u^*_{c,r}(\mathbf{p}_d)$ violating condition (11). Hence, the computed network-weight corrections do not yield a reduction of $\mathcal{E}^{*-h}_{c,r,d}$ for values $u^*_{c,r}(\mathbf{p}_d)$ violating condition (11).

In order to improve the training procedure, two modifications are proposed:

(1) The probability that training of the ANN by means of the GDM approaches a local minimum of $\mathcal{R}^{*-h}$ is reduced by employing a GA for determination of the initial choice of the network weights.
(2) A so-called "reduced-training" algorithm is used. It is characterised by reducing the contributions to the gradient of $\mathcal{R}^{*-h}$ which correspond to values $u^*_{c,r}(\mathbf{p}_d)$ satisfying condition (11).

3.4 Determination of initial network weights employing a real coded GA

A GA imitates the evolution of a population of biological individuals [5] following DARWIN's principle of the *survival of the fittest* [1]. At the beginning of the GA, a generation of 50 individuals [2], i. e., of 50 different choices of initial

[2] This is a very small number for a genetic algorithm.

weights, is generated randomly [3]. Hence, 50 different networks are obtained. The quality, i.e., the fitness of each individual of a certain generation is described by the respective value of the *fitness function* $\mathcal{F}$, which is obtained as follows: For each ANN, 10000 iteration steps are performed in the framework of the GDM. Based on the so-obtained networks, a feedforward procedure of all available input vectors $\mathbf{p}_d$ is carried out. This allows to determine the value of the error function $\mathcal{R}^{*-h}$ corresponding to each individual. The error function $\mathcal{R}^{*-h}$ is related to the fitness function $\mathcal{F}$ of the GA by $\mathcal{F} = 1/\mathcal{R}^{*-h}$. The "fittest" individual is characterised by the largest fitness function $\mathcal{F}$ and, hence, the smallest value of $\mathcal{R}^{*-h}$. Based on the obtained values of the fitness function, a new generation of network weights is computed by means of genetic operations:

Reproduction: Individuals are selected probabilistically and put in a so-called mating pool. Hereby, the probability to be selected is higher for individuals characterised by an above-average fitness. Hence, such individuals may be represented twice or even more often in the mating pool. Individuals characterised by a below-average fitness, on the other hand, might not be included in the mating pool. Hence, they will die out. Reproduction is terminated when the number of individuals in the mating pool is equal to the number of individuals in a generation, i. e., when it is equal to 50.

Crossover: Individuals in the mating pool are arranged randomly in couples, so-called parents. Each couple is designated to yield two offsprings by exchanging some of the parental parameters. In case of a two-point crossover, two crossing sites are determined randomly for each couple. These crossing sites subdivide the parental sets of parameters and represent borderlines. In between these borderlines the parameters of the two parents are exchanged, yielding two new individuals, i. e., two *offsprings* (see Table 1).

Mutation: The mutation operator is used in artificial genetic systems to protect the generations from losing some potentially useful genetic material [5]. With a small probability, network weights of offsprings get changed. E.g., one out of thousand weights is modified. The size of this modification is determined randomly.

After reproduction, crossover, and mutation, the individuals of the new generation are obtained. Finally, a randomly selected individual is replaced by the fittest individual of the old generation. Hence, the largest fitness value obtained in the new generation is at least equal to the largest fitness value of the old generation. In general, new generations are characterised by an increased

[3] Classical GAs employ individuals represented as binary strings. According, e.g., to [6], the basic sequence of a GA is the same for binary-coded individuals and for individuals represented as a set of real numbers. Hence, the use of binary-coded individuals is not necessary in the context of the proposed PI method.

Table 1. Two-point crossover for a couple of sets of network weights consisting of five weights w_j , $j = 1, 2, \ldots 5$

parent Ⓐ:	$w_1^A \mid w_2^A \; w_3^A \mid w_4^A \; w_5^A$
parent Ⓑ:	$w_1^B \mid w_2^B \; w_3^B \mid w_4^B \; w_5^B$
	⇑ ⇑
randomly determined crossing sites	
offspring ①:	$w_1^A \mid w_2^B \; w_3^B \mid w_4^A \; w_5^A$
offspring ②:	$w_1^B \mid w_2^A \; w_3^A \mid w_4^B \; w_5^B$

average fitness of the population. The fittest individual after a prespecified number of generations is the output of the GA. The network weights corresponding to this individual are the starting values for the "reduced-training" algorithm.

3.5 The "reduced-training" algorithm

Based on the network weights provided by the GA, the GDM is employed for determination of network weights that finally satisfy condition (11). In contrast to classical training algorithms, the "reduced-training" algorithm is characterised by the multiplicative reduction of contributions to the gradient of $\mathcal{R}^{*-h}$ corresponding to output values that already satisfy condition (11). For this purpose, an exponential scaling function $\rho(\mathcal{E}^{*-h}_{c,r,d})$ is introduced:

$$\rho(\mathcal{E}^{*-h}_{c,r,d}) = \begin{cases} 1 & \text{for } \mathcal{E}^{*-h}_{c,r,d} > \bar{\mathcal{E}}^{*-h} , \\ \exp\left[\ln(\alpha)\left(\frac{10}{9}\right)^2 \left(\frac{\mathcal{E}^{*-h}_{c,r,d} - \bar{\mathcal{E}}^{*-h}}{\bar{\mathcal{E}}^{*-h}}\right)^2\right] & \text{for } \mathcal{E}^{*-h}_{c,r,d} \leq \bar{\mathcal{E}}^{*-h} , \end{cases} \tag{13}$$

where $\alpha < 1$ is a constant scalar parameter. Figure 4 shows the scaling function $\rho(\mathcal{E}^{*-h}_{c,r,d})$ for two different values of α. A contribution to the gradient

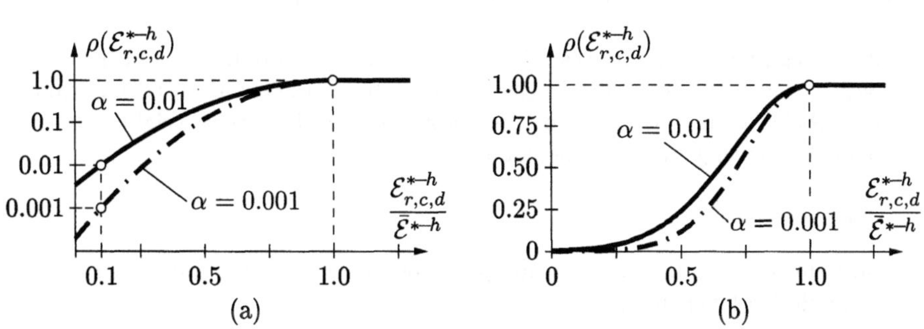

Fig. 4. Scaling function $\rho(\mathcal{E}^{*-h}_{c,r,d})$ used during "reduced training" of the ANN for two different values of α: (a) logarithmic scale, and (b) natural scale

of $\mathcal{R}^{*-h}$ corresponding to an output value $u^*_{c,r}(\mathbf{p}_d)$ which is characterised by a mapping accuracy of, e.g., $\mathcal{E}^{*-h}_{c,r,d} = \bar{\mathcal{E}}^{*-h}/10$ is reduced by the factor $\rho(\mathcal{E}^{*-h}_{c,r,d}) = \alpha$. A contribution to the gradient of $\mathcal{R}^{*-h}$ corresponding to an output value $u^*_{c,r}(\mathbf{p}_d)$ violating condition (11) is not affected by $\rho(\mathcal{E}^{*-h}_{c,r,d})$.

If, however, condition (11) is violated although a prespecified number of iteration steps within the framework of the GDM was performed, the training of the ANN will be terminated. The number of network weights, which is related to the number of neurons in the hidden layer (see (7)), must be increased in order to improve the approximation capability of the ANN. Accordingly, the number of neurons in the hidden layer is increased by one [4]. Training of the new ANN is started, determining the initial network weights by means of the GA.

3.6 Estimate and assessment of an optimal parameter set

The trained ANN provides an approximation of the n_d FE analyses that have been performed. Accordingly, a parameter set belonging to the precomputed n_d data sets is mapped by the ANN onto corresponding FE results with an accuracy greater than or equal to $\bar{\mathcal{E}}^{*-h}$. Parameter sets which do not belong to the n_d data sets are mapped onto interpolated values of FE results.

A parameter set yielding the smallest difference between corresponding FE results and the respective *in situ* measurements is an optimal parameter set. The ANN, providing only an approximation of the FE analysis, can be used to compute an estimate of an optimal parameter set. This estimate is obtained by means of minimisation of the error of the output values of the ANN (see (3):

$$\mathcal{R}^*(\mathbf{p}) = \sqrt{\frac{1}{n_m}\sum_{c=1}^{n_m}\frac{1}{n_{o,c}}\sum_{r=1}^{n_{o,e}}\left(\frac{u^*_{c,r}(\mathbf{p}) - u_{c,r}}{\bar{u}_c}\right)^2} \quad \rightarrow \quad \text{minimum.} \tag{16}$$

The minimisation problem (16) is solved in two steps:

[4] At the beginning of the process of PI, the initial number of neurons in the hidden layer, n_h, must be specified. In case all available data sets are used for training of the ANN, an upper bound for the initial number of network weights is recommended, following from the inequality

$$n_w \leq n_o \cdot n_d \,. \tag{14}$$

Substitution of (7) in (14) gives an upper bound for the initial number of neurons in the hidden layer n_h as

$$n_h \leq \frac{n_o \cdot (n_d - 1)}{1 + n_i + n_o} \,. \tag{15}$$

(1) A GA is employed to search for a near optimal solution in the entire parameter space. 100 000 different parameter combinations per generation are presented to the network. For each one of them, $\mathcal{R}^*$ is evaluated. The parameter combination yielding the largest value of the fitness function $\mathcal{F}$, with $\mathcal{F} = 1/\mathcal{R}^*$ and, hence, the smallest value of the error $\mathcal{R}^*$ within a prespecified number of generations represents the output of the GA.
(2) Starting from the near optimal solution provided by the GA, a GDM is employed to compute the parameters associated with the absolute minimum of $\mathcal{R}^*$. Iteratively, the input parameters are continuously updated. The increment of parameters related to one GDM iteration step, Δp_j, is a function of $\partial\mathcal{R}^*/\partial p_j$, $j = 1, 2, \ldots n_i$. The required partial derivatives of the error function $\mathcal{R}^*$ with respect to the unknown parameters p_j are given in [13].

In general, this two-step procedure yields the parameter set associated with the absolute minimum of $\mathcal{R}^*$.

An additional FE analysis on the basis of the estimated optimal parameter set gives the respective value of the error function $\mathcal{R}^h$. In general, because of inaccurate measurements and deficiencies of the numerical model, the attainable minimum of $\mathcal{R}^h$ is greater than zero. Traditional gradient-free PI methods are terminated if an obtained minimum of $\mathcal{R}^h$ cannot be outperformed anymore.

In contrast to such methods, the use of an ANN in the framework of the proposed PI method yields an estimate of the attainable minimum of $\mathcal{R}^h$. The output values obtained from traversing of the estimated optimal parameter set through the ANN can be interpreted as an estimate of numerical results as close as possible to the measurements. Hence, the computed minimum of $\mathcal{R}^*$ is an estimate of the attainable precision of reproducing the measurements by an FE simulation. Consequently, the difference between $\mathcal{R}^h$ and $\mathcal{R}^*$ corresponding to the estimated optimal parameter set, $|\mathcal{R}^h - \mathcal{R}^*|$, is used as an indicator of the quality of the obtained parameter set. If $|\mathcal{R}^h - \mathcal{R}^*|$ is smaller than a prespecified tolerance value, i.e., if

$$|\mathcal{R}^h - \mathcal{R}^*| \leq \text{tolerance}, \tag{17}$$

the PI is assumed to have converged. If, on the other hand, $|\mathcal{R}^h - \mathcal{R}^*| >$ tolerance, the PI will be restarted from the beginning. This time, however, the optimal parameter set just estimated and the respective FE results are added to the data sets. Hence, "$n_d = n_d + 1$". Based on the increased number of training data sets, a better approximation quality of the next ANN and, thus, a better estimate of optimal parameters can be expected.

3.7 Graphical interpretation of the proposed PI method

Figure 5 provides a graphical interpretation of the proposed PI method. In order to reduce the complexity of the illustration, a simplified problem involving a single unknown input parameter p is considered. The circles in Fig. 5a represent known values of $\mathcal{R}^h$. These values are computed from the difference between measurements and numerical results obtained from three FE analyses based on the parameters p_1, p_2, and p_3. The thick curve represents the unknown function of $\mathcal{R}^h(p)$. The PI method should give the parameter value that corresponds to the global minimum of $\mathcal{R}^h(p)$ within the prespecified parameter interval, i.e., it should provide the smallest possible difference between measurements and corresponding FE results. Hereby, the algorithm should not get trapped in a local minimum of $\mathcal{R}^h(p)$.

The first ANN is trained to reproduce the three known values of $\mathcal{R}^h$, i.e., $\mathcal{R}^h(p_1)$, $\mathcal{R}^h(p_2)$, and $\mathcal{R}^h(p_3)$. Hence, the approximation of $\mathcal{R}^h(p)$ by the ANN, $\mathcal{R}^*(p)$, is rather good at $p = p_1, p_2$, and p_3. For other values of p, $\mathcal{R}^*$ provides an approximation of $\mathcal{R}^h$ (see the thin curve in Fig. 5a). The global minimum of $\mathcal{R}^*(p)$ (the square in Fig. 5a) is determined. An estimate of an optimal parameter is obtained, with $p = p_4$. The quality of the parameter value p_4 is assessed by means of a FE simulation, giving $\mathcal{R}^h(p_4)$. As shown in Fig. 5a, the deviation of $\mathcal{R}^*$ from $\mathcal{R}^h$ at p_4 is rather large. Hence, the criterion (17) for terminating the PI is violated. Consequently, another PI iteration step is performed..

Training of the next ANN is based on the results from four FE analyses performed on the basis of p_1, p_2, p_3, and p_4. Consequently, the second approximation of $\mathcal{R}^h$ by $\mathcal{R}^*$ will give a good approximation of the four known values of $\mathcal{R}^h$, i.e., $\mathcal{R}^h(p_1)$ to $\mathcal{R}^h(p_4)$ (see the circles in Fig. 5b). In general, the quality of the approximation of $\mathcal{R}^h$ by $\mathcal{R}^*$ will be improved (see the thin curve in Fig. 5b). Again, the global minimum of $\mathcal{R}^*$ is determined. The respective estimate of an optimal parameter is assessed by means of another FE simulation. Again, the difference between the values of the errors $\mathcal{R}^h$ and $\mathcal{R}^*$ is too big. Hence, another PI iteration step is performed.

Step-by-step, the quality of subsequent approximations of $\mathcal{R}^h$ by $\mathcal{R}^*$ is increasing. Hence, the PI method will yield an optimal choice for the unknown parameter. The presented algorithm can escape local minima of $\mathcal{R}^h$. This is

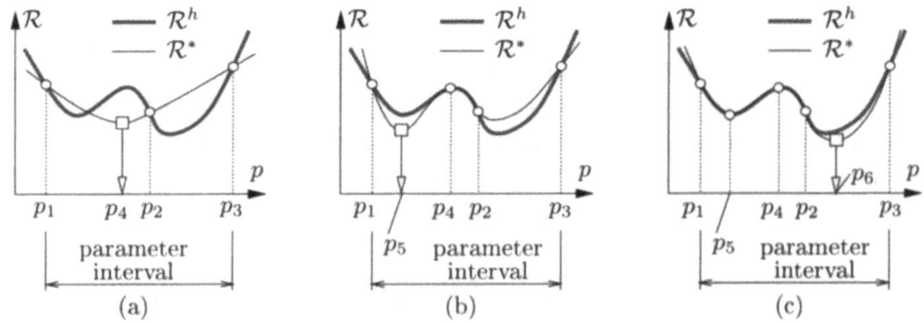

Fig. 5. 1D graphical interpretation of the PI method: error functions $\mathcal{R}^h(p)$ and $\mathcal{R}^*(p)$ for (a) the first, (b) the second, and (c) the third iteration step of the PI

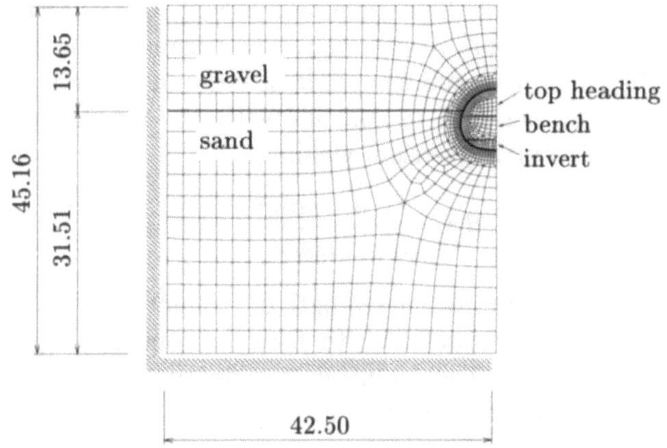

Fig. 6. MCS of underground station "Taborstraße": geometrical dimensions (in [m]) of the numerical model and FE mesh consisting of 991 finite elements

illustrated in Fig. 5b and c. The PI will be terminated if $\mathcal{R}^*$ is sufficiently close to $\mathcal{R}^h$ at the global minimum of $\mathcal{R}^h$.

4 Application to NATM tunnelling

In this section, the first steps towards genuine predictions of stress states in shotcrete tunnel shells will be described. By means of an example problem, unknown material parameters of two different soil layers are identified via back analysis. PI is based on "exact" and on inaccurate measurement data. Based on parameters identified at an already installed MCS, predictions of the structural behaviour at a future MCS become possible.

4.1 Numerical analysis of the excavation process

The example presented in this section refers to the planned excavation of the station tube "Taborstraße" of the Vienna underground line U2. The soil in this area of Vienna consists mainly of cohesionless materials such as gravel and sand. The geological conditions and the location of the station tube are illustrated in Fig. 6. The tunnel will be excavated stepwise. The three excavation steps are top heading, bench, and invert. After each excavation step, shotcrete will be applied onto the newly excavated tunnel surface. The numerical analyses are performed by means of the FEM. The employed FE discretisation is shown in Fig. 6.

The mechanical behaviour of soil is described by means of a multi-surface viscoplasticity model. It consists of the Drucker-Prager criterion and a tension-cut-off (see Fig. 7a), reading

$$f_D(\boldsymbol{\sigma}, \zeta_D) = \sqrt{J_2} + \alpha\, I_1 - \zeta_D(\chi_D)/\beta \quad \text{and} \quad f_T(\boldsymbol{\sigma}, \zeta_T) = I_1 - \zeta_T \ , \quad (18)$$

where ζ_D and ζ_T represent the hardening forces of the Drucker-Prager criterion and the tension-cut-off, respectively. The parameters α and β are computed from the cohesion c and the angle of internal friction φ by enforcing the Drucker-Prager meridian to coincide with the tensile meridian of the respective Mohr-Coloumb criterion, yielding

$$\alpha = \frac{2\sin\varphi}{\sqrt{3}(3+\sin\varphi)}, \quad \beta = \frac{f_c}{\sqrt{3}c}\frac{3+\sin\varphi}{2\cos\varphi} \quad \text{with} \quad f_c = \frac{6\,c\cos\varphi}{3-\sin\varphi} \tag{19}$$

denoting the uniaxial compressive strength of the material. An increase of the compressive strength ζ_D from $0.85 f_c$ to f_c in the context of strain hardening is considered, see Fig. 7b. The evolution laws for the viscoplastic strains and the hardening force ζ_D are given by [2],

$$\dot{\boldsymbol{\varepsilon}}^{vp} = \frac{1}{\tau}\mathbf{C}^{-1}(\boldsymbol{\sigma} - \boldsymbol{\sigma}^{\infty}) \qquad \text{and} \qquad \dot{\zeta}_D = -\frac{1}{\tau}(\zeta_D - \zeta_D^{\infty}), \tag{20}$$

where τ is the relaxation time describing the time-dependent behaviour of soil. $\boldsymbol{\sigma}^{\infty}$ and ζ_D^{∞} correspond to the solution for rate-independent elastoplasticity, i. e., to the solution for infinitely slow loading.

Shotcrete is modelled as a reactive porous medium. For a detailed description, see Chapt. 5.

4.2 "Measurement data" for the underground station "Taborstraße"

The construction of the station tube "Taborstraße" is scheduled for spring 2004. Hence, measurement data are not available. Therefore, "measurement data" are generated by a numerical analysis prior to parameter identification. The material parameters employed for the two soil layers in this analysis are given in Table 2. A very small value was assigned to the cohesion of the sand and the gravel. The material properties of shotcrete are taken from [8]. From the obtained numerical results, histories of surface settlements and displacements of three MPs of the shotcrete lining are recorded (see Figs. 8 and 9).

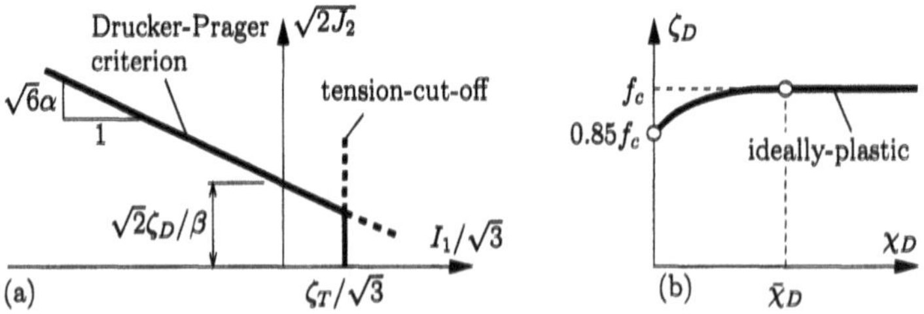

Fig. 7. Multi-surface plasticity model for soil: (a) yield surfaces in the $(I_1/\sqrt{3})$-$(\sqrt{2J_2})$-space, and (b) strain hardening within the framework of the Drucker-Prager criterion (f_c: uniaxial compressive strength)

Table 2. Material parameters of both soil layers employed for the generation of "measurement data" by means of a numerical analysis

	sand	gravel
Young's modulus E [N/mm^2]	65	80
Poisson's ratio ν [–]	0.350	0.316
angle of internal friction φ	36.5°	42.0°
relaxation time τ [h]	1.0	1.0

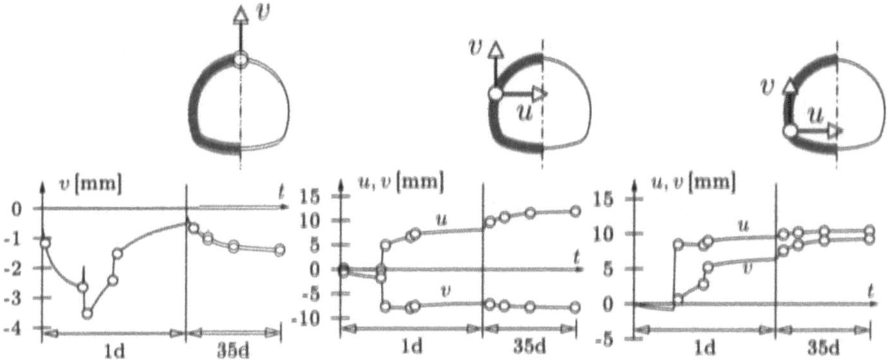

Fig. 8. "Measurement data" generated by numerical analysis: histories of horizontal and vertical displacement component, u and v, at selected points of the shotcrete lining

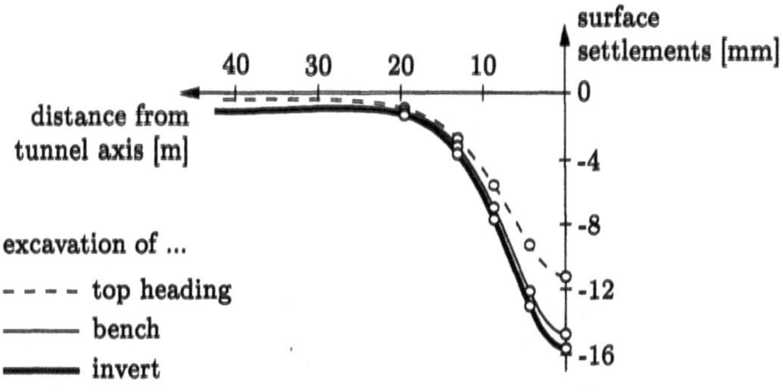

Fig. 9. "Measurement data" generated by numerical analysis: surface settlements

4.3 Parameter identification

During back analysis five material parameters listed in Table 2 are assumed to be unknown. They are collected in the vector $\mathbf{p}$, reading

$$\mathbf{p} = \lfloor E_{sand},\ \varphi_{sand},\ \tau,\ \varphi_{gravel},\ E_{gravel} \rfloor^T , \tag{21}$$

where $\tau_{sand} = \tau_{gravel} = \tau$. Accordingly, $n_i = 5$. The intervals for the unknown parameters are chosen as

$$\begin{aligned}
E_{sand} &\in [60.0\ \mathrm{N/mm^2},\ 80.0\ \mathrm{N/mm^2}]\ ,\\
\varphi_{sand} &\in [33.0°,\ 41.0°]\ ,\\
\tau &\in [0.15\ \mathrm{h},\ 1.50\ \mathrm{h}]\ ,\\
\varphi_{gravel} &\in [35.0°,\ 45.0°]\ ,\\
E_{gravel} &\in [70.0\ \mathrm{N/mm^2},\ 100.0\ \mathrm{N/mm^2}]\ .
\end{aligned} \tag{22}$$

For back analysis, the displacement histories illustrated in Figs. 8 and 9 are treated as measurement data provided by the construction site. 56 discrete values referring to different measurement curves are selected for PI (see circles in Figs. 8 and 9). They are collected in the vector $\mathbf{u}$, yielding $n_o = 56$.

Twelve FE analyses are performed in advance in order to provide twelve different data sets, i. e., $n_d = 12$. They are based on different parameter sets containing different combinations of the limit values of the defined parameter intervals (22), reading

$$\begin{aligned}
\mathbf{p}_1 &= \lfloor 60,\ 33,\ 0.15,\ 35,\ 70\rfloor^T \Rightarrow \mathcal{R}^h = 148.8\%\ ,\\
\mathbf{p}_2 &= \lfloor 80,\ 33,\ 0.15,\ 35,\ 70\rfloor^T \Rightarrow \mathcal{R}^h = 125.9\%\ ,\\
\mathbf{p}_3 &= \lfloor 60,\ 41,\ 0.15,\ 35,\ 70\rfloor^T \Rightarrow \mathcal{R}^h = 116.9\%\ ,\\
\mathbf{p}_4 &= \lfloor 60,\ 33,\ 1.50,\ 35,\ 70\rfloor^T \Rightarrow \mathcal{R}^h = 24.3\%\ ,\\
\mathbf{p}_5 &= \lfloor 60,\ 33,\ 0.15,\ 45,\ 70\rfloor^T \Rightarrow \mathcal{R}^h = 86.2\%\ ,\\
\mathbf{p}_6 &= \lfloor 60,\ 33,\ 0.15,\ 35,\ 100\rfloor^T \Rightarrow \mathcal{R}^h = 124.1\%\ ,\\
\mathbf{p}_7 &= \lfloor 80,\ 41,\ 1.50,\ 45,\ 100\rfloor^T \Rightarrow \mathcal{R}^h = 29.9\%\ ,\\
\mathbf{p}_8 &= \lfloor 60,\ 41,\ 1.50,\ 45,\ 100\rfloor^T \Rightarrow \mathcal{R}^h = 20.1\%\ ,\\
\mathbf{p}_9 &= \lfloor 80,\ 33,\ 1.50,\ 45,\ 100\rfloor^T \Rightarrow \mathcal{R}^h = 23.2\%\ ,\\
\mathbf{p}_{10} &= \lfloor 80,\ 41,\ 0.15,\ 45,\ 100\rfloor^T \Rightarrow \mathcal{R}^h = 28.0\%\ ,\\
\mathbf{p}_{11} &= \lfloor 80,\ 41,\ 1.50,\ 35,\ 100\rfloor^T \Rightarrow \mathcal{R}^h = 25.2\%\ ,\\
\mathbf{p}_{12} &= \lfloor 80,\ 41,\ 1.50,\ 45,\ 70\rfloor^T \Rightarrow \mathcal{R}^h = 23.3\%\ .
\end{aligned} \tag{23}$$

The PI is assumed to have converged if the indicator $|\mathcal{R}^h - \mathcal{R}^*|$ is smaller than or equal to 0.5 %. The required mapping accuracy for each output value, $\bar{\mathcal{E}}^{*-h}$, is set equal to 1.0 %. Training procedures still violating criterion (11) after 250 000 iteration steps in the framework of the GDM are viewed as having failed to converge. Based on these settings, six prognoses for the optimal parameter set were computed, giving $\mathbf{p}_{13}$ to $\mathbf{p}_{18}$:

$$\begin{aligned}
\mathbf{p}_{13} &= \lfloor 65.7,\ 38.6,\ 1.122,\ 42.6,\ 79.8\rfloor^T \Rightarrow \mathcal{R}^* = 2.0\%\ ,\ \mathcal{R}^h = 7.1\%\ ,\\
\mathbf{p}_{14} &= \lfloor 65.0,\ 35.0,\ 1.173,\ 40.2,\ 77.6\rfloor^T \Rightarrow \mathcal{R}^* = 1.0\%\ ,\ \mathcal{R}^h = 4.0\%\ ,\\
\mathbf{p}_{15} &= \lfloor 64.9,\ 36.5,\ 1.100,\ 42.0,\ 79.1\rfloor^T \Rightarrow \mathcal{R}^* = 1.1\%\ ,\ \mathcal{R}^h = 2.2\%\ ,\\
\mathbf{p}_{16} &= \lfloor 65.3,\ 36.3,\ 1.075,\ 42.0,\ 79.1\rfloor^T \Rightarrow \mathcal{R}^* = 0.7\%\ ,\ \mathcal{R}^h = 1.5\%\ ,\\
\mathbf{p}_{17} &= \lfloor 65.4,\ 36.3,\ 1.048,\ 41.8,\ 79.6\rfloor^T \Rightarrow \mathcal{R}^* = 0.4\%\ ,\ \mathcal{R}^h = 0.7\%\ ,\\
\mathbf{p}_{18} &= \lfloor 65.1,\ 36.4,\ 1.007,\ 42.0,\ 80.0\rfloor^T \Rightarrow \mathcal{R}^* = 0.4\%\ ,\ \mathcal{R}^h = 0.2\%\ .
\end{aligned} \tag{24}$$

Recall that the number of training data sets, n_d, increases by one after having obtained and assessed a new prognosis for an optimal parameter set. Parameter set $\mathbf{p}_{17}$, obtained from an ANN trained by means of 16 training data sets, yields $|\mathcal{R}^h - \mathcal{R}^*| = 0.3\,\% < 0.5\,\%$ (see Fig. 10a), signalling convergence of the PI by . Since the "measurement data" were generated by means of a numerical analysis, PI should finally give the parameters employed in this analysis, listed in Table 2. In fact, parameter set $\mathbf{p}_{17}$ represents a satisfactory solution of the PI problem. Hence, the indicator based on the difference between $\mathcal{R}^*$ and $\mathcal{R}^h$ (criterion (17)) provided a good estimate for the quality of the obtained parameters.

Figure 10(b) illustrates the estimated sensitivity of the parameters contained in the parameter set $\mathbf{p}_{17}$. The curves represent sections through the error function $\mathcal{R}^*$. They contain the estimated minimum of the error function, which is marked by circles. For each section four parameters were kept constant and set equal to the values of the parameter set $\mathbf{p}_{17}$, whereas the remaining parameter was varied over its entire interval (see (22)). The dash-dotted curve indicates that τ has the largest influence on $\mathcal{R}^*$. The remaining parameters of the bottom layer, E_{sand} and φ_{sand}, on the other hand, have the smallest influence on the error function $\mathcal{R}^*$.

4.4 Parameter identification based on inaccurate measurement data

In general, measurement data provided by the construction site are afflicted with inaccuracies. In order to assess the influence of these inaccuracies on the performance of the proposed PI method, the measurement data illustrated in Figs. 8 and 9 are modified:

$$u_{c,r} \rightarrow u_{c,r} \cdot (1+\delta)\,, \qquad c = 1,2,\ldots n_m, \quad r = 1,2,\ldots n_{o,c}\,, \tag{25}$$

where the value of δ is related to the inaccuracy of the measurement data. A negative value of δ, e. g., indicates that the surface settlements and displacements of the shotcrete tunnel shell provided by the construction site are too small.

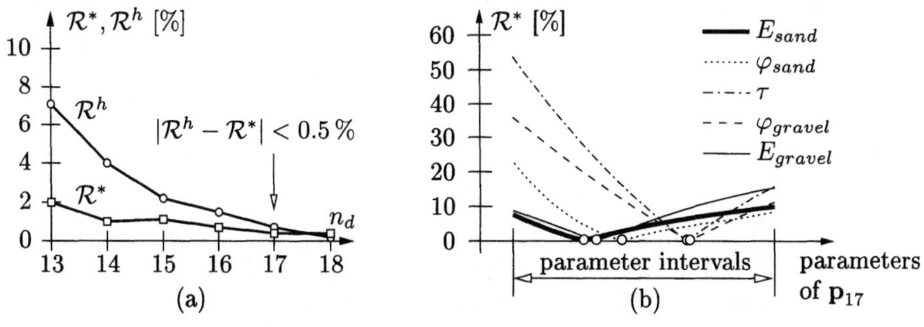

Fig. 10. Results from back analysis: (a) evolutions of $\mathcal{R}^*$ and $\mathcal{R}^h$ obtained in the course of PI and (b) ANN-estimated sensitivities of parameters (the minima of the respective functions are indicated by circles)

Six different values of δ are chosen (see Table 3) to obtain six sets of inaccurate measurement data. For each of them, PI is performed independently as described in Subsect. 4.3. These back analyses aim at determining parameter sets that yield numerical results which fit to the respective measurement data as well as possible. Consequently, $u_{c,r}$ needs to be replaced in (1) by $u_{c,r} \cdot (1+\delta)$, resulting in the following condition for an optimal parameter set:

$$\mathcal{R}^h(\mathbf{p},\delta) = \sqrt{\frac{1}{n_m}\sum_{c=1}^{n_m}\frac{1}{n_{o,c}}\sum_{r=1}^{n_{o,c}}\left(\frac{u^h_{c,r}(\mathbf{p}) - u_{c,r}\cdot(1+\delta)}{\bar{u}_c\cdot(1+\delta)}\right)^2} \rightarrow \min. \quad (26)$$

Table 3 contains the percentage difference between the optimal parameters obtained from these six PIs and the parameters listed in Table 2. In general, back analysis on the basis of inaccurate settlements and displacements of the tunnel shell, which are too small (triggered by $\delta < 0$), result in an increase of the values for the optimal parameter set. A positive value of δ, on the other hand, yields a decrease of the parameters. Notably, the percentage error of Young's modulus of both soil layers and of the relaxation time τ are closely related to the value of δ. The influence of δ on the angles of internal friction is significantly smaller.

4.5 Predictions of displacements and of the level of loading of the shotcrete tunnel shell at a future MCS

Based on the parameters obtained from six PIs of the previous subsection, PI 1 to PI 6 (see Table 3), the structural behaviour of the tunnel at a future MCS, i.e., of one that has not yet been installed, will be investigated in the following. The (changed) geological conditions and the (changed) location of the station tube at the future MCS are illustrated in Fig. 11. Characteristic results obtained from the FE analyses performed on the basis of the six parameter sets computed in Subsect. 4.4 are listed in Table 4. These results are predictions for the structural behaviour at the future MCS.

In order to assess these predictions, an additional FE analysis on the basis of the material parameters given in Table 2 is performed. The respective characteristic results are listed in Table 5.

Table 3. Influence of the value of δ on the parameter values obtained from PI

	δ	E_{sand}	φ_{sand}	τ	φ_{gravel}	E_{gravel}
PI 1	−10.0 %	+10.3 %	+0.6 %	+9.2%	+0.4 %	+11.3%
PI 2	−5.0 %	+4.7 %	+0.3 %	+5.4%	+0.1 %	+6.0%
PI 3	−2.5 %	+2.2 %	−0.2 %	+2.8%	+0.2 %	+2.2%
PI 4	+2.5 %	−2.1 %	−0.1 %	−2.2%	±0.0 %	−2.4%
PI 5	+5.0 %	−4.7 %	−0.4 %	−2.0%	−0.4 %	−5.0%
PI 6	+10.0 %	−8.6 %	−0.9 %	−4.8%	−0.7 %	−9.3%

Table 4. Prognoses of characteristic values describing the structural behaviour at a future MCS (for the definition of $\bar{\mathcal{L}}_2$ see Chap. 10)

material parameters obtained from ...	maximum values for ... surface settlements	vertical displacement at top of the shotcrete shell	level of loading $\bar{\mathcal{L}}_2$
PI 1 (δ=−10 %)	1.72 cm	1.67 cm	32.2%
PI 2 (δ=−5 %)	1.81 cm	1.68 cm	32.8%
PI 3 (δ=−2.5 %)	1.86 cm	1.70 cm	33.5%
PI 4 (δ=+2.5 %)	1.95 cm	1.72 cm	34.1%
PI 5 (δ=+5 %)	2.01 cm	1.73 cm	34.5%
PI 6 (δ=+10 %)	2.10 cm	1.75 cm	35.5%

Table 5. Characteristic values describing the structural behaviour at a future MCS obtained from an FE analysis based on the material parameters listed in Table 2 (for the definition of $\bar{\mathcal{L}}_2$ see Chap. 10)

maximum values for ... surface settlements	vertical displacement at top of the shotcrete shell	level of loading $\bar{\mathcal{L}}_2$
1.91 cm	1.70 cm	33.7%

A comparison of the values listed in Tables 4 and 5 allows an assessment of the prognoses of the structural behaviour at a future MCS. Deviations of surface settlements from the respective value given in Table 5 are closely related to the value of δ, whereas the influence of δ on the maximum displacement at the top of the tunnel shell is significantly smaller. The maximum level of

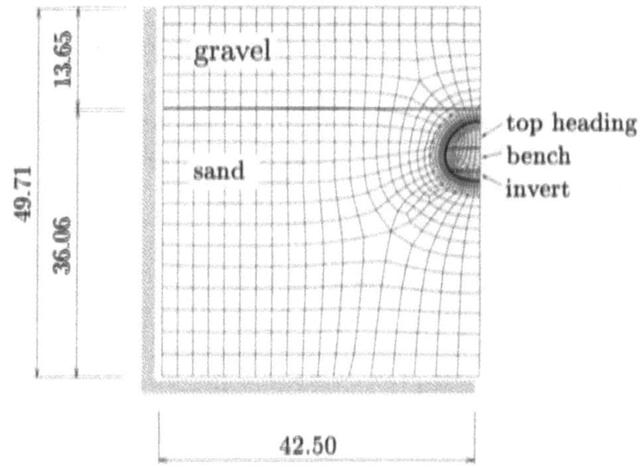

Fig. 11. Future MCS of underground station "Taborstraße": geometrical dimensions (in [m]) of the numerical model and FE mesh consisting of 1071 finite elements

loading of the shotcrete tunnel shell changes only slightly with δ. To summarise, the characteristic values of the structural behaviour at a future MCS were well predicted, taking into account the considered range of measurement inaccuracies of $\pm 10\,\%$. This result indicates that the presented mode of PI can be expected to be useful for real-life applications even in case of inaccurate measurement data, such as provided by the construction site.

References

[1] Darwin, C. (1995) Origin of species. Modern Library, Random House.

[2] Duvaut, G., Lions, J.L. (1971) Les inequations en mechanique et en physique [The inequations in mechanics and physics]. Dunod, Paris. In French.

[3] Feng, X.-T., Zhang, Z., and Sheng, Q. (2000) Estimating mechanical rock mass parameters relating to the Three Gorges Project permanent shiplock using an intelligent displacement back analysis method. *International Journal of Rock Mechanic and Mining Sciences*, **37**: 1039-1054.

[4] Gioda, G., Sakurai, S. (1987) Back analysis procedure for the interpretation of field measurements in geomechanics. *International Journal for Numerical and Analytical Methods in Geomechanics*, **11**: 555–583.

[5] Goldberg, D. E. (1989) Genetic algorithms in search, optimisation, and machine learning. Addison Wesley, Reading, Massachusetts.

[6] Gupta, J. N. D., Sexton, R. S. (1999) Comparing backpropagation with a genetic algorithm for neural network training. *Omega, The International Journal of Management Science*, **27**: 679-684.

[7] Huber, N. (2000) Anwendung Neuronaler Netze bei nichtlinearen Problemen der Mechanik [Application of neural networks to nonlinear problems of mechanics]. Habilitation-thesis, Forschungszentrum Karlsruhe GmbH, Karlsruhe, Germany. In German.

[8] Lackner, R., Hellmich, Ch., and Mang, H.A. (2002) Constitutive modelling of cementitious materials in the framework of chemoplasticity. *International Journal for Numerical Methods in Engineering*, **53**: 2357–2388.

[9] Mahnken, R., Stein, E. (1995) Parameter identification for viscoplastic models based on analytical derivatives of a least-squares functional and stability investigations. *International Journal of Plasticity*, **12**: 451-479.

[10] Ohkami, T., Ichikawa, Y. (1997) A parameter identification procedure for viscoelastic materials. *Computers and Geotechnics*, **21**: 255-275.

[11] Ohkami, T., Swoboda, G. (1999) Parameter identification of viscoelastic materials. *Computers and Geotechnics*, **24**: 279-295.

[12] Pichler, B., Lackner, R., and Mang, H.A. (2001) Soft-computing based structural back analysis of material model parameters for soil in the context of tunnelling according to the New Austrian Tunnelling Method. *In Valliappan, S. (editor) Proceedings of the 1st Asian Pacific Congress on Computational Mechanics*, **2**: 1101-1108.

[13] Pichler, B., Lackner, R., and Mang, H.A. (2002) Back analysis of model parameters in geotechnical engineering by means of soft computing. *International Journal for Numerical Methods in Engineering*, Accepted for publication.

[14] Pichler, B., Mang, H.A. (2001) Parameter identification for sophisticated material models by means of iterative back analyses based on soft computing. *In Waszczyszyn, Z., et al. (editors) CD-ROM Proceedings of the 2nd European Conference on Computational Mechanics*

[15] Rojas, R. (1996) Theorie der neuronalen Netze: eine systematische Einführung [Theory of neural networks: A systematic introduction]. Springer, Berlin. In German

[16] Sakurai, S., and Takeuchi, K. (1983) Back analysis of measured displacements of tunnels. *Rock Mechanics and Rock Engineering*, **16**: 173-180.

[17] Swoboda, G., Ichikawa, Y., Dong, Q., and Zaki, M. (1999) Back analysis of large geotechnical models. *International Journal for Numerical and Analytical Methods in Geomechanics*, **23**: 1455-1472.

Chapter 10
Quantification of stress states in shotcrete shells

Jürgen Macht, Roman Lackner, Christian Hellmich, and Herbert A. Mang
Institute for Strength of Materials, Vienna University of Technology

Abstract. As reported in Chap. 3, monitoring of deformations which occur during the excavation of tunnels is essential for the success of the NATM. Nowadays it is performed at every NATM construction site. The combination of these monitored displacements with the material model for shotcrete outlined in Chap. 5 has led to the development of a hybrid method for the analysis of closed shotcrete tunnel shells by means of nonlinear Finite Element (FE) analyses. It allows determination of the stress state in the shotcrete shell. Knowing this stress state a level of loading can be computed, amounting to 0% for the unloaded shell and to 100% for shotcrete (locally) loaded up to its compressive strength. The analysis of segmented tunnel shells required further development of this hybrid method. The desire for real-time monitoring of the level of loading led to the development of a shell-theory-based hybrid method avoiding time-consuming FE analyses.

1 Introduction

Quantification of stress states in tunnel shells made of shotcrete is of great importance for the assessment of the safety of tunnels during the excavation process. Different methods dealing with this aspect were reported in the literature. All of them are based on input data provided by *in situ* measurements. Recently, *in situ* measurements of displacements were employed to estimate the loading of the shell [15,13]. In 1995, Zachow [17] proposed to compare the displacements from *in situ* measurements with numerically-obtained displacements. In the numerical model, the tunnel shell was discretized by means of finite beam elements. For the simulations reported in [17], the material law of Rokahr and Lux [12] was used. The approach proposed by Zachow [17] was improved with respect to both material and structural modeling of shotcrete tunnel shells at Vienna University of Technology [1,2,6,8]. For an outline of the employed material model for shotcrete, the reader is referred to Chap. 5. This chapter focuses on improvements of the structural model. In contrast to Zachow [17], all displacement measurement data provided by the construction site enter the analysis. In general, 3D displacement vectors of optical targets, so-called measurement points (MPs) (see Fig. 1b), are recorded. MPs are arranged at measurement cross-sections (MCSs) (see Fig. 1a). The distance between adjacent MCSs ranges from 5 to 20 m.

Over the last years, different types of hybrid methods have been developed, starting from nonlinear FE analysis of closed tunnel shells. Further

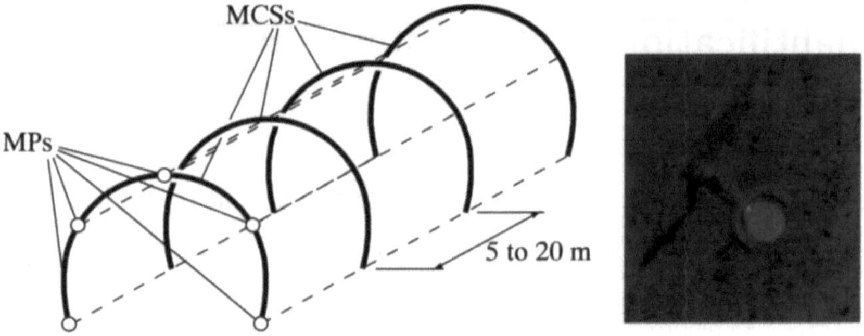

(a) (b)

Fig. 1. (a) Illustration of the location of MPs installed at MCSs and (b) optical target (MP) installed at the Lainzer Tunnel, Austria

development resulted in an increase of applicability (e.g., squeezing rock conditions). Finally, an increase of robustness and a reduction of the computational effort was achieved by replacing the time-consuming FE-analysis with a shell-theory-based model for the shotcrete shell. The different types of hybrid methods dealt with in this paper are based on continuous displacement fields in time and space. These fields are obtained from interpolation in time and space which is outlined in the following. The hybrid methods are applied to three Austrian tunnels in order to determine the level of loading of the shotcrete shell. First, the Sieberg tunnel is investigated by means of the hybrid method using 3D FE analyses developed for closed linings. Thereafter the hybrid method for segmented linings in squeezing rock is applied to the Semmering pilot tunnel. Finally, the Lainzer tunnel is analyzed with the shell-theory-based hybrid method.

2 Interpolation of displacements in time and space

Interpolation of displacements between MPs, based on total displacements measured at MPs, was first proposed in [1,2]. However, if a reflector breaks during the construction process and a new reflector is set, the interpolation of total displacements is no longer possible. To account for changes in the number of MPs during the construction process, the mode of interpolation described in [1,2] for total displacements was applied to displacement increments [9] as outlined in the following.

2.1 Interpolation in time

In general, displacement measurements are performed once a day. For the numerical simulation, however, a continuous displacement history is required.

This history is determined by means of linear interpolation of displacement increments in time intervals $[t_n; t_{n+1}]$ bounded by successive time instants t_n and t_{n+1} at which the measurements are performed:

$$\Delta \mathbf{u}_{MPk}(t) = \frac{\mathbf{u}_{MPk}(t_{n+1}) - \mathbf{u}_{MPk}(t_n)}{t_{n+1} - t_n}(t - t_n) \text{ for } t \in [t_n; t_{n+1}] \, , \tag{1}$$

where $\mathbf{u}^T_{MPk} = \lfloor u_{x,MPk}, u_{y,MPk}, u_{z,MPk} \rfloor$ denotes the three-dimensional displacement vector of the k-th MP. The total displacement of the k-th MP is given by

$$\mathbf{u}_{MPk}(t) = \mathbf{u}_{MPk}(t_n) + \Delta \mathbf{u}_{MPk}(t) \, . \tag{2}$$

2.2 Interpolation in space

As mentioned previously, measurements are only available at MPs. However, they form a relatively dense grid over the inner surface of the tunnel. Suitable interpolation functions are needed for the prescription of a continuous displacement field $\bar{\mathbf{u}}$ between these points. Therefore, a basic simplification in form of a so-called structural hypothesis is introduced to simplify the numerical model of the tunnel shell [1,2]:

H1: A *smooth displacement* field is assumed between the MPs. As long as the tunnel shell does not collapse this is a reasonable assumption. (In case of a collapse, localization phenomena would occur.)

Based on structural hypothesis H1, quadratic functions are used for the interpolation of the incremental displacements $\Delta u_{x,MPk}$, $\Delta u_{y,MPk}$, and $\Delta u_{z,MPk}$ in the circumferential direction of the MCS. Fig. 2 shows the interpolation of the incremental displacement in x-direction. The incremental displacement vector obtained from the interpolation is referred to as $\Delta \bar{\mathbf{u}} = \lfloor \Delta \bar{u}_x, \Delta \bar{u}_y, \Delta \bar{u}_z \rfloor^T$. It is given by

$$\Delta \bar{\mathbf{u}}^A(\varphi, t) = \left\{ \begin{matrix} \Delta \bar{u}^A_x \\ \Delta \bar{u}^A_y \\ \Delta \bar{u}^A_z \end{matrix} \right\} = \mathbf{a}^A(t) + \mathbf{b}^A(t)\varphi + \mathbf{c}^A(t)\varphi^2 \quad \text{with } A=I,II,\ldots \, , \tag{3}$$

with A denoting the index of the quadratic function (see Fig. 2). This index identifies the part of the shotcrete shell to which the respective interpolation refers. The coefficients $\mathbf{a}^A(t)$, $\mathbf{b}^A(t)$, and $\mathbf{c}^A(t)$ are either determined by means of the incremental displacements of three MPs (see, e.g., $\Delta \bar{u}^I_x(\varphi)$ in Fig. 2) or by means of the incremental displacements of two MPs and the continuity of $\Delta \bar{\mathbf{u}}' = \partial \Delta \bar{\mathbf{u}} / \partial \varphi$ between two adjacent quadratic interpolation functions (see, e.g., $\Delta \bar{u}^{II}_x(\varphi)$ and $\Delta \bar{u}^{III}_x(\varphi)$ in Fig. 2).

For the investigation of a MCS, in the following referred to as the *considered* MCS, the displacements of the adjacent, i.e., the *preceding* and the *succeeding* MCS, are required for consideration of the longitudinal deformations.

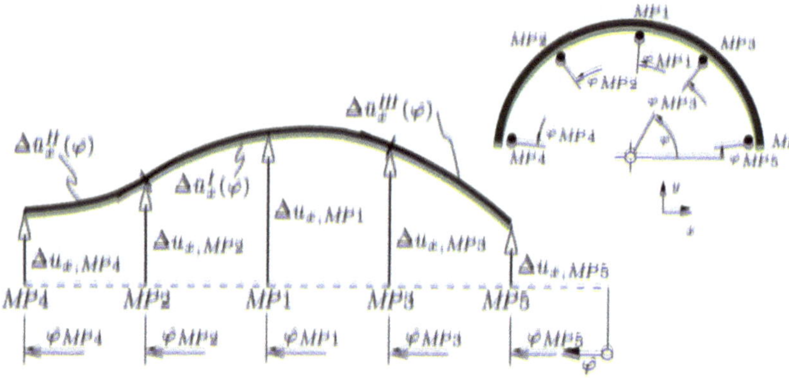

Fig. 2. Illustration of the use of quadratic functions for interpolation of the prescribed displacement $\Delta\bar{u}_x$ based on measured displacement increments $\Delta u_{x,MPk}$ at five MPs

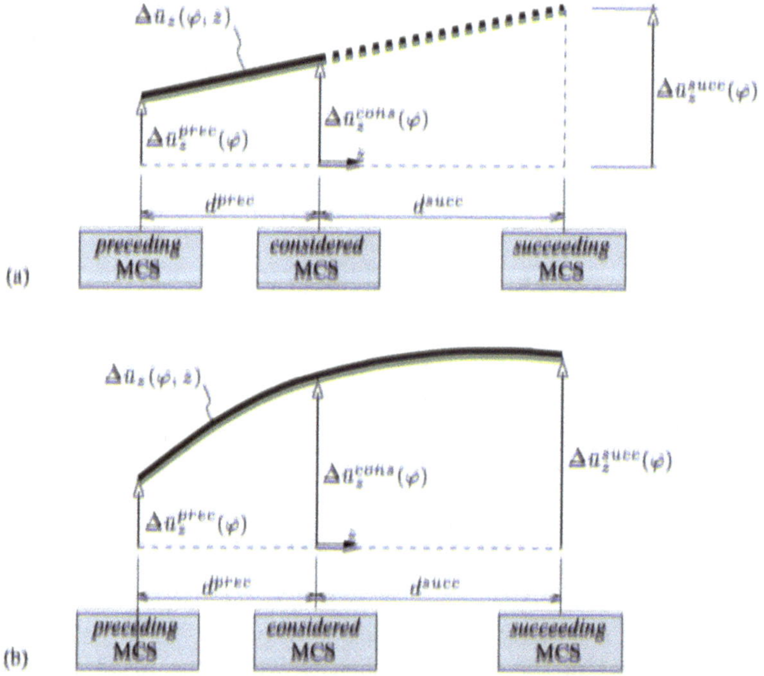

Fig. 3. On the evaluation of the incremental, longitudinal displacement $\Delta\bar{u}_z$ based on measured displacement increments $\Delta u_{z,MPk}$ at the *preceding*, the *considered*, and the *succeeding* MCS: (a) before installation of the *succeeding* MCS and (b) after installation of the *succeeding* MCS

After the (stress-free and strain-free) installation of the *considered* MCS, the longitudinal displacement between a point of the *preceding* MCS and the corresponding point of the *considered* MCS is approximated by a linear function (see Fig. 3a):

$$\Delta\bar{u}_z(\varphi, z, t) = a(\varphi, t) + b(\varphi, t)z \ . \tag{4}$$

The coordinate z is measured from the *considered* MCS in the direction of the tunnel advance. The coefficients $a(\varphi, t)$ and $b(\varphi, t)$ are determined from

$$\begin{aligned} z = -d^{prec} &: \Delta\bar{u}_z(\varphi, t) = \Delta\bar{u}_z^{prec}(\varphi, t) \ , \\ z = 0 \quad &: \Delta\bar{u}_z(\varphi, t) = \Delta\bar{u}_z^{cons}(\varphi, t) \ , \end{aligned} \tag{5}$$

where d^{prec} is the distance between the *considered* and the *preceding* MCS. In Eqn. (5), $\Delta\bar{u}_z^{cons}$ stands for the incremental longitudinal displacement obtained from interpolation in the circumferential direction according to Eqn. (3) at the *considered* MCS, and $\Delta\bar{u}_z^{prec}$ denotes the corresponding incremental displacement at the previously-installed MCS.

As soon as the *succeeding* MCS is established, displacements measured of all three MCSs, i.e., the *preceding*, the *considered*, and the *succeeding* MCS, are employed for the interpolation. A quadratic function is used for the interpolation of $\Delta\bar{u}_z$ between the three MCSs (see Fig. 3b):

$$\Delta\bar{u}_z(\varphi, z, t) = a(\varphi, t) + b(\varphi, t)z + c(\varphi, t)z^2 \ . \tag{6}$$

The coefficients $a(\varphi, t)$, $b(\varphi, t)$, and $c(\varphi, t)$ are determined from

$$\begin{aligned} z = -d^{prec} &: \Delta\bar{u}_z(\varphi, t) = \Delta\bar{u}_z^{prec}(\varphi, t) \ , \\ z = 0 \quad &: \Delta\bar{u}_z(\varphi, t) = \Delta\bar{u}_z^{cons}(\varphi, t) \ , \\ z = d^{succ} &: \Delta\bar{u}_z(\varphi, t) = \Delta\bar{u}_z^{succ}(\varphi, t) \ . \end{aligned} \tag{7}$$

In (7), $\Delta\bar{u}_z^{succ}$ denotes the incremental, longitudinal displacement obtained from interpolation in the circumferential direction of $\Delta\bar{u}_z$ at the *succeeding* MCS. d^{succ} is the distance between the *considered* and the *succeeding* MCS.

3 Determination of strain states by means of 3D FE analyses

A tunnel constructed according to the principles of the NATM is a complex structure. In order to define a proper finite element (FE) structural model, some additional structural hypotheses must be introduced [1,2]:

H2: The *change of the curvature in the longitudinal direction*, $\kappa_z = 12/h^3 \int_h \varepsilon_z(r - r_0)dr$, is negligible. In the expression for κ_z, h denotes the shell thickness and r_0 describes the radius of the middle surface of the shell. Depending on the magnitude of κ_z, $\kappa_z \neq 0$ may result in significant deviations of the tunnel from the prescribed track.

H3: During the deformation, the *thickness* of the shell is assumed to be approximately *constant*. This hypothesis is consistent with the Kirchhoff-Love shell theory. Because of the small thickness of the shell, this assumption is reasonable.

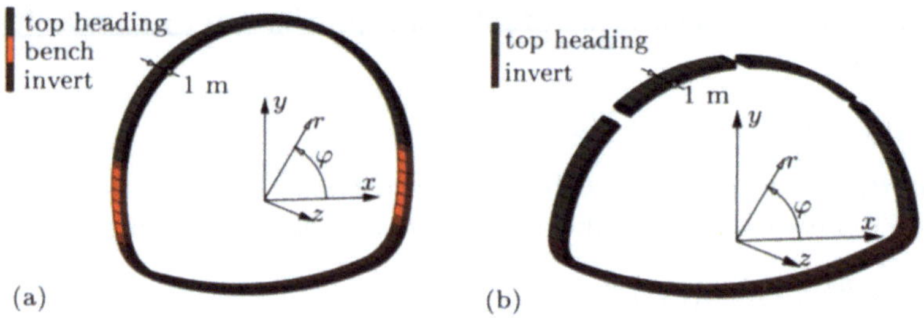

Fig. 4. 3D FE-meshes employed for hybrid analysis of (a) the Sieberg tunnel and (b) the Semmering pilot tunnel

H4: *Perfect bond is assumed between old and new parts* of the tunnel shell. In other words, as soon as a new part (e.g., a bench) is installed in a stress-free and strain-free state, bending moments can be resisted at the joint between the new part and the previously–installed parts (e.g., the top heading), which have already been subjected to stresses and strains. This assumption can be met by enforcing *continuous displacements* across joints.

H5: There is *no stress transfer* between the footings of the top heading or the benches and the adjacent rock during respective construction states. This assumption is made because of lack of respective design provisions within the framework of the NATM.

Because of these hypotheses it is adequate to consider a ring with a width of 1 m (0.5 m before and 0.5 m after the *considered* MCS) of the structure. This ring is discretized by means of 3D finite elements, allowing determination of the stress state in the *considered* MCS, see Fig. 4. The interpolated displacements, which are computed from measurements performed at the inner surface of the tunnel shell, are prescribed as boundary values on the outer surface of the shell, giving $\Delta\bar{u}_x(\varphi,t)$ and $\Delta\bar{u}_y(\varphi,t)$.

According to hypothesis H2, the increment of the longitudinal strain $\Delta\varepsilon_z$ is constant over the shell thickness. It is obtained from the Eqns. (4) and (6), respectively:

$$\Delta\varepsilon_z(\varphi,t) = \frac{d\Delta\bar{u}_z(\varphi,z,t)}{dz}\Big|_{z=0} \; . \tag{8}$$

The longitudinal displacement prescribed on the two surfaces perpendicular to the z-axis is computed from the longitudinal strain $\Delta\varepsilon_z(t)$ (Eqn. (8))

$$\Delta\bar{u}_z(\varphi, z=-0.5, t) = -\frac{1}{2}\Delta\varepsilon_z(\varphi,t) \tag{9}$$

for the surface with $z = -0.5$ m and

$$\Delta\bar{u}_z(\varphi, z=+0.5, t) = +\frac{1}{2}\Delta\varepsilon_z(\varphi,t) \tag{10}$$

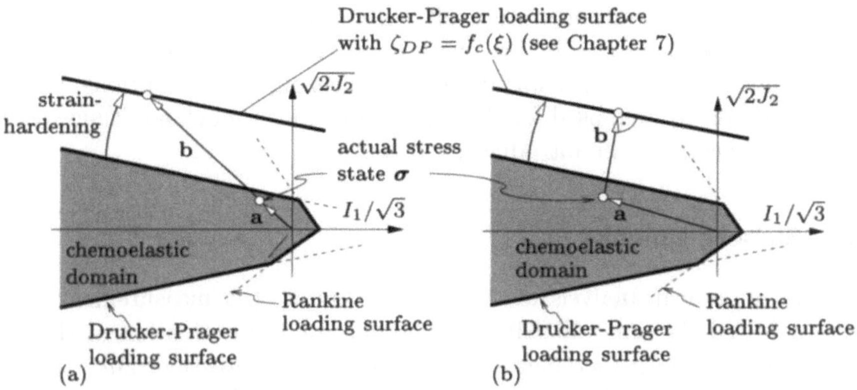

Fig. 5. Definition of the level of loading (a) $\mathcal{L}_1$ [1] and (b) $\mathcal{L}_2$ [7]

for the surface with $z = +0.5$ m.

Based on the material model for shotcrete outlined in Chap. 5, the FE analysis provides the stress state in the shotcrete shell. From the stress state, a level of loading or degree of utilization can be computed. It is defined by (see Fig. 5)

$$\mathcal{L} = 100 \frac{|\mathbf{a}|}{|\mathbf{a}| + |\mathbf{b}|} [\%] \; , \tag{11}$$

where the actual stress state is represented by the position of a vector **a** in the stress space. The vector **b** represents the distance from the actual stress state to the Drucker-Prager loading surface after completed hardening, given by $f_{DP}[\boldsymbol{\sigma}, \zeta_{DP} = f_c(\xi)] = 0$ (see Chap. 5). Accordingly, $\mathcal{L}$ amounts to 0% for stress-free states and to 100% when the (compressive) strength of shotcrete is reached. In the latter case, the stress state $\boldsymbol{\sigma}$ is located on $f_{DP}[\boldsymbol{\sigma}, \zeta_{DP} = f_c(\xi)] = 0$ and, hence, $\mathbf{b}=\mathbf{0}$. Depending on the determination of **b**, two different types of the level of loading will be used in the following:

- The first type of the level of loading, $\mathcal{L}_1$, is obtained by setting the direction of the vector **b** equal to the direction **a** (see Fig. 5a). Hence, the actual stress state is extrapolated towards the Drucker-Prager loading surface.
- As regards the second type of the level of loading, $\mathcal{L}_2$, the vector **b** represents the shortest distance between the actual stress state and the Drucker-Prager loading surface (see Fig. 5b). Accordingly, $\mathcal{L}_2 \geq \mathcal{L}_1$ and, hence, gives an upper bound for $\mathcal{L}_1$.

4 Applications

In this section, numerical results obtained from hybrid analyses on the basis of 3D FE meshes will be presented. Two Austrian tunnels, characterized by different ground conditions, are dealt with. The first tunnel is the Sieberg tunnel. It is characterized by low overburden. The second tunnel is the Semmering pilot tunnel, which was driven in squeezing rock conditions. For the tunnel support system used for the Semmering pilot tunnel, additional structural hypotheses will be introduced.

4.1 Sieberg tunnel

For the numerical analysis of the Sieberg tunnel, the measurement cross-section MCS1452 (km 156.990) was chosen. For the evaluation of the prescribed displacement increments in the longitudinal direction, $\Delta\bar{u}_z$, the *preceding* and the *succeeding* MCS, i.e., MCS1444 (km 156.998) and MCS1462 (km 156.980), were employed.

The geometric dimensions of the standard cross-section of the Sieberg tunnel are given in Fig. 6a. The investigated part of the tunnel shell was discretized by means of 648 trilinear brick elements, see Fig. 4a. The material properties for shotcrete employed in the numerical analysis are listed in Tab. 1 (see Appendix A).

The construction history for the part of the tunnel around MCS1452 is depicted in Fig. 6b. 120 h after construction of the top heading in MCS1452, the construction works were interrupted for 456 h (Christmas vacations). The left and the right bench were installed 288 h and 336 h, respectively, after resuming work.

The deformational state of the *considered* MCS is illustrated in Fig. 7. It is based on the circumferential stretches $\epsilon_\varphi = 1/h \int_h \varepsilon_\varphi dr$ and the longitudinal stretches $\epsilon_z = 1/h \int_h \varepsilon_z dr$ at the top of the cross section ($\varphi = 90°$) and in the middle part of the left bench ($\varphi = 182.6°$).

Figure 8a and b shows the evolution of the axial forces in the circumferential direction, $n_\varphi = \int_h \sigma_\varphi dr$, at the top of the cross section and in the middle part of the left bench, respectively. After application of the top heading, compressive forces develop (Fig. 8a). Simultaneously, the creep capacity of the

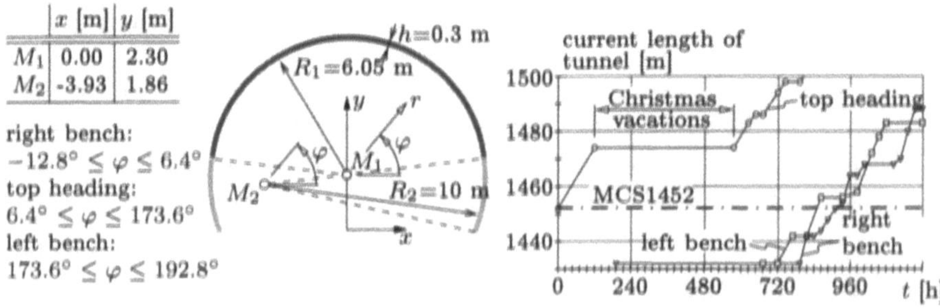

(a) (b)

Fig. 6. Sieberg tunnel: (a) cross section, and (b) construction history

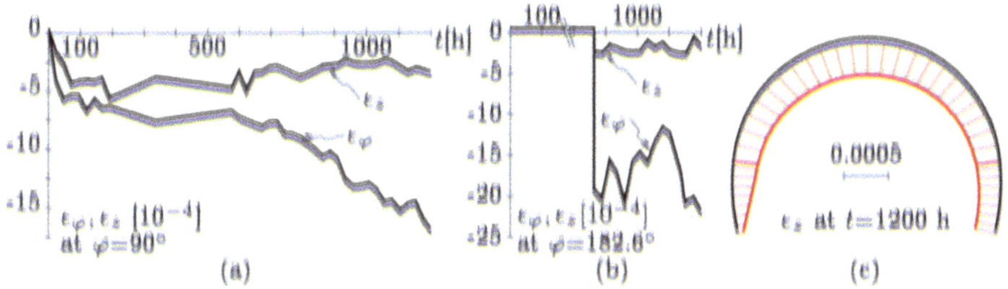

Fig. 7. Stretches at km 156.990 of the Sieberg tunnel: (a)(b) evolutions, and (c) spatial distribution

young shotcrete develops, resulting in a reduction of the compressive forces. During the Christmas vacation (see Fig. 6b), the stress relaxation continues ($200 < t < 330$ h). At $t = 330$ h, the compressive circumferential stretch ϵ_φ begins to decrease (Fig. 7a), temporarily resulting in tensile forces in the top heading. Continuation of the excavation after the Christmas vacations leads to compressive loading states in the circumferential direction of the shotcrete shell. In the left bench the reduction of the circumferential stretches shown in Fig. 7b between 960 and 1080 hours results in tensile loading (Fig. 8b). For $t > 1080$ hours, the increase of the circumferential compressive stretches leads to compressive loading of the left bench. As regards the axial force in the longitudinal direction, $n_z = \int_h \sigma_z dr$, chemical (autogeneous) shrinkage together with the decreasing evolution of the compressive longitudinal stretch ϵ_z between 200 and 920 hours, see Fig. 7a, leads to tensile stresses in the top heading (Fig. 9a). A similar response is observed in the left bench. Chemical shrinkage results in tensile longitudinal forces.

Figure 10 shows the evolution of the maximum value of the level of loading $\mathcal{L}_1$ averaged over the shell thickness, given by

$$\bar{\mathcal{L}}_{1,max} = \max \left\{ \bar{\mathcal{L}}_1(\varphi) = \frac{1}{h} \int_h \mathcal{L}_1(\varphi, r) dr, \; -12.8^\circ \leq \varphi \leq 192.8^\circ \right\} . \quad (12)$$

Right after the installation of the top heading, values of $\bar{\mathcal{L}}_{1,max}$ up to 74% are obtained. With increasing time, relaxation in consequence of creep of

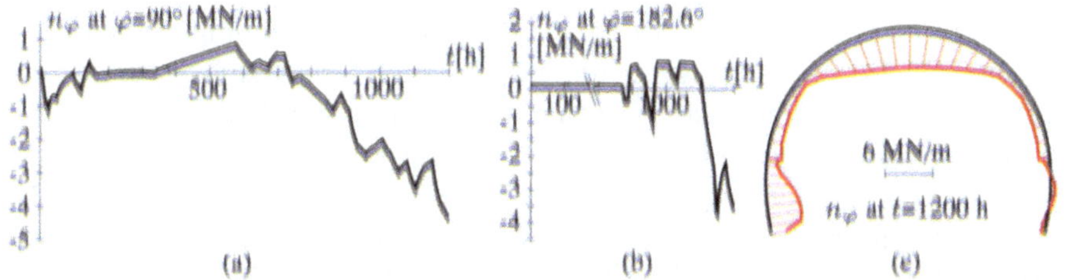

Fig. 8. Axial forces in the circumferential direction at km 156.990 of the Sieberg tunnel: (a)(b) evolutions, and (c) spatial distribution

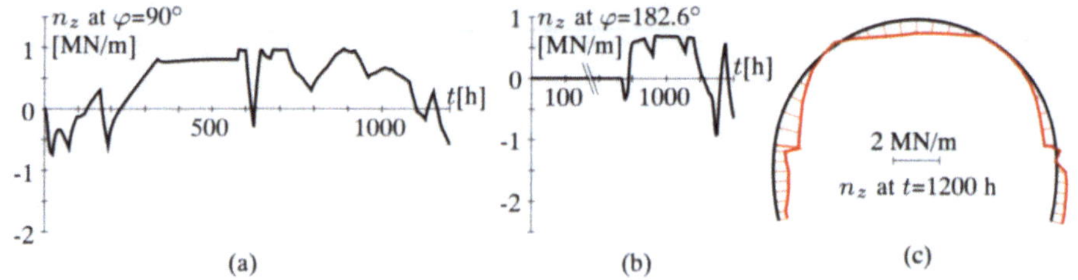

Fig. 9. Longitudinal axial forces at km 156.990 of the Sieberg tunnel: (a)(b) evolutions, and (c) spatial distribution

shotcrete results in a reduction of $\bar{\mathcal{L}}_{1,max}$ to about 20%. Installation of the benches causes further loading of the shell reflected by an increase of $\bar{\mathcal{L}}_{1,max}$. The increase of the compressive circumferential stretches ϵ_φ in the left bench for $t > 1090$ h (Fig. 7b) results in a high level of loading in this area of the cross section, see Fig. 10b.

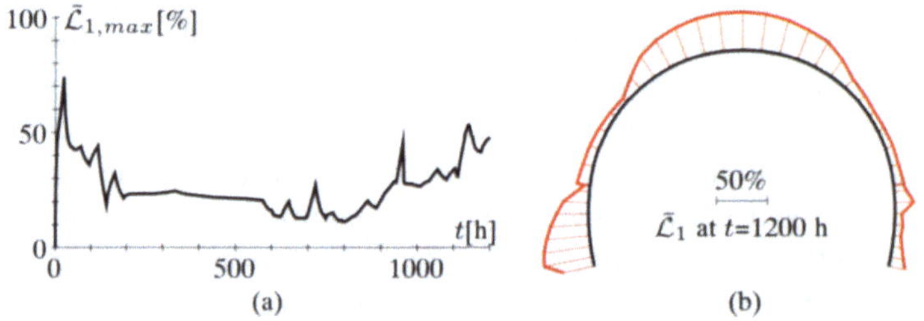

Fig. 10. Level of loading $\bar{\mathcal{L}}_1$ of the Sieberg tunnel shell at km 156.990: (a) evolution of $\bar{\mathcal{L}}_{1,max}$, and (b) spatial distribution

4.2 Semmering pilot tunnel

During the excavation of the Semmering pilot tunnel relatively large displacements of the rock mass were recorded. In order to avoid damage of the shotcrete shell, longitudinal gaps were left which resulted in an increase of the compliance of the shell (see Fig. 11). In the gaps, damping elements were installed [11,14]. In this way, circumferential axial forces could be transferred over the gaps increasing the utilization of the shotcrete shell. The mentioned damping elements, called Lining Stress Controllers (LSCs), are steel cylinders (see Fig. 11) with a nearly bilinear load-deformation behavior (see Fig. 13). They are supposed to function as follows: Before the strength of the shotcrete is reached in the process of loading of the tunnel shell, which, in principle, is a displacement–driven process, the LSCs are buckling. This results in a large circumferential shortening in the longitudinal gaps of the shell. According to

[11], from this point on the circumferential axial forces in the tunnel shell are controlled by the load–carrying behavior of the LSC.

For the support system used at the Semmering pilot tunnel, hypothesis H1 has to be reformulated. It reads [6]:

H1: As for the *parts of the shell made of shotcrete*, smooth displacement fields are assumed. With regards to the gaps, after buckling of the LSCs (see the local deformations in Fig. 11c), large displacement gradients occur in a very small region compared to the dimensions of the whole structure. This situation can be approximated by discontinuous displacement fields in the circumferential direction (see Fig. 12) whereas the displacement field in the radial direction is continuous (as the displacement field

(a) (b) (c)

Fig. 11. Semmering pilot tunnel: (a)(b) shotcrete shell with Lining Stress Controllers (LSCs) installed in longitudinal gaps and (c) prototype of LSC after buckling [11]

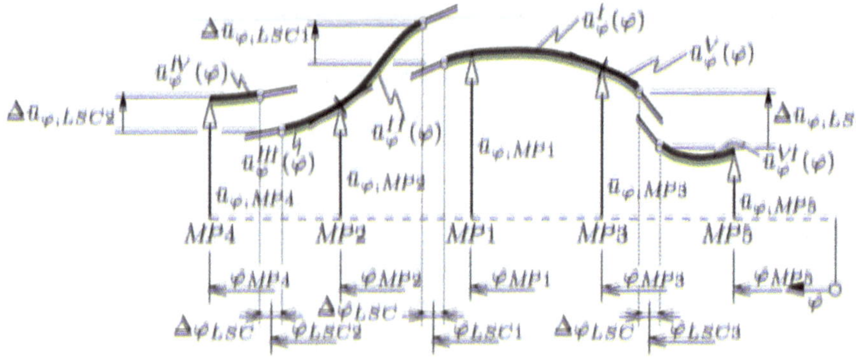

Fig. 12. Illustration of the total displacement in the circumferential direction, $\bar{u}_\varphi$, on the basis of displacements at the measurements points, displacement jumps and prescribed values of $\Delta\bar{u}'_\varphi = \partial\Delta\bar{u}'_\varphi/\partial\varphi$ at the LSCs

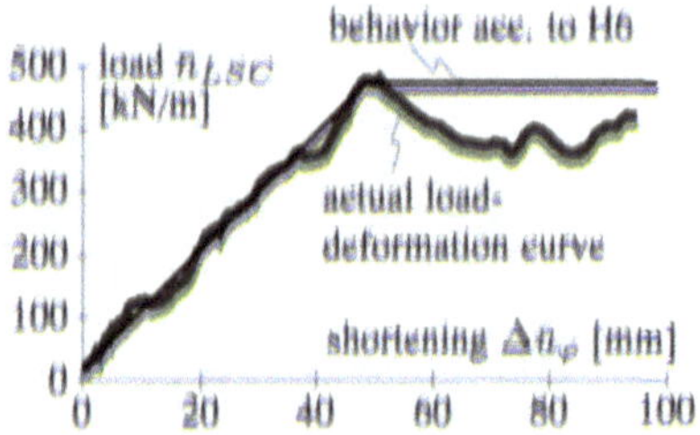

Fig. 13. Actual load-deformation curve of the LSC and behavior according to hypothesis H6

Δu_x depicted in Fig. 2). In contrast to the interpolation of displacement increments in the x- and the y-direction for the Sieberg tunnel, for the segmented shotcrete shells interpolation is performed for the circumferential and the radial displacement component.

Moreover, new hypotheses must be introduced for consideration of the behavior of the LSC [6]:

H6: The load-deformation characteristic of the LSCs is assumed to be linear-elastic – ideally-plastic as depicted in Fig. 13.

H7: The bending moment which can be taken by the LSCs is negligible in comparison to the one which can be carried by the shotcrete shell. Hence, the bending moment is assumed to be zero at the shotcrete-LSC interface.

The algorithmic realization of the structural hypotheses was accomplished by the development of an iterative calculation strategy. For a detailed description of this strategy, the reader is referred to [6]. The material properties for shotcrete employed in the numerical analysis are listed in Tab. 1 (see Appendix A).

For the numerical analysis of the Semmering pilot tunnel, the measurement cross-section MCS4274 (km 4.274) was chosen. For the evaluation of the prescribed displacements in the longitudinal direction, $\bar{u}_z$, the *preceding* and the *succeeding* MCS, i.e., MCS4269 (km 4.269) and MCS4281 (km 4.281), were employed.

The geometric dimensions of the standard cross-section of the Semmering pilot tunnel are given in Fig. 14. The investigated part of the tunnel shell was discretized by means of 312 trilinear brick elements, see Fig. 4b.

The interpretation of the results is based on three figures:

- Figure 15 shows the spatial distribution of the axial forces in the circumferential direction, n_φ, at selected time instants.
- Figure 17 shows the shear stresses $\sigma_{r\varphi}$ at the shotcrete-rock interface. They are obtained from the reaction forces at the nodes on the outer surface of the tunnel shell. It is noted that the influence of rock bolts on the circumferential displacement component was not considered in the respective interpolation scheme. Hence, the given distribution of the shear stresses does not reflect the increased shear capacity at rock bolts in consequence of radial compressive loading.
- Finally, Fig. 18 shows the spatial distribution of the level of loading $\bar{\mathcal{L}}_1$.

In order to better understand the stress states in the tunnel shell, it is instructive to study the displacement histories of the MPs shown in Fig. 16. *MP*2 and *MP*4 exhibit larger radial displacements than the remaining MPs. Therefore, large forces n_φ prevail in the left lower tunnel segment for the first 7 days, see Fig. 15a to Fig. 15c. These axial forces are induced at the shotcrete-rock interface by shear transfer from rock to shotcrete at the lower part of the segment (Fig. 17). The downward movement of *MP*2 with respect to *MP*1 results in tensile forces over the major part of the upper left segment. The values of the respective numerical quantities are bounded. This is a consequence of the employed Rankine failure criterion (see Chap. 5).

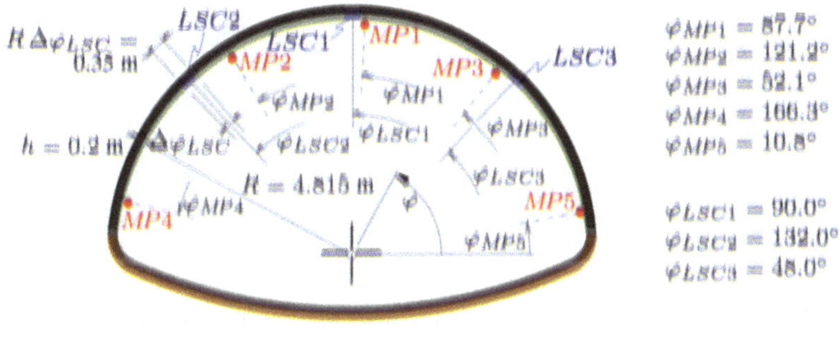

Fig. 14. Semmering pilot tunnel: cross section

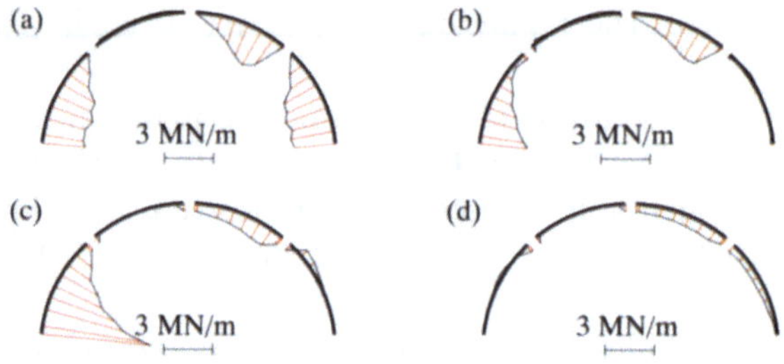

Fig. 15. Semmering pilot tunnel: distribution of the axial forces (in [MN/m]) at (a) $t = 12$ hours, (b) $t = 36$ hours, (c) $t = 7$ days, and (d) $t = 9$ days

The right upper segment is subjected to compressive forces n_φ mainly induced by the shear stresses at the shotcrete-rock interface, see Fig. 17. In contrast to the left lower segment, the less significant inward movement of the rock led to a continuous decrease of n_φ with time, resulting from creep deformations.

Finally, the right lower segment shows alternate compressive and tensile circumferential stress states.

The aforementioned creep deformations of shotcrete resulted in a low level of loading of the shell segments 9 days after shotcreting (see Fig. 18).

5 Determination of strain states from interpolated displacements only

To obtain sufficiently accurate results from FE analyses, rather fine (spatial) FE discretizations were employed (see the FE meshes depicted in Fig. 4). Moreover, the nonlinear material behavior of shotcrete required rather small time increments in nonlinear FE analyses (a numerical study illustrating the influence of the discretization in time is presented in [5]). Both the fine FE

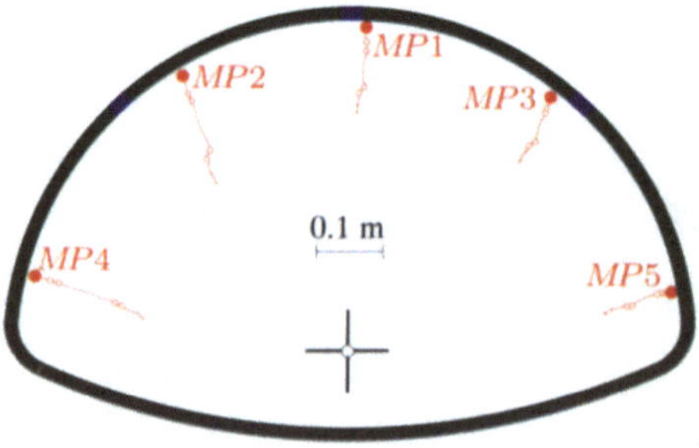

Fig. 16. Semmering pilot tunnel: displacement history of the five MPs during the first 35 days after shotcreting (circles indicate the location of the MPs 12 h, 36 h, 7 days, and 9 days after shotcreting)

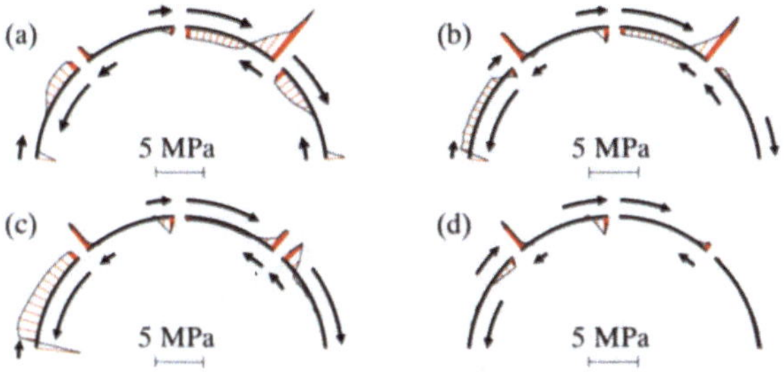

Fig. 17. Semmering pilot tunnel: distribution of the shear stresses (in [MPa]) at the shotcrete-rock interface at (a) $t = 12$ hours, (b) $t = 36$ hours, (c) $t = 7$ days, and (d) $t = 9$ days

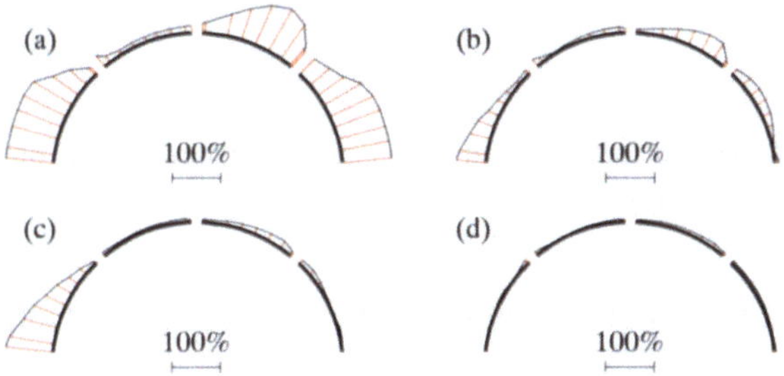

Fig. 18. Semmering pilot tunnel: distribution of the level of loading $\bar{\mathcal{L}}_1$ (in [%]) at (a) $t = 12$ hours, (b) $t = 36$ hours, (c) $t = 7$ days, and (d) $t = 9$ days

discretization and the small time increments have led to high computational effort including long computation times. In NATM tunneling the driving parameters are continuously adjusted during the construction process. Accordingly, quick information providing the basis for these adjustments is needed. This was the motivation for modification of the previously described hybrid method such that time-consuming FE analyses are not needed [8].

For this purpose hypothesis H3 is reformulated [8]:

H3: During the deformation the *thickness* of the shell is assumed to remain *constant*.

Furthermore, two additional hypotheses are adopted:

H6: Lines perpendicular to the inner surface of the tunnel shell immediately after installation, remain perpendicular to the inner surface during deformation. This hypothesis is consistent with the theory of thin shells (see, e.g., [16]). Because of the small thickness of the shell, this assumption is reasonable.

H7: The circumferential and the longitudinal direction are considered as principal directions of stress and strain. Since shotcrete shells are subjected predominantly to biaxial strain and stress states stemming from

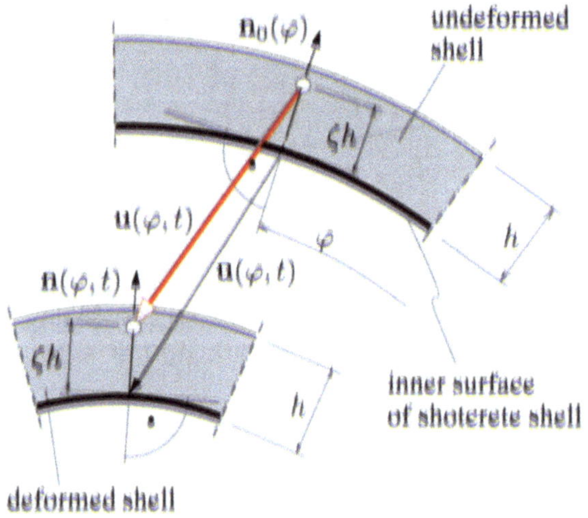

Fig. 19. On determination of the displacement $\mathbf{u}$ from known boundary value $\bar{\mathbf{u}}$, using hypotheses H3 and H6

circumferential and longitudinal deformations, this assumption is reasonable.

The displacement vector of an arbitrary point of the shotcrete shell, $\mathbf{u}$, can be determined from the displacement of the corresponding point on the inner surface, $\bar{\mathbf{u}}(\varphi, t)$, on the basis of hypotheses H3 and H6 (Fig. 19):

$$\mathbf{u}(\varphi, t, \zeta) = -\zeta h \bar{\mathbf{n}}_0(\varphi) + \bar{\mathbf{u}}(\varphi, t) + \zeta h \bar{\mathbf{n}}(\varphi, t) \ , \tag{13}$$

where ζ is a normalized coordinate perpendicular to the inner surface of the tunnel shell. $\bar{\mathbf{n}}_0$ defines the orientation of the shell section immediately after its installation whereas $\bar{\mathbf{n}}$ refers to the orientation of the deformed shell section. Both $\bar{\mathbf{n}}_0$ and $\bar{\mathbf{n}}$ are perpendicular to the inner surface of the tunnel shell (hypothesis H6), i.e., both can be computed from the given displacement field $\bar{\mathbf{u}}(\varphi, t)$ of the *considered* MCS.

According to hypothesis H7, only axial strains and stresses in the circumferential and longitudinal direction are considered. The circumferential strain ε_φ is a function of the displacement field $\mathbf{u}(\varphi, t, \zeta)$ given in (13). An efficient way for determination of ε_φ is to approximate the displacement field locally by means of a (linear) finite beam element. The length of the element, represented by $\Delta\varphi^e$, is arbitrary and has to be specified by the user.

In the thickness direction of the shell, the finite element is divided into n layers (Fig. 20). Hence, the axial strain in the circumferential direction of the k-th layer, $\varepsilon_{\varphi,k}$, can be determined from the displacements of the respective points at the two end points of the middle surface of this layer, see points P_1 and P_2 in Fig. 20. According to Eqn. (13), they are given by

$$\mathbf{u}_{1,k}(t) = -\zeta_k h \bar{\mathbf{n}}_0(\varphi_1) + \bar{\mathbf{u}}(\varphi_1, t) + \zeta_k h \bar{\mathbf{n}}(\varphi_1, t) \tag{14}$$

and

$$\mathbf{u}_{2,k}(t) = -\zeta_k h \bar{\mathbf{n}}_0(\varphi_2) + \bar{\mathbf{u}}(\varphi_2, t) + \zeta_k h \bar{\mathbf{n}}(\varphi_2, t) \ . \tag{15}$$

The strain $\varepsilon_{\varphi,k}$ is computed as

$$\varepsilon_{\varphi,k} = \frac{\ell_k - \ell_{0,k}}{\ell_{0,k}} \ , \tag{16}$$

where $\ell_{0,k}$ and ℓ_k denote the length of the k-th layer immediately after installation of the shotcrete shell and in the deformed configuration, respectively. They are computed from the coordinates of points P_1 and P_2 in the undeformed configuration, $\mathbf{x}_{1,k}$ and $\mathbf{x}_{2,k}$, and the respective displacements according to the Eqns. (14) and (15):

$$\ell_{0,k} = \mathrm{dist}(\mathbf{x}_{1,k}, \mathbf{x}_{2,k}) \tag{17}$$

and

$$\ell_k = \mathrm{dist}(\mathbf{x}_{1,k} + \mathbf{u}_{1,k}, \mathbf{x}_{2,k} + \mathbf{u}_{2,k}) \ . \tag{18}$$

The longitudinal strain at the *considered* MCS is computed with the help of Eqn. (8). ε_z is determined at the midpoint of the beam elements, given by $\varphi^e = (\varphi_1 + \varphi_2)/2$ (see Fig. 20). According to hypothesis H2, ε_z is constant for all layers of a beam element.

For determination of the axial stresses σ_φ and σ_z in the individual layers of the finite beam element, the material model for shotcrete is specified for a plane state of stress (see [8]). Hence, over the thickness of the tunnel shell, a piecewise constant stress distribution is obtained.

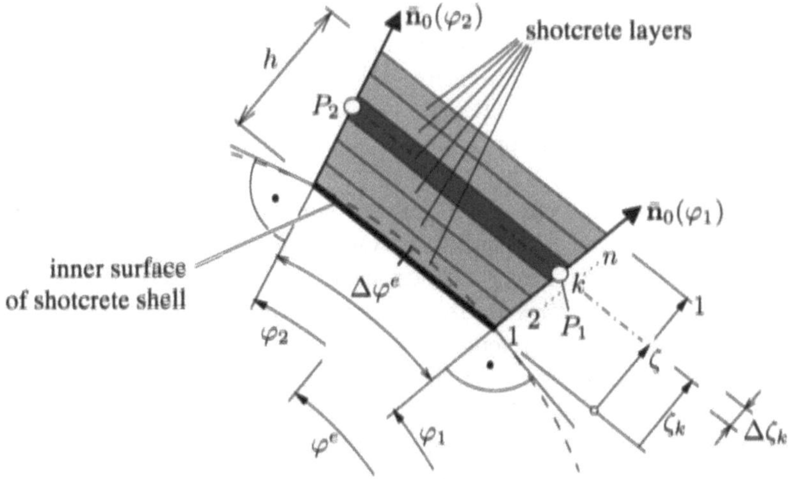

Fig. 20. Layered beam element in the undeformed configuration, used for determination of the axial strain in the circumferential direction

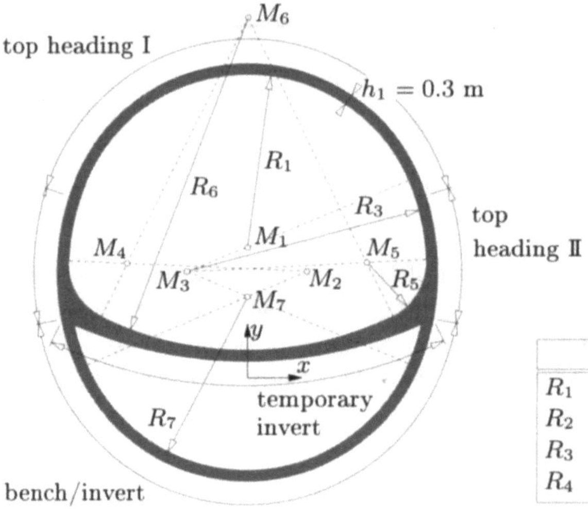

	x [m]	y [m]
M_1	0.00	3.40
M_2	1.51	2.62
M_3	-1.51	2.62
M_4	-2.76	2.87
M_5	2.76	2.87
M_6	0.00	9.30
M_7	0.00	1.73

	radius [m]		
R_1	4.08	R_5	1.50
R_2	5.78	R_6	8.50
R_3	5.78	R_7	4.03
R_4	1.50		

Fig. 21. Geometric dimensions of the standard cross-section of the Lainzer tunnel (Bierhäuselberg)

6 Application – Lainzer tunnel

The performance of the shell-theory-based hybrid method outlined in the previous section is demonstrated by a hybrid analysis of the Lainzer tunnel (Bierhäuslberg), which is a part of the high-capacity railway line between Vienna and Salzburg. The geometric dimensions of the tunnel cross-section are given in Fig. 21. The excavation is performed in three steps: top heading I, top heading II, and bench/invert (see Fig. 21). After the excavation of top heading II, a temporary invert is installed. This invert is removed at the beginning of the third excavation step.

During the construction of top heading I, top heading II, and the temporary invert, displacement data of five MPs (*MP*1 to *MP*5) were provided by the construction site. Three MPs were installed at top heading I, the remaining two MPs were placed approximately in the middle of top heading II (see Fig. 22a).

For the hybrid analysis of the Lainzer tunnel, 16 layered finite beam elements were employed (see Fig. 22a). Over the thickness each element was divided into 6 layers of 5 cm thickness each.

The material properties for shotcrete employed in the numerical analysis are listed in Tab. 1 (see Appendix A).

The computational efficiency of the shell-theory-based hybrid method allows to analyze several MCSs simultaneously. The respective results give insight into the spatial distribution of the level of loading in the circumferential and the longitudinal direction of the tunnel. In the following, this is illustrated for five MCSs: MCS00093 (km 8.322), MCS00100 (km 8.315), MCS00110 (km 8.305), MCS00119 (km 8.296), and MCS00130 (km 8.285).

Fig. 23 shows the distribution of the level of loading $\bar{\mathcal{L}}_2$ of the aforementioned MCSs on June 12, 2001. On that day, the excavation of top heading I and top heading II has reached km 8.264 and km 8.268, respectively (see Figure 22b and Fig. 23).

At MCS00093, $\bar{\mathcal{L}}_2$ reaches a value of 92% at the left footing, resulting from continuation of the excavation after the excavation break of 69 days (see Fig. 22b). Fig. 24 shows the evolution of $\bar{\mathcal{L}}_2$ for MCS00093 from February 5, 2001, to June 12, 2001. On continuation of the excavation on April 18, 2001, an increase of the level of loading in top heading II is observed. Whereas the loading of the right part decreases with increasing time, the loading of the left part remains almost unchanged, finally giving the distribution of $\bar{\mathcal{L}}_2$ for MCS00093 for June 12, 2001, shown in Fig. 23.

At MCS00100, the level of loading reaches 50% at top heading II at the left side on June 12, 2001 (see Fig. 23). Figure 25(b) shows the evolution of the maximum value of $\bar{\mathcal{L}}_2$, $\bar{\mathcal{L}}_{2,max}$, for this MCS. Immediately after the installation of top heading I, $\bar{\mathcal{L}}_2$ increases up to 60%. Whereas creep of shotcrete led to a reduction of $\bar{\mathcal{L}}_2$, the installation of top heading II caused a further increase of the loading of the shotcrete shell (see ① in Fig. 25). A few days later, an increase of the level of loading to 59% is observed at the top end of top heading II at the right side (see ② in Fig. 25). This increase is caused by the inward movement of the right side of the shotcrete shell (see displacement histories of *MP*1 and *MP*3 in Fig. 25b). The increase of $\bar{\mathcal{L}}_2$ at the left footing of the shotcrete shell shortly before June 12, 2001, is a consequence of longitudinal deformations. Results obtained by a 2D analysis, disregarding longitudinal deformations, did not show this increase (see ③ in Fig. 25).

Fig. 26 shows the history of $\bar{\mathcal{L}}_2$ for MCS00110. The installation of top heading I and top heading II resulted in an increase of $\bar{\mathcal{L}}_2$ up to 60% in the respective part of the shotcrete shell. Creep of shotcrete has led to a reduction of $\bar{\mathcal{L}}_2$ below 40%. At the end of the considered time span an increase of $\bar{\mathcal{L}}_2$

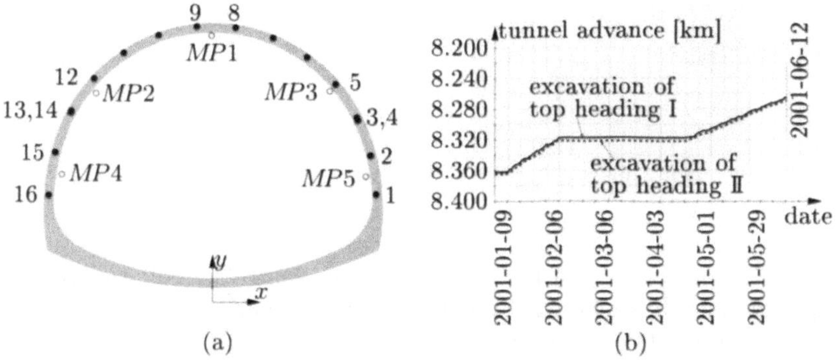

Fig. 22. (a) Location of the finite beam elements for the hybrid analysis and (b) construction history of the Lainzer tunnel (Bierhäuslberg)

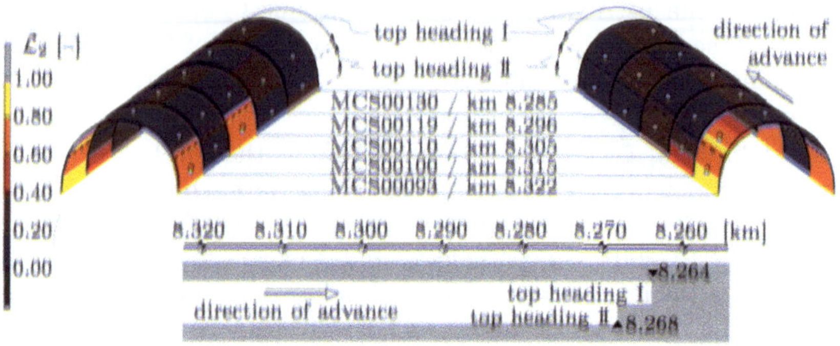

Fig. 23. Distribution of $\bar{\mathcal{L}}_2$ in five MCSs of the Lainzer tunnel (Bierhäuslberg) on June 12, 2001 (circles indicate the location of the MPs)

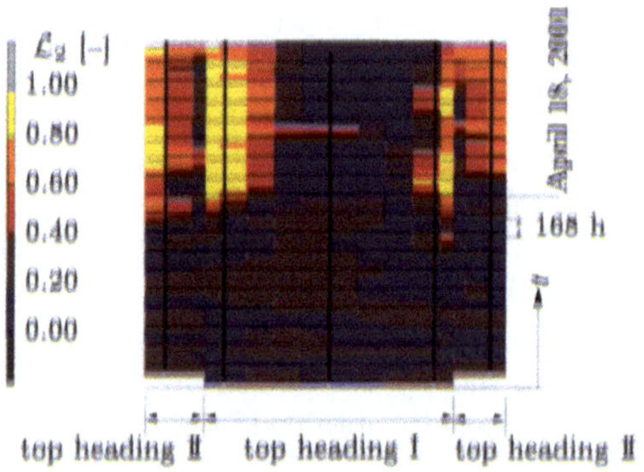

Fig. 24. History plot of $\bar{\mathcal{L}}_2$ for MCS00093 from February 5, 2001, to June 12, 2001 (top view)

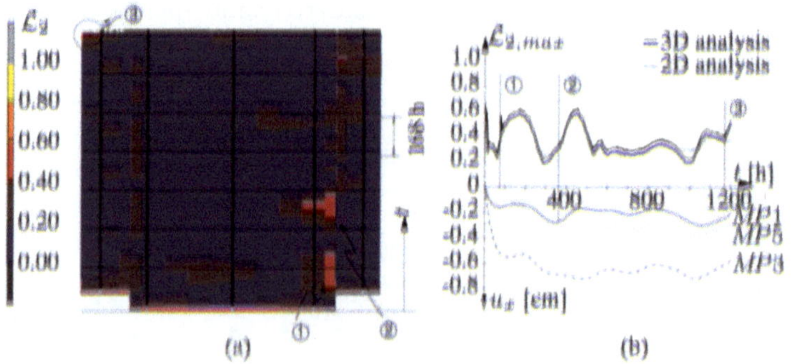

Fig. 25. (a) History plot of $\bar{\mathcal{L}}_2$ for MCS00100 from April 23, 2001, to June 12, 2001 (top view) and (b) evolution of the maximum value of $\bar{\mathcal{L}}_2$, $\bar{\mathcal{L}}_{2,max}$, and of the horizontal displacements u_x of $MP1$, $MP3$, and $MP5$

is observed at both footings of the shell. It is caused by the deformations of the shell as illustrated in Fig. 26b by the displacement history of *MP*4 and *MP*5. The observed displacements are characterized by oscillations beyond the measurement accuracy of ±1 mm. These oscillations and the rather large stiffness of the shotcrete shell have led to the increase of $\bar{\mathcal{L}}_2$ shown in Fig. 26.

7 Discussion

The experience with the hybrid tunnel analyses allow for some statements in a more general discussion related to the three tunnels analyzed in this chapter:

- Sieberg tunnel:
  The results of the hybrid method for closed linings have a quantitative meaning, but only when underlying a physically sound material model for shotcrete. The importance of realistic material modeling was pointed out by means of comparative studies in [1,2]. Consideration of dissipative phenomena changes the results of, e.g., an analysis of a classical creep test by a factor of 10 (see Chap. 5). The level of loading in the analyses of closed tunnel linings depends especially on realistic modeling of creep. Improved formulations lead to more desirable creep answers [10].
- Semmering pilot tunnel:
  The existence of the LSCs results in an increase of the circumferential forces at the gaps from zero (corresponding to a situation characterized by longitudinal gaps without LSCs) to a value below the compressive strength of the shotcrete. However, this value is by far *smaller* than the axial forces in the circumferential direction prevailing in some of the shotcrete segments. These forces are induced mainly by the shotcrete-rock interface at the footings of the shell and at the exterior surface of the shell and *not* by the LSCs. Therefore, as regards the analyzed tunnel shell, the LSCs do not control the "lining stress" as their name is

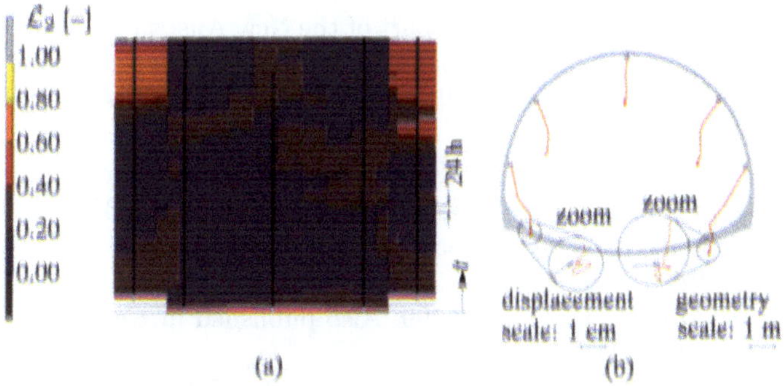

Fig. 26. (a) History plot of $\bar{\mathcal{L}}_2$ for MCS00110 for the period from May 5, 2001, to June 12, 2001, (top view) and (b) respective traces of the displacements of the MPs

insinuating. Their influence is rather local, i.e., restricted to the vicinity of the longitudinal gaps. The present lining system is rather acting as a sealing device for the surface of the tunnel which is composed of highly anisotropic and inhomogeneous rock. Remarkably, the level of loading is locally reaching values up to 100%. Since the present shell is a sealing device rather than a load-driven structure, this situation does not cause failure. Nevertheless, such large values of the level of loading is certainly not desirable. In order to avoid this, more than three longitudinal gaps would have to be installed.

- Lainzer tunnel:
  The development of a shell-theory-based hybrid method was a significant step towards real engineering practice. The computation time of one tunnel cross-section was reduced from typically 2 to 3 hours to less than 5 minutes on the same computer. The employed computer hardware was a Hewlett-Packard workstation (HP 9000 L2000) using HP PA-RISC 8500 440 MHz processors. Real-time determination of the level of loading gives the geotechnical engineer on the tunnel site useful additional information for in-time adjustment of the driving parameters of the excavation process.

Acknowledgement

This work has drawn great benefit from the close co-operation with Prof. Wulf Schubert (Institute for Rock Mechanics and Tunneling, Graz University of Technology), Dr. Bernd Moritz (Geoconsult ZT Gmbh, formerly Institute for Rock Mechanics and Tunneling, Graz University of Technology), and Dr. Harald Golser (Geoconsult Asia Singapore; formerly Institute for Rock Mechanics and Tunneling, Graz University of Technology).

References

[1] Hellmich Ch. (1999) Shotcrete as part of the New Austrian Tunneling Method. *from thermochemomechanical material modeling to structural analysis and safety assessment of tunnels*, PhD thesis, Vienna University of Technology, Vienna, Austria

[2] Hellmich Ch., Mang H.A., Ulm F.-J. (2001) Hybrid method for quantification of stress states in shotcrete tunnel shells: combination of 3D *in-situ* displacement measurements and thermochemoplastic material law. In W. Wunderlich, editor, *CD-ROM Proceedings of the 1st European Conference of Computational Mechanics*, Munich, Germany, 1999. Also published in *Computers and Structures*, 79: 2103 – 2115

[3] Hellmich Ch., Ulm F.-J., Mang H. A. (1999) Multisurface chemoplasticity I: Material model for shotcrete. *Journal of Engineering Mechanics (ASCE)*, 125(6):692–701

[4] Huber H.G. (1991) Untersuchungen zum Verformungsverhalten von jungem Spritzbeton im Tunnelbau [Investigations concerning the deformation behavior of young shotcrete in tunneling]. Master's thesis, University of Innsbruck, Innsbruck, Austria, In German.

[5] Lackner R., Hellmich Ch., Mang H.A. (2002) Constitutive modeling of cementitious materials in the framework of chemoplasticity. *International Journal for Numerical Methods in Engineering*, 53(10):2357–2388

[6] Lackner R., Macht J., Hellmich Ch., Mang H.A. (2002) Hybrid analyses of segmented shotcrete tunnel linings in squeezing rock. *Journal of Geotechnical and Geoenvironmental Engineering (ASCE)*, 128(4):298–308

[7] Lackner R., Macht J., Mang H.A. (2002) Projekt zur Erstellung eines Programmsystems zur praktischen Umsetzung eines hybriden Verfahrens zur Bestimmung der Beanspruchung von Tunnelschalen aus Spritzbeton [Project for the development of a program system for the practical realization of a hybrid method for determination of the loading of shotcrete tunnel shells]. *Technical report, University of Technology, Vienna*, In German

[8] Lackner R., Mang H.A. (2002) Cracking in shotcrete tunnel shells. In R. de Borst, J. Mazars, G. Pijaudier-Cabor, and J.G.M. van Mier, editors, *Fracture Mechanics of Concrete Structures, Proceedings of the 4th International Conference*: 857–870, Cachan, France To be published in *Engineering Fracture Mechanics*

[9] Lackner R., Mang H.A. (2002) On-line quantification of stress states in tunnel shells. *Géotechnique*, Submitted for publication.

[10] Lechner M., Hellmich Ch., Mang H.A. (2001) Short-term creep of shotcrete - thermochemoplastic material modeling and nonlinear analysis of a laboratory test and of a NATM excavation by the Finite Element Method. *Proceedings of the International Symposium on Continuous and Discontinuous Modeling of Cohesive Frictional Materials*, German Science Foundation, University of Stuttgart, Lecture Notes in Physics: 47-62, Springer, Berlin

[11] Moritz. B. (1999) *Ductile support system for tunnels in squeezing rock.* Ph.D. thesis, Graz University of Technology, Graz, Austria

[12] Rokahr R., Lux K.H. (1987) Einfluß des rheologischen Verhaltens des Spritzbetons auf den Ausbauwiderstand [Influence of the rheological behavior of shotcrete on the lining resistance]. *Felsbau*, **5**: 11–18 In German.

[13] Rokahr R.B., Stärk A., Zachow R. (2002) On the art of interpreting measurement results. *Felsbau*, **20**(2): 16–21

[14] Schubert W., Moritz B. (1998) Controllable ductile support system for tunnels in squeezing rock. *Felsbau*, **16**(4):224–227

[15] Schubert W., Steindorfer A., Button E.A. (2002) Displacement monitoring in tunnels – an overview. *Felsbau*, **20**(2):7–15

[16] Timoshenko S. (1940) *Theory of plates and shells.* McGraw-Hill, New York

[17] Zachow R. (1995) Dimensionierung zweischaliger Tunnel in Fels auf der Grundlage von *in-situ* Messungen [Dimensioning of bicoque tunnels in rock masses based on *in situ* measurements]. Technical Report 16, University of Hannover, Hannover, In German.

A Mixture characteristics and material parameters

Table 1 contains the mixture characteristics and the material parameter for two different shotcrete mixtures. The first mixture was used for the analyses of the Sieberg tunnel (Sect. 4.1) and the Semmering pilot tunnel (Subsect. 4.2, the second mixture was used for the analysis of the Lainzer tunnel (Subsect. 6).

Table 1. Mixture characteristics and material parameters for two different shotcrete mixtures

			shotcrete used at: Sieberg tunnel, Semmering pilot tunnel	shotcrete used at: Lainzer tunnel
water/cement-ratio		[–]	0.60	0.48
aggregate/cement-ratio		[–]	4.79	5.30
chemical affinity $\tilde{A}$	a_A	[1/s]	7.313	6.415
$\tilde{A}(\xi) = a_A \frac{1-\exp(-b_A \xi)}{1+c_A \xi^{d_A}}$	b_A	[–]	10.46	6.508
	c_A	[–]	169.3	48.358
	d_A	[–]	4.370	7.249
relaxation modulus H		[MPa]	$1/7 \cdot 10^6$	$1/7 \cdot 10^6$
compressive strength f_c	ξ_0	[–]	0.01	0.01
$f_c(\xi) = f_{c,\infty} \frac{\xi-\xi_0}{1-\xi_0}$	$f_{c,\infty}$	[MPa]	39.6	40
Young's modulus E_c				
$E_c(\xi) = E_{c,\infty}\sqrt{\xi}$	$E_{c,\infty}$	[MPa]	40 800	37 100
shrinkage parameter β *	a_s	[–]	$-4.08 \cdot 10^{-4}$	$-4.08 \cdot 10^{-4}$
$\varepsilon^s = a_s + b_s \xi$	b_s	[–]	$9.43 \cdot 10^{-4}$	$9.43 \cdot 10^{-4}$

* shrinkage parameters were determined from experimental data given in [4] (see [3])

Chapter 11
Compressed air tunnelling - determination of air requirement

Stephan Semprich[1], Yannick Scheid[1], and Jens Gattermann[2]

[1] Institute for Soil Mechanics and Foundation Engineering, Graz University of Technology

[2] Ed. Züblin AG

Abstract. In shallow tunnelling below the groundwater table compressed air can be used for preventing water inflow into the tunnel. When using this method air loss takes place through both the unsupported tunnel face and shrinkage cracks of the shotcrete lining. Until today it has been very difficult to correctly estimate the amount of air loss during the design phase of a project, although it could be a significant factor concerning the total costs of the tunnel construction. For solving this problem the multi-phase flow in the soil above the tunnel has to be considered. The aim of the project was to develop a new approach, based on existing design principles and unsaturated soil constitutive models. At the Institute for Soil Mechanics and Foundation Engineering in Graz, large scale laboratory tests were conducted to simulate the air-permeability of the shotcrete lining and the soil. Additionally, the experimental data are compared with results of numerical models. The models are based on existing constitutive laws to describe the mechanical behaviour of unsaturated soils. In this contribution results of the tests are discussed and a methodology is presented to estimate the amount of air loss during tunnel advance.

1 Introduction

Constructing urban traffic tunnels in soft ground and below the groundwater level remains a difficult task. A useful excavation method in water bearing soil is the NATM in combination with compressed air. In shallow tunnelling below groundwater table, compressed air can be used for preventing water inflows into the tunnel during construction. Compressed air in the working area results in air flow through the initially saturated and subsequently only partially saturated soil. The air escapes from the working space through the unsupported face as well as through shrinkage cracks in the shotcrete lining. The amount of air loss thus depends on the head of the air-pressure, on the air permeability of the ground, and on the chosen sequence of excavation and supporting methods.

Maintaining a certain air pressure requires that the air lost into the ground is constantly replaced by the use of compressors. The volume of air loss determines the dimensions of the compressor house and the energy costs for their operation.

An accurate method for the prediction of the air loss through the shotcrete cracks in the tunnel lining is not known. Therefore, at the Institute for Soil Mechanics and Foundation Engineering at Graz University of Technology a laboratory test for the investigation of the air permeability of a cracked shotcrete-soil element was developed and carried out. Results of the laboratory test and results of an evaluation of air loss measurements carried out within previous tunnel constructions will be presented in the current work. In particular the influence of the air pressure in the tunnel and the width of the shotcrete crack on the occurring air loss through the tunnel lining will be discussed.

An approach to use numerical methods to compute air loss through the cracks in the shotcrete lining will be given. At this stage of work, a model of the tests carried out at the laboratory was simulated and evaluated using the software code TOUGH2.

In Sect. 5, a tunnel advance simulation to estimate the air loss through the tunnel face and the cracks in the shotcrete lining is presented. The model is evaluated in a case study for a tunnel project in Siegaue (Germany). In addition, an application of the model is presented for a metro project in Vienna (Austria).

2 Compressed air in combination with NATM

Compressed air in shallow tunnelling is used to counteract the tendency of water to flow into the opened space. The invention goes back to Sir Thomas Cochrane and was patented in 1830 [1]. There is little doubt that it is related to the undertunnelling of the river Thames designed by Brunel from 1823-1843, who was struggling with constant inrushes of water into the tunnel. Cochrane knew that every meter in height of water exerts a pressure of 10 kN/m^2, therefore by introducing compressed air into the tunnel corresponding to the number of meters of water head acting upon the tunnel base would be possible to prevent the water inflow [2]. The principle is shown in Fig. 1. By this invention the method of excavation used in ground below ground water table could be switched to that used in ground above groundwater table and made it later possible to make use of the NATM, patented by Rabcewicz [3] in 1950, when tunnelling under the groundwater surface. The first application of Cochranes invention was by the French engineer Triger [4] in 1841 who published an account of the sinking of a colliery shaft by this method.

Construction using compressed air has gained new importance recently, because in urban areas it does not influence the surrounding groundwater. Another environmental aspect is that the transportation of existing contaminants into the direction of the tunnel can be prevented. The main advantage of tunnelling under compressed air is that this method does not affect the aquifer. One major disadvantage is that the excavation work takes place under severe conditions.

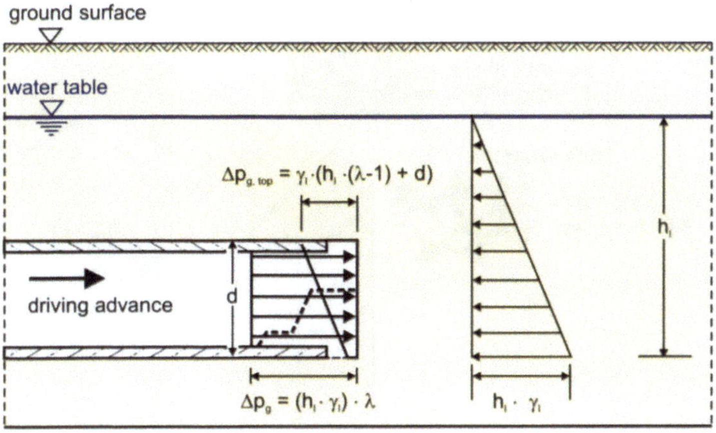

Fig. 1. Compressed air to prevent goundwater inflow

The required air pressure Δp_g is shown in Fig. 1. A safety factor λ is introduced. Since the pressure of the compressed air is constant over the height of the tunnel, while the water pressure increases constantly with depth, a disequilibrium at the tunnel crown and leads to air flow into the soil. Depending on the soil conditions, the geometry of the tunnel and the applied air pressure the amount of air loss varies. Air loss takes place through both the unsupported tunnel face (Q_F) and shrinkage cracks of the shotcrete lining (Q_S) (Fig. 2). While in homogenous ground the amount of air loss through the tunnel face is relatively constant the amount of air loss through the shrinkage cracks increases with the length of the tunnel section and becomes the determining factor. The air flow into the soil changes the primarily fully saturated conditions to unsaturated soil conditions. Therefore, the theoretical approaches of unsaturated soil mechanics have to be considered.

As mentioned before the use of compressed air is appropriate in combination with NATM. Under certain conditions the use of NATM in urban areas below the groundwater table might be more cost-effective than other methods e.g. shield tunnelling. Often the NATM is chosen when constructing short tunnels with variable diameters. Nevertheless, the most frequent use of compressed air is in slurry and earth pressure balanced shield tunnelling, when obstacles or maintenance and repair demand entering the shield area. Furthermore, compressed air might be used in pipe jacking. Table 1 shows some projects where compressed air has been used in combination with NATM. In Munich, a compressed air tunnel section is to be constructed in the years 2001 to 2003.

The amount of overpressure in the tunnel and the required volume of air to replace air loss through the tunnel face and the shotcrete lining is a

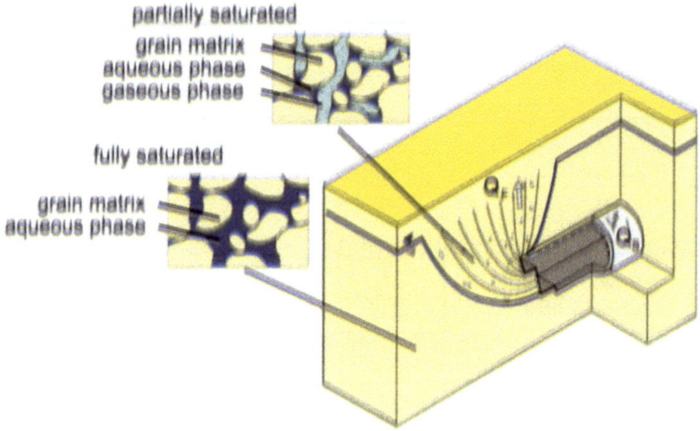

Fig. 2. Air loss due to compressed air tunnelling

Table 1. Compressed air projects

City	Length driven under compressed air	Air pressure	Air loss
Munich	6961 m	0,3 - 1,1 bar	25 - 580 m^3/min
Essen	1330 m	0,4 - 1,2 bar	52 - 250 m^3/min
Taipei	400 m	0,8 - 1,4 bar	50 - 180 m^3/min
Siegburg	240 m	0,6 - 1,2 bar	50 - 450 m^3/min

crucial point in scientific research. Table 2 shows some contributions based on judgement and empirical approaches.

In recent time various estimations of the flow of air and water in tunnel constructions using compressed air based on numerical models have been given for example by [9][10] extended by [11][12], [13] and [14].

These approaches need to be evaluated. On the one hand verifying, tunnel projects can do this, on the other hand laboratory tests give a great opportunity in fitting model approaches to test results. Strobl [15] conducted laboratory tests to analyse the flow of air and water through porous media. He also simulated the airflow in different excavation steps of the tunnel. Regarding the air loss due to cracks in the shotcrete elements of the tunnel lining an important contribution has been presented in [16].

An observation often reported is the control of settlement and ground heaving during compressed air tunnel advance [17]-[20]. Compressed air in the tunnel working area and air flow in the ground is supposed to cause displacements of the tunnel walls as well as in the ground and subsequently on ground level. Those displacements eventually counteract the deformations due to tunnel excavation and so exert a positive effect on the stability of the tunnel itself and on the settlement of adjacent buildings. Heaving has been effectively measured in a field test reported in [8].

The experiments conducted in Graz took consideration of the deformation behaviour. The testing apparatus was designed to measure surface deformation during the airflow phase. The results of the analyses are presented in Sect. 3.3.

Table 2. Analytical and empirical approaches for determining the amount of air flow

Author(s)	Equation(s)
Hewett and Johannesson (1922), [2]	$3,66d^2 \leq Q_g \leq 7,32d^2$ (1)
Schenck zu Schweinsberg and Wagner (1963), [5]	$Q_g = \frac{w \cdot k_g}{\gamma_l} \cdot \frac{(p_{g1} - p_{g2})}{l} \cdot \frac{(p_{g1} - p_{g2})}{2} \cdot \frac{A_S W}{p_{g2}} \cdot \frac{(d - h_k)}{d}$ (2)
Krabbe (1969), [6], modified by [7]	$Q_g = 2 \cdot k_g \cdot 70 \frac{(p_{g1} - p_{g2})}{\gamma_l} \cdot A_k \cdot \frac{p_{g1}}{p_{g2}} (n_a + n_b + n_c)$ (3)
Kramer and Semprich (1989) [8]	$Q_T = Q_F + Q_S + Q_L$ (4) $Q_F = C \cdot k_g \cdot A_F \cdot \frac{h_w}{h_g} \cdot (p_{g1} + 1) \cdot 60$ (5) $Q_S = k_{gL} \cdot A_S \cdot \frac{h_w}{h_g} \cdot (p_{g1} + 1) \cdot 60$ (6) $Q_L = 3 \div 10\ m^3/\min$ (7)

Q_g	[m³/h]	volumetric air flow
d	[m]	tunnel diameter
w	[-]	correction factor, typically 2
k_g	[m/s]	coefficient of air permeability
γ_l	[kN/m³]	unit weight of water
p_{g1}	[bar]	absolute pressure at tunnel face / Air pressure when entering the crack

p_{g2}	[bar]	absolute pressure at surface / Air pressure at top of the crack
l	[m]	length of airflow path from tunnel crown to a point of free flow
A_{SW}	[m^2]	flow effective section according to [5]
h_k	[m]	groundwater table above tunnel including capillary rise according to [5]
A_K	[m^2]	flow effective section according to [6]
n_a	[ ]	relates to the tunnel face. Varies from 1 for soft inhomogeneous ground to 0 for hard ground
n_b	[ ]	relates to the tail seal of the shield. In soft ground it is 0 with a tail seal and between 1 and 2 without a tail seal
n_c	[ ]	relates to a typical 500 m length of tunnel behind the face. For a grouted and caulked lining it is 1 and for grouted lining without caulking it is 2
Q_T	[m^3/h]	total air loss
Q_F	[m^3/h]	air loss through the unsupported tunnel face
Q_S	[m^3/h]	air loss through shrinkage cracks of the shotcrete lining
Q_L	[m^3/h]	air loss through locks and pipes
C	[-]	factor considering the three-dimensional spread of the air field behind the tunnel face
A_F	[m^2]	cross section of tunnel face
h_w	[m]	overpressure of air in the tunnel in m water height
h_g	[m]	depth of tunnel
k_{gL}	[m/s]	coefficient of air permeability of a layered soil
A_S	[m^2]	cross section of the shotcrete lining

3 Laboratory tests

From the implementations according to the NATM in combination with compressed air, stated in Sect. 2, it follows that with increasing surface area of the shotcrete lining the amount of air loss increases in relation to air loss at the tunnel face. With longer project units air loss through shotcrete lining is the principal part of the total air loss. The magnitude of air loss through the shotcrete lining depends essentially on the air permeability of the shotcrete lining and the ground adjacent to the shotcrete lining.

Until today the calculation of air loss through the shotcrete lining was exclusively based on empirical relationships. In order to develop calculation methods, based on theoretical considerations, experimental investigations were carried out during the project.

The air permeability of the shotcrete lining can be based on porosity of the shotcrete. However, the porosity-related air permeability of shotcrete can be neglected in comparison to the air permeability due to cracks. The latter problem is examined in this report.

Regarding air loss occurring over the tunnel wall the ground immediately behind the shotcrete lining is of equal importance. Behind the lining the air spreads in form of a spatial air flow field.

3.1 Development in experimental analyses

Influence of compressed air on shotcrete. The influence of the compressed air on shotcrete was analysed by Schreyer [21]. He focused on the porosity permeability of the shotcrete and neglected flow of air through cracks. Additionally he focused on the quality of the shotcrete under compressed air with respect to the compressive strength [21], the influence of the groundwater on the shotcrete and the possible detrimental effects of air pressure variation on the young shotcrete. The tests on the compressive strength were conducted in a pressure chamber. Tests on air permeability were made on specially prepared shotcrete specimens. Older specimens were impermeable below a pressure of 5 bar, but young shotcrete showed air permeability. Analytical solutions to estimate the permeability were derived from the tests. In conclusion in [21] it is recommended to apply water on young shotcrete under compressed air in order to prevent drying of the shotcrete due to the inflow of air. For shotcrete made by common mixtures no loss of compressive strength was observed. Loss of air pressure can lead to damage in the young shotcrete due to overpressure entrapped in the pores of the shotcrete. Sudden air loss may produce collapse on the shotcrete in the tunnel crown. Both effects may be prevented by use of early strengthening shotcrete.

One-dimensional air flow in porous media. Strobl conducted one-dimensional air flow tests on probes that had a volume of 0,11 m^3 [15]. Air and water permeability were determined. Since flow occurs under pressure gradients pressure distribution was measured in the soil probe. Furthermore, the influence of the upward acting force of the air flow was measured.

The testing apparatus, able to analyse flow of water or air, is shown in Fig. 3. The probe cylinder is fixed on a load cell. The air and the water flow downside up. A sieve separates the pressure chamber from the soil. No forces should be transmitted through the connection between the pressure chamber and soil. Therefore, the cylinder with the soil lays inside is a double wall system, which is filled with mercury. Pressure in the soil is measured using a piezometric system. Outflowing water at top of the probe is measured in a graduated cylinder. Atmospheric conditions exist on top of the soil. The diameter of the cylinder is 44 cm and the height is 72 cm.

The tests on dry, partially saturated and fully saturated sands with different grain distributions showed that there was no particular coefficient of air permeability. Instead, the coefficient of air permeability was dependent from the pressure gradient. Fine sands showed an increasing coefficient of air permeability with increasing pressure gradient, whereas the coefficient of air permeability decreased in coarse sands while the pressure gradient increased. Fine materials were supposed to be blown out during the test with fine sands and leading to a higher coefficient of air permeability in the probe. In coarse materials, the air was thought to be influenced by turbulent flow behaviour, becoming more important with increasing gradient. Another observation was that once air had flown through a soil, the coefficient of air permeability increased with respect to the first test. It was supposed, that permanent flow channels formed in the soil during the first analysis. Increasing density reduced the coefficient of air permeability, as well as a high level of water saturation. No results were presented with regard to the changes of weight in the cylinder.

In conclusion in reference [15] it was proposed to define a pressure dependent coefficient of air permeability and to distinguish saturated and partially saturated conditions.

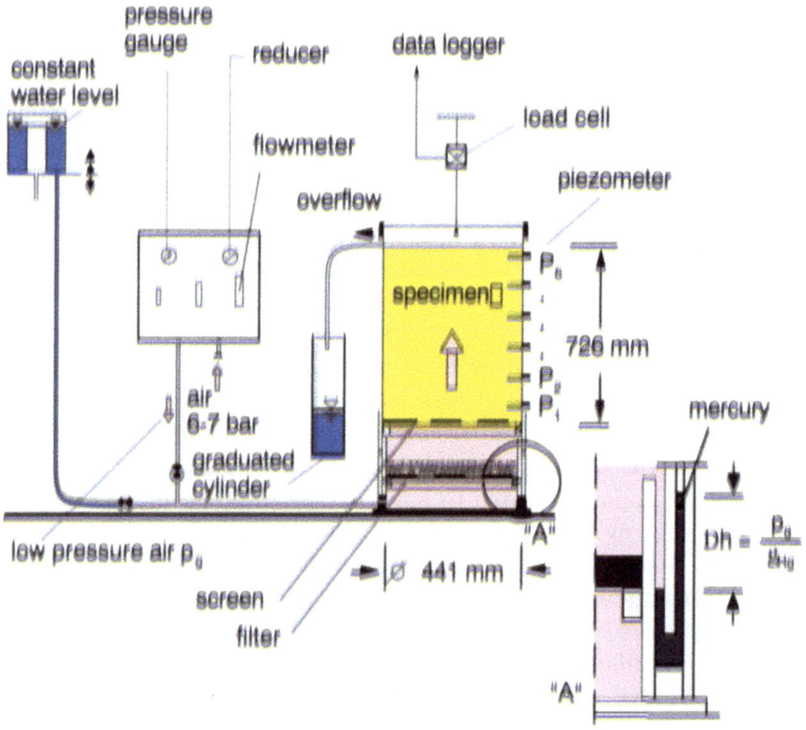

Fig. 3. One-dimensional testing apparatus [15]

Two-dimensional air flow in porous media. Two-dimensional experiments to analyse flow behaviour in a soil adjacent to a model tunnel were reported in [15]. The aim of experiment was to get more detailed information on the distribution of air pressure in the soil surrounding the tunnel. Figure 4 shows a scheme of the testing apparatus. The model includes a longitudinal section of a tunnel in the scale of 1:20. The adjacent soil has an equivalent length of 20 m and a maximum height up to 10 m. Water drainage is controlled by a filter and a drainage tube at the opposite end of the tunnel. An overpressure of air in the tunnel induces a flow field into the ground. The level of pressure in the tunnel, the permeability of the multiple soil layers, the height of soil above of the tunnel and the saturation degree control the size of the flow field and by this the distribution of the pressure in the soil. The front of the apparatus was made of acrylic glass to observe the pressure field. Piezometers in the soil allowed to measure pressures. The amount of air flow was also measured.

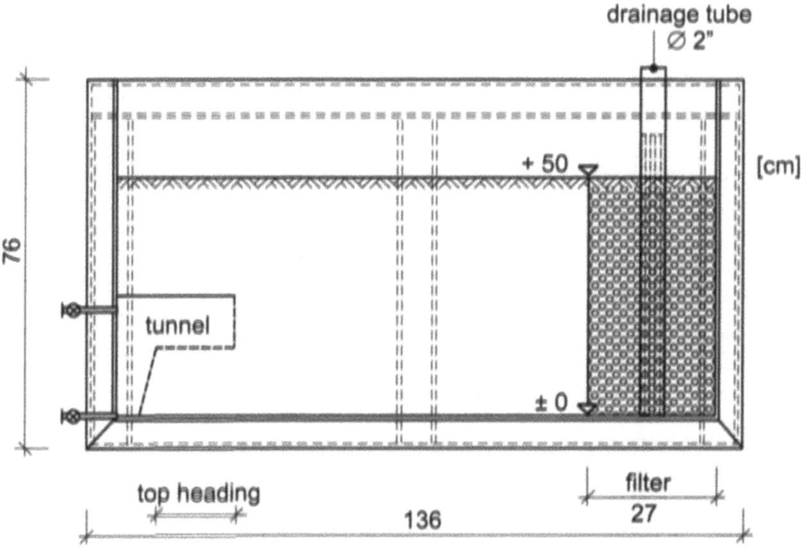

Fig. 4. 2-dimensional testing apparatus [15]

Tests were conducted on dry and partly saturated soils. During the tests it could be observed that in the beginning the flow occurred through pore channels with a width of 2-3 mm. With increasing time water saturation was reduced and the air used even smaller channels to escape the tunnel face. In a layered system with a layer of low permeability above of a layer of high permeability the air flowed nearly horizontally to the opposite end of the apparatus.

Laboratory test to analyse flow of air through cracked shotcrete and adjacent soil. Since the beginning of the Joint Research Initiative SITU in 1997, large-scale laboratory testing methods to simulate the air permeability of the cracked shotcrete lining and the soil have been developed. The testing apparatus was based on the knowledge of the results of in-situ tests, which were conducted for the first time by the German contractor Bilfinger+Berger Bauaktiengesellschaft in 1987 during the construction of the Metro in Essen [8][22] and on the results of experiments described above.

The influence of various parameters was analysed in laboratory tests. In particular, the amount of air pressure, the width of the cracks in the lining and the type of soil were varied. In the first step, the aim of the testing programme was to analyse the flow behaviour through the cracks of the shotcrete lining. In a second step, the soil was added to the system and the tests were repeated. Figure 5 shows the testing apparatus.

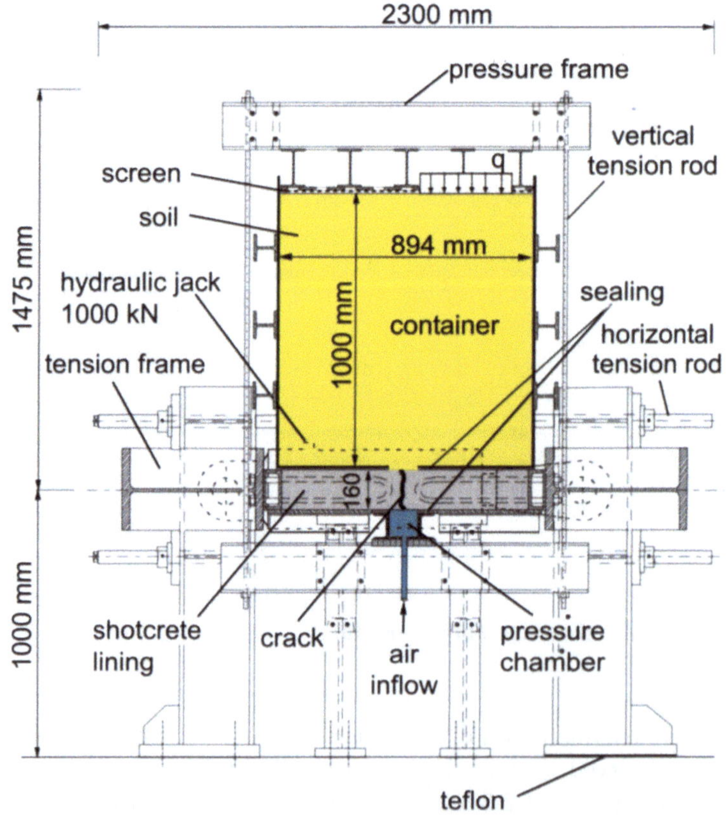

Fig. 5. Laboratory testing apparatus – flow of air through cracked shotcrete and soil [23]

Air flow tests through cracks. Flow of air through the cracks started with the generation of a controlled crack in the shotcrete as shown in Fig. 6. Four fissuremeters (Tab. 3) were fixed on the upper and lower sides of the shotcrete lining. The two hydraulic cylinders and the shotcrete lining were fixed in the frame and a force of total 80 kN was applied on the horizontal tension rods in order to pre-stress the lining. Next the hydraulic cylinders were extended until the specially developed software recognized the crack. Recognition was performed by evaluating the quotient of deformation and force in the cylinder for every time increment. When a defined value was exceeded the force in the cylinder was reduced to zero suddenly. This closed the crack.

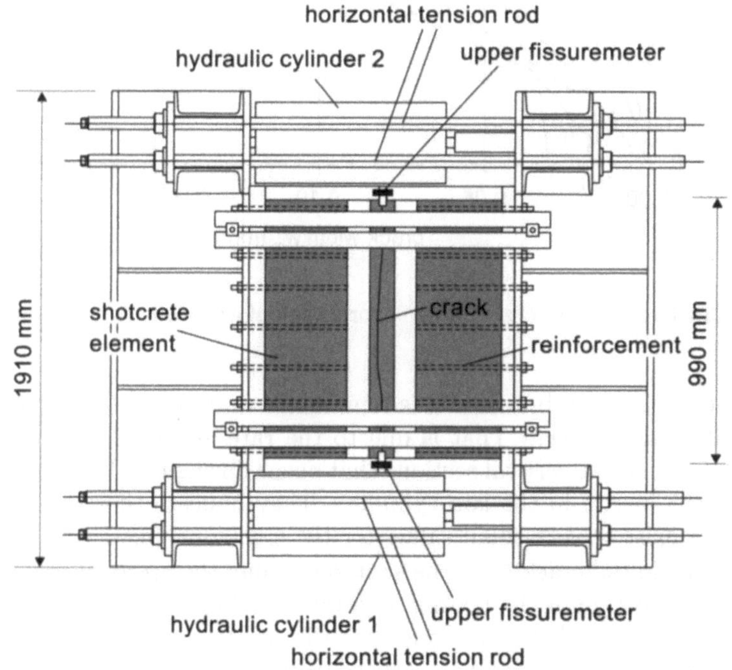

Fig. 6. Plan view of the cracking frame with shotcrete element

A typical deformation-force diagram is presented in Fig. 7. At the beginning (A), the cylinders and the fissuremeters are indicating zero force and deformation. During extension of the cylinders the force and deformation are increasing until cracking is recognized by the software(B). The total force in the cylinders at (B) is F_C= 420 kN. After reducing the force to zero the deformation drops back from w_m= 0,15 mm to w_m= 0,015 mm (C). In one fissuremeter the software did not recognize a crack. Therefore, the crack procedure was restarted. At a force of F_C= 380 kN (D) the cracking was indicated in all four positions. Deformation at (D) was about w_m= 0,18 mm. After unloading a deformation of w_m= 0,02 mm remained in the shotcrete element (E).

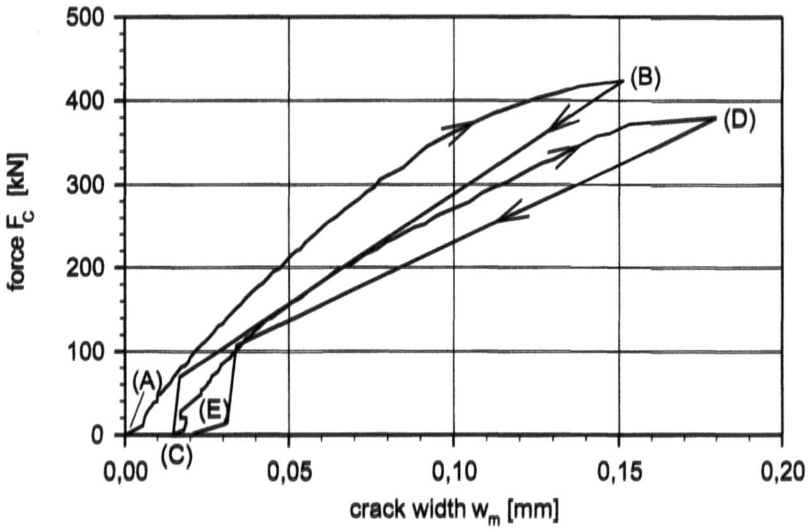

Fig. 7. Cracking the shotcrete lining – Force vs. deformation

On both discharge branches it is remarkable that only relatively few measured values are present. That is due to the rate at which the sample was relieved. The almost vertical sections that occur during the discharge are justified by the inertia in the two cylinders. They indicate that still relatively high forces exist at very small values for the crack opening.

After the initial cracking of the shotcrete lining the pressure chamber was fixed on the bottom of the testing apparatus and air flow tests through the cracks were conducted. Two parameters – the air pressure and the crack width – were varied. The air flow was measured using gas flow meters (Tab. 3). The results were used to compute the coefficient of air permeability as described in [16].

As can be seen in Fig. 8 the coefficient of air permeability of cracks in the 16 cm thick shotcrete lining is pressure dependent. Permeability decreases while the pressure is increasing. The crack width has nearly no influence on the coefficient of permeability. For small crack widths the values are more distributed. This is due to the accuracy of measuring values at low air flow rates. As mentioned before the coefficient of air permeability decreases with increasing pressure. This behaviour can be explained by the more turbulent flow of air at high air pressure.

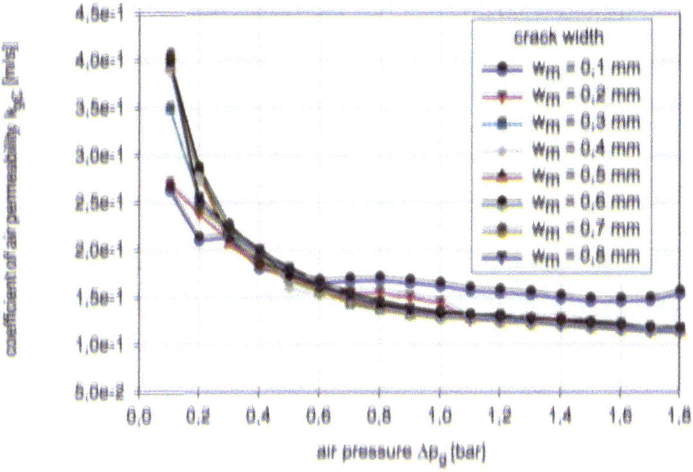

Fig. 8. Coefficient of air permeability of a crack in a shotcrete lining

Air flow tests through cracks and adjacent soil. During testing the overpressure in the pressure chamber was in the range of 100 mbar to 2,0 bar. The air flowed through a 16 cm thick shotcrete lining into the soil (Fig. 5). The crack width could be varied from 0,1 mm to 0,7 mm by the use of a hydraulic tension frame. The air passed through the crack and entered into a soil column with a height of 1 m. The range of experiments included different degrees of initial saturation and two types of soil, sand and silt. Full saturation was not analysed. At the top of the soil column the air could escape freely through a screen. An overload at the top surface representing an overburden was installed to simulate the state of stress according to the tunnel depth.

During the test the air pressure Δp_g was measured in the pressure chamber and at different locations in the soil column. Furthermore, the volumetric airflow Q_g based on standard conditions in the laboratory, the average saturation S_l of the soil column, the overburden applied on top of the soil column, and the temperature of the compressed air were measured.

Figure 9 shows the air pressure field in the sand column with an initial saturation of S_l= 0,52 when applying an overpressure of 990 mbar at a crack width of w_m = 0,30 mm at different time steps. During the test water appeared at top of the sample. After 45 min the water level at the top surface was about 40 mm and the saturation in the sample was reduced to S_l= 0,43. It can further be noticed, that the air pressure at the top of the crack was about 80 % lower than the applied pressure. At this moment, the volumetric air flow through the system of crack and soil was about Q_g= 8,08 m^3/h. After the first three minutes of the test it was Q_g= 7,85 m^3/h.

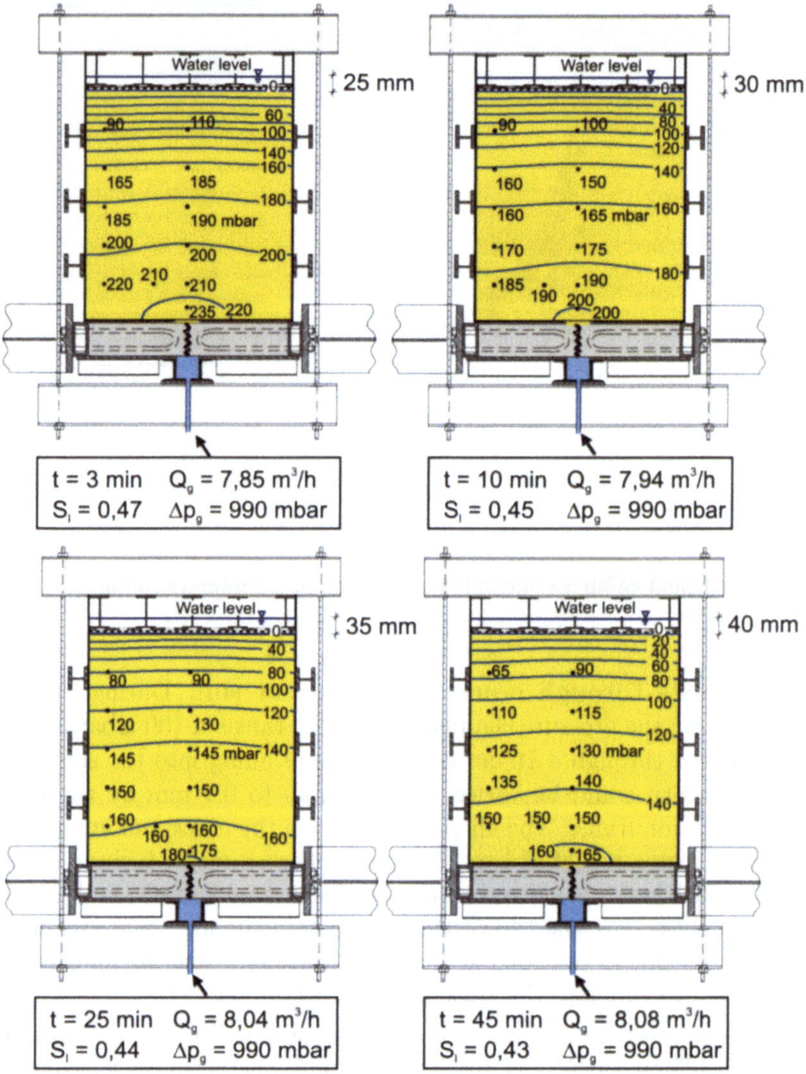

Fig. 9. Distribution of air pressure in the sand at different time steps [16]

Steady state conditions were reached after 45 min. At this time the volumetric airflow Q_g and the saturation S_l stagnated.

In addition, Fig. 9 shows that the highest pressure gradient was located at the top region of the soil column. This was due to a nearly saturated waterfront in this region, which had been reduced when reaching the steady state. At this state the water level above the soil column remained constant.

Tests where the sand was replaced by partially saturated silt did not show a characteristic air pressure potential distribution. The air rather flowed through preferred channels in the soil column.

3.2 Analytical approach

The evaluation of the test results in combination with the development of different theoretical models by [16] to describe the multiphase flow led to an improved calculation method for multiphase flow in partially saturated soils for tunnels driven under compressed air.

Airflow through a cracked shotcrete lining. The mathematical description of the air flow through cracks in the shotcrete lining is based on the laws of Newton on the mechanics of fluids. The law of Hagen-Poiseuille for steady state flow in long straight tubes with circular section extends these laws and these can be modified to describe the behaviour of steady state flow in thin long rectangular gaps. The rectangular gap in this model simulates the shrinkage cracks of the shotcrete support. In the formulation, the compressibility of air is considered. In conclusion, the law to describe the volumetric flow through the crack Q_g can be written as follows:

$$Q_g = \sqrt{\frac{2 \cdot b_c^2 \cdot w_m^3}{l_c \cdot \lambda_{\text{mod}}} \cdot \frac{(p_{g1})^2 - (p_{g2})^2}{\rho_g \cdot p_{atm}}}, \tag{8}$$

where b_c is the breadth of shotcrete lining, w_m is the measured width of the crack at the surface of the shotcrete lining, l_c is the length of the crack. The pressure in the pressure chamber is denoted as p_{g1} and pressure conditions at top of the crack are p_{g2}. p_{atm} is the atmospheric pressure. The density of air is ρ_g. The modified roughness coefficient λ_{mod} was calculated from the test results using (8) and (9) was applied to calculate the Reynolds' Number Re:

$$Re = \frac{2 \cdot \rho_g \cdot Q_g}{b_c \cdot \eta_g}, \tag{9}$$

where η_g is the viscosity of air. Both equations were used to develop the curve presented in Fig. 10, based on the test results. The lower limit line was proposed as the characteristic curve, which is to be used to calculate the volumetric airflow in an iteration process. A Reynolds' Number Re was chosen, which gives a modified roughness coefficient λ_{mod} when put into Fig. 10. This modified roughness coefficient λ_{mod} was inserted into (8) to calculate the volumetric airflow Q_g. Next the result was put into (9) in order to check the chosen Reynolds' Number Re.

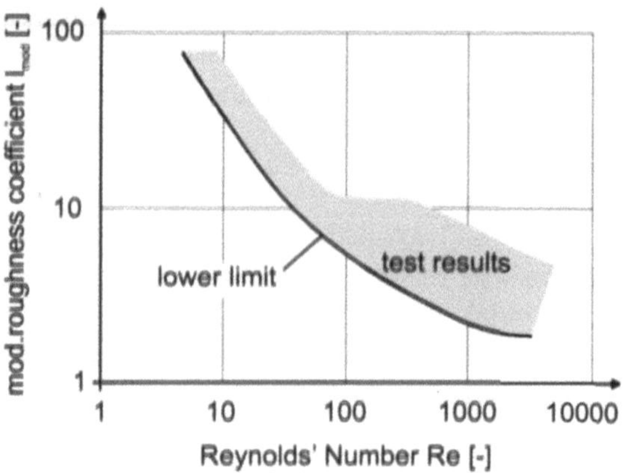

Fig. 10. Reynolds' Number versus modified roughness coefficient [16]

Airflow through partially saturated soil. The airflow in the soil is based on the Newtonian law to describe the mechanics of compressible fluids. This law was extended by the law of Hagen-Poiseuille for steady state flow in long straight tubes with circular section. The porous medium was hereby compared to a bundle of parallel cylindrical tubes. The soil is further idealized as a three-phase mixture, where the law of Darcy is applied for both phases gas and water. The coefficients of permeability $k_g$and k_l were connected by the intrinsic permeability K. The volumetric airflow was then calculated as

$$Q_g = k_{rg} \cdot K \cdot \frac{A_O}{\eta_g} \cdot \frac{(p_{g1} + p_{g2})}{2p_{atm}} \cdot \frac{(p_{g1} - p_{g2})}{l}. \tag{10}$$

Figure 11 shows the relative coefficient of air permeability k_{rg} versus water saturation for test results of the two soils, sand and silt. For a given water saturation the relative coefficient of air permeability may be taken from Fig. 11. A_O is the flow effective cross section at the tunnel face.

It is remarkable that the test results for sand matched the results of former investigations by [24] quite well (Fig. 11). For silt insufficient results were obtained. Consequently, no calibration curve can be proposed. Another phenomenon observed while conducting tests with silt, was that the air flowed through preferred channels. This observation might demand another model as a base for theoretical study of the flow of air through fine material.

Airflow through a cracked shotcrete lining with adjacent soil. Regarding the mathematical description of the air permeability of a crack in a shotcrete lining with adjacent soil the test results were analysed with the previously described theoretical models. It was assumed, that the field of airflow in the first meter behind the lining extends less than 1 m in each direction.

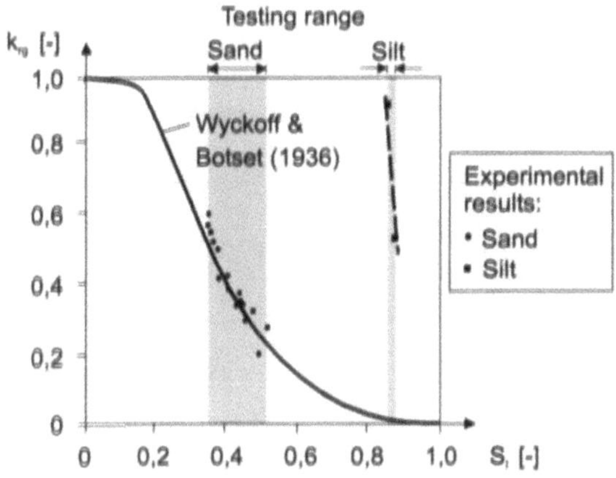

Fig. 11. Saturation versus relative coefficient of air permeability [16]

The airflow through the system of a shotcrete lining with a thickness of 16 cm was calculated as

$$Q_S = k_{gC} \cdot \frac{A}{\gamma_l} \cdot \frac{(p_{g1} + p_{g2})}{2p_{atm}} \cdot \frac{(p_{g1} - p_{g2})}{l} \cdot 60. \tag{11}$$

Equation (11) was then transformed in order to calculate the coefficient of air permeability of the combined system crack and soil. The results are presented in Fig. 12 for an applied overpressure of Δp_g= 100 mbar.

3.3 Air flow tests with improved testing method

The tests described in Sect. 3.1 have been re-conducted with an improved measuring programme focusing on unsaturated soil mechanics. Therefore, TDR-probes (Time Domain Reflectometry [25]-[27]) and tensiometers were installed into the soil. This measuring systems were used in combination with digital pressure gauges. In this testing series for the first time almost full saturated initial conditions could be achieved.

Measuring programme. The extended measuring programme is specified in Tab. 3. As in the tests before, the crack width was measured by means of fissuremeters and the flow measurements took place with glass tube volumetric flow meters. Thermometers, barometers and hygrometers were used to determine and control laboratory conditions.

Within the soil twelve sensors were inserted. They were fixed to the pressure gauges. The installation plan of the sensors is shown in Fig. 13. Water saturation and suction were measured by means of eight TDR-probes and eight tensiometers. The installation is shown in Fig. 13. The symmetry of the

Table 3. Measuring programme

value	instrument	number of instru-ments	range	accuracy	manufacturer
cylinder force	proportional valve to measure oil pressure	2	0 - 1 MN		Ulbrich Hydroautomatica d.o.o.(Slovenia)
crack width	fissuremeter	4	0 - 25 mm	$\pm 1 \mu m$	Heidenhain GmbH
air flow	gas flow meter	4 (different ranges)	0,195 - 880 m^3/h	$\pm 0,5\%$	Brooks Instruments
atmospheric pressure	barometer	1	0 - 1030 mbar		
temperature	thermometer	1	-50 $^\circ$C - +140 $^\circ$C	± 2 $^\circ$C	Elpro-Buchs
relative humidity	hygrometer	1			
pressure	pressure gauge PE 350	10	-1 – +1 bar	$\pm 0,2$ %	Hottinger Baldwin Messtechnik
	pressure gauge PE 350	2	0-4 bar	$\pm 0,2$ %	Hottinger Baldwin Messtechnik
	pressure gauge PE 300	3	0-20 bar	$\pm 0,3$ %	Hottinger Baldwin Messtechnik
saturation	TDR-probe	8	0 - 100 %	$\pm 2\%$	Easy Test (Poland)
suction	tensiometer	8	0 - 900 mbar	± 10mbar	Easy Test (Poland)
strain	strain gauge	4	-2 - 4 $\cdot 10^{-2}$ mm/m	$\pm 0,2$ %	Hottinger Baldwin Messtechnik

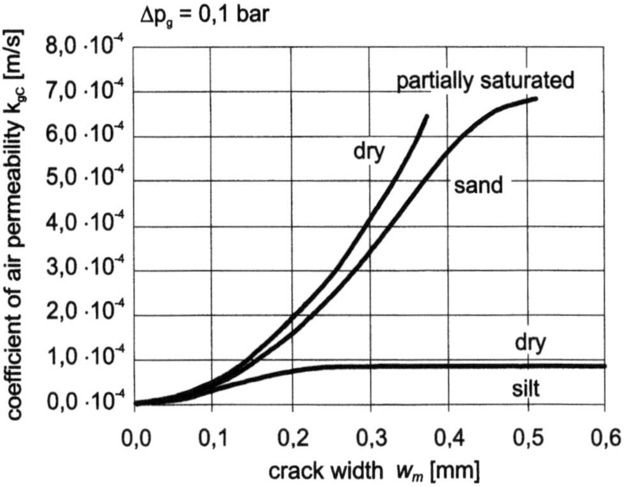

Fig. 12. Coefficient of air permeability of the system crack/adjacent soil versus crack width [16]

test was taken into consideration. The shown installation was chosen in order to get a surface resolution of the air flow field within the soil. Strain gauges were used to indicate deformation of the soil due to the forces of the flowing air. They were fixed on the vertical tension rods that can be seen in Fig. 5.

Performance of the test. A silty sand, common for compressed air application, was used for the experiment. Characteristics of the soil are given in Tab. 4. The soil was installed in layers of 5-10 cm with approximately optimum water content and compacted at proctor energy.

After installing the soil, a load was applied on top of the soil to simulate a shallow tunnel depth. The installation is shown in Fig. 5. Strain gauges were used to measure the deformation of the vertical tension rods and subsequently the forces acting on top of the soil. The force acting on top of the soil was about 60 kN.

Prior to air flow testing the soil was saturated with water from top down. No specified gradient was defined. The water inflow was driven by gravity. The TDR-probes were used to control the saturation. Figure 14 shows the development of water saturation during water infiltration phase. The numbering of the probes is corresponding to the scheme in Fig. 13.

During the saturation phase of the silty sand the saturation increased about 3 % at every location of the test. Unfortunately, data is missing at the beginning of the saturation phase. After five days no more change in saturation could be observed. After 7,5 days the saturation phase was stopped. Parallel to the saturation development the deformations of the soil were ob-

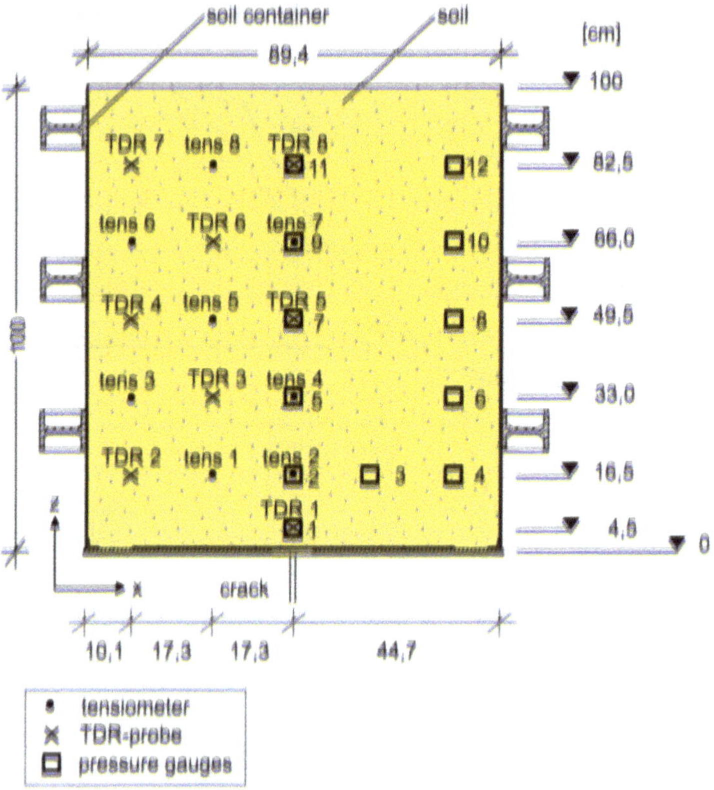

Fig. 13. Installation of pressure gauges, TDR-probes and tensiometers

served. A second criterion for stopping the wetting phase was the end of deformations due to consolidation.

Tensiometers were used to measure capillary pressure in the soil container. Capillary pressure of a soil depends strongly on the saturation conditions of an unsaturated soil. The capillary pressure during the wetting phase can be seen in Fig. 15.

It can be observed, that during saturation the capillary pressure decreases with increasing water saturation until negative capillary pressures are reached at approximately 90 % of water saturation. This indicates positive pore water pressure in the soil.

The air flow test was then performed. A compressor provided compressed air. The compressed air flowed through an air tight pressure chamber through the crack and into the soil above. The amount of air was measured by the flow meter before entering the pressure chamber (Tab. 3). The influence of the air flow on the deformation of the soil was registered in the strain gauges. Within the soil tensiometers, TDR-probes, and the pressure sensors registered the

Table 4. Characteristics of chosen silty sand

a) soil parameters

grain size distribution	clay	4,3
	silt	23,0 %
	sand	70,5%
	gravel	2,2%
effective grain-size	d_{10}	0,011 mm
coefficient of uniformity	U	18
grain density	ρ_s	2,67 g/cm^3

b) installation parameters

dry density	ρ_d	1,77 g/cm^3
porosity	n	0,33
void ratio	e	0,51
coefficient of water permeability	k_l	$4{\cdot}10^{-7}$ m/s

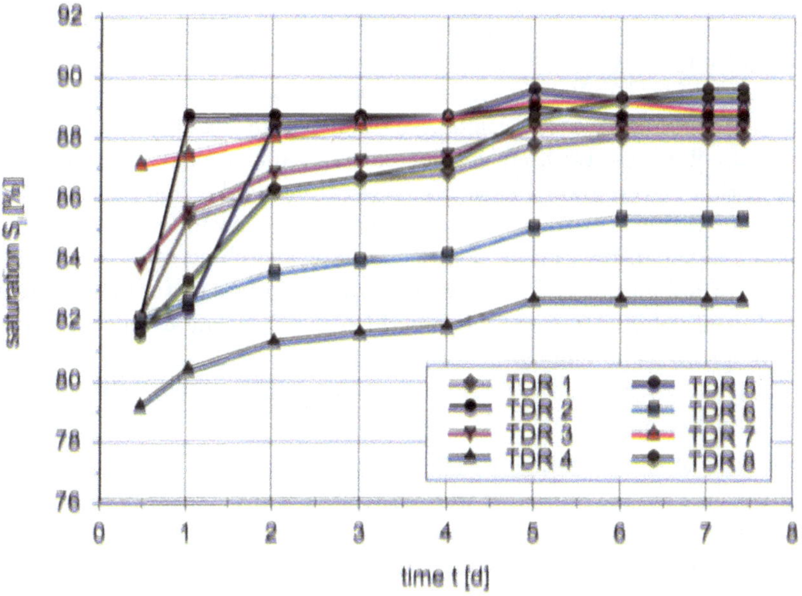

Fig. 14. Wetting phase – Saturation vs. time

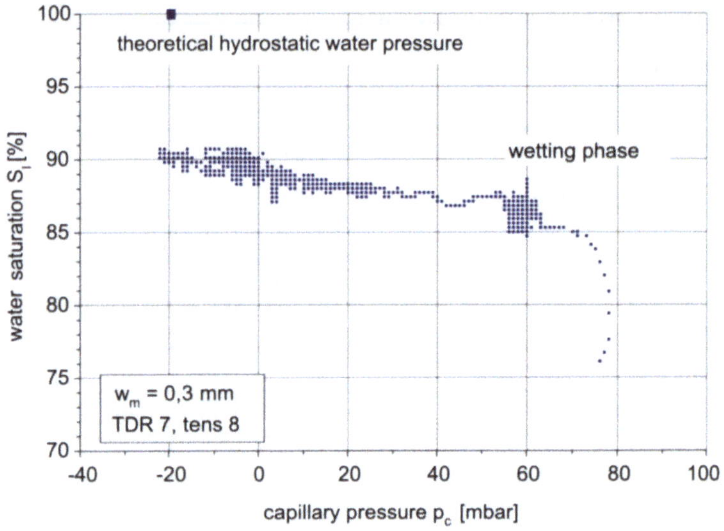

Fig. 15. Capillary pressure vs. water saturation

influence of the air flow on the soil. The test was performed with a crack width of w_m= 0,3 mm at a pressure level of Δp_g= 500 mbar.

Figure 16 shows the evolution of the gas flow rate during the air flow test. Initial flow rate of 15 m^3/h is reached soon after start of the air flow test. For almost six days the air flow rate is increasing, before steady state conditions were reached. Those results are compared to the water saturation measurement in Fig. 17. Values are presented for two probes on top and at the bottom of the container (positions are described in Fig. 13). With beginning of the air flow test both curves decrease instantaneously. The value of water saturation in the lower probe decreases about 20 %. For the next five days water saturation is decreasing with declining velocity in both probes. No additional change can be seen after 5 days. After switching off the air pressure the saturation measured at the upper probe increases instantaneously to almost its initial value. After one day a sudden increase can also be observed in the lower probe. The change in water saturation is about 30 %. For the last three days of re-saturation (second wetting phase) no more changes are measured in either probe.

An analysis of the pressures in the soil container at four different time steps is given in Fig. 18. Pressures were measured using digital pressure meters (Tab. 3). The thick line in the four plots represents the wall of the container. Atmospheric pressures reign on top of the container. At the beginning of the test, the pressure meters showed hydrostatic water pressure.

One hour after applying a pressure of Δp_g= 500 mbar in the pressure chamber a pressure of Δp_g= 431 mbar was measured in the lowest point of

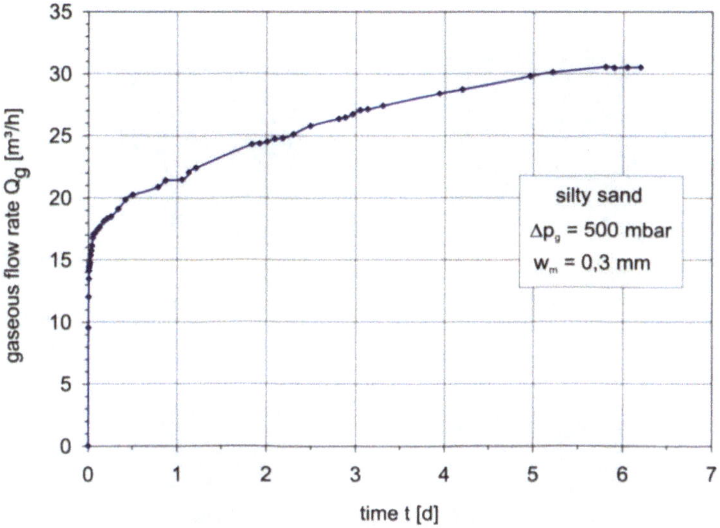

Fig. 16. Evolution of the gaseous flow rate during the air flow test

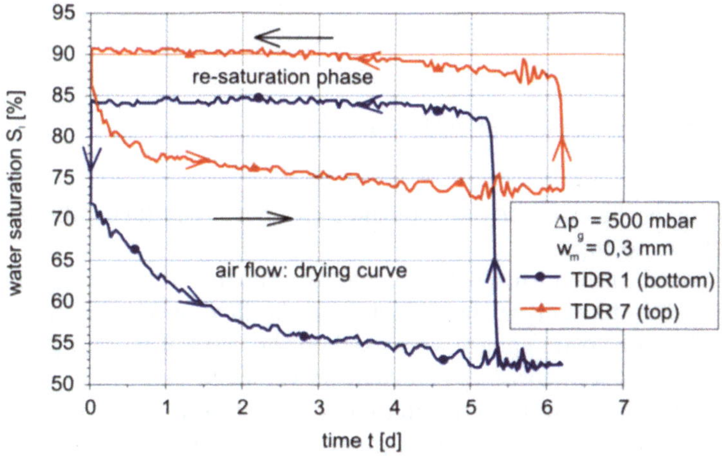

Fig. 17. Water saturation measurement in two probes vs. time

the box (Fig. 18, top-left). The difference to 500 mbar is the pressure that is relieved in the crack. After 0,5 days a decrease in pressure can be observed on the lowest point of the soil in the container. The remaining pressure is Δp_g= 404 mbar. This means that pressure gradient in the crack is becoming more important. A further decrease can be measured until the second day of testing. Almost no changes occur between the second and the sixth day. Hence, steady state conditions seems to appear in the soil box. A further

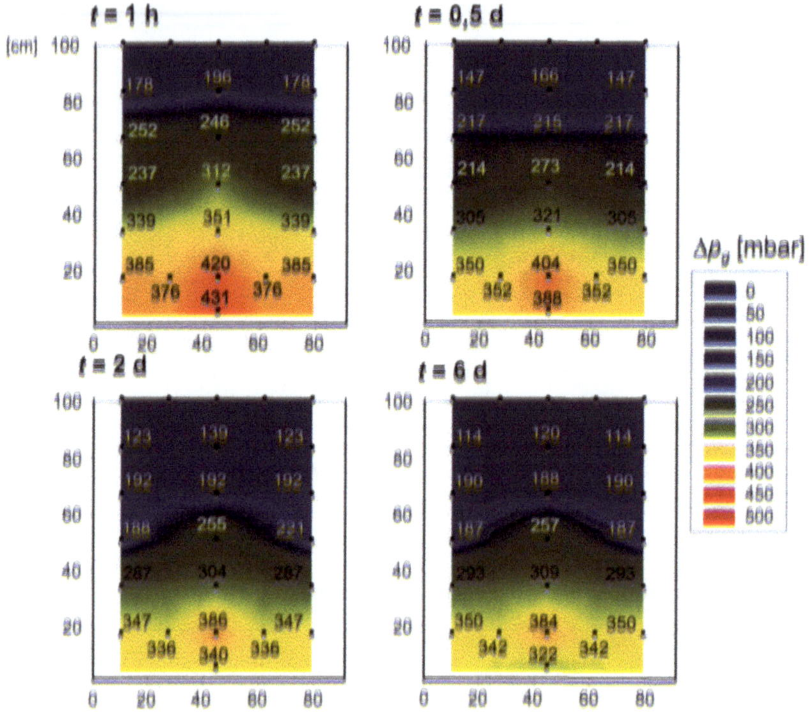

Fig. 18. Pressure distribution in the soil

observation is that the pressures in the container are reduced almost linearly from bottom to top.

Compared to the formerly conducted experiments at this stage of the tests it was possible to analyse the influence of the flow on the grain matrix. As described before strain gauges were fixed on the vertical tension rods (Fig. 5). Figure 19 shows a diagram of deformation versus time.

At the beginning of the flow phase the tension rods were pre-strained due to the force applied in the wetting phase. This strain was set to zero before the flow phase began. Directly after applying the pressure, the strains in the four tension rods increased to about $\varepsilon = 2{\cdot}10^{-5}$. This is due to the upward flow of air.

Before the end of the first day a maximum strain is reached in all four tension rods and strain is decreasing, until negative values are reached in every rod between the second and the fifth day of the testing. At the end of the analyses strain in the rods is lower than at the beginning. A possible explanation is given in the conclusions.

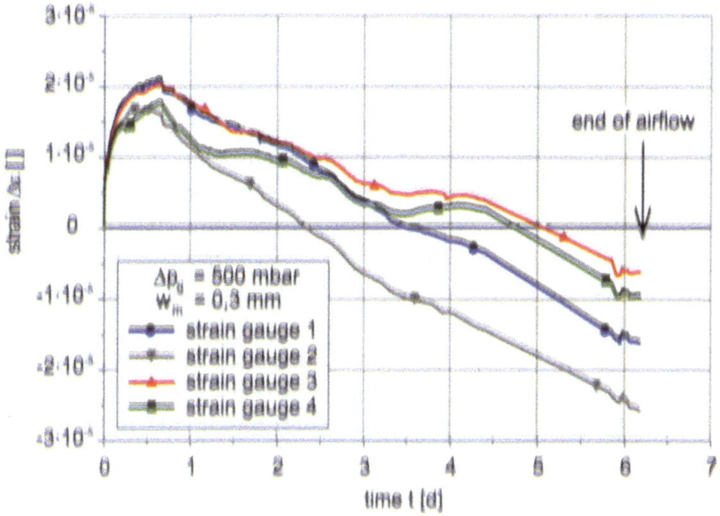

Fig. 19. Deformations in the laboratory test versus time

Conclusions.

Air flow. The flow tests at this stage have given a deeper knowledge on the behaviour of air flow through cracked shotcrete lining and adjacent soil due to NATM in combination with compressed air. The results allow better estimation of the amount of air flow through the cracks in the soil, that are crucial with respect to the total air flow in the compressed air area of the tunnel. The presented laboratory testing results are taken into consideration of the tunnel advance model presented in 5.

Deformation behaviour. In the past heaving of the soil could be observed during compressed air tunnel advance e.g. [28,19]. The deformation behaviour in the laboratory test has been described previously. Heaving of the soil was observed in the beginning of the air flow phase that turned to settlement after 2-5 days. Two reasons might explain this observation.

One possibility might be the loss of buoyancy of the water due to the dewatering process during air flow phase. The buoyancy of the water might not be replaced completely by the upward acting air that is flowing through distinct flow channels. It has to be considered that after dropping down the compressed air in the tunnelling area the uplifting forces of the air are missing completely. At the same time the soil is not fully saturated anymore.

Secondly, the de-saturation of the soil leads to an increase of capillary pressure. High values of capillary pressure are possible. The capillary pressures are acting on the grain matrix possibly leading to additional contraction. This has been reported in [29]. Since the observed deformations have been very small, further analyses should be performed.

4 Numerical Modelling

4.1 TOUGH2

One aim of the research project "Compressed Air" was to develop a numerical model that allows simulation of the flow behaviour in the laboratory test with the option of extending it to simulations of compressed air tunnel advance. TOUGH2 [30] was chosen to perform the numerical analyses. Besides its special relation to the multi-phase flow in porous media, the modular structure of TOUGH2 allows extensions to handle new fluid mixtures with various geotechnical and thermophysical properties [31].

TOUGH2 uses the Integrated Finite Difference Method (IFDM) to solve heat transfer problems in heterogeneous isotropic multidimensional flow regions [32]. Conductive heat transfer is conceptually similar to fluid flow in porous media. The central concept of IFDM is to discretize the total flow domain into conveniently small sub domains (elements) and evaluate the mass balance in each element.

Integrated Finite Difference Method (IFDM). Most of the computational work in the numerical simulation of fluid and heat flow in porous media arises from the solution of large systems of linear equations $\mathbf{Ax} = \mathbf{b}$, where $\mathbf{A}$ is a banded matrix of order N, $\mathbf{x}$ is the vector of the unknowns and $\mathbf{b}$ is the right hand side vector of known. These are solved using either direct or iterative methods. In the TOUGH2 general-purpose reservoir simulator the matrix $\mathbf{A}$ is a Jacobian with a consistent structure. For a flow problem with N_e mass-and-energy balance equations per grid block (N_e = NEQ in TOUGH2 nomenclature), the Jacobian consists of many N_e-dimensional sub-matrices. For a grid block n (Fig. 20), derivatives corresponding to accumulation terms will generate a sub-matrix in location (n,n). Each flow term between n and a neighbouring grid block m will give rise to two off-diagonal sub-matrices in locations (n,m) and (m,n), and will also contribute to the diagonal sub-matrices at (n,n) and (m,m). Thus, the incidence matrix will be symmetric, although for multi-phase and non-isothermal flow problems matrix $\mathbf{A}$ generally may be far from symmetric. Each grid block is typically connected only to a few other blocks so that $\mathbf{A}$ will be sparse [33].

Figure 20 shows an element in the IFDM network. As one can see there is no need for rectangular elements. The average properties of element 7 are associated with the representative nodal point $m = 7$. Nodal point $m = 7$ is connected to the 6 neighbouring nodes n. To ensure sufficient accuracy connection lines (interfaces) should be perpendicular to the borderlines of the elements and situated at an appropriate mean position (arithmetic mean of nondivergent coordinates, log mean of cylindrical radii, or geometric mean of spherical radii). In practice this might be difficult to achieve but should be approximated as closely as possible. The nodal points may be located anywhere within the element [32].

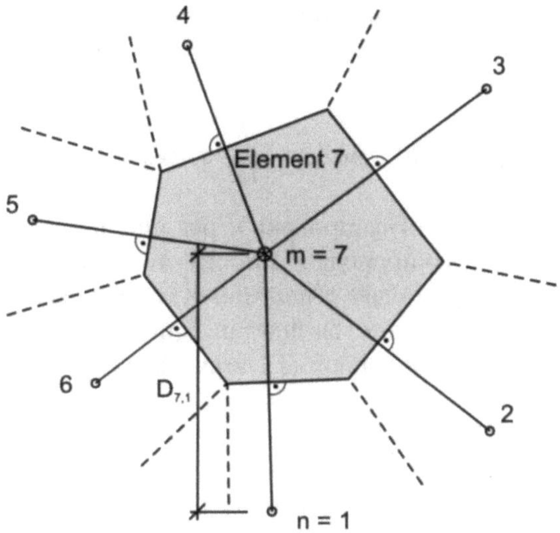

Fig. 20. Discretized Element with representative nodal point and neighbouring nodal points in the IFDM network

As mentioned before, $m = 7$ is connected to the adjoining nodes $n = 1$ - 6. The finite difference approximation can be written for one component condition as follows:

$$g_m V_m + \sum_n \bar{k}_{m,n} \frac{\phi_n - \phi_m}{D_{m,n}} A_{m,n} = c_m V_m \frac{\Delta \phi_m}{\Delta t}, \tag{12}$$

where g_m is the flow rate per unit volume of element m and V_m the volume of element m. $\bar{k}_{m,n}$ denotes the mean hydraulic conductivity between elements m and n and ϕ the hydraulic head. $D_{m,n}$ is the distance between nodal points m and n and $A_{m,n}$ is the area of interface between elements m and n. c_m is the specific storage for element m. The Integrated Finite Difference Method does not need to make reference to a global system of coordinates. Very irregular grid systems may be used, which may result in nearly random, but symmetric, sparsity structure. However, for regular grid systems involving global coordinates such as axisymmetric (r,z) and three-dimensional (x,y,z) grids, the IFDM is equivalent to conventional finite differences, giving rise to Jacobian matrices with regular banded structure [33].

Mass and energy balances. TOUGH2 [30][33] is a general-purpose simulator for non-isothermal flow of multiple fluid components (NK) distributed among multiple phases (NPH). Its modular structure was built on the recognition that the mass-and-energy balance equations for flow in porous media have the same form, regardless of the nature and number of fluid components and phases present. The balance equations for component κ (water, air, CO2, tracers, etc.) are written in integral form for an arbitrary flow region V_n with surface area A_n as follows [33]:

$$\frac{d}{dt}\int_{V_n} M^{\kappa} dV_n = \int_{A_n} \mathbf{F}^{\kappa} \cdot \mathbf{n} dA_n + \int_{V_n} q^{\kappa} dV_n \tag{13}$$

Here M^{κ} is the mass of component κ per unit porous medium volume, $\mathbf{F}^{\kappa}$is the mass flow of component κ into V_n, $\mathbf{n}$ is the inward unit normal vector and q^{κ} is the rate of mass generation of component κ per unit volume. The mass flow $\mathbf{F}$ is a sum over the flows in liquid and vapour phases, which are written as a multiphase version of Darcys' law expressed by

$$\mathbf{F}_{\beta} = -\mathbf{K}\frac{k_{r\beta}}{\mu_{\beta}}\rho_{\beta}(\nabla p_{\beta} - \rho_{\beta}\mathbf{g}), \tag{14}$$

where $\mathbf{K}$ denotes the intrinsic permeability tensor, $k_{r\beta}$ is the relative permeability of phase β, μ_{β} is the viscosity of phase β, p_{β} is the fluid pressure in phase β and $\mathbf{g}$ is the gravity vector. The fluid pressure in phase β is the sum of the pressure p (in the presented case taken to be the gas phase), and the capillary pressure $p_{c\beta}$ (≤ 0)

$$p_{c\beta} = p - p_{\beta}. \tag{15}$$

Capillary pressure-saturation function. TOUGH2 offers various possibilities to define empirically based capillary pressure-saturation function [34], for example the capillary pressure p_c may be defined as

$$p_c = -p_0\left([S_e]^{-1/\lambda} - 1\right)^{1-\lambda}, \tag{16}$$

where the effective water saturation S_e is

$$S_e = \frac{S_l - S_{rl}}{S_{sl} - S_{rl}}. \tag{17}$$

S_l denotes the water saturation and S_{rl} the residual water saturation, while S_{sl} is the water saturation at full saturation. The curve fitting parameter p_0 is defined by

$$\frac{1}{p_0} = \frac{\alpha}{\rho_l \cdot g}, \tag{18}$$

where α is another curve fitting parameter as λ and n are. ρ_l is the density of water and g the gravity vector.

$$\lambda = 1 - \frac{1}{n}. \tag{19}$$

The capillary pressure p_c is restricted to

$$-p_{\max} \leq p_c \leq 0. \tag{20}$$

Relative permeability-saturation function. The function used in this simulation is the van Genuchten-Mualem model [34] and [35]. The relative permeability regarding the aqueous phase k_{rl} appears as follows

$$k_{rl} = \begin{cases} \left\{ \sqrt{S_e} \left\{ 1 - \left(1 - [S_e]^{1/\lambda} \right)^{\lambda} \right\} \right\}^2 & if\ S_l < S_{rl} \\ 1 & if\ S_l \geq S_{rl} \end{cases} . \tag{21}$$

The relative permeability with respect to the gaseous phase k_{rg} can be chosen as one of the following two forms

$$k_{rg} = \begin{cases} 1 - k_{rl} & if\ S_{rg} = 0 \\ (1 - \hat{S})^2 (1 - \hat{S}^2) & if\ S_{rg} > 0 \end{cases} . \tag{22}$$

The term for $S_{rg} > 0$ is due to [36]. The equation is subjected to restriction $0 \leq k_{rl}, k_{rg} \leq 1$. S_e is to be calculated as shown above and $\hat{S}$ is determined by

$$\hat{S} = \frac{S_l - S_{rl}}{1 - S_{rl} - S_{rg}} . \tag{23}$$

4.2 Numerical model of the laboratory test

The first step of the numerical simulation is the evaluation of the laboratory test described in Sect. 3. The chosen Finite Difference model is a box in the xz-plane filled with a porous medium and bounded by two impermeable planes, on the left (BL 2) and on the right side (BL 1). The top of the model (BL 3) is of free-flow boundary type. The boundary conditions shown in Fig. 21 read

$$\begin{array}{ll} Q_g, Q_l = 0 & for\ BL\ 1\ and\ BL\ 2 \\ \left.\begin{array}{l} Q_g, Q_l \neq 0 \\ p_g = const = p_{atm} \\ p_l = const \end{array}\right\} & for\ BL\ 3 \end{array}$$

where Q_g denotes the flow of air and Q_l the flow of water. p_g is the pressure in the air phase and p_l the pressure in the water phase. The index atm indicates atmospheric pressure conditions.

Pressure conditions were defined as Δp_g= 500 mbar in the pressure chamber and reigning atmospheric conditions on BL 3. Soil characteristics are to be defined for the silty sand used in the simulation. It is the same material that has been used in the primary conducted laboratory test described in 3.3. Grain density ρ_s, porosity n and intrinsic permeability K are given in Fig. 21. An average value for the initial water saturation S_l was chosen.

The crack in the shotcrete lining is modelled as a continuum element with appropriate porosity, intrinsic permeability and saturation. The porosity is

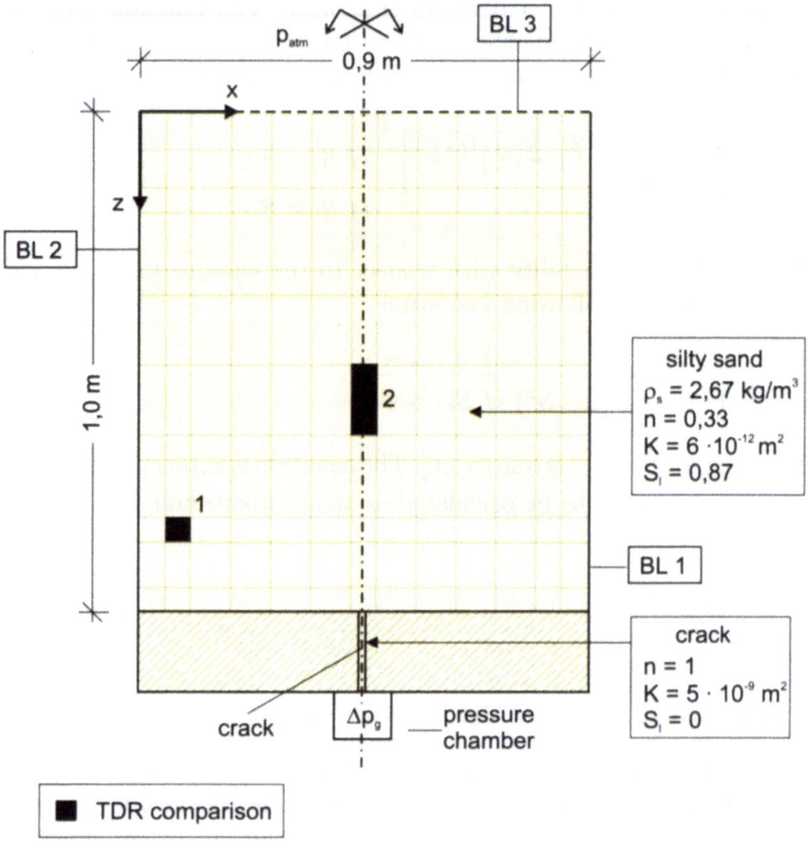

Fig. 21. Preliminary model of the laboratory test

n equal to one and saturation with respect to the water phase is zero S_l. Intrinsic permeability K is determined using the test results obtained by a flow analysis of air through the crack (Fig. 8). The values are shown in Fig. 21.

4.3 Simulation results

Figure 22 shows the results of the simulation with respect to the air flow for a given crack width of w_m= 0,3 mm and an over pressure of Δp_g= 500 mbar. Air flow into the soil is output at the interface between crack and soil. The parameters of the relative permeability-saturation function were fitted to show agreement between both curves in Fig. 22. Air flow is slightly underestimated in the beginning of the air flow test. A slight overestimation can be observed for the period of 0,5-5 days. Hence, the form of both curves is almost identically and differences are only small.

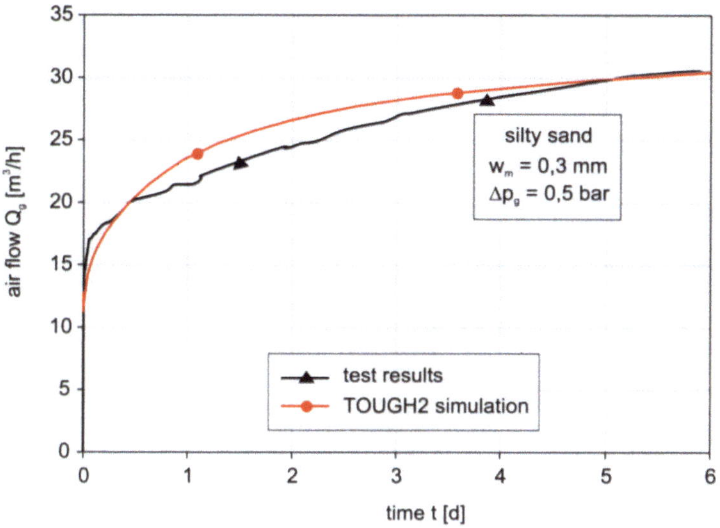

Fig. 22. Comparison of air flow in simulation and laboratory test

Next the results of the simulation are compared with the measured values of water saturation (Fig. 23). Two TDR-probes are chosen for the evaluation of the results. The probes are shown in Fig. 21. TDR1 is located in the lower part of the container close to a side wall and TDR2 is approximately in the middle of the container. Figure 23 shows different behaviour of test and simulation. In the test the decrease in water saturation stopped after short time. This is not represented in the simulation, where saturation is still decreasing after a time of 6 days. Until now the saturation behaviour was not considered in more detail, instead more importance is typically given to the amount of air flow through the system. Principally the correct behaviour of decreasing water saturation is shown. Further analyse will focus on the dewatering of the soil. A possible explanation for difference between test results and simulation is given in Sect. 4.4.

Finally, the air pressures in the soil are compared. Those are measured in the test with eleven digital pressure meters in the soil as described in 3.3. Six pressure meters are vertically positioned and five are horizontally placed in the centre of the box. The lowest position is not occupied. The comparison of simulation and test is presented in Fig. 24. Clearly, in the simulation the conditions are symmetric with respect to a vertical axis in the middle of the model. Symmetric conditions in the laboratory test as shown in Fig. 24 are assumed.

As an example a situation 12 h after the start of the analyses is represented in Fig. 24. Good agreement between the measured values of pressure and computed ones is obvious. This has been observed for the other time steps as well.

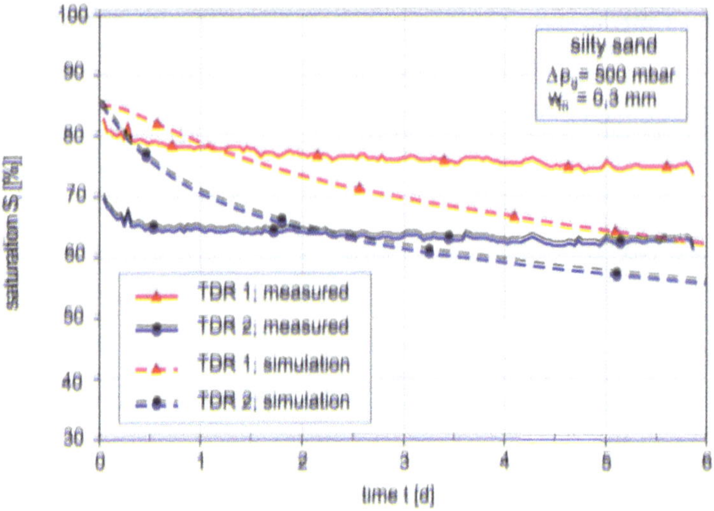

Fig. 23. Comparison of saturation (water phase) in simulation and laboratory test

4.4 Conclusions

Aim of this work was to collect experience from simulations in multi-phase flow and to analyse the plausibility of numerical results for further investigations. It was shown through a preliminary numerical simulation that the use of TOUGH2 gives reasonable results to describe the flow of air in a box filled with a porous medium. Various equations describing unsaturated flow behaviour of immiscible phases could be applied. They provide a wide range of possibilities to model the porous media. The results are strongly influenced by the choice of adequate parameters. In this study, the flow of the gaseous phase was identified and evaluated on the results of the laboratory tests. Compared to a preliminary study, presented in [37], the crack could be successfully implemented in the simulation.

The dewatering behaviour of the soil needs to be evaluated in more detail. It might be difficult to compare the results to measurements by TDR-probes. TDR-probes give result on very small areas, while the numerical model is based on a continuum. Under real conditions flow paths of air might develop beside of the TDR-probes without showing any influence on the probe measures.

5 Tunnel Advance Model

5.1 Introduction

This section describes the development of a model to calculate the air loss during the advance of a tunnel project driven by NATM. The model is in-

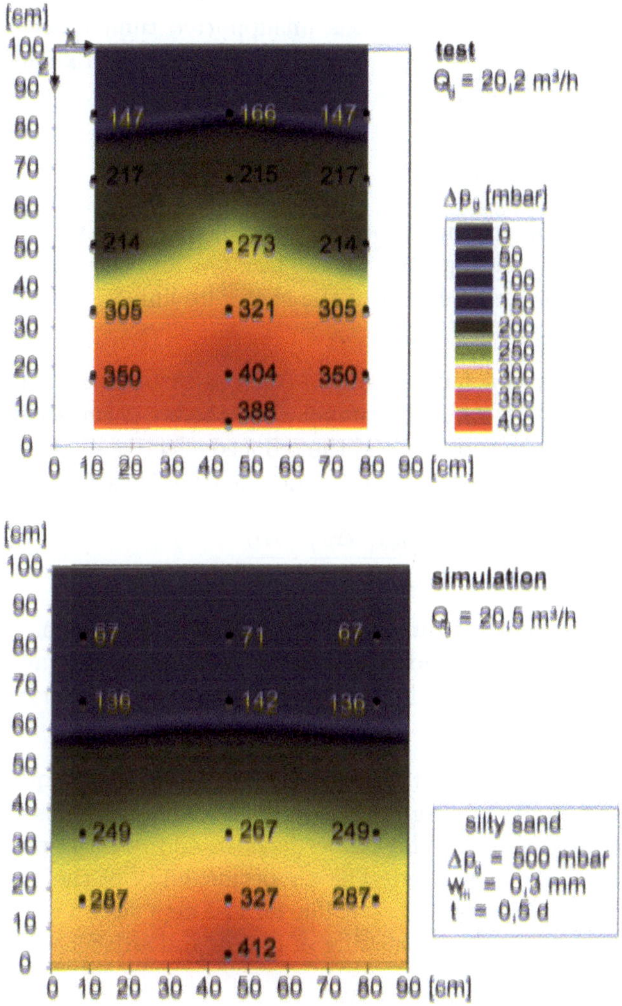

Fig. 24. Comparison of air pressure simulation and laboratory test

tended to serve as a tool to the planning engineer in order to get better knowledge on the total costs of the tunnel project during the design phase. It helps to plan the need of compressor power on the different phases of the tunnel advance. The two parameters governing the air loss are the position of the tunnel face and the height of water above of the tunnel face. Hence, in the presented model the total loss of air Q_g is a function of position x and of the height of water h_l above of the tunnel base at the face:

$$Q_g = f(x, h_l). \tag{24}$$

As described before [8] the total amount of air loss Q_T is the sum of the air loss that takes place through the unsupported tunnel face Q_F, through shrinkage cracks of the shotcrete lining Q_S and the approximately constant amount of air loss through air locks and pipes Q_L.

5.2 Governing equations

Governing equations are developed for two parts of the total air loss. One controls the air loss through the tunnel face Q_F and the other one controls air loss through shrinkage cracks of the shotcrete lining Q_S.

Air loss through the tunnel face. Air loss through the tunnel face depends on the air pressure in the tunnel Δp_g, the air permeability of the soil k_g, the length of the flow path l and the reduced effective section of the tunnel face A_O which is taken into account. The equation is as follows:

$$Q_F = C \cdot \frac{k_g}{\gamma_l} \cdot \frac{(\Delta p_g + 2p_{atm}) \cdot \Delta p_g}{2p_{atm}} \cdot \frac{A_O}{l}. \tag{25}$$

The factor itC controls the three-dimensional propagation of the air into the soil. Values between 2 and 4 have been used in former analyses [6,16]. γ_l is the unit weight of water and p_{atm} determines the atmospheric air pressure.

Air loss through cracks in the shotcrete lining. The equation to estimate air loss through shrinkage cracks in the shotcrete lining is similar to the one for the air loss through the tunnel face. In addition the permeability of the cracks has to be considered. The air permeability of the cracks k_{gC} has been determined in laboratory tests. The air loss depends on the permeability k_{gLC} of the layered system consisting of cracks and adjacent soil:

$$Q_S = \frac{k_{gLC}}{\gamma_l} \cdot \frac{(\Delta p_g + 2p_{atm}) \cdot \Delta p_g}{2p_{atm}} \cdot \frac{A_C}{l}. \tag{26}$$

Since the equation describes the air loss of the combined system crack and soil the equivalent section A_C corresponds to a cross section of 1 m^2. l is the length of the flow path. γ_l is the unit weight of water and p_{atm} determines the atmospheric air pressure.

5.3 Principles of modelling

The first step of the modelling is the subdivision of the tunnel into n blocks of the length d_{bl} (Fig. 25). The block length d_{bl} should be governed by the expected daily tunnel advance.

A block is first considered when it is reached by the tunnel face. Subsequently, the same block will be taken into account with regard to the air loss

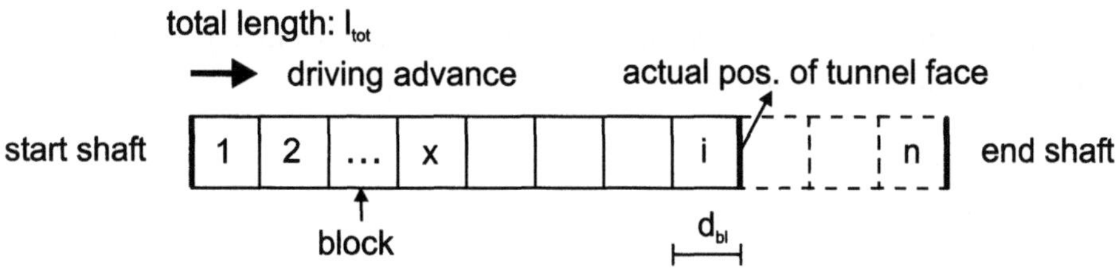

Fig. 25. Subdivision of the tunnel

through its shrinkage cracks. This block will be active until the total length is reached and compressed air is turned down. A specific air pressure $\Delta p_{g,i}$ is assigned to each block. The air pressure $\Delta p_{g,i}$ depends on the height of water above of the tunnel base h_l. With regard to the air loss through the tunnel face the air pressure $\Delta p_{g,i}$ in this block will be taken into account. Furthermore, it controls the air loss through the cracks of the shotcrete elements of all blocks behind of the tunnel face. Further parameters of the governing equations are assigned to every block as described below.

5.4 Determination of coefficient of air permeability

Coefficient of air permeability of the soil. It is assumed that flow occurs at constant degree of saturation. The parameters are defined for every block i.

Coefficient of air permeability of the crack. The coefficient of air permeability of the crack k_{gC} is obtained from tests conducted in the laboratory of the Institute for Soil Mechanics and Foundation Engineering at Graz University of Technology. Those results are described in 3.3. The parameters are defined for every block i.

Coefficient of air permeability of a layered system. Figure 26 shows a compressed air tunnel in a layered soil. The air that escapes at the tunnel face flows through the soil layers 1, 2 and 3 with different coefficients of air permeability. It is assumed that the length of the flow path l is equal to the sum of the thickness of the soil layers d_i:

$$l = \sum_{i=1}^{n} d_i. \tag{27}$$

The coefficient of air permeability of the layered system k_{gL} at the tunnel face is calculated as follows:

$$k_{gL} = \frac{\sum_{i=1}^{n} d_i}{\sum_{i=1}^{n} \frac{d_i}{k_{gi}}}, \tag{28}$$

where n corresponds to the total number of soil layers. The coefficient of permeability with respect to the air loss through the shrinkage cracks is only slightly different:

$$k_{gLC} = \frac{d_C + \sum_{i=1}^{n} d_i}{\frac{d_C}{k_{gC}} + \sum_{i=1}^{n} \frac{d_i}{k_{gi}}}. \tag{29}$$

The length of the flow path l of this system corresponds to the distance from the inner side of the lining to the surface of the groundwater table. Atmospheric conditions reign above of the groundwater table. The parameters are defined for every block i.

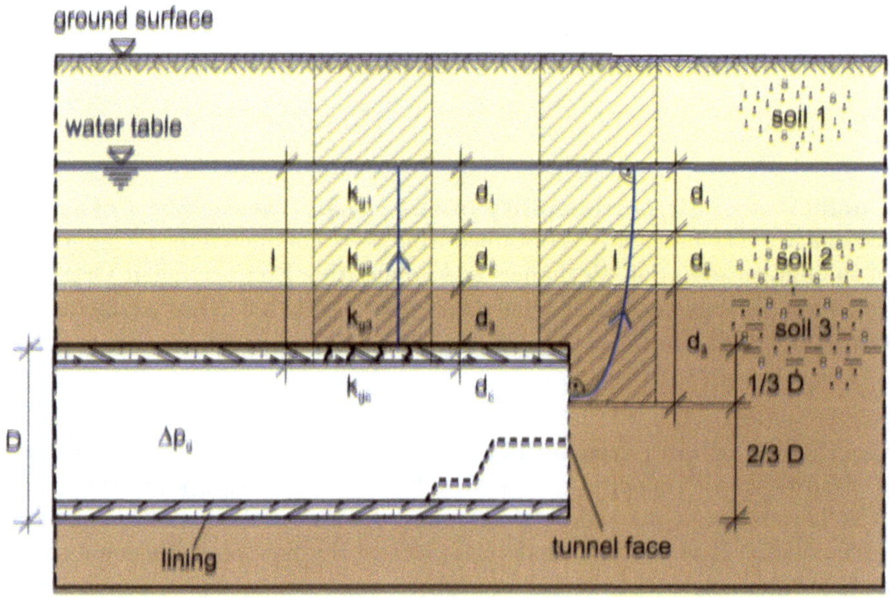

Fig. 26. Compressed air tunnel in a layered soil

5.5 Crack distribution

A crucial point for the amount of air loss through the cracks is their distribution and length. The distribution of the cracks and their length is determined by statistical analysis which have been done in [16][38][39].

The width of shrinkage cracks can be determined by temperature analyses. The temperature development during the hydration of the shotcrete is indicated in [40]. As is the case for the inner concrete lining the temperature is not constant over the cross section of the lining. Hence, the temperature is not constant from the beginning point of zero stress to the end of the hydration process. The average decrease of temperature however can be estimated as ΔT =20 K, this leads to an increase in strain of ε = 0,2 ‰[16].

On the assumption of impeded longitudinal deformation of an unreinforced shotcrete lining the cracks in ring direction will appear every 1,5 m with a crack width of 0,3 mm. Since the bonding between shotcrete lining and soil is substantially lower than between interior and outer tunnel lining, it is neglected in the context of the cracking. Thus the cracks in the shotcrete lining have a constant crack width contrary to the cracks in the tunnel interior lining [16].

For an reinforcement content higher than 0,4 % in the tunnels longitudinal direction the crack distance could furthermore be cut in half for a shotcrete lining. However, since the reinforcement content of a shotcrete lining is generally clearly below 0,4 %, the crack frequency is assumed as the same as for unreinforced concrete. Thus, with a crack distance of an average of 1,5 m in the longitudinal tunnel direction the crack width is 0,3 mm. In the circumferential direction a crack extending over 75 % of the tunnel circumference is assumed. This is due lower shrinkage in the tunnel base because of higher humidity [16].

With regard to the tunnel advance model the results of air loss from the laboratory test have to be scaled to the field condition. The relations are shown in Fig. 27.

The total air loss in a block $Q_{S,bl}$ is scaled by the following equation:

$$Q_{S,bl} = \frac{l_{CT}}{l_a b} \cdot \frac{d_{bl}}{d_{CT}} \cdot Q_S, l_{ub} \tag{30}$$

where l_{CT} is

$$l_{CT} = \tfrac{3}{4} U_T. \tag{31}$$

5.6 Processing of air loss

As mentioned before the tunnel is subdivided into n blocks of a length d_{bl}. The daily advance progress x is concerned as equal to the block length:

$$x(t) = d_{bl} \; for \; t = 1 \; day. \tag{32}$$

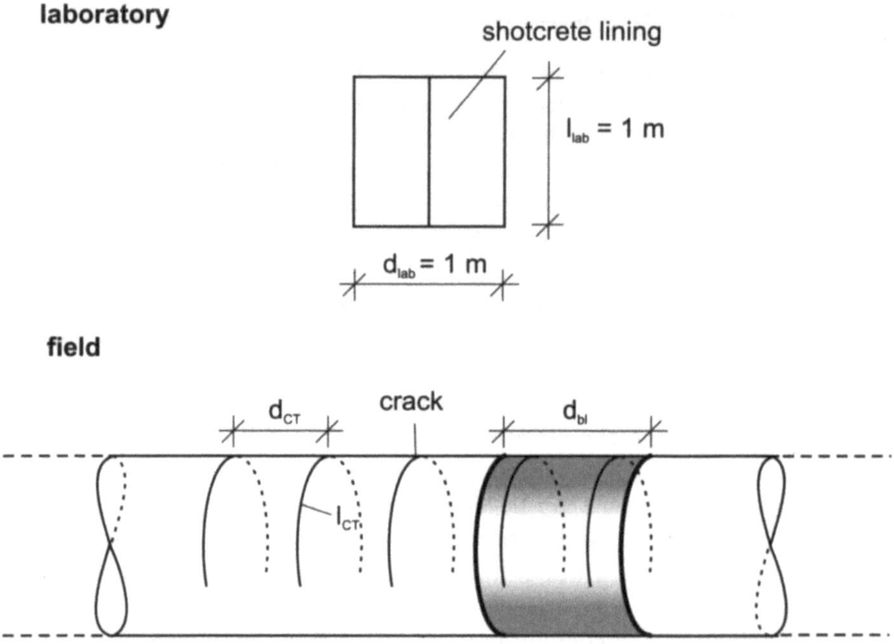

Fig. 27. Scaling of laboratory test dimension to field conditions

By this definition, time t and tunnel advance x are connected, as shown in Fig. 28. The air loss at the tunnel face will be:

$$Q_F(t) = C \cdot \frac{k_{gL}(t)}{\gamma_l} \cdot \frac{(\Delta p_g(t) + 2p_{atm}) \cdot \Delta p_g(t)}{2p_{atm}} \cdot \frac{A_O}{l(t)}, \tag{33}$$

where $k_{gL}(t)$, $\Delta p_g(t)$ and $l(t)$ depend on the position of the tunnel face.

The processing of the air loss through the shrinkage cracks of the lining is of larger magnitude. The actual total air loss through the shrinkage cracks is the sum of the air loss of all the blocks that have been excavated . For each of these blocks the actual pressure at the tunnel face has to be considered. According to Fig. 29 the equation is as follows:

$$Q_S(t) = \sum_{j=1}^{i} \left(\frac{k_{gLC,j}}{\gamma_l} \cdot \frac{(\Delta p_{g,i} + 2p_{atm}) \cdot \Delta p_{g,i}}{2p_{atm}} \cdot \frac{l_{CT}}{l_C} \cdot \frac{d_{bl}}{d_{CT}} \right) \tag{34}$$

where i denotes the block at the actual position of the tunnel face and j is linked to the blocks that have been previously excavated.

Total air loss can finally be determined by

$$Q_T(t) = Q_F(t) + Q_S(t) + Q_L. \tag{35}$$

The air loss through locks and pipes Q_L is a well known value based on experience and is between 3-10 m^3/min.

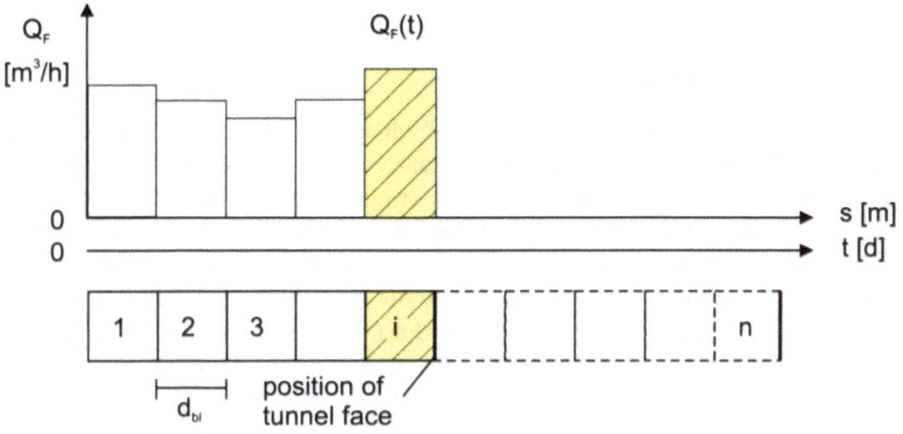

Fig. 28. Air loss at the tunnel face

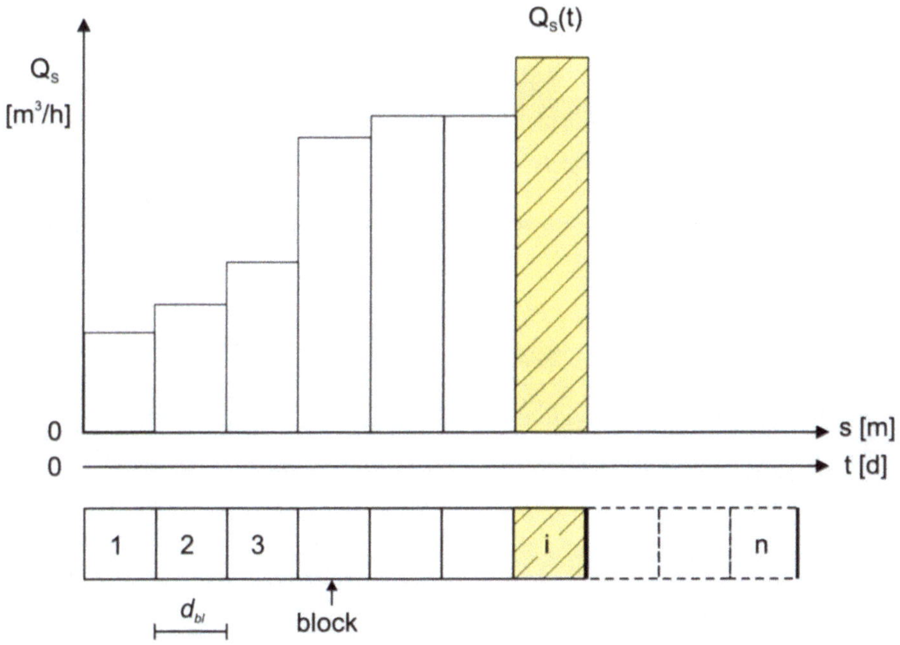

Fig. 29. Air loss through the shrinkage cracks

6 Applications

6.1 Tunnel Siegaue

Project description. A 219 km long high-speed railway line between Cologne and Rhein-Main was built in Germany. Due to its central position the new line is one of the most important transport projects in Germany and Europe. It was completed at the end of 2001 and has been in operation since mid 2002. Thirty tunnels with a total length of 47 km were constructed [41]. In Siegaue a 370 m long tunnel was built by means of NATM (Fig. 30). Compressed air was applied on the northern part of the tunnel. The ground showed a very complex groundwater situation. The construction method was chosen to prevent damage on nearby sensitive constructions in the area. The tunnel crosses a highway, and is situated below a church and a cemetery. Contractor for the project was the Ed. Züblin AG, Stuttgart.

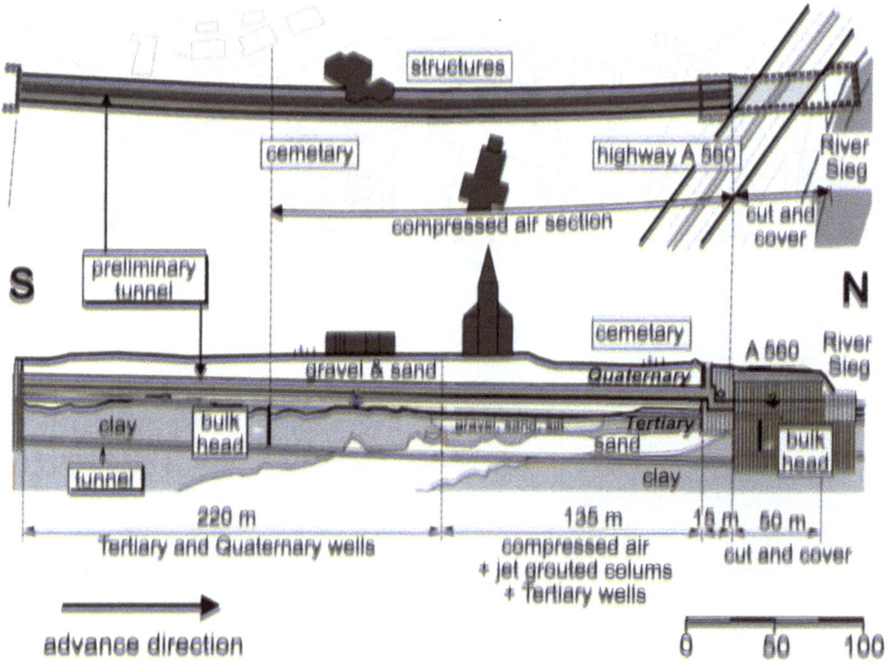

Fig. 30. Tunnel Siegaue – longitudinal section and plan view after reference [42]

Ground conditions. Three soil layers can be distinguished in the area [42]. The fill on top (not indicated in Fig. 30) has a thickness of 0-9 m and consists of sand, gravel and partly cohesive material. The Quaternary layers follow. They mostly consist of silty gravels and sand. The permeability of this layer varies strongly ($k_l = 10^{-1}$-10^{-7} m/s). The base of the Quaternary lies 10-19 m below the ground surface and close to the tunnel crown. The clay and silt layers of the Tertiary lie beneath. Their consistency varies from rigid to semi solid and partly soft. A thin band of gravel, sand and silt with sometimes very high permeabilities was found in the tunnel sections below Siegaue church. Some lignite bands with a thickness of 0,2-0.4 m are enclosed in the Tertiary layers. A thick layer of fine sand exists within the Tertiary with a length of 200 m and a maximum thickness of 7 m. The fine sands are hydrologically connected to the Quaternary sands. Their permeability is about $k_l = 10^{-4}$-10^{-7} m/s.

Construction techniques. The chosen tunnelling technique was governed by the large tunnel cross section with an area of 152 m^2 and the sensitive neighbouring structures. There was no opportunity to use large machines on ground surface. High ground water levels (maximum 6 m above tunnel crown) were present.

Figure 31 shows a cross section of the tunnel. Above the groundwater level a preliminary microtunnel with a diameter of d_i = 2,80 m was constructed by the pipe jacking method. Jet grouted columns were built out of the preliminary tunnel to build an impermeable roof above of the main tunnel. The columns reached into the Tertiary layer. The tunnel was then driven by means of NATM with divided sections of two side galleries and a core. Compressed air was applied to control groundwater inflow from the Quaternary and the Tertiary fine sands. The danger of sudden air loss into the Quaternary layers could be prevented by the jet grouted columns. In order to reduce the amount of air pressure to a level of Δp_g =0,8-0,9 bar the Tertiary aquifer was additionally lowered by a system of vacuum wells [42].

Tunnel advance simulation. An air flow simulation based on the model described in 5 was performed for the tunnel advance period of April to August 2000. The data provided by Ed. Züblin AG gave daily information on the positions of the faces of the two side galleries and the core. Furthermore, the varying amount of air pressure and the medium value of total air loss per day were given. Different plans indicated the ground situation and information on the construction.

The divided sections of the tunnel and the jet-grouted columns were taken into account for the simulation. The initial situation in April 2000 – when compressed air work began – is shown in Fig. 32. The position of the bulkhead is known from the given data as positions of the three tunnel faces are. It is assumed that the length of the core is constantly 40 m. The advance direction is south to north. The staged excavation is neglected in the simulation.

In this example the block length is about d_{bl} = 10 m. At the end of the simulation period the eastern gallery will be 222 m away from the bulkhead.

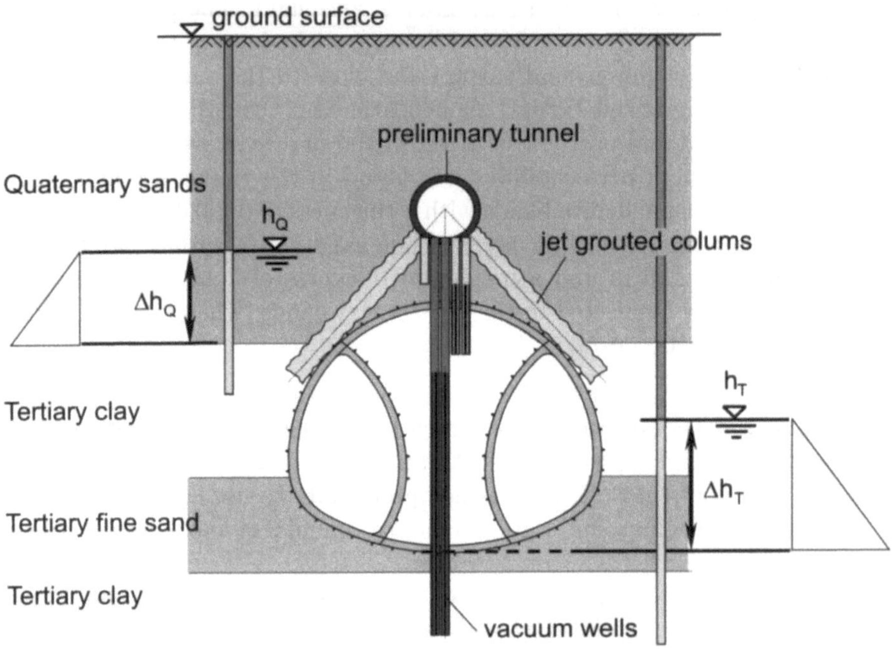

Fig. 31. Tunnel Siegaue – Cross section after reference [42]

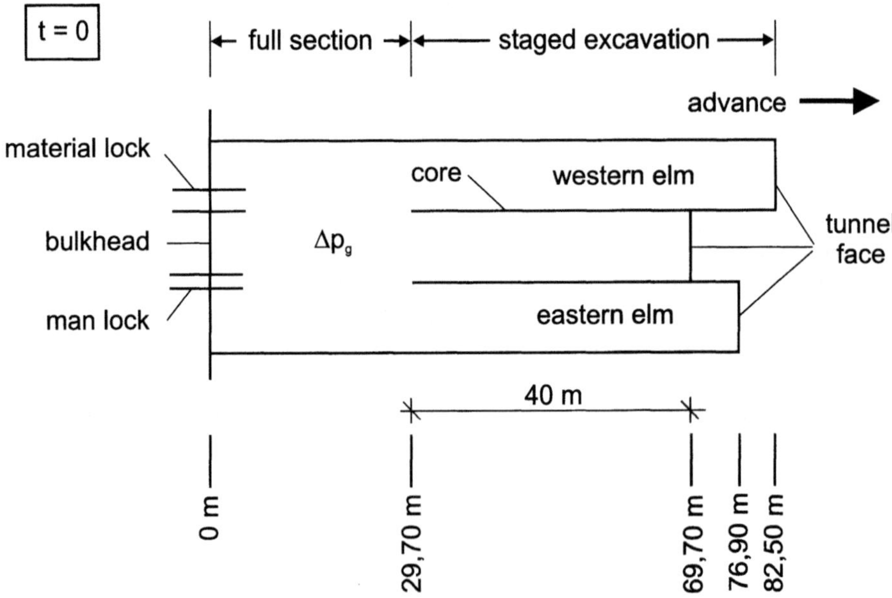

Fig. 32. Model at the beginning of compressed air advance

Therefore, the total number of blocks is $n = 23$. Parameters, as defined in Sect. 5, are given with respect to air flow through the tunnel face and the shrinkage cracks. Figure 33 shows the idealized flow path in the model. Atmospheric conditions reign in the preliminary tunnel. It is assumed that the air escaping the tunnel is drained by the jet grouted columns. A second assumption is that the jet-grouted columns lie horizontally in the ground.

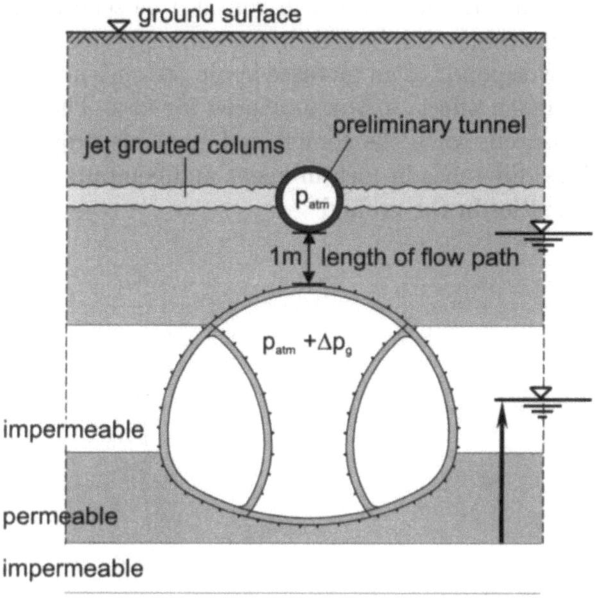

Fig. 33. Idealization of the cross section

As soil-water characteristic curves have not been investigated for the present soils, coefficients of air permeability have been estimated as shown in Tab. 5. Steady state conditions are assumed. Hence, no change in the coefficients of air permeability is computed.

Table 5. Coefficients of air permeability for the ground

Layer	k_g
Quaternary	10^{-5} m/s
Tertiary clay	10^{-8} m/s
Tertiary fine sands	10^{-4} m/s

The results of the simulation are shown in Fig. 34. Computed results are compared to the measured values. The horizontal axes show time and position of the galleries with respect to the bulkhead. The vertical axis shows the total air loss through locks and pipes, tunnel face and shrinkage cracks. The computed air loss is slightly larger than the measured ones. Both the measured and the computed values show high sensitivity to variation in the air pressure. This is reflected in the local maxima and minima of the curves. After 100 days the air pressure increased from 0,8 bar to 1,1 bar. This resulted in the highest increase in air loss, which can also be seen in the computed model. One local maximum can be seen in the computed model after 30 days. This maximum corresponds to an increase in air pressure at this time, which did not have a sudden effect on the measured air loss. The increase in air loss was first measured, when the compressed air value dropped back to its former value. This difference in measurement and computation may be due to local inhomogenities in the ground.

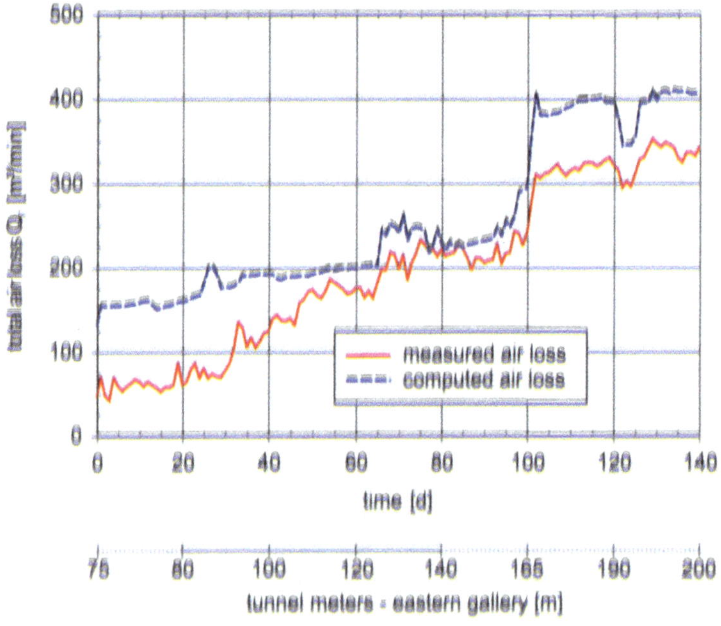

Fig. 34. Computation of air loss

6.2 Tunnel Section – U1/1 Vienna

Project description. Since the fall of 2001 a new metro tunnel is under construction in Vienna (Austria). The tunnel consists of two tubes with a total length of 435 m each. The inner diameter of the tubes is about $d_i = 5{,}85$ m. A shield machine is used to build the tunnel. The ground is grouted prior to tunnelling in order to reduce permeability of the ground and increase the strength of the ground [43]. The project is executed by the contractor Bilfinger Berger AG, Vienna.

During the tender phase the contractor investigated the alternative proposal of NATM under compressed air. SITU was asked to perform a simulation of air loss during tunnel advance. Due to the low price of a used shield machine the shield tunnelling method was chosen.

Ground conditions. The ground (Fig. 35) consists of a 2 m thick layer of fill at the top. Quaternary gravels extend to a depth of 10 m below ground surface. Sand, silt and clay layers lie beneath the gravels.

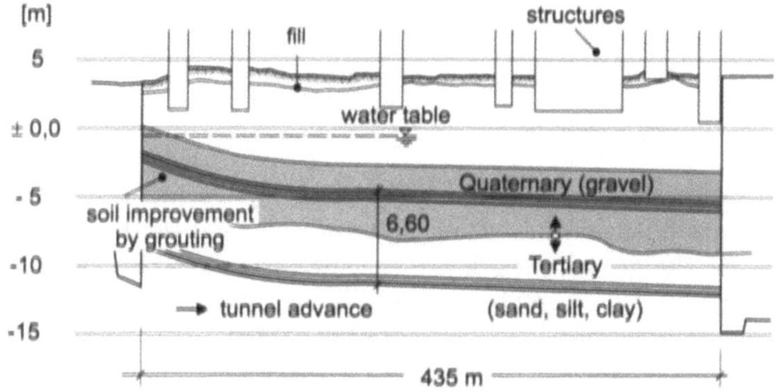

Fig. 35. Tunnel Section – U1/1 Vienna – Longitudinal section and ground conditions [44]

The Quaternary is constituted of sandy gravel. The content of sand varies widely (Fig. 36). Porosity is about $n = 28$ %. The water permeability ranges between $k_l = 1 \cdot 10^{-4}$ m/s and $k_l = 5 \cdot 10^{-2}$ m/s. The contact between the Quaternary and the Tertiary layers is encountered along the whole tunnel length at the level of the sidewall. The Tertiary is of relative low water permeability. Ground water level was measured about 1 m to 4 m above of the tunnel crown.

Construction techniques. The task given by the contractor was to evaluate a compressed air tunnel advance by means of NATM. Due to the relatively high water and air permeability of the gravel a compressed air advance without additional grouting of the gravel is not economical (Fig. 36). Therefore, the tunnel sections within the Quaternary and a 2 m thick neighbouring

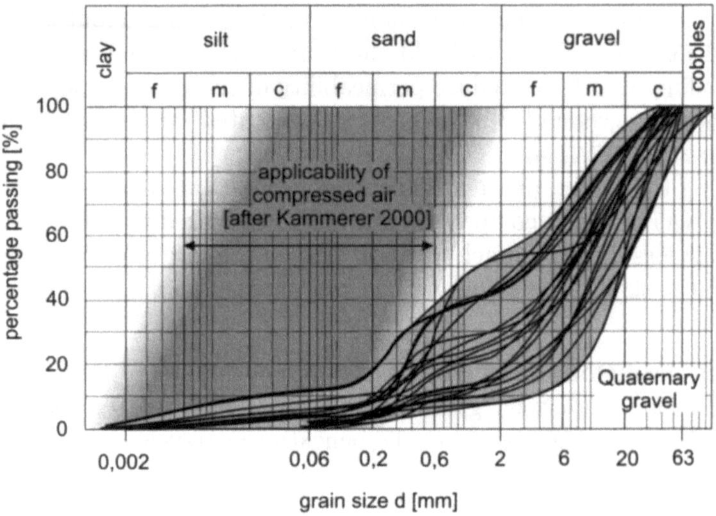

Fig. 36. Grain size distribution of Quaternary gravel [44]

area will be injected with grouting material. By this measure the pores in the gravel are filled with grouting material reducing the permeability. The drillholes for the grouting work are executed from shafts located along the tunnel (Fig. 37). The combination of both methods has been already applied successfully in metro tunnel projects in Duisburg/Germany [45] and in Munich/Germany [44].

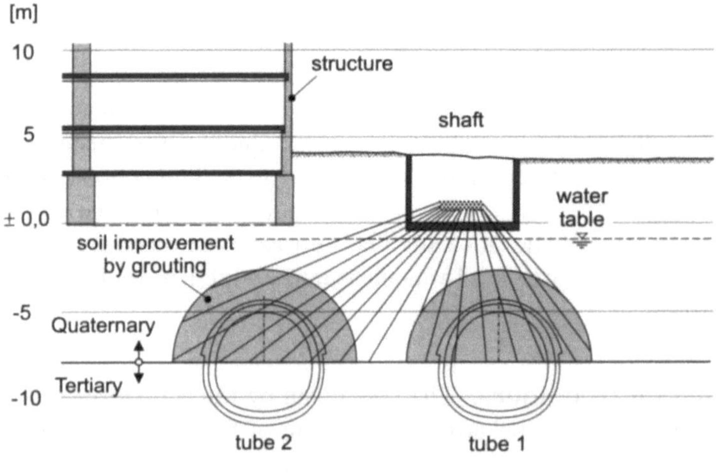

Fig. 37. Cross section of tunnel with shafts [44]

Tunnel advance simulation. An air flow simulation based on the model described in Sect. 5 was performed to estimate the total air loss. The position of the tunnel face of tube one was assumed to be always 5 m ahead of tube two. The grouted area in the Quaternary was taken into account during the simulation. In this model, the block length is about $d_{bl} = 5$ m, corresponding to the daily tunnel advance. The total number of blocks is $n = 87$. Subsequently, the compressed air work will be finished after $t = 87$ days. Parameters as defined in Sect. 5 are given with respect to air flow through the tunnel face and the shrinkage cracks.

The level of air pressure in the tunnel is governed by the inclination of the tunnel with respect to the water level (Fig. 35). The amount of air pressure is determined as described in 2. A safety factor of $\gamma = 1{,}1$ is taken into account.

As soil water characteristic curves have not been investigated for the present soils, coefficients of air permeability have been estimated as shown in Tab. 6. Steady state conditions are assumed. Hence, no change in the coefficients of air permeability is computed.

Table 6. Coefficients of air permeability for the ground

Layer	k_g
Fill (below groundwater)	10^{-4} m/s
Quaternary gravel	10^{-4} m/s
Quaternary grouted gravel	10^{-5} m/s
Tertiary sand	10^{-7} m/s
Tertiary silt and clays	10^{-9} m/s

The results of the simulation are shown in Fig. 38. At the beginning of the tunnel excavation a majority of the cross section was in the Quaternary. This led to high air loss through the tunnel face in the first ten days of the tunnel advance, as one can see in Fig. 38. The same effect is present concerning the air loss through the shrinkage cracks. Since, the air loss through the crack is increasing depending on the tunnel length, the higher air loss due to ground conditions cannot be seen.

The total amount of air loss is increasing every day. Figure 38 shows that the total amount of air loss is nearly completely represented by the flow through the shotcrete lining. As a result of the present calculation the compressor station can be dimensioned more accurate. Furthermore, the need of air for each stage during the construction period can be foreseen and the costs could be estimated during the design phase of a tunnel project.

This application has shown, that modelling the air loss can help to foresee costs of a compressed air tunnel project. Soils with large grain diameters are

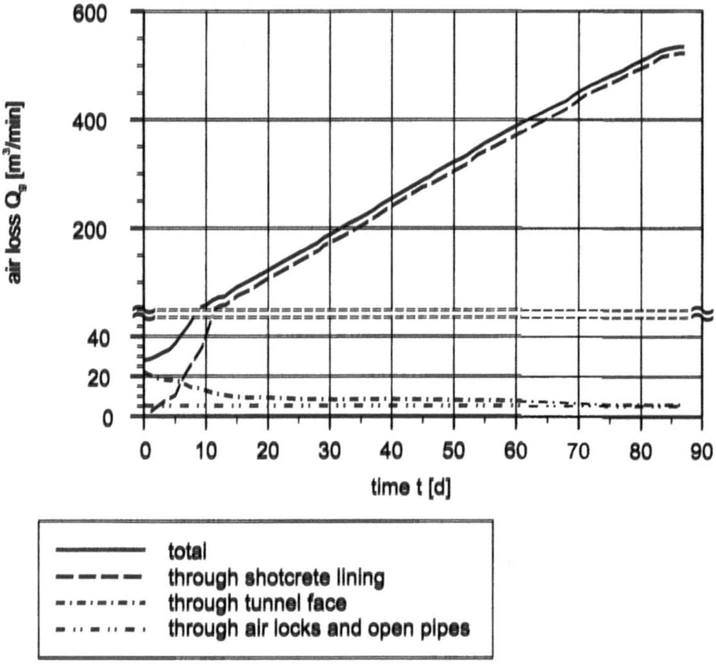

Fig. 38. Air loss during tunnel advance [44]

not strictly unsuitable for compressed air, but grouting measures have to be considered. An analyse of the cost effectiveness has to follow the analyses of technical feasibility.

7 Recommendations for practice

The state of the art in compressed air tunnelling has been developed in 1989 [8]. The approach takes into consideration that the amount of air loss is a function of the pressure in the tunnel and the coefficient of air permeability of the soil. The total air losses in the tunnel are defined as the sum of the air escaping at the tunnel face, through the shotcrete lining and through locks and pipes.

Principally this approach is sufficient. Nevertheless some major influences had been neglected. On one hand the compressibility of air was not taken into consideration in a more simple way. On the other hand all influences due to unsaturated soil mechanics such as relative permeability and capillary pressures were neglected. Furthermore airflow through cracks in the shotcrete lining was not analysed and a simulation of tunnel advance could not be calculated. In consequence the prognostics are of reduced reliability. Compared to the state of the art compressed air tunnel advance analyses should take into account the above mentioned influences.

For example an tunnel advance simulation should consider the increasing height of water above of the tunnel in a declining tunnel section. An application is given in Sect. 6.

Until today the grain matrix of the soil is considered as not deformable.

Compressed air is used when the tunnel is driven below the groundwater surface. The initial conditions are fully saturated with water. Due to a gradient in air pressure the air flows upward where atmospheric air pressure reigns. The air displaces the water in the soil and reduces saturation of the water phase. Subsequently the saturation of the air phase increases. The coefficient of air permeability increases respectively until steady state conditions are reached. Steady state conditions are governed by the dewatering capabilities of the soil. This process is time dependent. However, the advance velocity of the tunnel is controlling the level of saturation at the tunnel face too. This dependency should be taken into account in future simulation of tunnel advance. One-dimensional standard testing devices could be developed to describe the time and pressure dependency of the dewatering capabilities of the soil.

Crack permeability of the shotcrete lining is of crucial importance when tunnel length is increasing. In practice crack permeability should be determined using the analytical approach given in this contribution. It was shown [16] that the permeability of the crack depends on the pressure in the tunnel, the thickness of the lining and the distribution and length of the cracks. The air losses through the shotcrete lining into the soil are described using a layered soil composite permeability.

Different measures can be applied in order to reduce the air losses into the soil. The thickness of the shotcrete lining may be increased or polymers can be sprayed on the lining to seal it. The mixture of the sprayed concrete might be optimised, eventually by applying fibre reinforced shotcrete. Accompanying soil injection to reduce air permeability might be useful. For long tunnel section it might be useful to replace the bulkhead or to install a second one.

Air losses at the tunnel face depend strongly on the ground conditions. Inhomogenities of the ground can lead to sudden air losses. Those inhomogenities have to be considered when estimating the reliability of given air loss prognostics. A sophisticated measuring programme is crucial to the success of a compressed air tunnelling project. Ground conditions should be known very well prior to estimating the air losses. During tunnel advance prognostics and conditions in the tunnel should be adapted and eventual modifications should always be possible. To achieve a cost-effective project different parameters should be measured. Those were first of all the air losses and the pressure in the tunnel. In addition the groundwater conditions and ground conditions at the tunnel face should be observed and evaluated. Measurement of surface heaving and settlements can give additional information. Conducting preliminary field tests might help to collect data related to the compressed air tunnel advance. With respect to the parameters related to unsaturated soil mechanics the use of TDR-probes, tensiometers or pressure gauges could give accurate information to the behaviour of the soil when air flows.

References

[1] Glossop R. (1976) The Invention and Early Use of Compressed Air to Exclude Water from Shafts and Tunnels during Construction. *Géotechnique*, Vol. 26, No. 2: 253-280

[2] Hewett B.H.M., Johannesson S. (1922) Shield and Compressed Air Tunnelling. McGraw Hill, New York, USA

[3] Rabcewicz L. von (1950) Verfahren zum Ausbau von unterirdischen Hohlräumen, insbesondere von Tunneln. Österreichisches Patentamt, Patentschrift No. 165573, Austria

[4] Triger J. (1841) Mémoire sur un Appareil à Air Comprimé pour le Percement des Puits de Mines et autres Travaux sous les Eaux et dans les Sables Submergés. *Comte Rendu des Séances de l'Académie des Sciences*, Vol. 13: 884-896

[5] Schenck zu Schweinsberg W.-R., Wagner H. (1963) Luftverbrauch und Überdeckung beim Tunnelvortrieb mit Druckluft, *Bautechnik*, Vol. 40, No. 2: 41-47

[6] Krabbe W. (1969) Die Wasserhaltung beim Schildvortrieb durch Grundwasserabsenkung und Druckluft. Vorträge der Baugrundtagung in Hamburg, Herausgegeben von der Deutschen Gesellschaft für Erd- und Grundbau e.V., Essen, Germany: 285-317

[7] Javadi A.A., Farmani R., Toropov V.V., Snee, C.P.M. (1998) Identification of parameters for air permeability of shotcrete tunnel lining using a genetic algorithm. *Computers and Geotechnics*, Vol. 25, No. 1: 1-24

[8] Kramer J., Semprich S. (1989) Erfahrungen über Druckluftverbrauch bei der Spritzbetonbauweise. *Taschenbuch für den Tunnelbau*, Glückauf Verlag, Essen, Germany: 91-153

[9] Chen Z.-S., Hofstetter G., Mang H. (1991) 3D-Boundary Element Analysis of the Lowered Groundwater Level for Tunnels Driven Under Compressed Air. *Numerical and Analytical Methods in Geomechanics*, Vol.15: 735-752

[10] Gülzow H.G. (1994) Dreidimensionale Berechnung des Zweiphasenströmungsfeldes beim Tunnelvortrieb unter Druckluft in wassergesättigten Böden, Veröffentlichungen des Instituts für Grundbau, Bodenmechanik, Felsmechanik und Verkehrswasserbau der RWTH Aachen, Germany, No. 25

[11] Hochgürtel T. (1998) Numerische Untersuchungen zur Beurteilung der Standsicherheit der Ortsbrust beim Einsatz von Druckluft zur Wasserhaltung im schildvorgetriebenen Tunnelbau. Veröffentlichungen des Instituts für Grundbau, Bodenmechanik, Felsmechanik und Verkehrswasserbau der RWTH Aachen, Germany, No. 32

[12] Javadi A.A. (2001) Estimation of Air Losses from Tunnels Driven under Compressed Air. *Proc.* 10th *Int. Conf. on Computer Methods and Advances in Geomechanics*, Desai, Kundu, Harpalani, Contractor and Kemeny (eds.), Vol. 1: 207-211

[13] Perau E. (1999) Flow of Water and Air in Soils due to Tunnelling under Compressed Air. Proc. 12th European Conf. Soil Mechanics Geotechnical Engineering (ECSMGE), *Geotechnical Engineering for Transportation Infrastructure*, Barends, Lindenberg, Luger, de Quelerij and Verruijt (eds.), Balkema, Rotterdam, Netherlands, Vol. 3: 2093-2099
[14] Oettl G., Stark R.F., Hofstetter G. (2000) Verification of a Fully Coupled FE Model for Tunneling under Compressed Air. GAMM Annual Meeting
[15] Strobl B. (1991) Die NATM in Böden in Kombination mit Druckluft. Dissertation, Institut für Bodenmechanik, Felsmechanik und Grundbau, Technische Universität Graz, Austria
[16] Kammerer G. (2000) Experimentelle Untersuchungen von Strömungsvorgängen in teilgesättigten Böden und Spritzbetonrissen im Hinblick auf den Einsatz von Druckluft im Tunnelbau. Mitteilungshefte der Gruppe Geotechnik Graz, Technische Universität Graz, Austria, No. 8
[17] Distelmaier H. (1981) Anwendung von Spritzbetonbauweisen im Tunnelbau unter Druckluftbedingungen. *Tunnel*, Vol. 81, No. 1: 199-240
[18] Snee C.P.M., Javadi A.A. (1996) Prediction of Compressed Air Leakage from Tunnels. Tunnelling and Underground Space Technology, Vol. 11, No. 2: 189-195
[19] Javadi A.A., Snee C.P.M. (1997) The Contribution of Air Flow from Compressed Air Tunnelling to Surface Settlement. Tunnels for People, Golser, Hinkel and Schubert (eds), Vol. 2: 483-489
[20] Kammerer G., Semprich S. (1998) Settlements Due to Tunnelling Under Compressed Air. *Int. Conf. on Soil-Structure Interaction in Urban Civil Engineering*, Darmstadt Geotechnics, Vol. 1, No. 4: 85-96
[21] Schreyer J. (1981) Spritzbeton unter Druckluft. STUVA-Forschungsbericht 15/81, Köln, Germany
[22] Semprich S. (1988, un-publ.) Luftdurchlässigkeitsversuch Essen (Tunnelvortriebe unter Druckluft). F+E Bericht Nr. 87-55, Bilfinger+Berger Bauaktiengesellschaft, Mannheim, Germany
[23] Scheid Y., Semprich S., Kammerer G. (2001) Using Compressed Air in Shallow Tunnelling – A Method to Reduce Detrimental Effects on the Environment. Proc. of the AITES-ITA 2001 World Tunnel Congress, Progress in Tunnelling after 2000, Teuscher and Colombo (eds), Milan, Italy: 665-679
[24] Wyckoff R.D., Botset H.G. (1936) The Flow of Gas-liquid Mixtures through Unconsolidated Sands. General and Applied Physics, Vol. 7: 325-345
[25] Roth C.H., Malicki M.A., Plagge R. (1992) Empirical Evaluation of the Relationship Between Soil Dielectric Constant and Volumetric Water Content as the Basis for Calibrating Soil Moisture Measurements by TDR. *Journal of Soil Science*, Vol. 43: 1-13
[26] Stacheder M. (1996) Die Time Domain Reflectometry in der Geotechnik, Messung von Wassergehalt, elektrischer Leitfähigkeit und Stofftransport. Schriftenreihe Angewandte Geologie, Universität Karlsruhe, Germany, No. 40
[27] Scheid Y. (2000) Die TDR-Methode zur Bestimmung des Sättigungsgrades teilgesättigter Böden. 2nd Workshop Teilgesättigte Böden, Schanz and Witt (eds.), Schriftenreihe Geotechnik, Bauhaus-Universität Weimar, Germany, No. 4: 35-55
[28] Kramer J. (1987) U-Bahn-Bau in Essen, Baulos 30, Spritzbetonbauweise unter Druckluft – Luftverbrauch, Spannungsumlagerungen im Boden, Senkungen. Forschung + Praxis, Vol. 32: 193-199

[29] Croney D., Coleman J.D. (1960) Pore Pressure and Suction in Soil. Proc. of the Conf. Pore Pressure and Suction in Soils: 31-37

[30] Pruess K., Oldenburg C., Moridis G. (1999) TOUGH2 User's Guide, Version 2.0. Earth Sciences Division, Ernest Orlando Lawrence Berkeley National Laboratory

[31] Finsterle S., Moridis G.J., Pruess K. (1994) A TOUGH2 Equation-of-state Module for the Simulation of Two-Phase Flow of Air, Water, and a Miscible Gelling Liquid. Report LBL-36086 UC-400, Earth Sciences Division, Ernest Orlando Lawrence Berkeley National Laboratory, University of California

[32] Narasimhan T.N., Witherspoon P.A. (1976) Integrated Finite Difference Method for Analyzing Fluid Flow in Porous Media. *Water Resources Research*, Vol. 12, No. 1: 57-64

[33] Moridis G.J., Pruess K. (1998) T2SOLV: An Enhanced Package of Solvers for the TOUGH2 Family of Reservoir Simulation Codes. *Geothermics*, Vol. 27, No. 4: 415-444

[34] Genuchten M.T. van (1980) A Closed-Form Equation for Predicting the Hydraulic Conductivity of Unsaturated Soils. *Soil Sciences Society*, Vol. 44: 892-898

[35] Mualem Y. (1976) A New Model for Predicting the Hydraulic Conductivity of Unsaturated Porous Media. *Water Resources Research*, Vol. 12, No. 3: 513-522

[36] Corey A.T. (1954) The Interrelation between Gas and Oil Relative Permeabilities. Producers Monthly: 38-41

[37] Semprich S., Scheid Y. (2002) An Approach to Describe the Flow of Air and Water in Shallow Tunnelling using Compressed Air. *Proc. of the 3rd Int. Conf. on Unsaturated Soils (UNSAT2002)*, Jucá, de Campos and Marinho (eds), Recife, Brazil, Vol. 2: 787-792

[38] Semprich S. (1985) Untersuchungsergebnisse an Innenschalen aus wasserundurchlässigem Beton bei Tunnelbauwerken im Fels. Teil 1: Theoretische Überlegungen zur Rissbildung sowie Ausführungserfahrungen vom Baulos 14 des Hasenbergtunnels Stuttgart. *Forschung+Praxis*, Vol. 30: 97-102

[39] Deutscher Beton-Verein (1994) DBV-Sachstandsbericht Stahlbetoninnenschalen im U-Bahnbau, Germany

[40] Hellmich Ch., Mang H.A. (1999) Influence of the Dilatation of Soil and Shotcrete on the Load Bearing Behaviour of NATM-Tunnels. *Felsbau 17*, No. 1: 35-43

[41] Sternath R. (2000) Tunnels for HS Railways in Germany. *Tunnels and Tunnelling International*, Vol. 32, No. 10: 53-59

[42] Sänger C. (2000) Neubaustrecke Köln-Rhein/Main Los 2.4 – Siegauen Tunnel. Vorträge der Baugrundtagung 2000 in Hannover, Deutsche Gesellschaft für Geotechnik (ed.), Verlag Glückauf GmbH, Cloppenburg, Germany: 375-392

[43] Semprich S., Sochatzy G., Gutzeit M., Fuchs R. (2002) Stabilisierungsinjektionen der quartären Kiese – ein Qualitätsmerkmal für den Schildvortrieb Baulos U1/1 der U-Bahn Wien. Beiträge zum 17. Christian Veder Kolloquium, Injektionen in Boden und Fels, Mitteilungshefte der Gruppe Geotechnik Graz, Technische Universität Graz, Austria, No. 13: 53-64

[44] Semprich S., Lesnik M., Scheid Y. (2002, in print) Tunnelling under Compressed Air using Grouting Techniques. *Proc. 4th Int. Symp. Geotechnical Aspects of Underground Construction in Soft Ground*, IS-Toulouse 2002

[45] Weiler A., Misch V. (1982) Tunnelvortrieb unter Druckluft im Schutze einer chemischen Injektion (Baulos 6/7, Stadtbahn Duisburg, TA 5A). *Geotechnik*, No. 4: 169-177

Chapter 12
A coupled FE-model for tunnelling by means of compressed air

Gerhard Öttl, Rudolf F. Stark, Robert Stelzer, and Günter Hofstetter

Institute for Structural Analysis and Strength of Materials, University of Innsbruck

Abstract. This chapter deals with the development and the application of a coupled numerical model for tunnelling below the groundwater table, taking into account compressed air as a means for displacing the groundwater in the vicinity of the tunnel face. The coupled solid-fluid model is characterized by treating the soil as a three-phase material, consisting of the deformable soil skeleton and the fluid phases water and compressed air. It contains a number of special cases, which are of interest in geotechnical engineering. In particular, a two-phase formulation for dewatering of soils under atmospheric conditions and a two-phase formulation for fully saturated conditions, applicable to consolidation problems, are included in the complete model.

1 Introduction

The construction of shallow tunnels, located below the groundwater level, is an important topic in geotechnical engineering. In this case deformations of the ground and surface settlements are not only caused by the advance of the tunnel face but also by dewatering of the soil. Especially in urban areas the minimization of surface settlements during the construction of the tunnel plays a predominant role in order to avoid damage of existing buildings and infrastructure. For tunnels driven in aquifers the groundwater has to be displaced from the working area at the tunnel face. Compared to lowering the groundwater table with pumping wells and driving the tunnel under atmospheric conditions, the application of compressed air for displacing the groundwater yields smaller settlements as shown in a study conducted during the subway construction in Essen, Germany [19]. This advantage is due to the air pressure and the drag forces of the airflow in the soil which counteract the deformations caused by dewatering and tunnel excavation.

This chapter deals with the development and the application of a coupled numerical model for tunnelling below the groundwater table which allows to take into account compressed air as a means for dewatering the soil in the vicinity of the tunnel face. The schematic diagram in Fig. 1 shows the loss of compressed air at the tunnel face and through cracks in the shotcrete lining as well as the flow of compressed air in the adjacent soil.

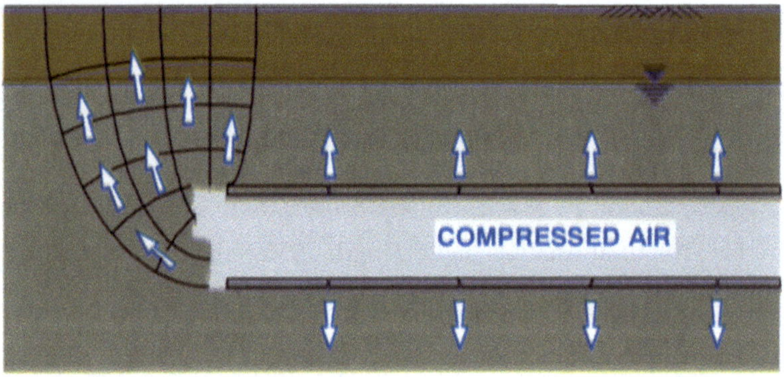

Fig. 1. Tunnelling under compressed air

Basically, there exist two different solution strategies for the numerical simulation of tunnelling under compressed air, an uncoupled and a coupled approach. In the uncoupled approach the flow of water and compressed air in the soil and the deformations of the soil caused by dewatering as well as by the advance of the tunnel face are treated in two consecutive steps. In the first step the fluid flow in the soil is determined, assuming a rigid soil skeleton. To this end, numerical models based on the boundary element method [7] or on the finite element method [8] have been developed. In the second step the deformations of the soil are computed by making use of the results of the flow analysis. Consequently, interactions between the fluid flow through the pores and the deformations of the soil skeleton are neglected.

On the other hand, in the coupled solution procedure adopted in this study the fluid flow in the soil and the deformations of the soil are determined simultaneously, treating the soil as a three-phase material. From a theoretical point of view, only this approach allows to properly take into account the intrinsic coupling of the process of dewatering with the deformations of the soil in a physically consistent manner. In addition, from a practical point of view the coupled model offers the advantage that all quantities are computed simultaneously on the basis of one coupled numerical model without switching between flow analysis and deformation analysis.

The chapter is organized as follows: At the outset the governing equations of the three-phase soil model are presented on the basis of respective mathematical models available in the literature [21],[27],[28] and special cases, which are contained in the three-phase formulation and are of interest for geotechnical engineering, are described. Then the weak formulations of the governing equations are derived, which form the basis for a numerical solution strategy. These equations are discretized in space and time within the framework of the finite element method (FEM). Subsequently, an adaptive FEM is presented, which allows to enhance the efficiency and accuracy of the numerical simulations. The last part of this chapter is devoted to the validation and application of the coupled model. First, the three-phase formulation is applied to the numerical simulation of laboratory tests dealing with the flow of compressed air through dry and partially saturated soil. Such tests have been performed at the Institute for Soil Mechanics and Foundation Engineering at Graz University of Technology within the framework of the Joint Research Initiative on Numerical Simulation in Tunnelling [2]. Subsequently, the numerical simulation of a large-scale in-situ test for displacing the groundwater by means of compressed air, which was carried out at a subway construction site in Essen, Germany [20], is presented. Finally, the coupled model is applied to the numerical simulation of tunnelling by means of compressed air.

2 Governing equations of the three-phase model

The coupled numerical model for displacing the groundwater in the soil by means of compressed air relies on a soil model treating the soil as a three-phase medium consisting of the deformable soil skeleton and the two fluid phases water and compressed air. Generally there exist various approaches for the mathematical description of multi-phase media [5]:

(i) Mixture theory starts from the macroscopic level and can be viewed as an extension of the classical continuum mechanics theory [6],[11]. Employing the concept of volume fractions allows to account for the different constituents of the soil;

(ii) Averaging theory, which is characterized by applying the balance laws of classical continuum mechanics to the actual microstructure of the porous medium and using so-called averaging operators to arrive at the macroscopic level, yields the same governing equations for the multi-phase medium as shown in [5];

(iii) Alternatively, the basic equations can also be derived by extending Biot's theory [3] as outlined , e.g., in [21] and [32].

The mathematical description of the three-phase medium, presented in this chapter, follows [21],[27],[28]. At the outset averaged quantities for the density of the three-phase mixture, for the stresses acting on the surfaces of a differential volume element of the mixture and for the velocities of the individual fluid phases relative to the soil skeleton are defined. Following this elementary definitions the governing equations for the three-phase formulation, which consist of

- the kinematic relations,
- the equilibrium equations for the three-phase mixture,
- the mass balance equations,
- the constitutive equations for each phase and
- a constitutive relationship between the capillary stress and the degree of water saturation

are presented.

2.1 Definition of averaged quantities for the three-phase medium

A soil element can be thought of as a mixture of the three phases: soil grains, water and compressed air which continuously fill the entire volume according to their percentile share. Of paramount importance for deriving the governing equations of the model is the definition of averaged values for the density, the stresses and the fluid velocities relative to the soil skeleton. Figure 2 shows a section of a representative elementary volume (REV) of the soil (a) before, and (b) after averaging.

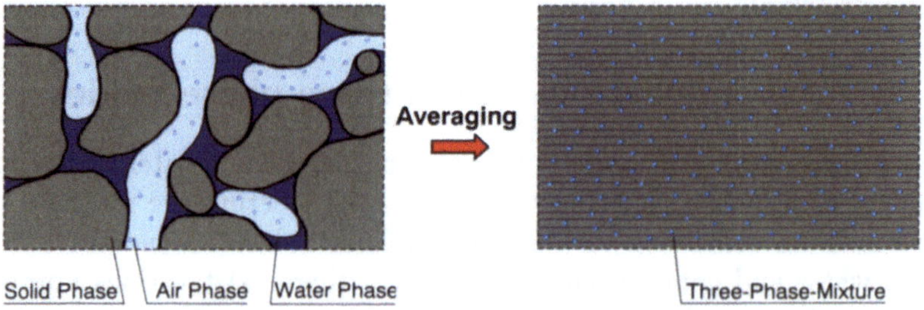

Fig. 2. Representative elementary volume of the soil: (a) before, and (b) after averaging

The three constituents solid phase (superscript s), water (superscript w) and compressed air (superscript a) occupy the parts dV^s, dV^w and dV^a of the volume dV of the REV. Introducing the porosity n of the soil, defined as the ratio of the pore volume to the total volume of the REV, and the degree of saturation of the respective fluid phase S^f, $f = w, a$, indicating the part of the void space filled by the particular fluid, yields the following definitions for the volume fractions:

$$\frac{dV^s}{dV} = (1-n) \ , \quad \frac{dV^w}{dV} = nS^w \ , \quad \frac{dV^a}{dV} = nS^a \ . \tag{1}$$

For the degrees of saturation the condition $S^w + S^a = 1$ holds. The partial surface elements of the individual constituents, dA^s, dA^w and dA^a, are assumed to be related to the surface element dA of the three-phase mixture in the same way as the respective partial volumes of the different phases to the volume of the REV.

Assuming each constituent to fill the entire volume dV of the soil element according to its volume fraction (1) yields the averaged density $\tilde{\rho}$ of the three-phase mixture as

$$\tilde{\rho} = \rho_s + \rho_w + \rho_a = (1-n)\rho^s + nS^w\rho^w + nS^a\rho^a \ , \tag{2}$$

with ρ^s, ρ^w and ρ^a being the intrinsic densities of the three phases and

$$\rho_s = (1-n)\rho^s \ , \quad \rho_w = nS^w\rho^w \ , \quad \rho_a = nS^a\rho^a \ , \tag{3}$$

representing the densities of the three phases averaged over the volume of the REV, denoted as partial densities.

Using the ratios for the partial surface elements of the individual constituents the averaged stress tensor for the three-phase material can be defined as

$$\tilde{\boldsymbol{\sigma}} = \boldsymbol{\sigma}_s + \boldsymbol{\sigma}_w + \boldsymbol{\sigma}_a = (1-n)\boldsymbol{\sigma}^s + nS^w\boldsymbol{\sigma}^w + nS^a\boldsymbol{\sigma}^a \ , \tag{4}$$

where $\boldsymbol{\sigma}^s$, $\boldsymbol{\sigma}^w$ and $\boldsymbol{\sigma}^a$ denote the stress tensors of the individual phases and

$$\boldsymbol{\sigma}_s = (1-n)\boldsymbol{\sigma}^s \ , \quad \boldsymbol{\sigma}_w = nS^w\boldsymbol{\sigma}^w \ , \quad \boldsymbol{\sigma}_a = nS^a\boldsymbol{\sigma}^a \ , \tag{5}$$

are the stresses of the three phases averaged over the surface elements of the REV, i. e., the partial stresses. In this chapter tensile stresses are defined as positive quantities for all phases. The averaged fluid density and the averaged hydrostatic fluid stresses follow from the second and the third equation of (3) and (5) as

$$\rho_f = nS^f\rho^f \ , \quad \boldsymbol{\sigma}_f = nS^f\boldsymbol{\sigma}^f \ , \quad f = w, a \ . \tag{6}$$

Since the stress state in the fluid phases is considered as a hydrostatic one, the fluid stresses are given as

$$\boldsymbol{\sigma}^f = p^f\,\mathbf{I} \ , \quad f = w, a \ , \tag{7}$$

where $\mathbf{I}$ denotes the second-order unit tensor and p^f is the hydrostatic stress in the fluid phase f. Equation (7) can be used to rewrite Equation (4) as

$$\tilde{\boldsymbol{\sigma}} = (1-n)\boldsymbol{\sigma}^s + n(S^w p^w + S^a p^a)\,\mathbf{I} \ . \tag{8}$$

In the context of dewatering by means of compressed air it seems to be reasonable to assume isothermal conditions and to consider the water in the pores of the soil being present as condensed liquid, i.e., the pores only contain a mixture of water and air. Hence, water vapour is not taken into account. The difference between the hydrostatic stress in the air phase p^a and the hydrostatic stress in the water phase p^w is defined as the capillary stress [21],[27],[28]:

$$p^c = p^a - p^w \ . \tag{9}$$

Due to surface tension the air pressure in the pores is larger than the water pressure. Hence, p^c can be viewed as an additional pressure acting on the soil skeleton.

The velocity of a fluid f, flowing through the pores of the soil, relative to the velocity $\mathbf{v}$ of the soil skeleton is given as

$$\mathbf{v}^{fr} = \mathbf{v}^f - \mathbf{v} \ , \tag{10}$$

where $\mathbf{v}^f$ is the intrinsic velocity of the fluid. Since the flow of the fluid phase f occurs through the partial surface dA^f of dA, the relative velocity averaged over dA, the so-called artificial velocity, is obtained as

$$\tilde{\mathbf{v}}^{fr} = nS^f\mathbf{v}^{fr} \ . \tag{11}$$

2.2 Kinematic equations

For tunnelling under compressed air the assumption of small displacements and small strains seems to be justified. Hence, the relationship between the displacements $\mathbf{u}$ of the soil skeleton and the strains ε in the soil skeleton is given as

$$\varepsilon = \frac{1}{2}\left[\nabla \mathbf{u} + (\nabla \mathbf{u})^T\right] . \tag{12}$$

2.3 Equilibrium equations

The linear momentum balance equations for a single-phase material are given as [23]

$$\operatorname{div} \boldsymbol{\sigma} + \mathbf{b} = \rho \mathbf{a} , \tag{13}$$

where $\mathbf{b}$ and $\mathbf{a}$ denote the body force vector and the acceleration vector, respectively.

Substituting (3_1) for the density ρ, (5_1) for the stress tensor $\boldsymbol{\sigma}$ and $\mathbf{b}_s = \rho_s \mathbf{g}$ for the body forces into (13) yields the linear momentum balance equations for the soil skeleton of the three-phase material as

$$\operatorname{div} \boldsymbol{\sigma}_s + \rho_s \mathbf{g} + \mathbf{f}_s = \rho_s \mathbf{a} , \tag{14}$$

where $\mathbf{a}$ is the acceleration vector of the soil skeleton and $\mathbf{f}_s$ represents the vector of interaction forces of the soil skeleton with the fluid phases, referred to the volume of the REV.

Substituting ρ_f according to (6_1) for the density ρ, $\boldsymbol{\sigma}_f$ according to (6_2) for the stress tensor $\boldsymbol{\sigma}$ and $\mathbf{b}_f = \rho_f \mathbf{g}$ for the body forces into (13) yields the linear momentum balance equation for the fluid phases $f = w, a$ as

$$\operatorname{div} \boldsymbol{\sigma}_f + \rho_f \mathbf{g} + \mathbf{f}_f = \rho_f \mathbf{a}^f , \tag{15}$$

where $\mathbf{a}^f$ represents the acceleration vector for the respective fluid phase and $\mathbf{f}_f$ denotes the vector of interaction forces of the respective fluid phase with the other phases, referred to the volume of the REV.

The interaction force $\mathbf{f}_f$ for the fluid phase f accounts for viscous effects due to the fluid flow and can be expressed using the artificial velocity as [21]

$$\mathbf{f}_f = -nS^f \frac{\gamma^f}{k^f} \mathbf{v}^{fr} - p^f \nabla \left(nS^f\right) , \tag{16}$$

where k^f denotes the permeability of the soil with respect to the fluid phase f and γ^f is its unit weight. Inserting (16) into (15), making use of (6) together with (7) for the density and the hydrostatic stress state in the fluid phase

f and neglecting the acceleration term yields Darcy's law for the artificial velocity of a fluid relative to the soil skeleton

$$\tilde{\mathbf{v}}^{fr} = \frac{k^f}{\gamma^f} \left(\nabla p^f + \rho^f \mathbf{g} \right) \ . \tag{17}$$

For a soil fully saturated with one fluid phase (either water or air) the coefficient of permeability k^f can be considered as a constant. If two fluid phases are present in the pores k^f depends on the degree of saturation S^f such that k^f decreases with decreasing values of S^f. This is due to the facts that water and air are assumed to be immiscible and each fluid can only flow in the part of the pores filled by the respective phase and that in the case of partial saturation the fluid paths are characterized by some tortuosity.

Defining relative permeability coefficients k^{rw} and k^{ra}, the permeabilities k^w and k^a of the soil skeleton with respect to the fluid phases water and air can be formulated as functions of the degree of water saturation:

$$k^w(S^w) = k^{ow}\, k^{rw}(S^w) \ , \quad k^a(S^w) = k^{oa}\, k^{ra}(S^w) \ . \tag{18}$$

In (18) k^{ow} and k^{oa} are the permeabilities for the fully saturated states, i.e. k^{ow} refers to fully water saturated conditions ($S^w = 1$) and k^{oa} to a dry soil ($S^w = 0$), respectively. k^{rw} and k^{ra} are introduced in order to consider partially saturated states. They are scalar quantities, varying between zero and one, depending on the degree of water saturation S^w. Empirical relations $k^{rw}(S^w)$ and $k^{ra}(S^w)$ can be found in the literature [13],[14],[24]. Figure 3 shows graphs of the two curves for a typical soil.

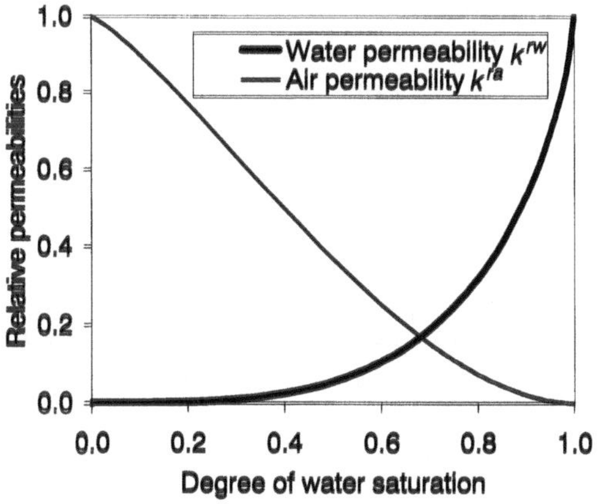

Fig. 3. Relative permeability curves for the fluid phases water and air

The linear momentum balance equations for the three-phase mixture are obtained by adding up Equation (14) and Equation (15), the latter for both the water phase and the air phase, as

$$\operatorname{div}(\boldsymbol{\sigma}_s + \boldsymbol{\sigma}_w + \boldsymbol{\sigma}_a) + (\rho_s + \rho_w + \rho_a)\mathbf{g} = \rho_s \mathbf{a} + \rho_w \mathbf{a}^w + \rho_a \mathbf{a}^a \ . \tag{19}$$

In (19) use of the fact has been made that the interaction forces must vanish by summation of the three phases, i.e. $\mathbf{f}_s + \mathbf{f}_w + \mathbf{f}_a = \mathbf{0}$.

Since for tunnelling under compressed air, the air pressure in the tunnel is increased slowly up to the desired value, the flow of water and air as well as the pressure in the soil develop rather slowly. Hence, it is justified to assume quasi-static conditions. Consequently, using the averaged density (2) and the averaged stresses (4) of the three-phase mixture, Equation (19) can be rewritten as

$$\operatorname{div}\tilde{\boldsymbol{\sigma}} + \tilde{\rho}\mathbf{g} = \mathbf{0} \ . \tag{20}$$

2.4 Mass balance equations

In each configuration the total mass of the solid phase, given as

$$m^s = \int_V (1-n)\rho^s \, dV \ , \tag{21}$$

is constant; hence, $Dm^s/Dt = 0$ holds, yielding

$$\frac{Dm^s}{Dt} = \int_V \left\{ \frac{D[(1-n)\rho^s]}{Dt} + (1-n)\rho^s \operatorname{div} \mathbf{v} \right\} dV = 0 \ , \tag{22}$$

$\mathbf{v}$ being the intrinsic velocity of the soil skeleton [21]. Equation (22) is satisfied, if the integrand vanishes. Performing the material time derivative, taking into account the incompressibility of the solid phase (*not* the soil skeleton, which is of course deformable) and using the volumetric strain rate of the soil skeleton $\dot{\varepsilon}_{vol}$ ($\dot{\varepsilon}_{vol} = \dot{\varepsilon}_{ij}\,\delta_{ij}$) instead of the divergence of $\mathbf{v}$, Equation (22) can be rewritten as

$$\frac{Dn}{Dt} = (1-n)\,\dot{\varepsilon}_{vol} \ . \tag{23}$$

Considering a domain without any wells or sinks the mass balance equation for a fluid phase f can be derived in an analogous manner. The fluid mass is given as

$$m^f = \int_V n S^f \rho^f \, dV \tag{24}$$

and its material time derivative again has to vanish, resulting in [21],[27]

$$\frac{Dn}{Dt} + \frac{n}{S^f \rho^f} \frac{D\left(S^f \rho^f\right)}{Dt} + n \,\mathrm{div}\, \mathbf{v} + \frac{1}{S^f} \,\mathrm{div}\, \tilde{\mathbf{v}}^{fr} + \frac{1}{S^f \rho^f} \nabla \rho^f \cdot \tilde{\mathbf{v}}^{fr} = 0. \quad (25)$$

Substitution of the mass balance equation for the solid phase (23) into the mass balance equation for the fluid phase (25) and specialization of the resulting equation for quasi-static conditions by neglecting the convective terms yields

$$\rho^f S^f \dot{\varepsilon}_{vol} + n S^f \dot{\rho}^f + n \rho^f \dot{S}^f = -\rho^f \mathrm{div}\, \tilde{\mathbf{v}}^{fr} - \nabla \rho^f \cdot \tilde{\mathbf{v}}^{fr} \,. \quad (26)$$

Due to the assumption that each phase continuously fills the entire domain, the mass balance equations for the individual constituents are referred to the same volume element. Hence, Equation (26) can be interpreted as follows: The terms on the right hand side, which can be written more concise as $-\mathrm{div}\left(\rho^f \tilde{\mathbf{v}}^{fr}\right)$, represent the inflow of fluid mass into a given volume element which has to be balanced by the terms on the left hand side of this equation; i.e. the inflow can be stored in the volume element by an increase of the volumetric strain of the soil skeleton $\dot{\varepsilon}_{vol}$ (being associated with an increase of porosity n as in Equation (23)), by an increase of the density $\dot{\rho}^f$ and by an increase of the degree of saturation $\dot{S}^f$ of the respective fluid phase.

2.5 Constitutive equations

Simple models for considering nonlinear material behaviour of unsaturated soils are characterized by extending the effective stress concept for water saturated soils to unsaturated soils. Following this approach, Equation (8) can be reformulated as

$$\tilde{\boldsymbol{\sigma}} = \tilde{\boldsymbol{\sigma}}' + \left(S^w p^w + S^a p^a\right) \mathbf{I} \,, \quad (27)$$

where

$$\tilde{\boldsymbol{\sigma}}' = (1-n)\left[\boldsymbol{\sigma}^s - \left(S^w p^w + S^a p^a\right) \mathbf{I}\right] \quad (28)$$

denotes the effective averaged stress tensor. According to (27), the total averaged stress tensor $\tilde{\boldsymbol{\sigma}}$ is partitioned into the effective averaged stresses $\tilde{\boldsymbol{\sigma}}'$ and the fluid stresses. Since the solid phase is assumed to be incompressible, only the effective stresses cause deformations of the soil skeleton. Therefore, the nonlinear constitutive law for the soil skeleton is formulated relating the rate of the effective averaged stresses to the strain rate in the soil skeleton:

$$\dot{\tilde{\boldsymbol{\sigma}}}' = \mathbf{C}_T \, \dot{\boldsymbol{\varepsilon}} \,. \quad (29)$$

In Equation (29) $\mathbf{C}_T = d\tilde{\boldsymbol{\sigma}}'/d\boldsymbol{\varepsilon}$ denotes the tangent material matrix of the soil skeleton. A comparison of several elastic-plastic soil models, including simple models based on the Drucker-Prager and the Mohr-Coulomb yield criterion and a more sophisticated cap model, in the context of 2D FE-analyses of tunnelling can be found in [26].

More complex constitutive models for unsaturated soils contain two independent stress state variables, e.g., the total stress in excess of pore air pressure, denoted as net stress, and the capillary stress [4],[10],[29]. The use of two stress state variables allows, e.g., to take into account the effect of the capillary stress on the shear strength of unsaturated soils. The model, described in [12],[13], was employed for the numerical simulation of tunnelling under compressed air on the basis of an uncoupled approach [15].

The constitutive equation for a compressible barotropic fluid phase (i.e., the state of thermodynamic equilibrium is independent of the temperature) is given as

$$\frac{\dot{\rho}^f}{\rho^f} = -\frac{\dot{p}^f}{K^f} \tag{30}$$

with K^f denoting the bulk modulus of the respective fluid phase f. The minus sign on the right hand side of (30) is a consequence of having defined tensile stresses as positive quantities. Assuming the air phase at constant temperature, it follows from (30), using Boyle-Mariotte's law $pV = const.$ and the principle of conservation of mass, that the bulk modulus K^a is equal to the magnitude of the air pressure p^a.

Finally, a constitutive relationship between the capillary stress p^c and the degree of water saturation S^w is required:

$$S^w = S^w(p^c) \; . \tag{31}$$

Inherent in Equation (31) is the assumption of isothermal conditions. Otherwise, the degree of water saturation would not only be a function of the capillary stress but also of the temperature. Equation (31) usually is obtained from laboratory experiments for determining the relation between matrix suction s and the degree of water saturation S^w as shown in Fig. 4b [1]. For stresses in the air phase different from the atmospheric pressure the matrix suction s has to be replaced by the capillary stress p^c according to (9). Following curve D_0 in Fig. 4a, departing from the state of a fully water saturated soil (i.e., from $S^w = 1$ and $s = 0$) for decreasing values of s (i.e., increasing absolute values of s) decreasing values of S^w are obtained. Approaching the residual degree of water saturation S^w_r, it is not possible to further withdraw water from the pores by increasing the absolute value of s. Subsequent wetting follows curve W until the maximum value of the degree of water saturation S^w_s is reached, being perhaps somewhat smaller than 1 because of the inclusion of air bubbles. Repeated dewetting yields curve D approaching curve D_0. As can be seen from Fig. 4a dewetting followed by wetting exhibits hysteresis which, to

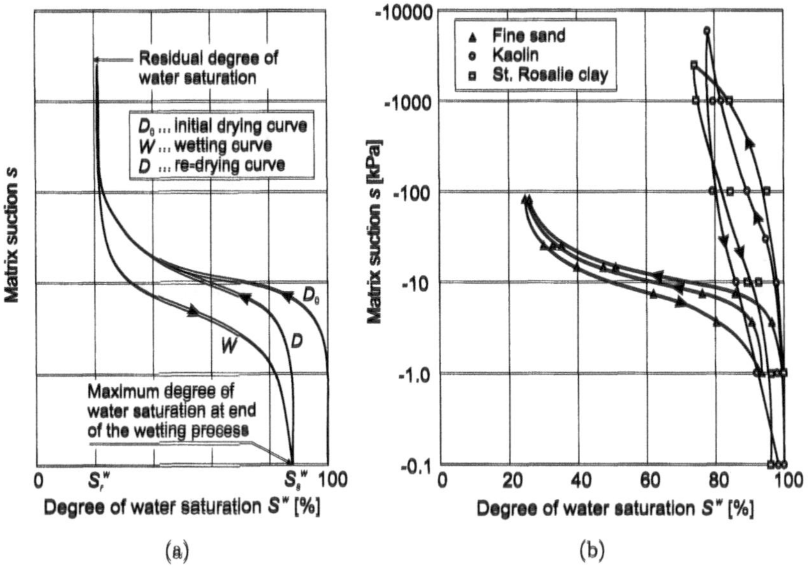

Fig. 4. Relationship between the degree of water saturation S^w and matrix suction s: (a) principal sketch, (b) examples for different soils

some extent, can be viewed similar to the active and passive capillary rises encountered in a capillary tube. However, while this hysteresis effect is rather pronounced in fine-grained soils, it is of minor importance for more coarse-textured ones and it may even be neglected when dealing, e.g., with coarse sands or gravel [15]. Moreover, the shape of the relationship between S^w and s and the magnitude of s strongly depend on the soil type. For fine-grained soils the latter is considerably larger than for coarse textured soils (Fig. 4b) [1].

In the literature several more or less sophisticated approximations for the relationship between matrix suction s and the degree of water saturation S^w have been suggested. A relatively simple empirical equation was developed by Brooks and Corey (Fig. 5a) which is described in more detail in [13] and is used in the approach presented in [21]. An improved approximation proposed by van Genuchten [14] is depicted in Fig. 5b.

The approximation of relation (31) proposed by Brooks and Corey is given as

$$S^w = S_r^w + (S_s^w - S_r^w)\left(\frac{p_b^c}{p^c}\right)^{\lambda} \tag{32}$$

and the one developed by van Genuchten as

$$S^w = S_r^w + (S_s^w - S_r^w)\left[1 + \left(\frac{p^c}{p_b^c}\right)^{n}\right]^{-m} . \tag{33}$$

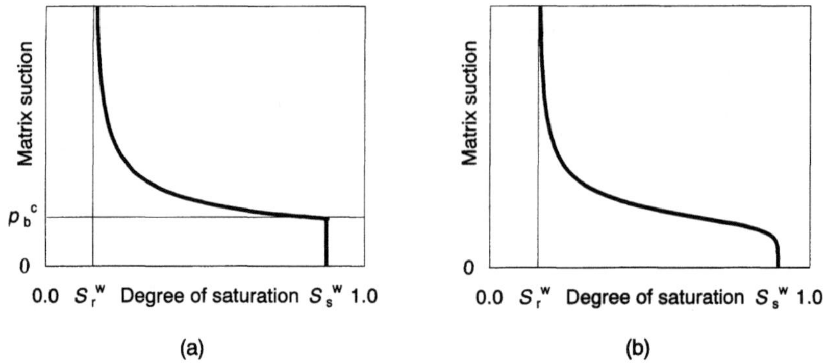

Fig. 5. Approximations of the $S^w p^c$-relationship according to (a) Brooks and Corey, and (b) van Genuchten

In (32) and (33) p_b^c denotes the air entry value, also referred to as bubbling pressure, which can be viewed as a characteristic pressure that has to be reached before the air actually enters the pores. λ, m and n are empirical constants to fit the curves to experimental data. Mualem determined these parameters for a large number of soils which in extracts are given in [24]. In the present model the approximation of the relationship between S^w and p^c by van Genuchten is used. Note that according to Fig. 4b no definite value of p_b^c can be identified for very fine grained soils, such as kaolin and clays.

3 Special cases contained in the three-phase formulation

The theoretical framework of the three-phase formulation can be employed for a broad range of problems encountered in geotechnical engineering. However, not only the complete model may be useful, but also different special cases forming certain subsets of the full three-phase formulation. In particular, a two-phase formulation for dewatering of soils under atmospheric conditions and a two-phase formulation for fully saturated conditions (i.e., consolidation) are contained in the complete model. Additionally, the uncoupled approach can be deduced and of course the familiar equations for drained conditions.

3.1 Dewatering under atmospheric conditions

Starting from the three-phase formulation, assuming the hydrostatic stress in the air phase to be equal to the atmospheric pressure yields a model for dewatering of soils under atmospheric conditions. In this case any changes of the hydrostatic air stresses are neglected. Consequently, the mass balance equation for the air phase can be omitted and the original three-phase model is reduced to a two-phase formulation. However, unsaturated soil behaviour is still possible, i.e. the degree of water saturation may vary.

Since the air stresses are considered to be atmospheric, the capillary stress (9) is equal to the negative hydrostatic water stress. Hence, from (31) the degree of water saturation is obtained as

$$S^w = S^w(-p^w) \,. \tag{34}$$

From (8) the averaged stress tensor for this special case follows as

$$\tilde{\boldsymbol{\sigma}} = (1-n)\,\boldsymbol{\sigma}^s + nS^w p^w\,\mathbf{I} \,. \tag{35}$$

The equilibrium equations (20) are now formulated with the averaged density according to (2) and the averaged stresses according to (35). The mass balance equation (26) and Darcy's law (17) have to be specified only for the water phase.

3.2 Consolidation

If the soil is water saturated and under atmospheric air pressure, then the three-phase formulation degenerates to the consolidation theory. Similar to the case of dewatering under atmospheric conditions the hydrostatic air stresses are equal to the atmospheric pressure and thus the mass balance equation for the air phase can be omitted. Consequently, this special case is a two-phase formulation as well. However, the degree of water saturation is now equal to 1 and kept constant during the analysis which means that the soil stays fully saturated. Therefore, the coefficient of permeability k^w with respect to the water phase, given in the first of the two equations of (18), may be considered as a constant as well. Thus, the relations

$$S^w = 1 = const. \,, \qquad S^a = 0 = const. \,, \qquad k^{rw} = 1 = const. \tag{36}$$

are valid when dealing with consolidation problems.

Because of (36) the averaged density $\tilde{\rho}$ (2) and the averaged stress tensor $\tilde{\boldsymbol{\sigma}}$ (8) are now given as

$$\tilde{\rho} = (1-n)\rho^s + n\rho^w \qquad \text{and} \qquad \tilde{\boldsymbol{\sigma}} = (1-n)\,\boldsymbol{\sigma}^s + np^w\,\mathbf{I} \,, \tag{37}$$

in the equilibrium equations (20). Since the degree of water saturation is constant, its time derivative is zero and consequently the last term on the left hand side in (26) vanishes. Hence, insertion of (36) into (26) yields the following form of the mass balance equation for the water phase

$$\dot{\varepsilon}_{vol} = -\operatorname{div} \tilde{\mathbf{v}}^{wr} \,, \tag{38}$$

where the water phase has been assumed as incompressible, i. e., $\rho^w = const.$

3.3 Uncoupled approach

The uncoupled approach is characterized by treating the fluid flow in the soil in a first step, assuming a rigid soil skeleton, and determining the deformations of the soil in a second step, making use of the results of the flow analysis. Consequently, interactions between the fluid flow in the soil and the deformations of the soil are neglected in this simplified approach. Nevertheless, it is rather popular in engineering practice.

The two-phase model for consolidation problems ($S^w = 1 = const., S^a = 0$) serves as a starting point for the subsequent derivations. The linear momentum balance equations for the soil skeleton are given in (14). For quasi-static conditions they read as

$$\mathrm{div}\boldsymbol{\sigma}_s + \rho_s \mathbf{g} + \mathbf{f}_s = \mathbf{0} \,. \tag{39}$$

From Equation (28) the averaged effective stress tensor can be obtained for $S^w = 1$ ($S^a = 0$), considering (5_1), as

$$\tilde{\boldsymbol{\sigma}}' = (1-n)\boldsymbol{\sigma}^s - (1-n)p^w \mathbf{I} = \boldsymbol{\sigma}_s - (1-n)p^w \mathbf{I} \,. \tag{40}$$

When calculating the divergence of (40), the resulting equation can be substituted into the first term of the equilibrium equations (39) yielding

$$\mathrm{div}\tilde{\boldsymbol{\sigma}}' + (1-n)\,\nabla p^w + \rho_s \mathbf{g} + \mathbf{f}_s = \mathbf{0} \,, \tag{41}$$

where n has been assumed as a constant quantity. For quasi-static conditions the linear momentum balance law for the water phase follows from (15), employing the relationship $\boldsymbol{\sigma}_w = n\,\boldsymbol{\sigma}^w = n\,p^w \mathbf{I}$ for the hydrostatic water stress, which follows from (5_2) and (7), as

$$n\,\nabla p^w + \rho_w \mathbf{g} + \mathbf{f}_w = \mathbf{0} \,. \tag{42}$$

From (42) the interaction force $\mathbf{f}_w$ can be calculated as

$$\mathbf{f}_w = -n\,\nabla p^w - \rho_w \mathbf{g} \,. \tag{43}$$

When considering the water saturated porous medium, the interaction forces for the soil skeleton and the water phase, $\mathbf{f}_s$ and $\mathbf{f}_w$, have to sum up to zero, that is

$$\mathbf{f}_s + \mathbf{f}_w = \mathbf{0} \qquad \text{or} \qquad \mathbf{f}_s = -\mathbf{f}_w \,. \tag{44}$$

Hence, when employing the relationship

$$\nabla p^w = \frac{\gamma^w}{k^w}\,\tilde{\mathbf{v}}^{wr} - \rho^w \mathbf{g} \,, \tag{45}$$

which is obtained from Darcy's law (17) for the water phase, and taking into account (3_2) with $S^w = 1$, the interaction force for the soil skeleton can be expressed as

$$\mathbf{f}_s = n\frac{\gamma^w}{k^w}\,\tilde{\mathbf{v}}^{wr} \,. \tag{46}$$

Substituting the Equations (45) and (46) into the linear momentum balance law for the soil skeleton (41) and making use of (3_1) yields

$$\operatorname{div}\tilde{\boldsymbol{\sigma}}' + \frac{\gamma^w}{k^w}\,\tilde{\mathbf{v}}^{wr} + (1-n)(\rho^s - \rho^w)\mathbf{g} = \mathbf{0}\ . \tag{47}$$

Equation (47) implies the subsequent interpretation. The first term represents the divergence of the effective averaged stress tensor of the soil skeleton; only this term causes deformations of the soil skeleton when considering an incompressible solid phase. The second term describes the drag forces arising from the flow of water in the porous soil skeleton. The third term accounts for the weight of the soil skeleton due to the presence of the water phase and is known as the buoyant weight.

Hence, in an uncoupled approach, after dealing with the flow of water in a rigid soil skeleton yielding the drag forces, the deformations of the soil skeleton are calculated from the linear momentum balance equation (47) taking into account these drag forces and the buoyant weight of the soil skeleton.

3.4 Drained conditions

To obtain the special case for drained conditions, the fluid phases are omitted and thus the fluid stresses and consequently also the mass balance equations for the fluids can be neglected. Apart from the kinematic relations and the constitutive equations for the soil skeleton only the equilibrium equations are required in this case, relating the internal forces of the soil skeleton to any present external forces consisting of body forces and surface tractions.

The definitions for the averaged density $\tilde{\rho}$ (2) and the averaged stress tensor $\tilde{\boldsymbol{\sigma}}$ (8) are then reduced to

$$\tilde{\rho} = (1-n)\rho^s \qquad \text{and} \qquad \tilde{\boldsymbol{\sigma}} = (1-n)\,\boldsymbol{\sigma}^s\ . \tag{48}$$

Using these relations in combination with the linear momentum balance equations (20) yields the equilibrium conditions for a dry soil with the porosity n.

4 Finite element formulation

4.1 Weak formulation of the governing equations

Weak formulations of the equilibrium equations for the three-phase mixture (20) and of the mass balance equations for the fluid phases (26) form the basis for the numerical solution of problems involving a three-phase medium.

Using the principle of virtual displacements and taking into account the partitioning of the total stress tensor as described in (27), the weak formulation of the equilibrium equations for quasi-static conditions (20) is obtained in vector and matrix notation as

$$-\int_V \delta\boldsymbol{\varepsilon}^T \tilde{\boldsymbol{\sigma}}' \, dV - \int_V S^w p^w \, \delta\boldsymbol{\varepsilon}^T \mathbf{1} \, dV - \int_V S^a p^a \, \delta\boldsymbol{\varepsilon}^T \mathbf{1} \, dV$$

$$+ \int_V \delta\mathbf{u}^T \, \tilde{\mathbf{b}} \, dV + \int_S \delta\mathbf{u}^T \, \tilde{\mathbf{t}} \, dS = 0 \; , \tag{49}$$

where $\mathbf{1} = \lfloor 1 \;\; 1 \;\; 1 \;\; 0 \;\; 0 \;\; 0 \rfloor^T$ denotes a vector similar to the Kronecker-Delta and the stresses and strains are given as $\boldsymbol{\sigma} = \lfloor \sigma_{11} \;\; \sigma_{22} \;\; \sigma_{33} \;\; \sigma_{12} \;\; \sigma_{23} \;\; \sigma_{31} \rfloor^T$ and $\boldsymbol{\varepsilon} = \lfloor \varepsilon_{11} \;\; \varepsilon_{22} \;\; \varepsilon_{33} \;\; 2\varepsilon_{12} \;\; 2\varepsilon_{23} \;\; 2\varepsilon_{31} \rfloor^T$, respectively; $\delta\mathbf{u}$ and $\delta\boldsymbol{\varepsilon}$ denote virtual displacements and virtual strains; V and S are the volume and the surface of the domain under consideration; $\tilde{\mathbf{b}} = \tilde{\rho}\,\mathbf{g}$ is the vector of the body forces and $\tilde{\mathbf{t}}$ denotes the vector of the surface tractions.

Neglecting in (26) the gradient of the fluid density according to [21], dividing the resulting equation by ρ^f, substituting Darcy's law (17) and the constitutive relationship for the fluid phase f (30) into this equation, multiplying the resulting equation by a virtual hydrostatic stress δp^f, integrating over the domain under consideration and applying the divergence theorem to the term containing the divergence of $\tilde{\mathbf{v}}^{fr}$ yields the following weak formulation of the mass balance equation for the fluid phase f:

$$\int_V \delta p^f \left\{ S^f \dot{\varepsilon}_{vol} - \frac{n S^f}{K^f} \dot{p}^f + n \dot{S}^f \right\} dV$$

$$- \int_V \frac{k^f}{\gamma^f} \left(\delta \nabla p^f\right)^T \left(\nabla p^f + \rho^f \mathbf{g}\right) dV + \int_S \delta p^f \, \tilde{q}^f \, dS = 0 \tag{50}$$

with $\tilde{q}^f$ denoting the averaged flux of the respective fluid phase through the boundary of the domain under consideration.

4.2 Discretization

Spatial discretization of (49) and (50), the latter for both the water phase ($f = w$) and the air phase ($f = a$), within the framework of the finite element method (FEM) involves subdivision of the domain V under consideration into finite elements. The displacements of the soil skeleton $\mathbf{u}$ and the hydrostatic stresses in the water- and the air phase, p^w and p^a, are chosen as primary variables. Within a single finite element e they are approximated in terms of a matrix of suitable shape functions $\mathbf{N}_u$ and of a vector of suitable shape functions $\mathbf{N}_p$ and of the respective vectors of the nodal values of the displacements, $\mathbf{U}_e$, and of the hydrostatic fluid stresses, $\mathbf{P}^w_e$ and $\mathbf{P}^a_e$, as

$$\mathbf{u}_e = \mathbf{N}_u \, \mathbf{U}_e \; , \quad p^w_e = \mathbf{N}_p^T \, \mathbf{P}^w_e \; , \quad p^a_e = \mathbf{N}_p^T \, \mathbf{P}^a_e \; . \tag{51}$$

From (51) the strains of the soil skeleton and the gradients of the fluid stresses within a single finite element e are obtained as

$$\boldsymbol{\varepsilon}_e = \mathbf{B}_u \, \mathbf{U}_e \, , \quad \nabla p_e^w = \mathbf{B}_p^T \, \mathbf{P}_e^w \, , \quad \nabla p_e^a = \mathbf{B}_p^T \, \mathbf{P}_e^a \, , \tag{52}$$

where the matrix $\mathbf{B}_u$ contains the derivatives of the shape functions $\mathbf{N}_u$ with respect to the coordinates and the vector $\mathbf{B}_p$ contains the derivatives of the shape functions $\mathbf{N}_p$ with respect to the coordinates.

Robust finite elements for coupled solid-fluid problems are characterized by different orders of interpolation for the displacements of the soil skeleton and the fluid stresses, e.g., for plane strain and axisymmetric problems the displacements and the hydrostatic stresses are interpolated with quadratic and bilinear shape functions, respectively (Fig. 6) [31].

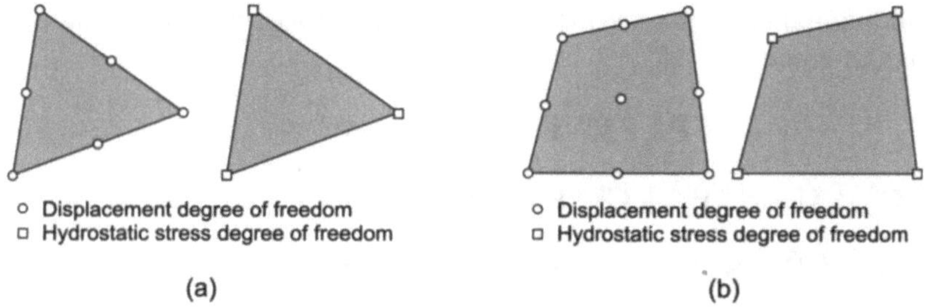

Fig. 6. 2D and axisymmetric finite elements for a three-phase formulation: (a) triangular elements, (b) quadrilateral elements

Assuming first the discretization to consist only of a single finite element, substitution of (51) and (52) into the weak formulation of the equilibrium equations for the three-phase mixture (49) and into the weak formulation of the mass balance equation for a fluid phase f (50) yields

$$\begin{aligned} -(\delta\mathbf{U}_e)^T \int_{V_e} \mathbf{B}_u^T \, \tilde{\boldsymbol{\sigma}}' \, dV_e - (\delta\mathbf{U}_e)^T \int_{V_e} \mathbf{B}_u^T \, \mathbf{1} S^w \, \mathbf{N}_p^T \, dV_e \, \mathbf{P}_e^w \\ - (\delta\mathbf{U}_e)^T \int_{V_e} \mathbf{B}_u^T \, \mathbf{1} S^a \, \mathbf{N}_p^T \, dV_e \, \mathbf{P}_e^a + (\delta\mathbf{U}_e)^T \int_{V_e} \mathbf{N}_u^T \, \tilde{\mathbf{b}} \, dV_e \\ + (\delta\mathbf{U}_e)^T \int_{S_e} \mathbf{N}_u^T \, \tilde{\mathbf{t}} \, dS_e = 0 \end{aligned} \tag{53}$$

and

$$(\delta\mathbf{P}_e^f)^T \int_{V_e} \mathbf{N}_p \, S^f \mathbf{1}^T \, \mathbf{B}_u \, dV_e \, \dot{\mathbf{U}}_e - (\delta\mathbf{P}_e^f)^T \int_{V_e} \mathbf{N}_p \frac{n S^f}{K^f} \mathbf{N}_p^T \, dV_e \, \dot{\mathbf{P}}_e^f$$

$$+ (\delta \mathbf{P}_e^f)^T \int_{V_e} \mathbf{N}_p \, n \, \frac{dS^f}{dp^c} \, \mathbf{N}_p^T \, dV_e \left(\dot{\mathbf{P}}_e^a - \dot{\mathbf{P}}_e^w \right)$$

$$- (\delta \mathbf{P}_e^f)^T \int_{V_e} \mathbf{B}_p \, \frac{k^f}{\gamma^f} \left[\mathbf{B}_p^T \, \mathbf{P}_e^f + \rho^f \mathbf{g} \right] dV_e + (\delta \mathbf{P}_e^f)^T \int_{S_e} \mathbf{N}_p \, \tilde{q}^f \, dS_e = 0 \, . \tag{54}$$

In (53) and (54) V_e is the volume and S_e the surface of the finite element e.

Extending the discretization in Equations (53) and (54) to a domain with an arbitrary number of finite elements, making use of the fact that the resulting equations hold for arbitrary virtual displacements $\delta\mathbf{U}$ and arbitrary virtual hydrostatic stresses $\delta\mathbf{P}^f$ and assembling the vectors and matrices for the individual elements according to standard procedures yields the coupled system of equations [25]

$$\begin{aligned} \mathbf{f}^{in} + \mathbf{C}^w \, \mathbf{P}^w + \mathbf{C}^a \, \mathbf{P}^a &= \mathbf{f}^{ex} \, , \\ (\mathbf{C}^w)^T \, \dot{\mathbf{U}} - \mathbf{S}^w \, \dot{\mathbf{P}}^w + \mathbf{C}^{wa} \, \dot{\mathbf{P}}^a - \mathbf{H}^w \, \mathbf{P}^w &= \dot{\mathbf{f}}^w \, , \\ (\mathbf{C}^a)^T \, \dot{\mathbf{U}} - \mathbf{S}^a \, \dot{\mathbf{P}}^a + \mathbf{C}^{wa} \, \dot{\mathbf{P}}^w - \mathbf{H}^a \, \mathbf{P}^a &= \dot{\mathbf{f}}^a \, . \end{aligned} \tag{55}$$

In (55) $\mathbf{U}$, $\mathbf{P}^w$ and $\mathbf{P}^a$ denote the global vectors of the nodal values for the displacements of the soil skeleton and for the hydrostatic stresses in the fluid phases water and air; $\dot{\mathbf{U}}$, $\dot{\mathbf{P}}^w$ and $\dot{\mathbf{P}}^a$ are the respective time derivatives. The contributions of a single finite element e to the vectors of the internal and the external forces of the soil skeleton, $\mathbf{f}^{in}$ and $\mathbf{f}^{ex}$, are given as

$$\mathbf{f}_e^{in} = \int_{V_e} \mathbf{B}_u^T \, \tilde{\boldsymbol{\sigma}}' \, dV_e \, , \quad \mathbf{f}_e^{ex} = \int_{V_e} \mathbf{N}_u^T \, \tilde{\mathbf{b}} \, dV_e + \int_{S_e} \mathbf{N}_u^T \, \tilde{\mathbf{t}} \, dS_e \, . \tag{56}$$

The matrices $\mathbf{C}^w$ and $\mathbf{C}^a$ describe the coupling between the water phase and the soil skeleton and between the air phase and the soil skeleton, respectively; for a single finite element they are given as

$$\mathbf{C}_e^w = \int_{V_e} \mathbf{B}_u^T \, \mathbf{1} \, S^w \, \mathbf{N}_p^T \, dV_e \, , \quad \mathbf{C}_e^a = \int_{V_e} \mathbf{B}_u^T \, \mathbf{1} \, S^a \, \mathbf{N}_p^T \, dV_e \, . \tag{57}$$

The coupling between the two fluid phases is contained in matrix $\mathbf{C}^{wa}$, which is obtained from the contributions for the individual finite elements

$$\mathbf{C}_e^{wa} = \int_{V_e} \mathbf{N}_p \, n \, \frac{dS^w}{dp^c} \, \mathbf{N}_p^T \, dV_e \, . \tag{58}$$

The matrices $\mathbf{S}^w$ and $\mathbf{S}^a$ contain the compressibility of the particular fluid and the constitutive relationship between the degree of water saturation and the capillary stress; the respective element matrices are given as

$$\mathbf{S}_e^w = \int_{V_e} \mathbf{N}_p \left(\frac{nS^w}{K^w} + n\frac{dS^w}{dp^c} \right) \mathbf{N}_p^T \, dV_e$$

$$\mathbf{S}_e^a = \int_{V_e} \mathbf{N}_p \left(\frac{nS^a}{K^a} + n\frac{dS^w}{dp^c} \right) \mathbf{N}_p^T \, dV_e \ , \tag{59}$$

where use of $\dot{S}^a = -\dot{S}^w$ has been made. The permeability matrices $\mathbf{H}^w$ and $\mathbf{H}^a$ are assembled from the respective element matrices

$$\mathbf{H}_e^w = \int_{V_e} \mathbf{B}_p \frac{k^w}{\gamma^w} \mathbf{B}_p^T \, dV_e \ , \quad \mathbf{H}_e^a = \int_{V_e} \mathbf{B}_p \frac{k^a}{\gamma^a} \mathbf{B}_p^T \, dV_e \ . \tag{60}$$

The vectors $\dot{\mathbf{f}}^w$ and $\dot{\mathbf{f}}^a$ are associated with the flow of the respective fluid due to gravity and the fluid flow through the surface of the domain under consideration. They are obtained from the contributions of the individual elements

$$\dot{\mathbf{f}}_e^w = \int_{V_e} \mathbf{B}_p \frac{k^w}{g} \mathbf{g} \, dV_e - \int_{S_e} \mathbf{N}_p \, \tilde{q}^w \, dS_e \ ,$$

$$\dot{\mathbf{f}}_e^a = \int_{V_e} \mathbf{B}_p \frac{k^a}{g} \mathbf{g} \, dV_e - \int_{S_e} \mathbf{N}_p \, \tilde{q}^a \, dS_e \ . \tag{61}$$

Since (55) contains time derivatives of the nodal values, a numerical integration in the time domain has to be performed. Assuming the solution to be known for the discrete point of time t_n and Δt_{n+1} being the current time step size, application of the implicit and unconditionally stable Euler backward integration scheme results in

$$\mathbf{X}_{n+1} = \mathbf{X}_n + \dot{\mathbf{X}}_{n+1} \Delta t_{n+1} \quad \text{with} \quad \mathbf{X} = \lfloor \mathbf{U} \;\; \mathbf{P}^w \;\; \mathbf{P}^a \rfloor^T \ , \tag{62}$$

where $\mathbf{X}_n$ and $\mathbf{X}_{n+1}$ denote the known and unknown solution for the nodal values at the discrete points of time t_n and t_{n+1}, respectively. Making use of (62) in (55) yields the following coupled set of equations for the incremental nodal values:

$$\begin{bmatrix} \mathbf{K} & \mathbf{C}_{n+1}^w & \mathbf{C}_{n+1}^a \\ (\mathbf{C}_{n+1}^w)^T & -\mathbf{S}_{n+1}^w - \Delta t_{n+1} \mathbf{H}_{n+1}^w & \mathbf{C}_{n+1}^{wa} \\ (\mathbf{C}_{n+1}^a)^T & \mathbf{C}_{n+1}^{wa} & -\mathbf{S}_{n+1}^a - \Delta t_{n+1} \mathbf{H}_{n+1}^a \end{bmatrix} \begin{Bmatrix} \Delta \mathbf{U}_{n+1} \\ \Delta \mathbf{P}_{n+1}^w \\ \Delta \mathbf{P}_{n+1}^a \end{Bmatrix}$$

$$= \begin{Bmatrix} \mathbf{f}_{n+1}^{ex} - \mathbf{f}_n^{in} - \mathbf{C}_{n+1}^w \mathbf{P}_n^w - \mathbf{C}_{n+1}^a \mathbf{P}_n^a \\ \Delta t_{n+1} (\dot{\mathbf{f}}_{n+1}^w + \mathbf{H}_{n+1}^w \mathbf{P}_n^w) \\ \Delta t_{n+1} (\dot{\mathbf{f}}_{n+1}^a + \mathbf{H}_{n+1}^a \mathbf{P}_n^a) \end{Bmatrix} . \tag{63}$$

To arrive at (63) in (55) the vector of internal forces was partitioned as $\mathbf{f}_{n+1}^{in} = \mathbf{f}_n^{in} + \Delta \mathbf{f}_{n+1}^{in}$. Then for linear elastic material behaviour of the soil skeleton $\Delta \mathbf{f}_{n+1}^{in}$ can be expressed as $\Delta \mathbf{f}_{n+1}^{in} = \mathbf{K}\,\Delta \mathbf{U}_{n+1}$ with $\mathbf{K}$ representing the stiffness matrix of the soil skeleton. Although (63) refers to the special case of linear elastic constitutive behaviour of the soil skeleton the equations are still nonlinear. This follows from the dependence of the matrices $\mathbf{C}$, $\mathbf{S}$ and $\mathbf{H}$ and of the vectors $\dot{\mathbf{f}}$ on the nodal values of the fluid stresses, because $\mathbf{C}$, $\mathbf{S}$, $\mathbf{H}$ and $\dot{\mathbf{f}}$ are functions either of the degree of water saturation (31) or of the permeability coefficients (18).

Equation (63) can be extended to take into account nonlinear material behaviour of the soil skeleton by evaluating the soil stresses and, consequently, the vector of the internal forces $\mathbf{f}^{in}$ and the stiffness matrix $\mathbf{K}$ according to the employed constitutive model for the soil skeleton.

The system of equations (63) can be written in concise form as

$$\mathbf{A}\left(\mathbf{X}_{n+1}\right) \Delta \mathbf{X}_{n+1} = \mathbf{f}\left(\mathbf{X}_{n+1}\right) \; . \tag{64}$$

$\mathbf{X}_{n+1}$ can be determined, e.g., by means of a direct iteration procedure as

$$\Delta \mathbf{X}_{n+1}^{i} = \left[\mathbf{A}(\mathbf{X}_{n+1}^{i-1})\right]^{-1} \mathbf{f}\left(\mathbf{X}_{n+1}^{i-1}\right) \; , \quad i = 1, 2, 3, \ldots \tag{65}$$

with

$$\mathbf{X}_{n+1}^{i} = \mathbf{X}_{n} + \Delta \mathbf{X}_{n+1}^{i} \; , \quad \mathbf{X}_{n+1}^{0} = \mathbf{X}_{n} \; , \tag{66}$$

and i denoting the iteration step within the time step Δt_{n+1}.

5 Adaptive FE-formulation for three-phase problems

Adaptive FE-analyses allow to enhance the accuracy and reliability of numerical simulations [30]. To this end the discretization error of a FE-solution is estimated a posteriori and – if a chosen error threshold is exceeded – an improved FE-mesh is generated and the analysis is continued by employing the improved mesh.

The application of an adaptive solution strategy to nonlinear problems, which are solved by an incremental iterative solution procedure, requires the estimation of the incremental discretization error. On the basis of the incremental error an improved FE-mesh is generated for the current time step (or load step) and only the current incremental step is re-calculated. Hence, a time-consuming re-calculation starting with the first time step (or load step), is avoided. However, an algorithm for transferring the state variables from the old mesh to the improved mesh is required. Thus, an adaptive nonlinear FE-analysis is computationally more expensive than a conventional nonlinear FE-analysis with a fixed FE-mesh.

Nevertheless, the development of an adaptive solution strategy for three-phase problems can be motivated by the following facts. Commonly, engineers employing the FEM for the numerical simulation of geotechnical problems are familiar with the solution of nonlinear single-phase problems. They have, however, less experience with time-dependent FE-analyses of multi-phase problems. Hence, especially for multi-phase problems it is often difficult to determine a suitable FE-mesh a priori, because the regions of the discretized domain, which are characterized by large gradients of the stresses in the soil skeleton and of the velocities of the fluid phases and, thus, requiring a sufficiently fine discretization, may change during the analyses. For instance, in consolidation problems, shortly after load application, large pore water pressures may be generated in the soil in the vicinity of the applied load and, thus, pore water pressures will govern mesh refinement at the earlier stages of the analysis. At later stages, however, the pore water pressures are decreasing, whereas the displacements are increasing. Thus, the latter will gain a stronger influence on mesh refinement.

Since the coupled system of equations of the three-phase model (55) contains the nodal values of the displacements of the soil skeleton and of the hydrostatic stresses in the fluid phases as well as their time derivatives, within the framework of the finite element method in addition to the spatial discretization a discretization in the time domain is required. Since the size of suitable time steps commonly show large variations over the different stages of a time-dependent FE-analysis, the time step size is determined also on the basis of an adaptive procedure, controlling the error due to numerical integration in the time domain.

5.1 Error estimation for the spatial discretization

In general, for the estimation of the error with respect to the spatial discretization one can choose between error estimators and error indicators [30],[33]. The former are formulated on a mathematical basis, yielding an objective measure of the actual error. However, usually they are more expensive to evaluate and error estimators for linear problems cannot be automatically applied to nonlinear problems. On the other hand, error indicators may be based on relatively simple intuitive ideas. If they are based on already available results of the finite element analysis, they may have the advantage of an easy and efficient evaluation.

The widely employed Zienkiewicz-Zhu error estimator [34] for linear problems, based on the superconvergent patch recovery technique, has been employed successfully as error indicator for nonlinear problems and, in particular, for consolidation analysis (see, e.g. [35],[36]). For instance, the superconvergent patch recovery technique enhanced by incorporating equilibrium and boundary conditions, proposed in [37], has been applied to consolidation problems in [36] by including the pore pressure in the error estimation. In [38], departing from the Zienkiewicz-Zhu error indicator for the effective stresses in the soil skeleton, the error estimation is extended in order to include elastic-plastic material behaviour of the soil by incorporating the plastic strains as well as the plastic strain rate. A further extension in [38] concerns saturated soils, for which the seepage velocity is included in the error estimation.

From this brief survey it follows that suitable error indicators of the three-phase formulation for tunnelling under compressed air can be obtained by applying the well-known superconvergent patch recovery technique, proposed by Zienkiewicz and Zhu [34]. For consolidation analysis the superconvergent patch recovery technique is commonly applied to the effective stresses [35],[36]. Since within the framework of the three-phase formulation material models for unsaturated soils may be based on different stress state variables, the superconvergent patch recovery technique is applied to the strains in the soil skeleton and, in addition, to the gradients of the hydrostatic stresses in the two fluid phases water and air. The latter are related to the averaged velocities of the fluid phases by means of Darcy's law according to (17). The error indicators can be computed in terms of the L_2-norms for a single finite element e as

$$\| \mathbf{e}_\varepsilon \|_e = \sqrt{\int_{V_e} \mathbf{e}_\varepsilon \cdot \mathbf{e}_\varepsilon \, dV_e} \ , \quad \| \mathbf{e}_w \|_e = \sqrt{\int_{V_e} \mathbf{e}_w \cdot \mathbf{e}_w \, dV_e} \ ,$$

$$\| \mathbf{e}_a \|_e = \sqrt{\int_{V_e} \mathbf{e}_a \cdot \mathbf{e}_a \, dV_e} \ . \tag{67}$$

In (67) the estimated errors of the strains $\boldsymbol{\varepsilon}$ and of the gradients of the hydrostatic stresses in the fluid phases, ∇p^w and ∇p^a, at an arbitrary point within element e are obtained as the difference between the smoothed values, denoted by the superscript *, and the respective values of the finite element solution, denoted by the superscript h:

$$\mathbf{e}_\varepsilon = \boldsymbol{\varepsilon}^* - \boldsymbol{\varepsilon}^h \ , \quad \mathbf{e}_w = \nabla p_w^* - \nabla p_w^h \ , \quad \mathbf{e}_a = \nabla p_a^* - \nabla p_a^h \ . \tag{68}$$

The smoothed values are computed from the FE-solution by applying the superconvergent patch recovery technique [34]. In order to account for elastic-plastic material behaviour of the soil skeleton, in addition, the plastic strain rate can be included in the superconvergent patch recovery, yielding the error indicator for a single finite element e

$$\| \dot{\mathbf{e}}_p \|_e = \sqrt{\int_{V_e} \dot{\mathbf{e}}_p \cdot \dot{\mathbf{e}}_p \, dV_e} \tag{69}$$

with the estimated error of the plastic strain rate at an arbitrary point within element e

$$\dot{\mathbf{e}}_p = \dot{\boldsymbol{\varepsilon}}_p^* - \dot{\boldsymbol{\varepsilon}}_p^h \ . \tag{70}$$

From the estimated element errors global error measures are computed:

$$\| \mathbf{e}_\varepsilon \| = \sqrt{\sum_{e=1}^{m} \| \mathbf{e}_\varepsilon \|_e^2} \ , \quad \| \mathbf{e}_w \| = \sqrt{\sum_{e=1}^{m} \| \mathbf{e}_w \|_e^2} \ ,$$

$$\| \mathbf{e}_a \| = \sqrt{\sum_{e=1}^{m} \| \mathbf{e}_a \|_e^2}\,, \qquad \| \dot{\mathbf{e}}_p \| = \sqrt{\sum_{e=1}^{m} \| \dot{\mathbf{e}}_p \|_e^2}\,, \tag{71}$$

where m denotes the number of finite elements of the actual mesh. From (71) relative global error measures can be obtained as

$$\eta_\varepsilon = \frac{\| \mathbf{e}_\varepsilon \|}{\| \boldsymbol{\varepsilon}^h \|}\,, \quad \eta_w = \frac{\| \mathbf{e}_w \|}{\| \nabla p_w^h \|}\,, \quad \eta_a = \frac{\| \mathbf{e}_a \|}{\| \nabla p_a^h \|}\,, \quad \eta_p = \frac{\| \dot{\mathbf{e}}_p \|}{\| \dot{\boldsymbol{\varepsilon}}_p^h \|}\,, \tag{72}$$

where the L_2-norms

$$\| \boldsymbol{\varepsilon}^h \| = \sqrt{\sum_{e=1}^{m} \left(\int_{V_e} \boldsymbol{\varepsilon}^h \cdot \boldsymbol{\varepsilon}^h \, dV_e \right)}\,,$$

$$\| \nabla p_w^h \| = \sqrt{\sum_{e=1}^{m} \left(\int_{V_e} \nabla p_w^h \cdot \nabla p_w^h \, dV_e \right)},$$

$$\| \nabla p_a^h \| = \sqrt{\sum_{e=1}^{m} \left(\int_{V_e} \nabla p_a^h \cdot \nabla p_a^h \, dV_e \right)}\,,$$

$$\| \dot{\boldsymbol{\varepsilon}}_p^h \| = \sqrt{\sum_{e=1}^{m} \left(\int_{V_e} \dot{\boldsymbol{\varepsilon}}_p^h \cdot \dot{\boldsymbol{\varepsilon}}_p^h \, dV_e \right)}\,, \tag{73}$$

serve as reference values.

5.2 Generation of the improved mesh

If the computed relative global error measures (72) differ from a user specified threshold $\bar{\eta}$, then the mesh will be modified. Actually, two threshold values, namely an upper threshold value $\bar{\eta}_{max}$ and a lower threshold value $\bar{\eta}_{min}$, have to be defined. The former represents the chosen acceptable error, whereas the latter serves as the target value for the generation of an improved mesh, if the estimated error exceeds $\bar{\eta}_{max}$. It has turned out to be advantageous to meet the condition $\bar{\eta}_{min} \leq 0.90\ \bar{\eta}_{max}$ [40]. In the following, for brevity, no distinction between the two threshold values is made. Several strategies for mesh refinement have been proposed in the literature. As shown in [33], the requirement of a uniform distribution of element errors, proposed in [39], is very efficient, yielding meshes that minimize the number of elements for a given acceptable error. Hence, for instance, the global error of the strains can be expressed as

$$\| \mathbf{e}_\varepsilon \|^2 = m \cdot \bar{e}_{m,\varepsilon}^2\,, \tag{74}$$

where $\bar{e}_{m,\varepsilon}$ denotes the error threshold of the strains for a single finite element e. The latter can be determined by substituting (74) into the first equation (72) and setting $\eta_\varepsilon = \bar{\eta}$ as

$$\bar{e}_{m,\varepsilon} = \frac{\parallel \boldsymbol{\varepsilon}^h \parallel}{\sqrt{m}} \bar{\eta} \, . \tag{75}$$

Employing the h-version of the available mesh refinement techniques, which is characterized by adjusting the size h_e of the finite elements, on the basis of the estimated error of the strains a new element size $\bar{h}_{e,\varepsilon}$ is computed according to [30]

$$\bar{h}_{e,\varepsilon} = \frac{h_e}{\xi_{e,\varepsilon}^{1/p}} \quad \text{with} \quad \xi_{e,\varepsilon} = \frac{\parallel \mathbf{e}_\varepsilon \parallel_e}{\bar{e}_{m,\varepsilon}} \, , \tag{76}$$

where p denotes the order of the error of the finite element solution for the strains, provided that no singularities are present. By analogy to (74), (75) and (76), the required element size can be computed on the basis of the estimated errors of the gradients of the fluid stresses and of the plastic strain rate. Either the smallest value for the element size, computed on the basis of the estimated errors, can be employed for the local modification of the mesh or, alternatively, a weighted element error [38] can be computed from the different element error measures, which serves as the basis for the determination of the new element size. This procedure is repeated for all elements of the actual mesh. Hence, employing the actual mesh as a background grid, it provides the new element size for a mesh generator at each point of the domain under consideration.

The three-phase formulation is based on quadratic interpolation of the displacements and bilinear interpolation of the hydrostatic stresses in the fluid phases (Fig. 6). A mesh generator for bilinear 2D quadrilateral elements [40], which is based on the advancing front method [41], has been extended to generate 2D quadrilateral finite elements with quadratic interpolation of the displacements. The improved FE-mesh is determined from scratch, employing the mesh density function, computed on the basis of the a posteriori error estimation. Generating the updated mesh from scratch instead of subdividing or combining elements of the actual mesh, yields a better graded mesh and a more uniform distribution of the element errors. The mesh generation on the basis of the advancing front method is started with the determination of boundary nodes (on basis of the mesh density function), followed by the generation of typically one or several rows of offset elements, proceeds with the active front method by generating element by element until the whole domain is covered by elements and is finished by the application of mesh improvement algorithms [40].

5.3 Incremental strategy for nonlinear adaptive FE-analyses

On the basis of the improved mesh the analysis can either be restarted from the beginning or be continued from the time step, for which the error criterion was violated. The former strategy is simpler, however, it requires to repeat all load steps and time steps, which have been applied so far. A further disadvantage of this approach is given by the fact that meshes, which are very appropriate at later stages of an analysis are not necessarily appropriate at early stages of the analysis. This is demonstrated in [35] by means of the so-called Mandel consolidation problem, which is characterized by pore pressures governing mesh refinement at the earlier stages of the analysis whereas at later stages pore pressures have dissipated and the displacements control the refinement. Hence, it is advantageous to continue the analysis from the previous load step or time step with the improved mesh. To this end the errors are estimated on the basis of incremental quantities instead of total quantities. If the mesh is modified, then the variables are transferred from the old mesh to the improved mesh by means of the respective smoothed values.

5.4 Adaptive procedure in the time domain

From the analysis of consolidation problems it is known that at early stages of the analysis relatively small time increments are required to obtain results with the desired accuracy, whereas for later stages much larger time increments may be chosen. Hence, an efficient multi-phase formulation also requires an adaptive procedure in the time domain. The employed automatic time stepping algorithm is characterized by adjusting the time step size according to the norm of the solution vector. If, for a given time step, a user specified threshold for the convergence tolerance is not met within a certain number of iteration steps, then the step size is reduced. Conversely, time step sizes are increased, if for a given time increment only small changes of the solution vector occur, i.e., if only few iteration steps are required to meet a specified tolerance.

6 Application of the three-phase formulation

The three-phase formulation was implemented into the finite element program AFENA [9]. For the validation of the coupled model, experiments both reported in the literature and conducted to this end were used. To check the two-phase formulation several consolidation analyses were performed. In this section, the numerical simulation of a laboratory test is presented, dealing with the flow of compressed air in dry soil, which was carried out at the Institute for Soil Mechanics and Foundation Engineering at Graz University of Technology [17] within the Joint Research Initiative on Numerical Simulation in Tunnelling [2].

The applications of the three-phase formulation, described in this section, contain the numerical simulation of the flow of compressed air in a partially saturated soil specimen. The respective laboratory test [17] was also conducted at the Institute for Soil Mechanics and Foundation Engineering at

Graz University of Technology within the framework of the Joint Research Initiative. Then the numerical simulation of a full-scale in-situ test for dewatering of soil by means of compressed air, which was carried out at a subway construction site in Essen, Germany [19],[20], is presented. Finally, the three-phase model is applied to the numerical simulation of a tunnel excavation where compressed air is used for displacing the groundwater during the construction process.

6.1 Flow of compressed air in dry and partially saturated soil specimens

In the laboratory tests performed at the Institute for Soil Mechanics and Foundation Engineering at Graz University of Technology [17] the air loss through cracks in the shotcrete lining and the flow of compressed air in the adjacent soil were investigated. The numerical simulations of two of these experiments, studying the air flow through a dry sand and through a partially saturated sand, are described subsequently.

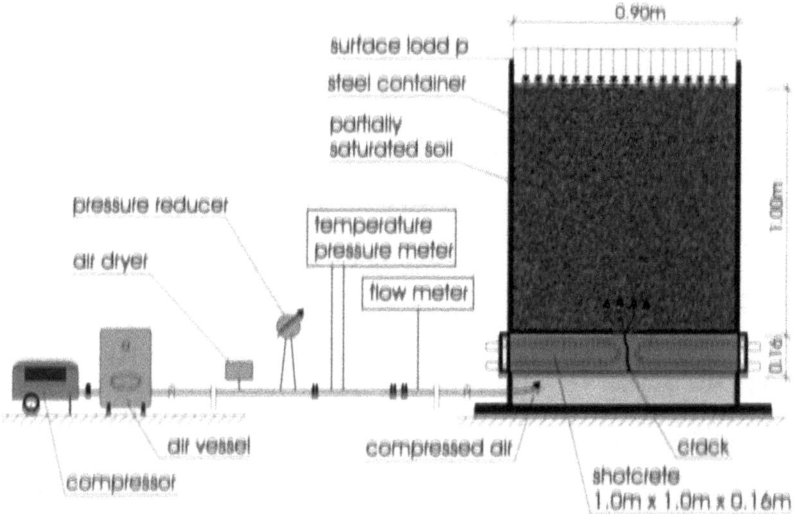

Fig. 7. Set-up of the laboratory tests conducted at Graz University of Technology

In Fig. 7 a schematic diagram of the test set-up is shown. A steel container filled with dry sand or with partially saturated sand was placed on top of a shotcrete element containing a crack of defined width. To simulate the overburden above the crown of the tunnel a constant surface load of $p = 250\ kN/m^2$ was applied on top of the soil using a grid system of girders and beams which was vertically prestressed and fixed by four spindle rods. At the bottom of the shotcrete element an excess air pressure was applied causing compressed air to flow through the crack in the slab as well as through the adjacent soil. During the experiments the development of the air pressure in the soil was measured at selected points.

The material parameters for the sand and the fluid phases water and air are summarized in Tab. 1. Young's modulus of the sand has been determined by taking into account the overburden pressure of the soil specimen. Since the changes of the soil stresses during the experiment are small compared to the initial soil stresses due to the applied overburden, linear elastic material behaviour of the soil skeleton is an adequate assumption for the numerical simulation. The parameters for the relationship between the degree of water saturation and the capillary stress according to Equation (33) are given in Tab. 2.

Table 1. Material properties for the sand and the fluid phases water and air

Parameter	Notation	Unit	Value
Young's modulus	E	kN/m^2	24 048
Poisson's ratio	ν	–	0.23
Porosity	n	–	0.37
Permeability water	k^{ew}	m/s	$3.0 \cdot 10^{-5}$
Permeability air	k^{ea}	m/s	$2.0 \cdot 10^{-6}$
Density grains	ρ^s	t/m^3	2.62
Density water	ρ^w	t/m^3	1.00
Density air	ρ^a	t/m^3	$1.295 \cdot 10^{-3}$
Bulk modulus water	K^w	kN/m^2	$2.0 \cdot 10^6$
Bulk modulus air (at atmospheric conditions)	K^a	kN/m^2	100.0

Table 2. Parameters for the $S^w p^c$-relationship according to Equation (33)

Parameter	Unit	Value
p_b^c	kN/m^2	2.0
S_r^w	–	0.05
S_s^w	–	1.0
n	–	5.0
m	–	0.80

For the numerical simulations of the experiments the soil specimen is discretized with a regular mesh, consisting of 20 × 20 quadrilateral finite elements for plane strain conditions. The boundary conditions for the soil skeleton and the fluid phases used in the analyses are given as follows: The domain is fully constrained at the bottom, vertically constrained at the top and horizontally constrained at the vertical boundaries. For the two fluids water and air the vertical boundaries and the lower horizontal boundary are assumed to be impermeable. Atmospheric air pressure prevails along the upper horizontal boundary of the soil element.

In the experiment, studying the air flow in dry sand, the excess air pressure of 8.5 kN/m^2, which enters the soil, is applied at the middle of the lower horizontal boundary of the soil specimen. The material parameters for the soil skeleton are taken according to Tab. 1; the air permeability is chosen as $k^{oa} = 2.23 \cdot 10^{-3}$ m/s, whereas the parameters for the water phase are not required in this case. The initial conditions are given as follows: The overburden of the soil specimen is accounted for in the primary stress state. At the beginning of the experiment the excess hydrostatic air stress in the soil is given as $p^a = 0$. For the numerical simulation of this test a two-phase formulation, consisting of the soil skeleton fully saturated with the compressible air phase, is sufficient.

Figure 8 shows the time dependent development of the air pressure in the soil. Immediately after the excess air pressure is applied, only a small region of the entire soil specimen close to the location of the air injection at the bottom boundary is affected by the excess air pressure (Fig. 8a). The computed air pressure distribution in the soil at steady state conditions is shown in Fig. 8d. For a comparison of computed and measured values of the air pressure in the soil, the latter have been included as dots and printed values in Fig. 8d.

In order to assess the accuracy of the numerical solution, the experiment was re-analyzed employing the adaptive solution procedure and choosing the error threshold of $\bar{\eta} = 20\%$. Shortly after the application of the air pressure the relative global errors for the strains of the soil skeleton and for the gradients of the air pressure are estimated according to (72) as $\eta_\varepsilon = 6.8\%$ and $\eta_a = 40.1\%$ on the basis of the results from the uniform FE-mesh (Fig. 8a). The respective relative element errors, which are computed from (67) und (73) as

$$(\eta_\varepsilon)_e = \frac{\| \mathbf{e}_\varepsilon \|_e}{\| \boldsymbol{\varepsilon}^h \|} \,, \quad (\eta_a)_e = \frac{\| \mathbf{e}_a \|_e}{\| \nabla p_a^h \|} \,, \tag{77}$$

are shown for the lower left half of the uniform FE-mesh in Fig. 9a. Hence, an improved FE-mesh is generated by means of the mesh density function (76). It is shown in Fig. 10a. Using the latter mesh, a reduction of the estimated relative global errors to $\eta_\varepsilon = 1.0\%$ and $\eta_a = 16.7\%$ is achieved. The respective relative element errors are shown for the lower left half of the improved FE-mesh in Fig. 9b. In spite of the improved FE-solution the number of elements is only increased from 400 to 474. This is due to the fact that large gradients of the air pressure are only present in the vicinity of the location of the air injection, where the element size is reduced, whereas for the remaining part of the domain the element size is increased. Figure 10b shows the FE-mesh at

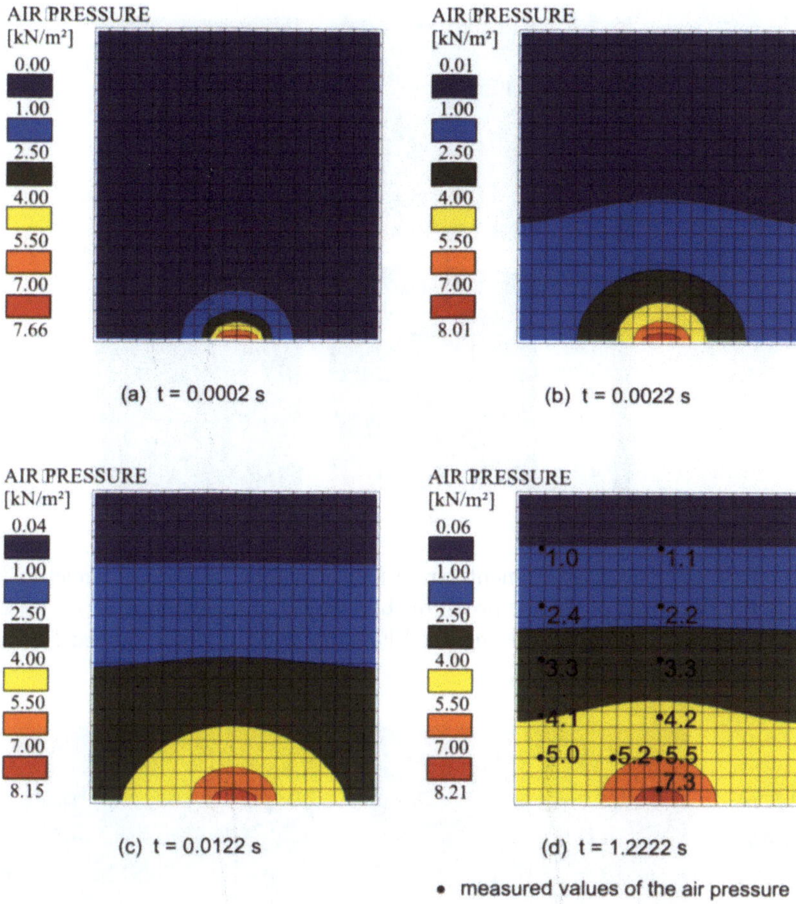

Fig. 8. Computed air pressure distribution in a dry soil specimen

steady state conditions, which consists of 424 elements with estimated errors of $\eta_\varepsilon = 0.4\%$ and $\eta_a = 8.3\%$. Although the error is smaller than the chosen error threshold, because of a prescribed upper limit of the element size of 65 mm, no further coarsening of the mesh was performed.

In the experimental study of the air flow in a partially saturated sand, the excess air pressure of 17.5 $\mathrm{kN/m^2}$, which enters the soil, is applied at the middle of the lower horizontal boundary of the soil specimen. In the numerical model at the top of the soil specimen a row of finite elements of 4 cm height is added to consider a small layer of water as observed in the experiment. The material parameters for the sand and the fluid phases and the parameters for the $S^w p^c$-relationship are chosen according to Tab. 1 and Tab. 2. The initial conditions are given as follows: The overburden of the soil specimen is accounted for in the primary stress state. At the beginning of the experiment the excess hydrostatic air stress in the soil is given as $p^a = 0$

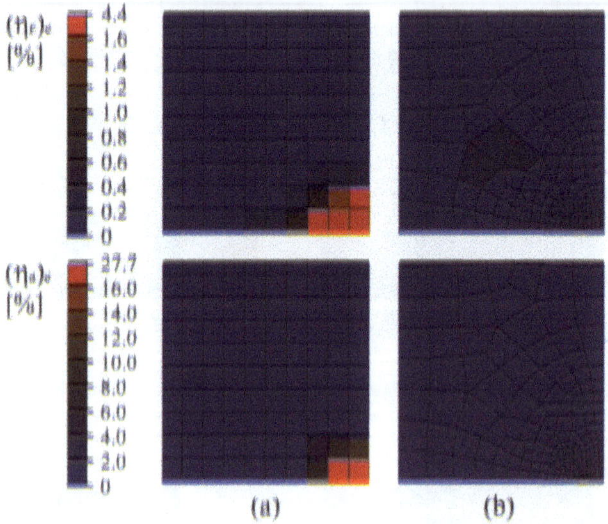

Fig. 9. Estimated relative element errors for the strains of the soil skeleton (top) and for the gradients of the air pressure (bottom) at the beginning of air injection for the lower left half of (a) the regular FE-mesh, and (b) the improved FE-mesh

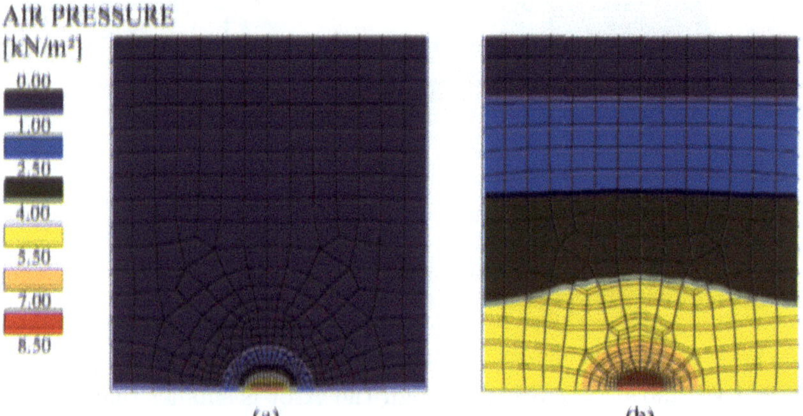

Fig. 10. Air pressure distribution in a dry soil specimen obtained by means of the adaptive FE-procedure: (a) at the beginning of air injection, (b) at steady state conditions

and the hydrostatic water stress p^w is computed from Equation (33) for the initial degree of water saturation of the partially saturated soil of 52 %, i.e., for $S^w = 0.52$.

The development of the hydrostatic fluid stresses in the soil are shown in Fig. 11 and Fig. 12 for two instants of time. As expected, shortly after the excess air pressure is applied only a small region of the entire soil specimen close to the location of the air injection at the bottom boundary is affected by the air pressure (Fig. 11a). Figure 11b shows the computed air pressure distribution in the soil at the end of the experiment after 45 minutes.

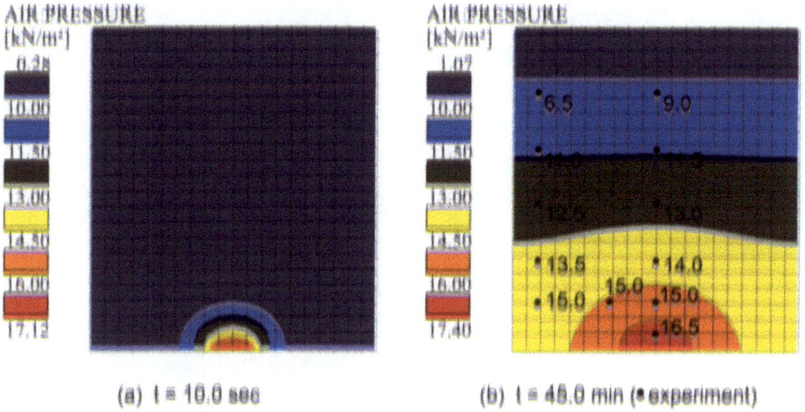

Fig. 11. Computed air pressure distribution in the partially saturated soil

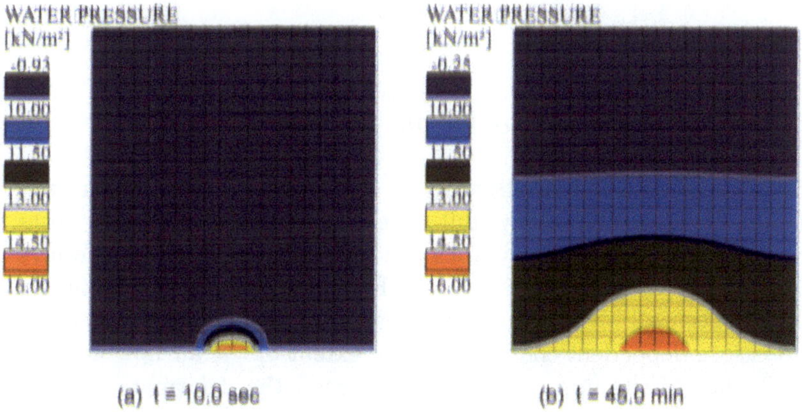

Fig. 12. Computed water pressure distribution in the partially saturated soil

Considering horizontal sections of the soil column in Fig. 11b the computed distribution of the air pressure at a fixed height is quite uniform in the upper part of the soil specimen. For comparison, the values of the excess air pressure, which have been measured in the test, are included (dots and printed values in Fig. 11b). The water pressure, shown in Fig. 12, is related to the air pressure of Fig. 11 by the $S^w p^c$-relationship according to Equation (33).

The displacements of the soil skeleton at steady state conditions are shown in Fig. 13. As one would expect the fluid flow from the bottom to the top of the soil column causes upwardly directed displacements of the soil. Due to the displacement constraint applied at the top of the soil column the maximum values of the displacements at steady state conditions are obtained about half-way between the lower and upper horizontal boundaries.

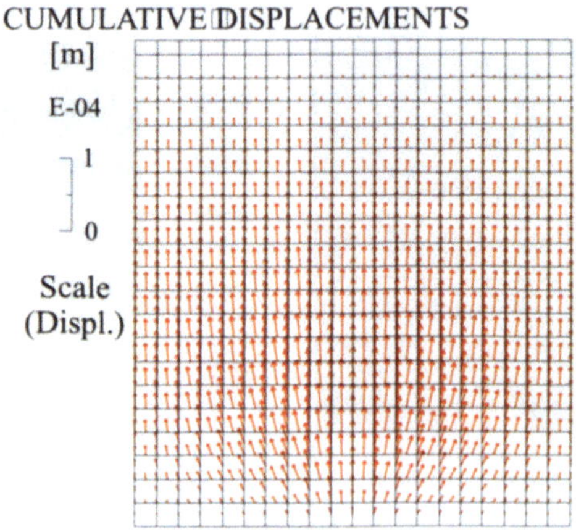

Fig. 13. Computed displacements of the soil specimen at steady state conditions

6.2 Dewatering of soil by means of compressed air

In connection with the application of compressed air for the subway construction in Essen, Germany, full-scale in-situ air permeability tests were carried out [20]. The primary aims of these tests focussed on the investigation of (i) the air permeability of the Essen soil, i.e., on the determination of the air permeability coefficients of the various soil layers and their time- and pressure-dependent changes, (ii) the extent of the air flow field and (iii) the influence of the flow of compressed air on the deformations of the soil skeleton.

Figure 14 contains a diagram of the test set-up. In total, three different sets of tests were performed, in which compressed air was injected into the ground via a large bore hole of 1.5 m diameter at different depths below the ground surface according to the various types of outcropping soil. In test 1A compressed air was pumped into the ground at a depth where rather weathered marl was encountered and thus the joints were less pronounced while in test 1B undisturbed and jointed marl was present at the air injection level. In experiment 2 the soil was grouted and compressed air was injected in the transition zone between marl and silt. In a number of smaller bore holes located in the vicinity of the main one the air pressure in the soil was measured at different distances from the point of air injection and at various depths below the ground surface.

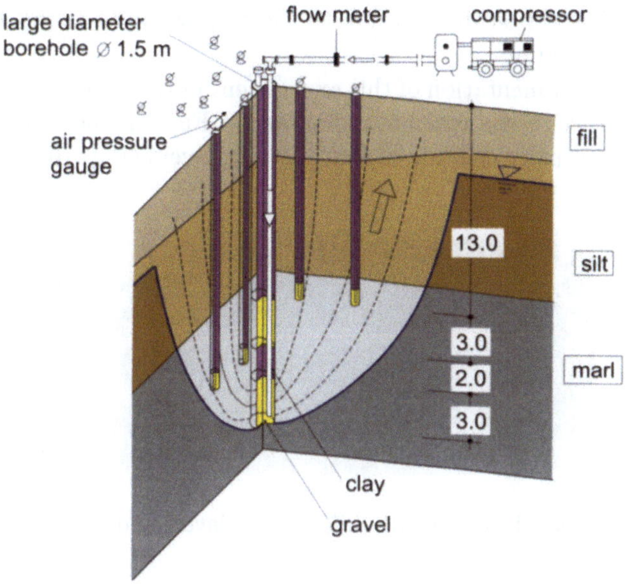

Fig. 14. Diagram of the in-situ air permeability test in Essen (dimensions in [m])

In experiment 1B, which is considered here, compressed air was injected into the ground at a depth between 18 and 21 m below the ground surface with an air pressure of up to 2.35 bar. The pressure was applied in three steps (1.60, 2.20 and 2.35 bar) keeping it constant for about one day at each of the three levels. Figure 15 shows a diagram of the three air pressure levels and the air loss versus time. It follows from Fig. 15 that at each pressure level the air loss increased with time. This can be attributed to the increase of the air permeability of the soil which is associated with the reduction of

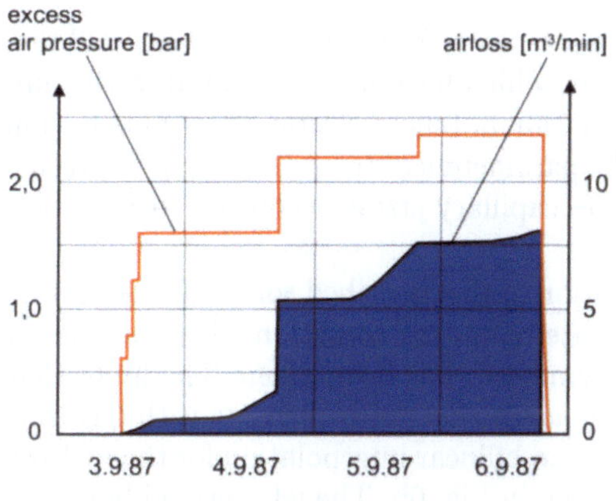

Fig. 15. Different excess air pressure levels versus time (date (D.M.Y))

the degree of water saturation due to the displacement of the groundwater by the compressed air.

From the documentation of this experiment in the literature [20] it follows that, in contrast to the schematic diagram in Fig. 14, the soil profile at the test location actually consists of four distinct layers (Fig. 16).

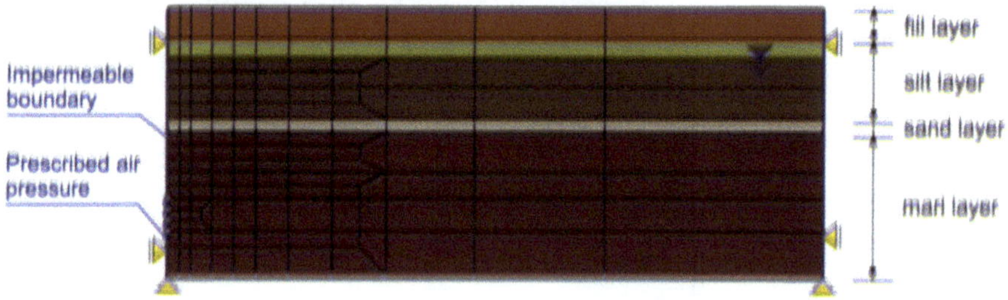

Fig. 16. Employed FE model with different soil layers and boundary conditions

At the top a fill layer of about 3 m thickness is encountered mainly composed of sand and silt but also containing parts of construction debris. Next is a 7 to 8 m thick silt layer with portions of sand in the upper parts and organic additions in the lower parts. Its permeability both with respect to water and air is rather low. In this layer also the groundwater table is located at a depth of about 4.7 m below the ground surface. Underneath the silt there is a thin, highly permeable sand-gravel layer also containing portions of silt. At the depth where the compressed air is injected a thick layer of marl is encountered which is rather weathered in the upper region while showing pronounced joints further down. Considering the granulometric composition the marl corresponds to a silt containing high portions of sand and clay. According to [20] averaged values of the permeabilities are used for the marl. Details about the material parameters of the various layers can also be found in [20].

The material parameters for the four soil layers and the fluid phases water and compressed air, which have been employed in the numerical simulation, are summarized in Tab. 3. Linear elastic material behaviour was assumed for all soil layers. The parameters of the relationship between the degree of water saturation and the capillary stress according to Equation (33) are given in Tab. 4.

The discretization of the described soil profile is shown in Fig. 16 together with the appropriate boundary conditions. For the numerical simulation advantage of axial symmetry has been taken. The finite element mesh consists of axisymmetric quadrilateral finite elements with quadratic interpolation for the displacements and bilinear interpolation for the hydrostatic stresses in the fluid phases as shown in Fig. 6b. The left vertical boundary of the domain to be discretized is given by the wall of the bore hole whereas the right vertical

Table 3. Material properties for the four layers of the Essen soil and the fluid phases water and air

Parameter	Notation	Unit	Value
Fill layer			
Young's modulus	E	kN/m^2	20 000
Poisson's ratio	ν	–	0.33
Porosity	n	–	0.36
Permeability water	k^{ow}	m/s	$5 \cdot 10^{-5}$
Permeability air	k^{oa}	m/s	$3 \cdot 10^{-4}$
Density grains	ρ^s	t/m^3	2.72
Silt layer			
Young's modulus	E	kN/m^2	5 200 – 12 500
Poisson's ratio	ν	–	0.35
Porosity	n	–	0.42
Permeability water	k^{ow}	m/s	$5 \cdot 10^{-6}$
Permeability air	k^{oa}	m/s	$4.5 \cdot 10^{-5}$
Density grains	ρ^s	t/m^3	2.72 – 2.90
Sand-gravel layer			
Young's modulus	E	kN/m^2	21 200
Poisson's ratio	ν	–	0.32
Porosity	n	–	0.36
Permeability water	k^{ow}	m/s	$1 \cdot 10^{-4}$
Permeability air	k^{oa}	m/s	$4.5 \cdot 10^{-4}$
Density grains	ρ^s	t/m^3	2.72
Marl layer			
Young's modulus	E	kN/m^2	14 300
Poisson's ratio	ν	–	0.40
Porosity	n	–	0.33
Permeability water	k^{ow}	m/s	$2.5 \cdot 10^{-5}$
Permeability air	k^{oa}	m/s	$3.7 \cdot 10^{-4}$
Density grains	ρ^s	t/m^3	2.79
Fluid phases			
Density water	ρ^w	t/m^3	1.0
Density air	ρ^a	t/m^3	$1.295 \cdot 10^{-3}$
Bulk modulus water	K^w	kN/m^2	$2.0 \cdot 10^6$
Bulk modulus air (at atmospheric conditions)	K^a	kN/m^2	100.0

boundary and the lower horizontal boundary have to be chosen such that the error induced by these artificial boundaries remains small.

The boundary conditions are given as follows: As indicated in Fig. 16, with respect to the displacements the domain is fully constrained at the bottom and horizontally constrained at the vertical boundaries. With respect to the two fluid phases the following boundary conditions are applied: The wall of the bore hole, except the filter part where the compressed air is injected, as

Table 4. Parameters of the $S^w p^c$-relationship for the four soil layers according to Equation (33)

Parameter	Unit	Value
Fill layer		
p_b^c	kN/m^2	−4.0
S_r^w	–	0.2
S_s^w	–	0.99
n	–	5.0
m	–	0.8
Silt layer		
p_b^c	kN/m^2	−30.0
S_r^w	–	0.15
S_s^w	–	0.99
n	–	2.5
m	–	0.6
Sand-gravel layer		
p_b^c	kN/m^2	−3.0
S_r^w	–	0.05
S_s^w	–	0.99
n	–	2.86
m	–	0.65
Marl layer		
p_b^c	kN/m^2	−9.0
S_r^w	–	0.15
S_s^w	–	0.99
n	–	2.5
m	–	0.6

well as the lower horizontal boundary and the right vertical boundary are assumed to be impermeable. Along the filter part of the bore hole the excess air pressure, which has been applied in the experiment, is prescribed. At the upper horizontal boundary, i.e., at the ground surface, atmospheric air pressure prevails.

The initial conditions are given as follows: The primary soil stresses are computed from the self-weight of the soil taking into account the buoyancy forces below the groundwater level. The hydrostatic stresses in the two fluid phases are determined according to the given groundwater table, whereas the excess hydrostatic air stresses are equal to zero at the beginning of the experiment.

The results of the numerical simulation of the field test 1B concerning the first pressure level of 1.60 bar are presented in Fig. 17 to Fig. 19. In these figures only the part of the discretized domain in the vicinity of the bore hole is shown.

Figure 17 shows the computed distributions of the air pressure in the soil at four different instants of time after the air pressure of 1.60 bar has been applied in the bore hole. In order to compare the numerical results with experimental data, isolines composed from the recorded air pressure gauge values have been included in Fig. 17d, showing the air pressure distribution at the end of the first pressure level after approximately one day. As can be seen from this figure, the computed air pressure distribution agrees quite well with the measurements.

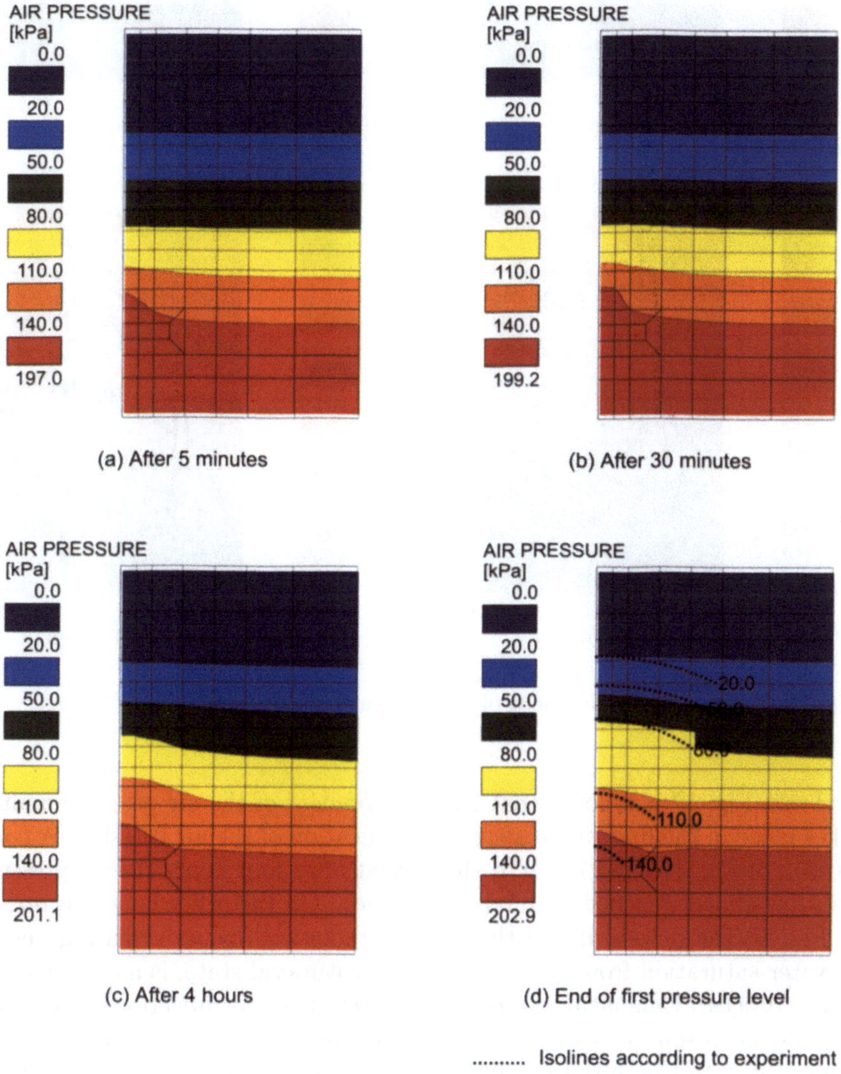

Fig. 17. Computed development of the air pressure in the soil

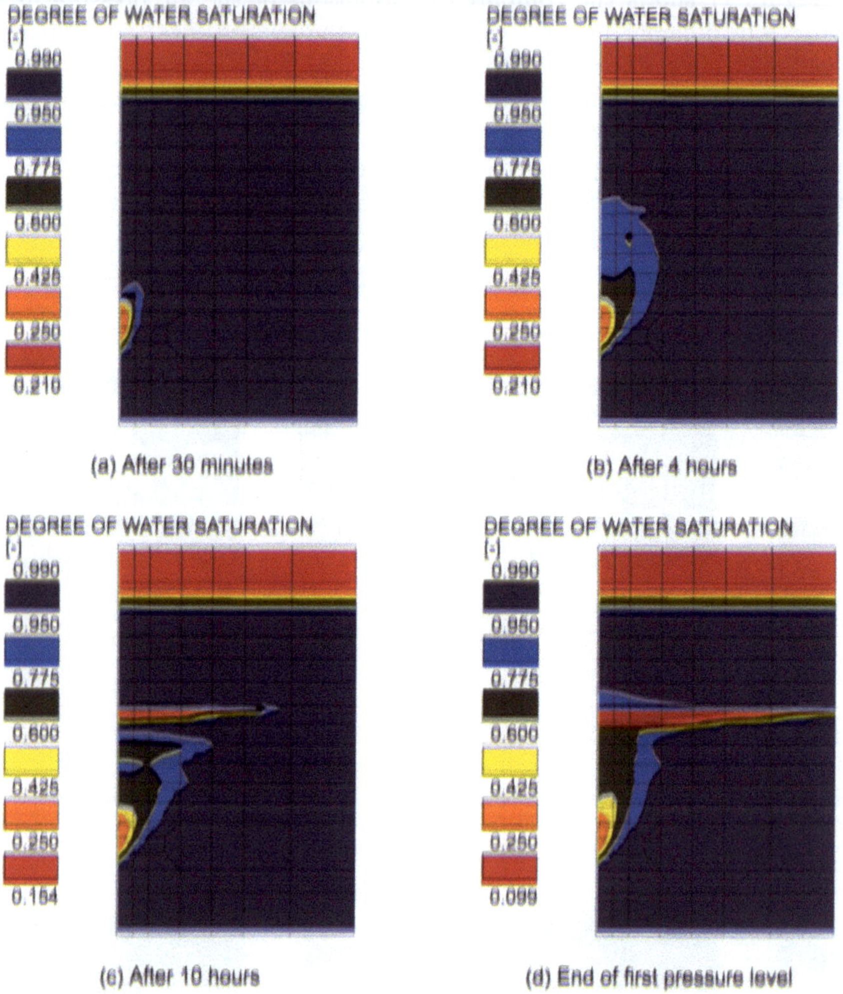

Fig. 18. Computed development of the degree of water saturation in the soil

From Fig. 17 it follows, that the development of the air pressure mainly takes place in the first phase after the application of the air pressure. A comparison of Fig. 17c and Fig. 17d shows that after four hours the air pressure field has almost reached its final distribution. On the other hand, according to Fig. 18 the dewatering of the soil, characterized by a decreasing degree of water saturation from the initially fully saturated state, is a considerably slower process. Emanating from the location of the air injection in the bore hole a cone of soil mass with a considerably lower degree of water saturation compared with the initial conditions is formed, indicating the extent of the air flow field, and the part of the thin highly permeable sand-gravel layer, which is close to the borehole, is almost completely dewatered at the end of the first pressure level. However, the applied air pressure of 1.60 bar is not large enough to dewater the silt layer, located above the sand-gravel layer. Consequently, also the soil above the groundwater level represented by the upper two rows of elements remains at the initial conditions, which are characterized by a decrease of the water saturation above the groundwater level, representing the capillary fringe, and the assumed degree of water saturation of 20 % in the fill layer.

Figure 19 contains vector plots of the displacements of the soil skeleton for two points of time at the beginning and at the end of the first pressure level. These plots clearly show the effect of the flow of compressed air on the deformations of the soil skeleton. They also confirm the findings reported by Kramer [19] that by the application of compressed air as a means for dewatering the soil in the working area of the tunnel face the overall ground settlements due to tunnel construction tend to be smaller compared with dewatering by pumping wells and driving the tunnel under atmospheric conditions.

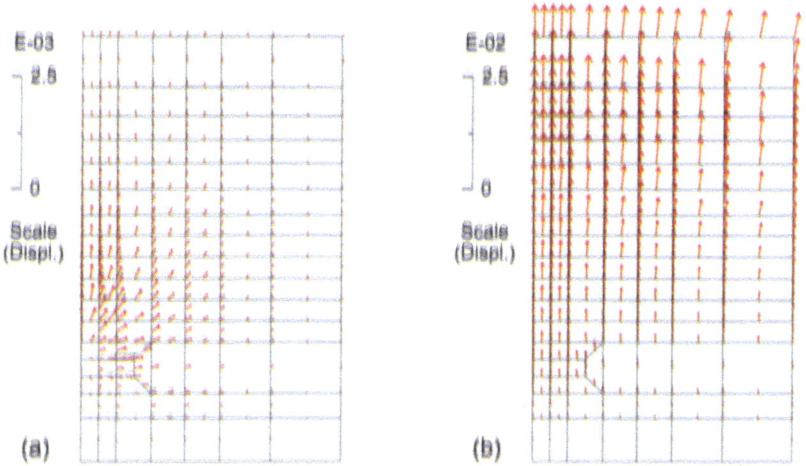

Fig. 19. Computed soil displacements (a) at the beginning and (b) at the end of the first pressure level (note the different scaling of the displacements)

Since shortly after the application of the compressed air only the soil region close to the location of the air injection is affected by the excess air pressure, only very small displacements of the soil occur primarily in this region (Fig. 19a), whereas at the end of the first pressure level larger displacements in a considerably larger soil region are obtained (Fig. 19b).

6.3 2D numerical simulation of tunnelling by means of compressed air

Finally, the three-phase formulation is applied to the coupled numerical simulation of the excavation of a tunnel by means of compressed air, which is used for displacing the groundwater in the working area at the tunnel face. The aim of this subsection is not a case study, but to demonstrate the principal steps of a coupled numerical analysis of tunnelling.

The crown of the tunnel is located about 12 m below the ground surface and the tunnel diameter is approximately 6.5 m (Fig. 20).

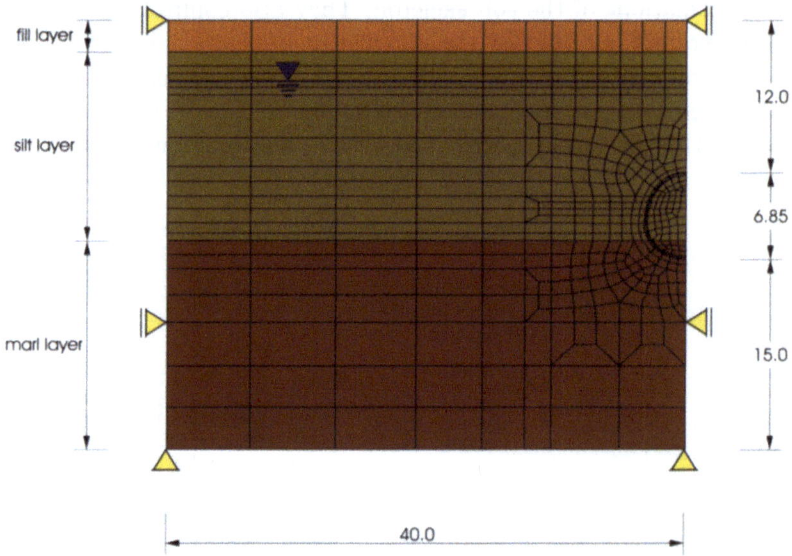

Fig. 20. 2D FE-mesh for the numerical simulation of tunnelling by means of compressed air

The soil consists of three layers. At the top there is a fill layer of about 2.5 m thickness, followed by a silt layer, ranging down to 17.4 m below the surface. The groundwater table is located in this layer at about 4.7 m below the surface. Underneath the silt a thick layer of marl is encountered. The material parameters for the three soil layers and the fluid phases water and compressed air and the parameters for the relationship between the degree of water saturation and the capillary stress according to Equation (33) are similar to those given in the respective Tab. 3 and Tab. 4 of the previous example. Young's modulus and Poissons's ratio of the shotcrete are given as 15 000 MN/m^2 and 0.20, respectively. The permeability of the lining, resulting from cracks in the shotcrete, is assumed to be characterized by the permeability coefficients $k^{ow} = 2.7 \cdot 10^{-5}$ and $k^{oa} = 2.0 \cdot 10^{-3}$ [17]. The thickness of the shotcrete lining is 25 cm.

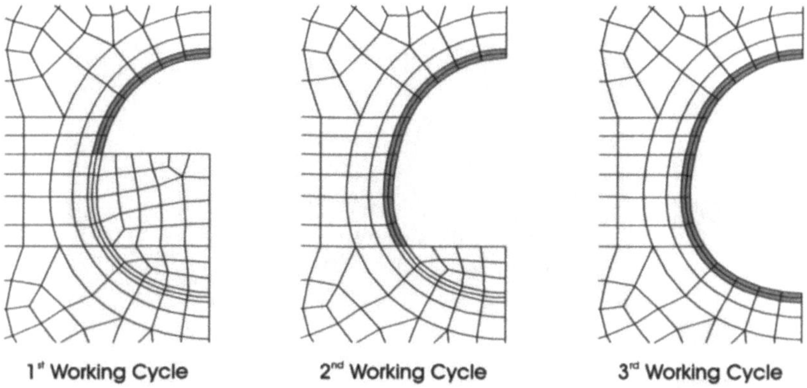

Fig. 21. Construction sequences of the tunnel

The tunnel is constructed in three working cycles, dealing sequentially with the crown, the bench and the invert, as indicated in Fig. 21.

The numerical model relies on the simplifying assumption of plane strain conditions, resulting in a two-dimensional discretization. Taking advantage of symmetry, the finite element mesh with the boundary conditions shown in Fig. 20 is employed. The discretized domain is fully constrained at the bottom and horizontally constrained at the vertical boundaries. Additionally, the vertical and the lower horizontal boundaries are assumed to be impermeable for the two fluid phases water and air and atmospheric conditions are assumed at the surface.

In a first step the primary stress state in the soil and in the two fluid phases is determined, taking into account the given groundwater table. Subsequently, the excess air pressure is applied at the tunnel face. The latter is chosen such that the air pressure balances the water pressure at the invert of the tunnel section, requiring an air pressure in the tunnel, which exceeds the atmospheric air pressure by about 1.4 bar. Figure 22 shows the distribution of the air pressure in the soil one day after the beginning of air injection at the tunnel face and Fig. 23 contains the resulting vertical displacements of the soil, which are characterized by a surface heave of approximately 2.5 cm.

The applied excess air pressure initiates dewatering of the soil in the vicinity of the tunnel face. Because of the relatively low permeability of the silt, the dewatering is a rather slow process. The development of the degree of water saturation after the beginning of air injection at the tunnel face is plotted in Fig. 24 for two points of time. The sharp tip of the isolines of the degree of water saturation close to the invert results from the common boundary of the silt layer and the marl layer, which are characterized by different values of the permeabilities and of the parameters of the $S^w p^c$-relationship, which govern the fluid flow within the respective type of soil.

Subsequently, the numerical simulation of the three working cycles is performed. Each working cycle is modelled by two computation steps, simulating

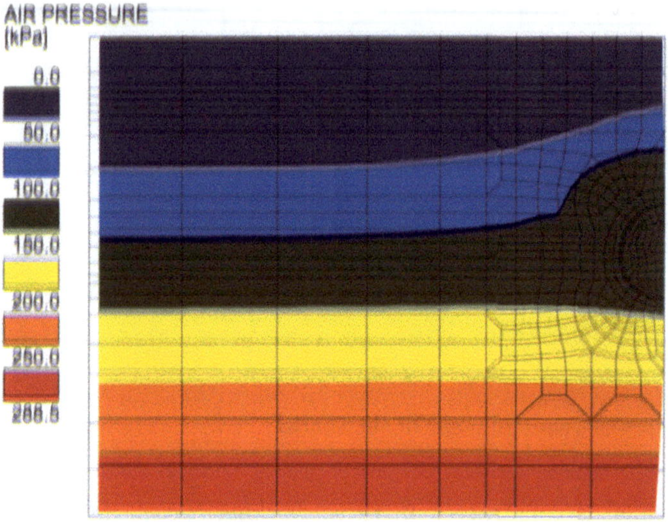

Fig. 22. Air pressure in the soil one day after the beginning of air injection at the tunnel face

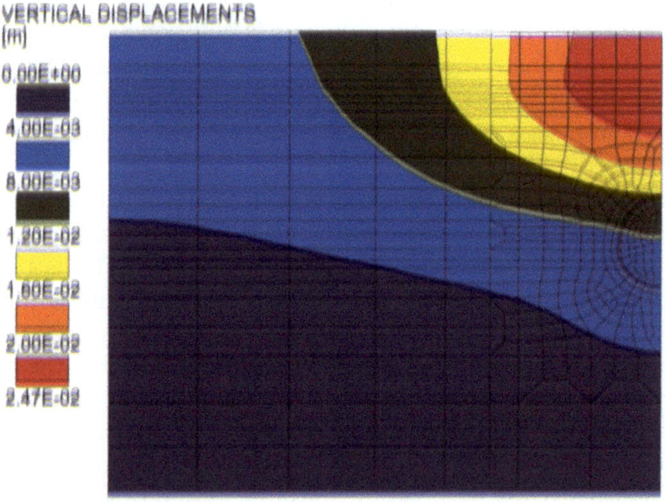

Fig. 23. Vertical displacements of the soil one day after the beginning of air injection at the tunnel face

the excavation of the soil in the first step and the placement of the shotcrete lining in the second step. To account for 3D-effects in the 2D analysis in an approximate manner a partial initial stress relief of 50% is assumed, allowing some deformations in the soil before the lining is installed.

The vertical stresses and the displacements after each of the three working cycles of the construction sequence are shown in Fig. 25 and Fig. 26, respec-

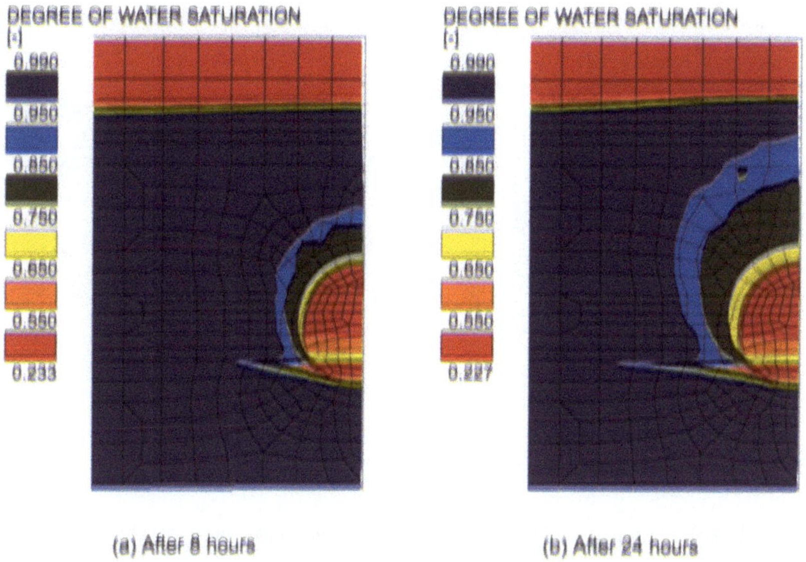

Fig. 24. Degree of water saturation in the soil at two instants of time after the beginning of air injection at the tunnel face

tively. Note that because of the application of the coupled numerical model, the deformations and the stress changes of the soil, caused by displacing the

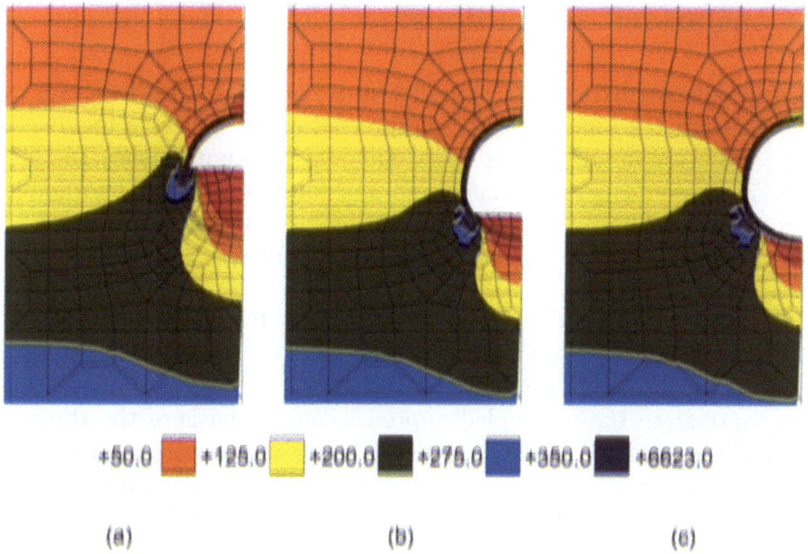

Fig. 25. Vertical stresses [kN/m^2] of the soil (positive values refer to compressive stresses) after completion of each of the three working cycles

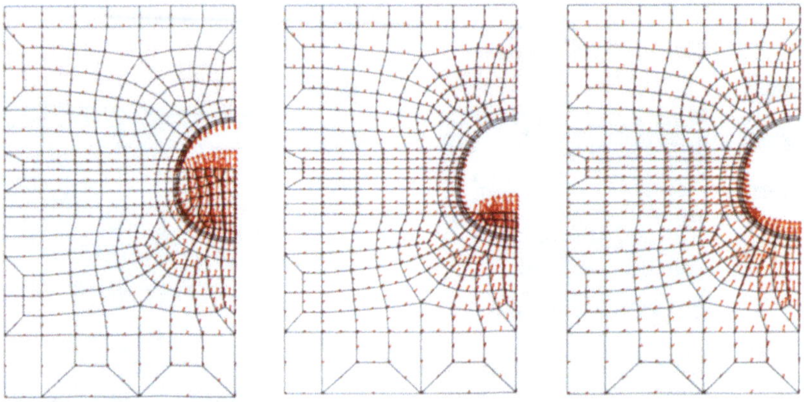

(a) (b) (c)

Fig. 26. Displacements of the soil after completion of each of the three working cycles

groundwater by means of compressed air, are included in the displacements and stresses after completion of each working cycle. The surface heave of 25 mm due to injection of compressed air at the tunnel face is reduced to 5 mm after the first working cycle. The maximum values of the displacements of 60 mm occur at the invert of the excavated part of the tunnel section.

7 Conclusions and outlook

The proposed coupled numerical model for tunnelling below the groundwater table allows to take into account compressed air as a means for displacing the groundwater in the soil in the vicinity of the tunnel face. To this end the soil is treated as a three-phase medium, consisting of the deformable soil skeleton and the fluid phases water and compressed air. Appealing features of the presented coupled approach are given as follows:

- Interactions between the fluid flow in the pores of the soil and the deformations and stresses of the soil skeleton are taken into account in a physically consistent manner;
- In contrast to the uncoupled approach, on the basis of the three-phase model the pressures in the fluid phases water and compressed air as well as the deformations and stresses of the soil skeleton are computed simultaneously without the need to switch between flow analysis and deformation analysis;
- Apart from the numerical simulation of tunnelling by means of compressed air the three-phase model can be employed for the numerical

simulation of a broad range of problems encountered in geotechnical engineering. This is possible, because models for dewatering of soils under atmospheric conditions and for consolidation of soils are contained in the three-phase formulation as special cases.

The potential of the three-phase model was demonstrated by the numerical simulation of

(i) laboratory tests, which had been conducted at the Institute for Soil Mechanics and Foundation Engineering at Graz University of Technology within the framework of the Joint Research Initiative, dealing with the flow of compressed air in dry and partially saturated soil [17];
(ii) a full-scale in-situ test for displacing the groundwater by means of compressed air, which had been conducted at a subway construction site in Essen [20], Germany, and
(iii) the excavation of a tunnel below the groundwater table, employing compressed air for displacing the ground water in the vicinity of the tunnel face, on the basis of a simplified 2D numerical model.

Current developments concentrate on the application of the 2D adaptive solution strategy for three-phase problems to tunnelling [42] and on the extension of the 2D coupled model to a fully three-dimensional model [43]. Although adaptive FE-analyses allow to enhance the efficiency and reliability of numerical solutions, they have rarely been applied to tunnelling so far, even in the case of single phase problems characterized by drained conditions. With respect to the extension of the 2D coupled model to three dimensions it is emphasized that in many cases three-dimensional models allow a better representation of the actual physical situation. Figure 27 contains results of a 3D three-phase FE-analysis of the injection of compressed air into water saturated soil. The dimensions of the discretized domain are $7 \times 7 \times 7$ m; the upper horizontal surface represents a river bed, which is located 7 m below the water level. Hence, the initial water pressure at the bottom of the discretized domain is 140 kN/m^2. An excess air pressure of 50 kN/m^2 is applied at a corner of the lower horizontal surface, acting on an area of 1×1 m, resulting in the excess air pressure field and the total air pressure field in the soil, shown in Fig. 27a and Fig. 27b. The excess air pressure yields dewatering of the soil, shown in Fig. 27c, and displacements of the ground (Fig. 27d).

However, for the application of the 3D coupled model to tunnelling, because of the large number of degrees of freedom, a very efficient equation solver has to be implemented. Further complexities are encountered formulating the boundary conditions for the fluid phases. Since the latter change in each excavation step, they have to be adapted in each analysis step. In contrast to two-dimensional FE-analyses, for three-dimensional FE-analyses these modifications can only be handled in an automated manner.

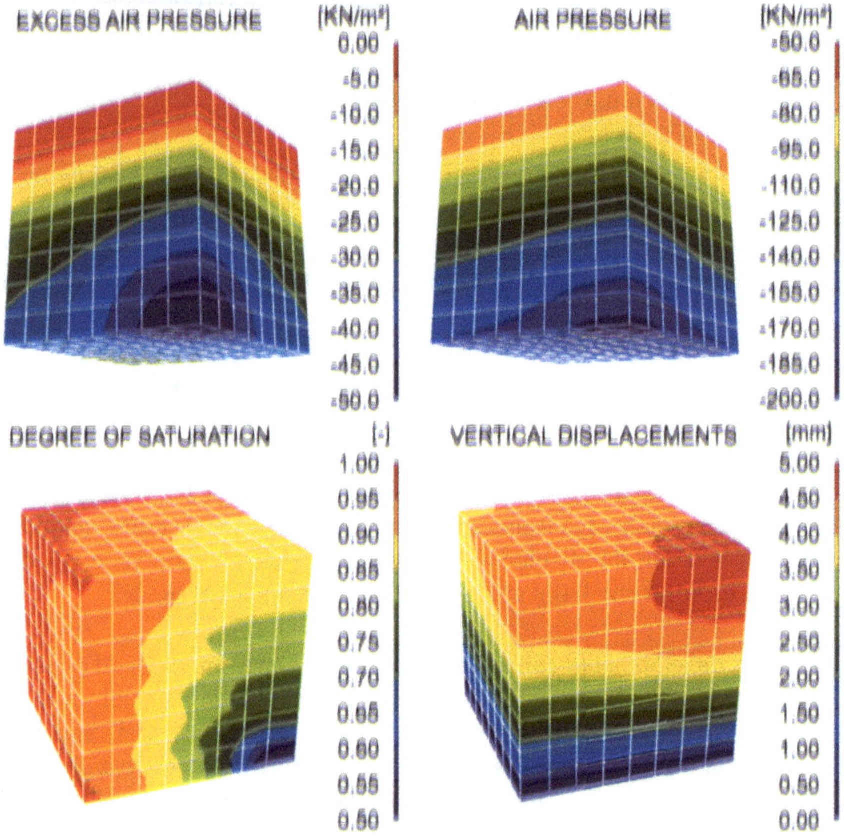

Fig. 27. 3D three-phase FE-analysis of dewatering by means of compressed air: (a) excess air pressure field, (b) total air pressure field, (c) degree of water saturation, (d) vertical displacements of the ground (positive values in opposite direction of gravity)

References

[1] Alonso E., Gens A., Hight D. (1987) Special problems soils: General report. *Proceedings of the 9th European Conference on Soil Mechanics and Foundation Engineering*, **3**: 1087-1146

[2] Beer G., Plank J. (1999) Joint research initiative on numerical simulation in tunnelling. *Felsbau*, **17**(1): 7–9

[3] Biot M.A. (1941) General theory of three-dimensional consolidation. *Journal of Applied Physics*, **12**(2): 155–164

[4] Bishop A.W., Blight G.E. (1963) Some aspects of effective stress in saturated and partly saturated soils. *Geotéchnique*, **13**(3): 177–197

[5] de Boer R., Ehlers W., Kowalski S., Plischka J. (1991) Porous Media: A Survey of Different Approaches. *Forschungsberichte aus dem Fachbereich Bauwesen*, **54**, Universität - Gesamthochschule Essen, Essen, Germany

[6] de Boer R. (1996) Highlights in the historical development of the porous media theory: Toward a consistent macroscopic theory. *Applied Mechanics Reviews*, **49**(4): 201-261

[7] Chen Z.S., Hofstetter G., Mang H.A. (1991) 3D Boundary Element, Analysis of the Lowered Groundwater Level for Tunnels Driven Under Compressed Air. *International Journal for Numerical and Analytical Methods in Geomechanics*, **15**: 735-752

[8] Gülzow H.G. (1994) Dreidimensionale Berechnung des Zweiphasenströmungsfeldes beim Tunnelvortrieb unter Druckluft in wassergesättigten Böden. *Veröffentlichungen des Instituts für Grundbau, Bodenmechanik, Felsmechanik und Verkehrswasserbau der RWTH-Aachen*, **25**, Germany

[9] Carter J.P., Balaam N.P. (1995) *AFENA User's Manual, Version 5.0.* Centre for Geotechnical Research, University of Sydney, Sydney

[10] Coleman J.D. (1962) Stress strain relations for partly saturated soil. *Geotéchnique*, **12**(4): 348–350

[11] Ehlers W. (1989) *Poröse Medien, ein kontinuumsmechanisches Modell auf der Basis der Mischungstheorie.* Forschungsberichte aus dem Fachbereich Bauwesen, **47**, Universität - Gesamthochschule Essen, Essen, Germany

[12] Fredlund D.G., Morgenstern N.R., Widger R.A. (1978) The shear strength of unsaturated soils. *Canadian Geotechnical Journal*, **15**(3): 313–321

[13] Fredlund D.G., Rahardjo H. (1993) *Soil Mechanics for Unsaturated Soils*, John Wiley and Sons, New York

[14] van Genuchten M.Th. (1980) A closed-form equation for predicting the hydraulic conductivity of unsaturated soils. *Soil Science Society of America Journal*, **44**: 892–898

[15] Hochgürtel T. (1998) *Numerische Untersuchungen zur Beurteilung der Standsicherheit der Ortsbrust beim Einsatz von Druckluft zur Wasserhaltung im schildvorgetriebenen Tunnelbau.* PhD thesis, Institut für Grundbau, Bodenmechanik, Felsmechanik und Verkehrswasserbau, RWTH-Aachen, Aachen

[16] Hofstetter G., Oettl G., Stark R.F. (1999) A soil model for tunneling under compressed air. *Proceedings of the European Conference on Computational Mechanics, ECCM'99*, Ed.: W. Wunderlich, Lehrstuhl für Statik, TU München, München, CD-ROM

[17] Kammerer G. (2000) *Experimentelle Untersuchungen von Strömungsvorgängen in teilgesättigten Böden und in Spritzbetonrissen im Hinblick auf den Einsatz von Druckluft zur Wasserhaltung im Tunnelbau.* Dissertation, Institut für Bodenmechanik und Grundbau, Technische Universität Graz, Graz

[18] Klubertanz G. (1999) *Zur hydromechanischen Kopplung in dreiphasigen porösen Medien.* PhD thesis, École Polytechnique Fédérale de Lausanne, Lausanne

[19] Kramer J. (1987) U-Bahn Bau in Essen, Baulos 30 - Spritzbetonbauweise unter Druckluft: Luftverbrauch und Spannungsumlagerungen im Boden, Senkungen. *Proceedings der STUVA Tagung, Essen, Forschung + Praxis*, STUVA, Essen, Vol. **32**: 193–199

[20] Kramer J., Semprich S. (1989) Erfahrungen über Druckluftverbrauch bei der Spritzbetonbauweise. *Taschenbuch für den Tunnelbau*, **13**: 91–153

[21] Lewis R.W., Schrefler B.A. (1998) *The Finite Element Method in the Static and Dynamic Deformation and Consolidation of Porous Media.* John Wiley and Sons, Chichester, 2nd edition

[22] Liakopoulos A.C. (1965) *Transient Flow Through Unsaturated Porous Media.* PhD thesis, University of California, Berkeley

[23] Malvern L.E. (1969) *Introduction to the Mechanics of a Continuous Medium.* Prentice-Hall, Englewood Cliffs, New Jersey

[24] Mualem Y. (1976) A new model for predicting the hydraulic conductivity of unsaturated porous media. *Water Resources Research,* **12**(3): 513–522

[25] Oettl G. *A coupled FE-model for dewatering of soils.* Dissertation, University of Innsbruck, in preparation.

[26] Oettl G., Stark R.F., Hofstetter G. (1998) A comparison of elastic-plastic soil models for 2D FE analyses of tunnelling. *Computers and Geotechnics,* **23**: 19-38

[27] Schrefler B.A., Xiaoyong Z. (1993) A fully coupled model for water flow and airflow in deformable porous media. *Water Resources Research,* **29**(1): 155–167

[28] Schrefler B.A., Simoni L. (1995) Numerical Solutions of Thermo-Hydro-Mechanical Problems. *Modern Issues in Non-saturated Soils,* Eds.: A. Gens, P. Jouanna, B.A. Schrefler, CISM Courses and Lectures, **357**: 213-275

[29] Wheeler S.J., Sivakumar V. (1995) An elasto-plastic critical state framework for unsaturated soil. *Geotéchnique,* **45**(1): 35-53

[30] Wriggers P. (2001) *Nichtlineare Finite-Element-Methoden.* Springer Verlag Berlin Heidelberg

[31] Zienkiewicz O.C., Taylor R.L. (1989) *The Finite Element Method,* 4. ed., Vol. 1: Basic Formulation and Linear Problems, Vol. 2: Solid and Fluid Mechanics, Dynamics and Non-linearity, 1991, McGraw Hill, UK.

[32] Zienkiewicz O.C., Xie Y.M., Schrefler B.A., Ledesma A., Biĉaniĉ N. (1990) Static and dynamic behaviour of soils: A rational approach to quantitative solutions. ii. semi-saturated problems. *Proceedings of the Royal Society of London,* **A**(429): 311–321

[33] Huerta A., Rodriguez-Ferran A., Diez P., Sarrate J. (1999) Adaptive finite element strategies based on error assessment. *International Journal for Numerical Methods in Engineering,* **46**: 1803-1818

[34] Zienkiewicz O., Zhu J. (1992) The superconvergent patch recovery and a posteriori error estimates. *International Journal for Numerical Methods in Engineering,* **33**: 1331-1382

[35] El-Hamalawi A., Bolton M. D. (1999) An a posteriori error estimator for plane-strain geotechnical analyses. *Finite Elements in Analysis and Design,* **33**: 335-354

[36] Hicks M. A. (2000) Coupled computations for an elastic-perfectly plastic soil using adaptive mesh refinement. *International Journal for Numerical and Analytical Methods in Geomechanics,* **24**: 453-476

[37] Wiberg N. E., Abdulwahab F., Ziukas S. (1994) Enhanced superconvergent patch recovery incorporating equilibrium and boundary conditions. *International Journal for Numerical Methods in Engineering,* **37**: 3417-3440

[38] Ehlers W., Ellsiepen P., Ammann M. (1999) Localization phenomena in saturated and empty frictional porous materials computed by time- and space-adaptive methods. *Proceedings of the European Conference on Computational Mechanics (ECCM),* Munich

[39] Li L. Y., Bettess P. (1995) Notes on mesh optimal criteria in adaptive finite element computations. *Communications in Numerical Methods in Engineering,* **11**: 911-915

[40] Huemer T. (1998) *Automatische Vernetzung und adaptive nichtlineare statische Berechnung von Flächentragwerken mittels vierknotiger finiter Elemente.* Ph.D. Thesis, University of Technology, Vienna, Austria
[41] Zhu J., Zienkiewicz O., Hinton E., Wu J. (1991) A new approach to the development of automatic quadrilateral mesh generation. *International Journal of Numerical Methods in Engineering*, **32**: 849-866
[42] Stelzer R. *2D adaptive FE-analysis of three-phase problems.* Dissertation, University of Innsbruck, in preparation.
[43] Hächl M. *3D numerical simulation of three-phase problems.* Dissertation, University of Innsbruck, in preparation.

Chapter 13
A multilaminate model for finite element analysis of tunnel excavation

Helmut F. Schweiger[1] and Hartmut Schuller[2]

[1] Institute for Soil Mechanics and Foundation Engineering, Computational Geotechnics Group, Graz University of Technology
[2] Garber and Dalmatiner, Consulting Engineers

Abstract. A constitutive model formulated within the framework of the multilaminate concept for soils is presented. It is shown that anisotropic material behaviour is easily accounted for with these types of models. The potential of the formulation to predict plastic volumetric strains purely caused by rotation of principal stress axes is demonstrated. Furthermore the model is able to capture the formation of shear bands due to an enhanced strain softening formulation including both frictional and cohesive softening behaviour. Due to a simple regularisation technique mesh independent results are obtained with sufficient accuracy for practical purposes. The latter feature will be highlighted by solving the practical problem of a shallow tunnel excavation constructed using the principles of the NATM. It is shown that the development of plastic shear strains leading to a failure mechanism that involves shear banding is realistically predicted by the proposed formulation.

1 Introduction

Constitutive modelling is still an issue in geotechnical engineering, in particular if solutions to complex practical problems have to be found under the financial and time restraints usually associated with this kind of work. Although a large number of sophisticated constitutive models has been proposed in the literature more advanced formulations are still not routinely employed for solving practical problems. The reason for this is either that these models are sometimes very difficult to understand and therefore not accepted in practice or that the computational effort is too high or both. However, as commercial software packages include more advanced models together with fast solution algorithms the advantages of powerful constitutive models over simple linear elastic - perfectly plastic models in predicting the deformation behaviour of soils and rocks under working load conditions and in defining possible failure mechanisms are increasingly appreciated in practice. By doing so the limitations of higher order models also become obvious because it is found that some deficiencies in modelling soil behaviour remain not only from a scientific but also from a practical point of view. Many important features of soil behaviour are still not included in most of the advanced models available in commercial software, possibly with a few exceptions.

These include

- anisotropic behaviour,
- small strain stiffness behaviour,
- the behaviour of partially saturated soils,
- the effects of rotation of principal stress axes,
- formation of shear bands including post peak behaviour.

There is a strong need therefore to formulate constitutive models for practical engineering which at one hand include all relevant features of soil behaviour and on the other hand are not too complex to be accepted by non-specialists in numerical modelling.

Some of the above mentioned aspects, namely anisotropy, rotation of principal stress axes and formation of shear bands will be addressed in the first part of this paper. A conceptually very simple model based on the multi-laminate framework for soils is presented. Only drained behaviour will be considered here thus neither consolidation problems nor partially saturated soils will be dealt with. Also the stiffness at very small strains will not be considered although extension of the concept to account for small strain stiffness seems to be possible.

In the second part of this paper the developed framework is applied to solve the practical boundary value problem of a tunnel excavation.

When excavating tunnels with low overburden it is of particular interest to identify possible failure mechanisms involving strain localisation. Especially under difficult geotechnical conditions the construction process can cause the formation of extensive shear zones in the ground. In this case an inappropriate combination of excavation sequence, support measures and excavation speed could lead to a progressive failure mechanism which, consequently, could result in the collapse of the tunnel. In addition, the development of shear bands can occur suddenly and with little warning.

Efficient numerical methods could provide the geotechnical engineer with an additional tool for getting better insight into the behaviour of the ground. Consequently, a numerical model for predicting possible collapse mechanisms and the deformation behaviour of the ground near failure could support the engineer in avoiding dangerous situations during tunnelling. However, standard finite element formulations are not capable of predicting soil behaviour that includes the formation of shear bands and, therefore, they have to be enhanced. Various strategies have been proposed so far for modelling the localisation of strains in a zone of limited thickness, e.g. a shear band, with numerical methods. These include enhanced finite element models ([1], [2], [3]), non-local models ([4], [5]), gradient plasticity models ([6]), micropolar continua ([7], [8], [9]), viscoplastic models ([10]) and mesh adaptivity techniques ([11], [12]). However, none of these models have been successfully applied to investigate the deformation behaviour and collapse mechanisms when excavating a tunnel in soil, at least not to the authors' knowledge.

A model that includes strain softening behaviour at the level of constitutive modelling has been developed by Cividini and Gioda ([13]). Shear band formation in a plane strain laboratory test of a tunnel excavation ([14], [15]) and a small scale test investigating the face stability ([16]) have been numerically simulated with this model.

A more pragmatic concept based on the multilaminate model for soils is adopted for this work. Basically, all material parameters required by the model have a clear physical meaning and are familiar to geotechnical engineers. The capabilities of the formulation in capturing the deformation behaviour of the ground near failure is demonstrated by numerical simulations of NATM tunnel excavations with low overburden. At first, the performance of the model with respect to varying initial stress states and the possibility to model both frictional softening and cohesive softening is shown. Finally, an application to a NATM tunnel excavation in staged excavation sequence is presented.

Of course, tunnelling involves three-dimensional stress redistributions in the ground, especially at the tunnel face. For the sake of simplicity all analyses presented here are performed under plane strain conditions and the problem of face stability is not addressed at all at this stage. However, the model discussed in the following can be extended to 3-D analyses in a straightforward manner because the constitutive equations are formulated on contact planes and only number and position of these contact planes have to be adapted according to integration rules valid for the three-dimensional space.

2 Multilaminate framework

The concept of multilaminate models for geomaterials is based on so-called integration planes or contact planes. It is assumed that the overall deformation behaviour of geomaterials can be obtained from an evaluation of deformations on these contact planes.

When load is applied to a soil mass, contact forces between adjacent particles develop which can be split into normal and tangential components. In the initial stress state these contacts and bounds depend on the history the soil has undergone prior to construction activities, such as tunnelling, which will be considered in this paper. The overall deformation behaviour of soils results from both, deformation of individual particles and relative sliding between particles. The inter-particle sliding which is governed by the shear resistance at the contacts between the particles makes the major contribution to the overall strain and this contribution is accounted for in the multilaminate model. Since the individual particles are not modelled in a discrete way, the interactions between particles are considered in an averaged form on contact planes. Rotation of individual grains cannot be explicitly modelled. Number, direction and orientation of the planes are governed by integration rules. It is assumed that the individual planes do not interact and usually the same mathematical relations hold for all planes. However, this assumption is not strictly required and initial anisotropy may be easily introduced by varying parameters over the planes prior to loading.

Depending on the stress path while loading, certain integration planes will become active leading to plastic deformations due to strain hardening or softening while other planes remain inactive. Therefore, anisotropy depending on the stress path is automatically considered. Effects of rotations of principal stress axes, which generally takes place in boundary value problems, can be accounted for in a simple and physically meaningful way.

When implementing a multilaminate model into a finite element code, the contact planes are located at each integration point of the finite element mesh. The constitutive relations for describing the material behaviour are formulated on the planes in local coordinates. Therefore, the stress state $\{\sigma\}$ at an integration point is transformed into normal and shear stresses (σ_n, τ in Fig. 1) on each plane utilising a transformation matrix $[\mathbf{T}_\sigma]$ (1).

$$\left\{ \begin{matrix} \sigma_n \\ \tau \end{matrix} \right\} = [\mathbf{T}_\sigma] \{\sigma\} . \tag{1}$$

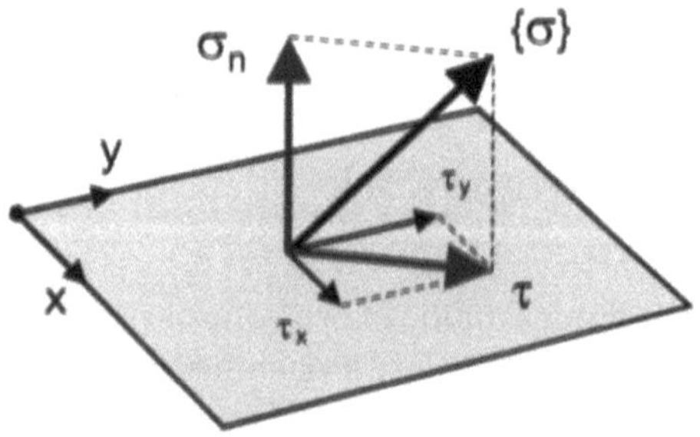

Fig. 1. Transformation of global stress $\{\sigma\}$ in integration point into local stresses σ_n and τ on a contact plane

By applying a yield criterion and a plastic potential function plastic volumetric and deviatoric strain rates ($\dot{\varepsilon}^p, \dot{\gamma}^p$) can be calculated on each of the planes. Since the constitutive relations are defined on the level of the integration planes and formulated in terms of σ_n and τ relatively simple mathematical expressions can be used, even to describe complex material behaviour that includes strain hardening and softening. The global plastic strain rates $\{\dot{\varepsilon}^p\}$ are obtained by numerical integration of the strain rates calculated on the individual planes using a transformation matrix $[\mathbf{T}_\varepsilon]$ (2).

$$\{\dot{\varepsilon}^p\} = \sum_{i=1}^{n_{pl}} w_i \, [\mathbf{T}_\varepsilon] \left\{ \begin{matrix} \dot{\varepsilon}^p \\ \dot{\gamma}^p \end{matrix} \right\} . \tag{2}$$

The number of planes n_{pl}, their orientation and the weight factors w_i are chosen according to integration rules (e.g. [17]).

In the formulation presented here only plastic behaviour is considered on contact planes, elastic strains are calculated globally. The multilaminate Model has been implemented into the finite element / boundary element code BEFE ([18]) which was developed for geomechanical analyses and employs a (pseudo-) viscoplastic algorithm for solving non-linear equation systems. Although 3-D analyses can be performed with BEFE only a 2-D (plane strain) version of the multilaminate model has been implemented as yet.

The first multilaminate model for rock has been proposed by Zienkiewicz and Pande ([19]). They assumed that a pronounced anisotropy of rock can be represented by introducing a limited number of planes of weakness, typically two or three. On these planes the constitutive model can be formulated in local coordinates describing the amount of sliding and opening which in turn governs the overall behaviour of jointed rock. This approach has been implemented into finite element codes also by others and is successfully used in solving problems in practical rock engineering.

Pande and Sharma ([20]) introduced the multilaminate concept for clays and formulated a critical state model on contact planes. A multilaminate Model for sand including shear and volumetric hardening has been presented by Krajewski ([21]), a different formulation for sands has been suggested by Sadrnejad and Pande ([22]). Above formulations have been implemented into finite element codes employing a (pseudo)-viscoplastic algorithm, i.e. viscoplasticity is used for solving the non-linear equation system but no true time dependent behaviour has been included. A plasticity formulation of the multilaminate model has been presented by Pietruszczak and Pande ([23]). More recent developments of the multilaminate model include research on strain localisation in dense sands by Karstunen ([24]) where a plasticity formulation of the multilaminate model with deviatoric hardening and a non-associated flow rule has been used. A similar approach developed to model fracture propagation in concrete as well as the behaviour of geomaterials is the microplane model first presented by Bazant and Prat ([25]).

As mentioned above a numerical integration is required to arrive at global strains based on local strains on contact planes. Number, distribution and orientation of these planes are governed by integration rules. Pande and Sharma ([20]) and Krajewski ([21]) used a 9-plane integration rule for plane strain analyses and a 13-plane integration rule for 3-D analyses, which to the authors' opinion is not sufficient for obtaining accurate results. Due to developments in hardware, higher order integration rules are feasible today as far as run times on computers are concerned. In the two-dimensional (plane strain) version of the multilaminate model, the planes are evenly distributed about a common axis (Fig. 2) and the number of planes can be freely chosen, a higher number of planes leading to a higher accuracy ([26]). Throughout the work presented here 64 contact planes have been used.

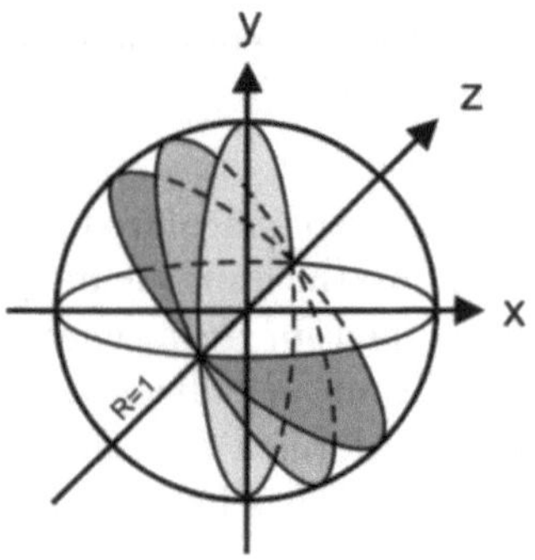

Fig. 2. Contact planes in integration point distributed about common axis for 2-D plane strain calculations

3 Constitutive relations

Compared to linear elastic – perfectly plastic constitutive laws, which are still widely used in geotechnical engineering, more advanced constitutive relations have been utilised in the multilaminate model presented here. Based on Mohr-Coulomb's failure criterion deviatoric hardening is introduced in the pre-peak range via mobilisation of the friction angle. Starting from an initial value the friction angle is increased with plastic shear strains until the peak friction angle is reached. A non-associated flow rule governs the development of volumetric and deviatoric strains. The post-peak range is characterised by a strain softening formulation. With increasing plastic shear strains the shear parameters, friction angle and cohesion, are reduced to residual values.

It has to be emphasised again that the constitutive relations presented below are formulated on the level of the contact planes of the multilaminate Model. Thus, they are formulated in terms of normal effective stress σ_n and shear stress τ.

3.1 Yield criterion including deviatoric hardening

The well known Mohr-Coulomb failure criterion was extended by introduction of a mobilised friction angle φ_m to account for deviatoric hardening (Fig. 3a).

$$F = |\tau| + \sigma_n' \tan\varphi_m - c = 0 \tag{3}$$

In (3), c represents the cohesion and tensile stresses have positive sign.

Governed by plastic shear strain accumulation on the respective planes the friction angle φ_m is increased until the peak friction angle φ is reached. In (4) the relation between mobilised friction angle and plastic shear strains is formulated:

$$\sin\varphi_m = \sin\varphi_0 + \left[\frac{2\sqrt{\gamma^{vp}\gamma_p^{vp}}}{\gamma^{vp} + \gamma_p^{vp}}\right](\sin\varphi - \sin\varphi_0)\,. \tag{4}$$

φ_0 and φ are initial friction angle and peak friction angle, respectively, γ^{vp} is the accumulated plastic shear strain on the plane, γ_p^{vp} represents the shear strain where the mobilised friction angle φ_m reaches its peak value φ. The initial friction angle φ_0 defines the elastic region prior to loading. This expression for the deviatoric hardening is based on Vermeer ([27]) and de Borst ([28]) but their invariant formulation has been modified to be suitable for implementation into the multilaminate framework.

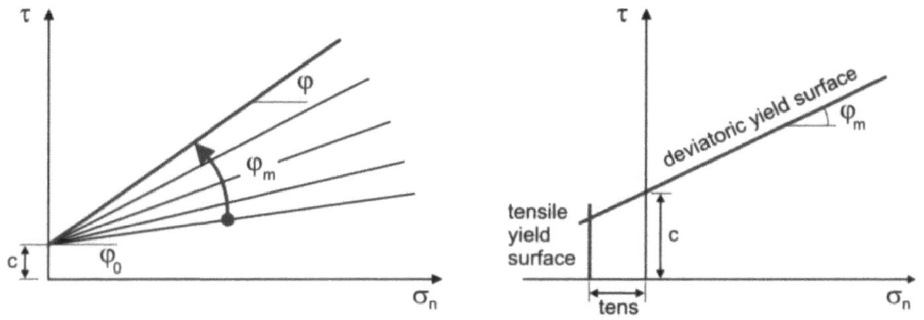

Fig. 3. (a) Mobilisation of friction angle φ_m starting from initial friction angle φ_0 to peak friction angle φ; (b) Tensile and deviatoric yield surfaces

In Fig. 3a, it can be observed that the cohesion c remains at a constant value during mobilisation of the friction angle. Strictly speaking the shear strength, i.e. c and φ, should be mobilised but the formulation utilised here has an advantage for modelling the post-peak behaviour since independent formulations for strain softening of friction angle and cohesion can be introduced. However, in the pre-peak range this does not have a significant influence on the material behaviour.

The tensile strength of the material is limited in the constitutive law by implementation of a so-called tension cut off criterion (Fig. 3b).

3.2 Plastic potential

Plastic flow and dilatancy are governed by a non-associated flow rule. Since the plastic potential is formulated on the contact planes, it is a function of normal and shear stresses and the mobilised dilation angle ψ_m on the respective plane. The mobilised dilation angle ψ_m is obtained as a function of the mobilised friction angle φ_m as described in (5) and (6).

$$\sin\psi_m = \frac{\sin\varphi_m - \sin\varphi_{cv}}{1 - \sin\varphi_m \sin\varphi_{cv}}, \tag{5}$$

$$\sin\varphi_{cv} = \frac{\sin\varphi - \sin\psi}{1 - \sin\varphi \sin\psi}. \tag{6}$$

φ_m, φ and ψ are mobilised friction angle, peak friction angle and peak dilation angle, respectively. φ_{cv} is the critical state friction angle or friction angle of constant volume. This formulation corresponds to the stress-dilatancy theory proposed by Rowe ([29]).

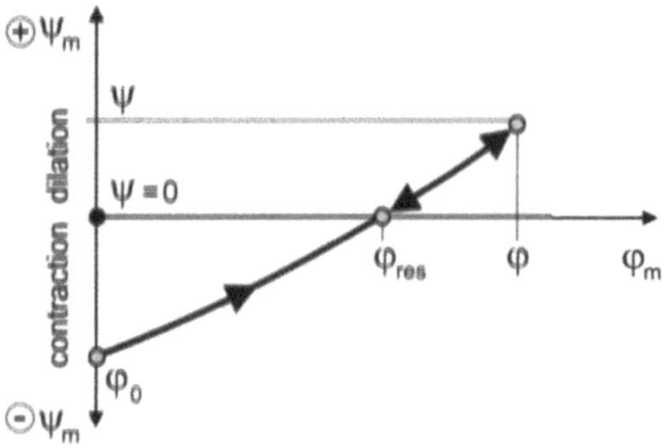

Fig. 4. Relation between mobilised friction angle φ_m and mobilised dilation angle ψ_m

As demonstrated in Fig. 4, the dilation angle starts with a negative value which leads to contractive volumetric strains. With mobilisation of the friction angle the dilation angle becomes positive leading to dilatant material behaviour. The peak dilation angle ψ is reached together with the peak friction angle φ. During softening the dilation angle is reduced until a state of constant volume ($\psi = 0$) is reached together with the residual value of the friction angle.

It has to be emphasized that (5) and (6) are formulated in terms of various parameters on contact planes. If the peak shear resistance is reached on one of the planes, the dilation angle is fully mobilised on the respective plane. At the same time, due to different levels of mobilisation of the friction angle, other planes will show less dilatant or even contractant behaviour. The volumetric behaviour in the respective integration point is obtained by a summation of the contributions of all planes. In an equivalent invariant formulation, the plastic volumetric behaviour is entirely governed by the peak dilation angle when the peak shear resistance is reached in the integration point. Thus, when comparing results of both formulations, the multilaminate framework leads to a less dilatant behaviour of the respective integration point if compared to an invariant formulation.

3.3 Strain softening

If the accumulated plastic shear strains γ^{vp} approach γ_p^{vp} at one or more of the integration planes, the shear strength of the material is fully mobilised on these planes and the peak friction angle φ is reached. If shearing continues on the respective planes, friction angle and cohesion are reduced in a linear relation until residual values φ_{res} and c_{res} are reached (Fig. 5). This formulation is similar to what has been proposed by Cividini and Gioda ([13]). The softening rates can be controlled individually for the friction angle φ_m and the cohesion c, respectively.

Since in multilaminate models sliding on certain contact planes is the main mode of deformation, the results of direct shear tests could be utilised for determination of the required softening rates. Another approach has been proposed by Marcher and Vermeer ([5]) who established a relation between mobilisation of the friction angle and changes of the void ratio, which is based on a scalar softening modulus and an internal length depending on the mean element size.

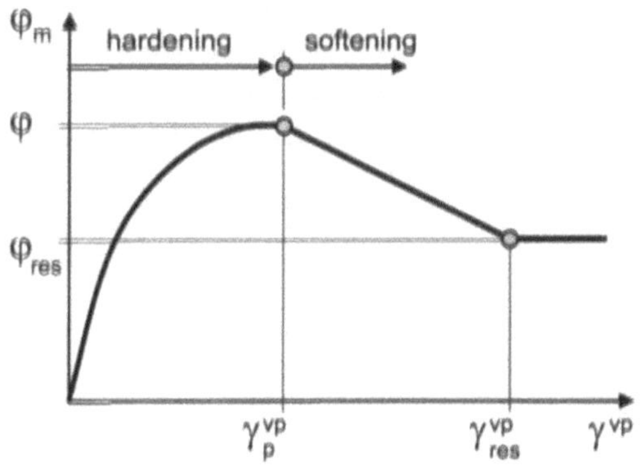

Fig. 5. Strain hardening and softening; relation between mobilised friction angle φ_m and accumulated plastic shear strains γ^{vp} on integration plane level

The degree of strain softening, namely the amount of plastic shear strains on the plane γ^{vp}_{res} where the residual friction angle is reached, is determined by a so-called softening factor F_{soft} defined as

$$F_{soft} = \frac{\gamma^{vp}_{res}}{\gamma^{vp}_{p}}. \tag{7}$$

It is well known that considering strain softening behaviour at constitutive level as suggested here may lead to severe mesh dependence of results in the post peak range if implemented with standard finite element techniques. In addition, the size of load increments and in the multilaminate model the number of contact planes have some influence. In order to avoid this a scaling factor F_{sc} is introduced, arriving at an internal parameter for defining the softening rate F_{int}. Equation (7) becomes

$$F_{int} = F_{soft} \cdot F_{sc}. \tag{8}$$

Determination of F_{sc} is described in the following. Strain softening behaviour, e.g. the reduction of the friction angle to a residual value, is governed by the accumulation of plastic shear strains. The process of shear strain accumulation happens on each contact plane independently during the iteration process. However, the overall shear resistance of an integration point is primarily governed by the shear resistance of the weakest contact plane and a high shear strain concentration on the weakest plane, therefore, leads to more pronounced softening of the integration point in the post peak behaviour. Thus, the key factor for handling strain softening behaviour in the multilaminate framework is the concentration of plastic shear strains on certain integration planes, which is represented by the scaling factor F_{sc}.

The development of plastic shear strains in an iteration step depends on the value of the yield function F on the planes of an integration point. By relating the maximum value of all contact planes, F_{max}, to the sum over all planes, ΣF_i, the scaling factor F_{sc} is determined (9). In the calculation of ΣF_i only planes which experience plastic straining are included. By doing so the influence of the localised plane is smeared over the integration point leading to results which are no longer dependent on the size of the load increment or the number of contact planes. F_{sc} is determined for each integration point independently in each step of the iteration procedure towards convergence.

$$F_{sc} = \frac{F_{\max}}{\frac{1}{n_{pl}} \sum_{i=1}^{n_{pl}} F_i} . \tag{9}$$

In (9), F_{max} and F_i represent the maximum value of the yield function and the value of the yield function on each contact plane, n_{pl} is the number of contact planes used.

Distributions of F on the integration planes normalised by the maximum value F_{max} are shown in Fig. 6 for 3 different load increments $\Delta\varepsilon_1 = 0.25\%$, $\Delta\varepsilon_1 = 0.5\%$ and $\Delta\varepsilon_1 = 2.0\%$ respectively. It is obvious that different values of F_{sc} will be obtained by the curves.

4 Simulation of stress paths

4.1 Stress path in biaxial test

A strain controlled biaxial test was simulated to demonstrate the development of plastic strains on individual contact planes in a two-dimensional (plane strain) multilaminate model. The hardening and softening response as well as the development of anisotropy is monitored on 64 integration planes which are evenly distributed about a common axis. To keep calculations simple the biaxial test was modelled using only one single eight-noded quadrilateral finite element with parabolic shape functions. The boundary conditions consisted of horizontal constraints at the left side together with vertical constraints

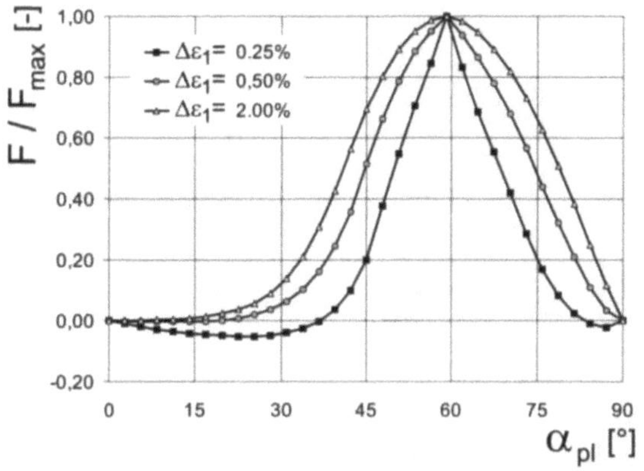

Fig. 6. Possible distributions of F over integration planes normalised by F_{max} (α_{pl} denotes the orientation of the contact planes)

at the bottom of the element. Strain controlled loading of the sample was simulated by nodal displacements applied to the top of the element (Fig. 7). Starting from an initial friction angle $\varphi_0 = 0\,^\circ$ on all planes the friction angle was mobilised until the peak friction angle $\varphi = 30\,^\circ$ was reached at certain integration planes at an accumulated plastic shear strain $\gamma_p^{vp} = 0.50$. With further loading, the respective planes switched into softening and the friction angle was reduced to a residual value of $\varphi_{res} = 20\,^\circ$. In this case, the residual value was reached at an accumulated plastic shear strain of $\gamma_{res}^{vp} =$ 2.50. It has to be emphasised, that in this configuration a homogeneous stress distribution is maintained within the sample. The aim here was to monitor the development of plastic strains on the planes. Clearly, effects of strain localisation and shear band formation cannot be captured with this simple model.

In Fig. 8, the development of strain hardening on the integration planes is presented by monitoring the accumulation of plastic shear strains and the development of the mobilised friction angle. The orientation of the planes with respect to the horizontal direction can be identified by the angle α_{pl} (Fig. 7). In Fig. 8 and Fig. 9, the location of the planes is indicated by the small circles. Due to loading in a fixed direction no rotation of principal stress axes occurs and hardening and softening develops symmetric with respect to the vertical axis which is equal to the direction of maximum principal stress. Therefore, only the planes located between $\alpha_{pl} = 0\,^\circ$ and $\alpha_{pl} = 90\,^\circ$ are included in the diagrams.

In the left diagram in Fig. 8 accumulation of plastic shear strains on the integration planes is presented. Each line represents the distribution of plastic shear strains for a certain load step. Starting from an almost equal distribution over the planes with a slight maximum located at planes close to

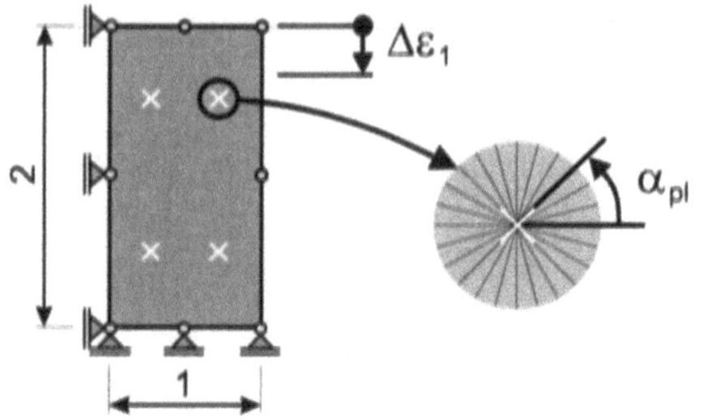

Fig. 7. Geometry, boundary conditions and load increments of 1-element biaxial test; orientation of integration planes in an integration point (plane strain version of multilaminate model) is given by the angle α_{pl}

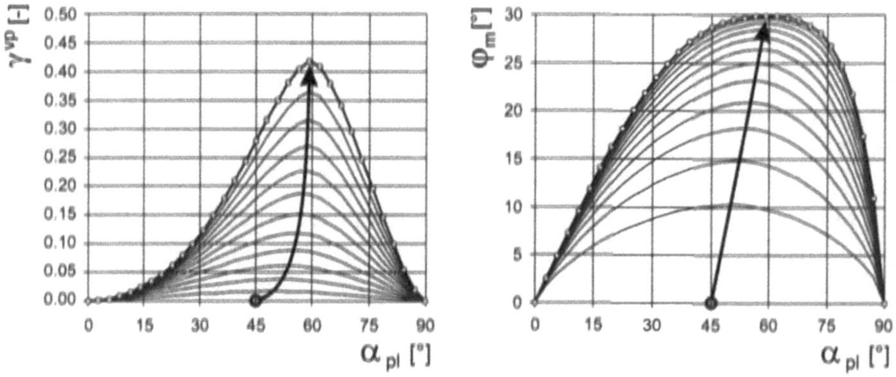

Fig. 8. Accumulation of plastic shear strains and mobilisation of friction angle monitored on integration planes located within one quadrant ($\alpha_{pl} = 0°$ to $\alpha_{pl} = 90°$)

$\alpha_{pl} = 45°$, plastic shear strains tend to concentrate on certain planes with increased loading. When loading gets close to the peak load as depicted by the top line, the peak value of plastic shear strains $\gamma_p^{vp} = 0.50$ is approached at a plane located at approximately $\alpha_{pl} = 60°$. This corresponds to the angle of $45° + \varphi/2$ with respect to the direction of minimum stress and is, therefore, in accordance with Mohr-Coulomb's failure criterion. In the right diagram in Fig. 8 mobilisation of the friction angles on the integration planes is presented. The mobilised friction angle φ_m is a function of accumulated plastic shear strains γ^{vp}, therefore it develops on the corresponding planes.

It can be observed that on the planes $\alpha_{pl} = 0°$ and $\alpha_{pl} = 90°$, respectively, no plastic shear strains develop because these planes are oriented normal to the principal stress axes.

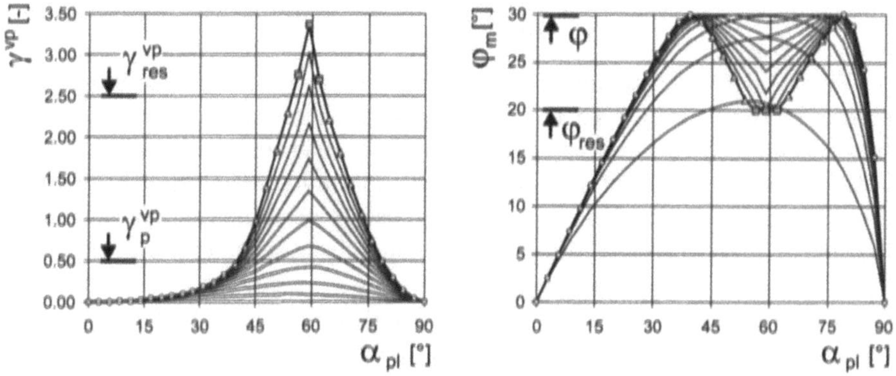

Fig. 9. Accumulation of plastic shear strains and mobilisation of friction angle on integration planes in pre-peak strain hardening and post-peak strain softening

In Fig. 9, the relations presented in Fig. 8 for the pre-peak range are extended to the post-peak strain softening behaviour. Again, each line represents the results of one particular load stage. The tendency of shear strain concentration on a few integration planes is further increased when the respective planes switch into softening. In the diagrams in Fig. 9 the relation between accumulated plastic shear strains and mobilisation of the friction angle in hardening and softening is demonstrated: Planes where the "peak" plastic shear strain γ_p^{vp} (in this case γ_p^{vp}= 0.50) is exceeded switch into softening behaviour and the friction angle is reduced. For the last load step presented in the diagrams this holds for the planes between approximately α_{pl} = 42° and α_{pl} = 76°. When the "residual" plastic shear strain γ_{res}^{vp} = 2.50 is exceeded, as well, the residual friction angle is reached. In the diagrams three planes located between α_{pl} = 56° and α_{pl} = 62° are at residual state. Furthermore it is demonstrated that within the same load step the friction angle can be reduced on certain planes due to softening while other planes still show strain hardening behaviour.

4.2 Stress path in biaxial test with exchange of principal stress axes

In this test example ([30]) a stress controlled biaxial test was simulated using one eight-noded quadrilateral finite element and the same geometry and boundary conditions as previously described. After isotropic consolidation at σ_0 = -100 kN/m², first stresses in horizontal direction were increased up to σ_x = -300 kN/m² until the mobilised friction angle approached its peak value at the critical planes. Horizontal stresses were then reduced in steps of $\Delta\sigma_x$ = 10 kN/m² while vertical stresses were increased maintaining a constant mean stress p'. During this process, the maximum and minimum principal stresses exchange. Finally, the shear resistance of the material was fully mobilised again at a stress state of σ_x= -100 kN/m^2 / σ_y= -300 kN/m^2.

Fig. 10 displays accumulation of plastic shear strains on the integration planes. In the diagrams, each dot represents an integration plane. The location of the dot with respect to the center of the diagram gives the orientation of the plane α_{pl}. The distance between the dot and the center of the diagram indicates the amount of plastic shear strains γ_{pl}^{vp} accumulated on the respective plane and, therefore, the degree of mobilisation of the friction angle. When the horizontal stress approaches the peak value of σ_x = -300 kN/m^2, strains concentrate on certain planes located at an angle of 45° - $\varphi/2$ with respect to the direction of maximum principal stress σ_x. When σ_x and σ_y stresses are changed in increments of $\Delta\sigma$ = 10 kN/m^2 maintaining a constant mean stress p', the model responds elastically at first. However, at a certain stress state, namely σ_x = -120 kN/m^2, σ_y = -280 kN/m^2, different planes become "active" and plastic shear strains start to develop. When the maximum stress state at σ_x = -100 kN/m^2, σ_y = -300 kN/m^2 is reached, again the largest amount of plastic shear strains has accumulated on planes located at an angle of 45° - $\varphi/2$ to the direction of maximum stress.

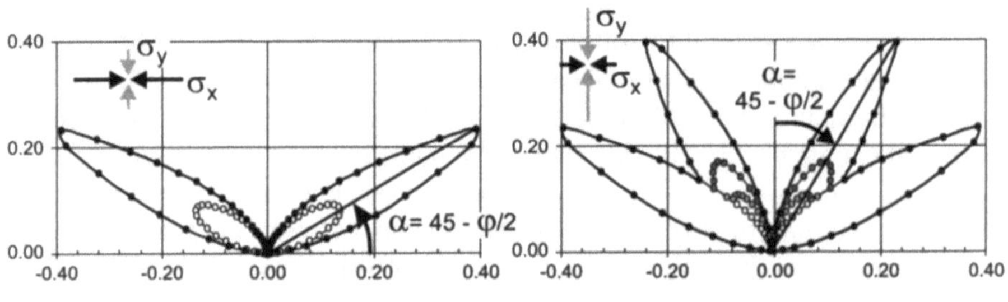

Fig. 10. Accumulation of plastic shear strains on integration planes due to exchange of principal stress axes; left diagram: σ_x = -300 kN/m^2, σ_y = -100 kN/m^2; right diagram: σ_x = -100 kN/m^2, σ_y = -300 kN/m^2

4.3 Stress paths including rotation of principal stress axes

Engineering works, such as excavations, the loading of foundations or construction of embankments result in complex stress paths in the ground including continuous rotation of the principal stress axes. To evaluate the influence of stress rotation on the constitutive behaviour, a number of laboratory tests have been developed which allow controlled rotation of principal stress axes, such as the directional shear cell ([31]), the hollow cylinder apparatus ([32]) or more recently the "1γ2ε" apparatus ([33]). Results of these tests indicate that principal stress axes rotation can have significant influence on the stress

– strain behaviour of soils. However, most of the proposed constitutive models are expressed in terms of stress and strain invariants. Thus, they respond to changes in these invariants but they cannot predict deformations correctly if the principal stress axes rotate while stress invariants remain unchanged.

When employing the proposed multilaminate model this disadvantage of invariant formulations can be overcome. To show the performance of the multilaminate model when capturing the effects of rotation of the principal stress axes, a very simple numerical model consisting of only one integration point and 64 integration planes has been used, i.e. in this section results shown are obtained with a small programme integrating the constitutive equations and not by a finite element calculation.

In the first calculation step anisotropic consolidation up to principal stress values of $\sigma_1 = -240\,\mathrm{kN/m^2}$ and $\sigma_3 = -160\,\mathrm{kN/m^2}$ is simulated. Then, the stress axes are rotated under constant magnitudes of σ_1 and σ_3 and, consequently, under constant mean principal stress p' and stress ratio k. As a consequence of utilising a plane-strain formulation of the multilaminate Model, the intermediate principal stress σ_2 is not considered because it does not influence the behaviour of the model. All integration planes remain within the pre-peak range at all load steps.

$$p' = \frac{\sigma_1 + \sigma_3}{2} = -200\mathrm{kN/m^2} = constant, \tag{10}$$

$$k = \frac{\sigma_1 - \sigma_3}{\sigma_1 + \sigma_3} = 0.2 = constant. \tag{11}$$

During the test the stress path is circular as presented in the stress diagram in Fig. 11: $(\sigma_y - \sigma_x) / (\sigma_y + \sigma_x)$ versus $2\tau_{xy} / (\sigma_y + \sigma_x)$. σ_x, σ_y and τ_{xy} are stresses with respect to the chosen co-ordinate system. During the numerical simulation, stress axes are rotated in steps of $\Delta\alpha_\sigma = 5^\circ$ up to a total rotation of $\alpha_\sigma = 180^\circ$. In addition, in the diagram the orientation of the principal stresses is indicated for rotations of $\alpha_\sigma = 0^\circ / 180^\circ$, $\alpha_\sigma = 45^\circ$, $\alpha_\sigma = 90^\circ$ and $\alpha_\sigma = 135^\circ$, respectively.

The numerical simulations using the multilaminate model result in compressive volumetric strains exclusively caused by rotation of the principal stress axes (Fig. 12). During the anisotropic consolidation plastic shear strains and strain hardening have occurred on certain integration planes. Due to stress axes rotation, different planes become active and both, plastic shear strains and volumetric strains develop contributing to the global volumetric strains. After a principal stress axes rotation of $\alpha_\sigma = 90^\circ$, no further compressive volumetric strains develop with this formulation of the multilaminate model. At this stage, the friction angle is mobilised to the same degree on all planes corresponding to the actual stress state. If stress axes are further rotated no additional activation of planes takes place.

A closer look at the integration planes is required to point out the mechanism of activation and deactivation of planes due to rotation of stress axes.

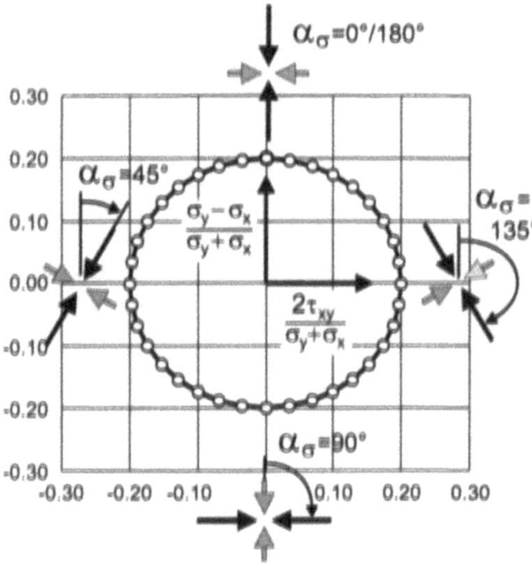

Fig. 11. Rotation of principal stress axes at constant mean stress p' and constant stress ratio k

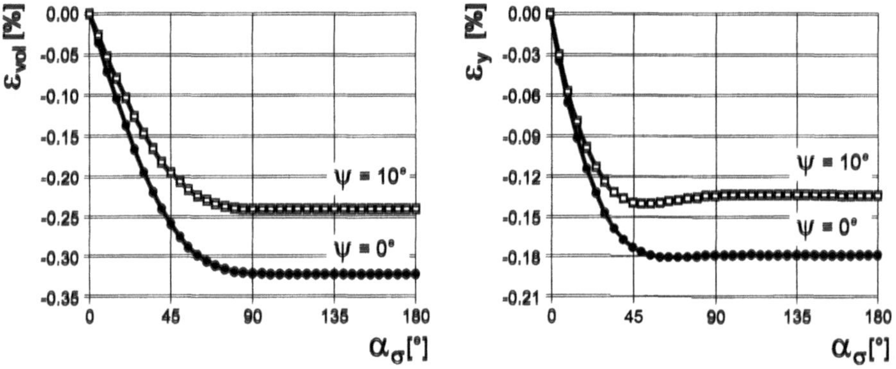

Fig. 12. Volumetric strains ε_{vol} and vertical strains ε_y exclusively due to stress rotation for different dilation angles ψ

In Fig. 13 accumulated plastic shear strains on the integration planes are presented for a sequence of rotation steps. As previously stated, accumulated plastic shear strains are a direct measure for mobilisation of the friction angle on the respective plane. Each dot represents an integration plane, the location of the dot with respect to the centre of the diagram gives the orientation of the plane α_{pl}. The distance between the dot and the centre of the diagram indicates the amount of plastic shear strains γ_{pl}^{vp} accumulated on the plane and, therefore, the degree of mobilisation of the friction angle.

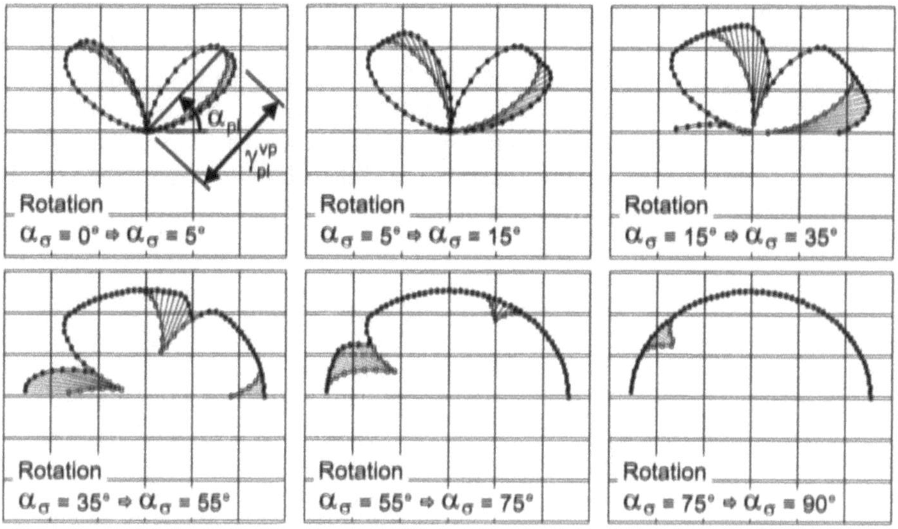

Fig. 13. Accumulation of plastic shear strains on integration planes during rotation of principal stress axes

After the anisotropic consolidation phase at $\alpha_\sigma = 0^\circ$ plastic shear strains have developed as depicted by the circles on the thin line in the first diagram in Fig. 13. The maximum values are obtained at planes located at $\alpha_{pl} = 45^\circ + \varphi_m/2$ and $\alpha_{pl} = 135^\circ - \varphi_m/2$. The distribution of shear strains is symmetric about the axis of maximum principal stress, in this case the vertical axis. When the first step of principal stress axes rotation to $\alpha_\sigma = 5^\circ$ takes place, additional plastic shear strains develop on certain planes represented by the thin radial lines and the distribution indicated by black dots is obtained. The strain distribution on the planes does not remain symmetric. The following diagrams in Fig. 13 describe the evolution of plastic shear strains with further rotation of principal stress axes. It becomes clear that additional plastic shear strains develop on certain planes until, finally, the same amount of strains has been accumulated on all planes after a stress rotation of $\alpha_\sigma = 90^\circ$. Consequently, at this moment the previously anisotropically distributed mobilised friction angle becomes isotropic again. A more detailed analysis of development of stresses and stress paths followed on different planes is given in Schweiger and Schuller ([34]).

A slightly different formulation for the deviatoric hardening and an extension of the model to include volumetric hardening by an additional cap yield surface has been presented by Wiltafsky et al. ([35]). With this model a comparison with experimental data by Akagi and Saitoh ([36]) on remoulded clay has been performed (Fig. 14). It follows that the qualitative behaviour is captured very well by the model, a perfect quantitative match hovever could not be achieved. This is at least to some extent due to the fact that not all required material parameters for the model have been available and thus some parameters had to be estimated based on the information given by Akagi and Saitoh ([36]). It is notable though that with this advanced formulation plastic volumetric strains are predicted up to a rotation of 180 ° as compared to 90 ° with the formulation not including the cap yield surface. In Fig. 14 results from a multilaminate Model including creep behaviour is also included, details of which can be found in Neher et al. ([37]).

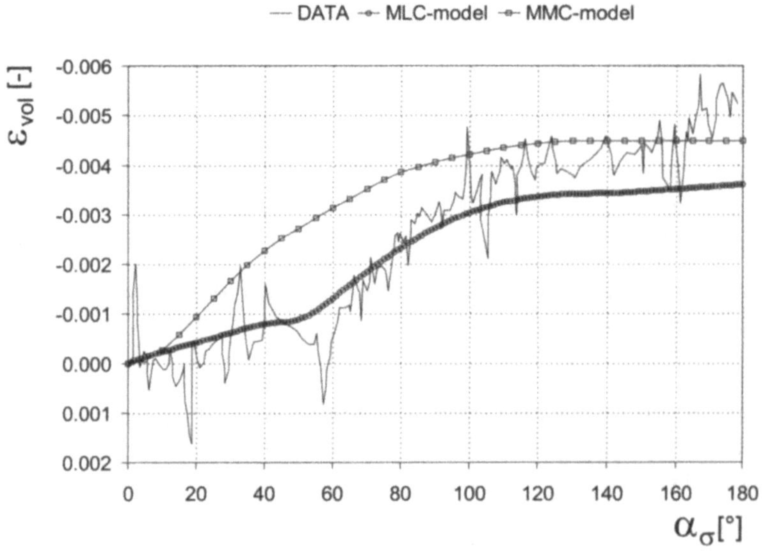

Fig. 14. Rotation of principal stress axes vs. volumetric strains for reconstituted clay with $k = 0.2$.

5 Numerical simulation of a biaxial test

A strain controlled biaxial (plane strain compression) test on dense Hostun RF sand as performed by Desrues and Hammad ([38]) in Grenoble was numerically simulated. To evaluate possible mesh dependencies in the post-peak range, four different meshes consisting of 126, 224, 504 and 880 elements, respectively, were analysed. The slender specimens have a width to height ratio of 1:3.4. In the analyses the bottom of the sample was vertically constrained but free to move in horizontal direction. The top of the sample is bounded by a stiff top platen which is horizontally constrained at its boundaries. Full friction between top platen and the sample introduces an inhomogeneous stress distribution which, at a certain load stage, initiates strain localisation. A double biased mesh leads to a slightly non-symmetric stress state and, therefore, to strain localisation in a single shear band.

The material parameters to model Hostun RF sand are included in Tab. 1. The calculation sequence consists of isotropic consolidation to an initial stress of σ_0 = -100 kN/m^2, then strain controlled loading is simulated by stepwise application of vertical nodal displacements to the stiff top platen.

In Fig. 15, curves of the maximum stress σ_1 in vertical direction versus vertical strains ε_1 are presented. The stress-strain curves for all meshes, consisting of 126, 224, 504 and 880 elements, respectively, are congruent up to the peak stress which is reached at a vertical strain ε_1 of approximately 3.0%. In the post peak range some deviation of the curves can be observed during the formation of shear bands but these differences are not relevant from a practical point of view, thus, the scaling of the softening rate as described in section 3.3 proves to be successful.

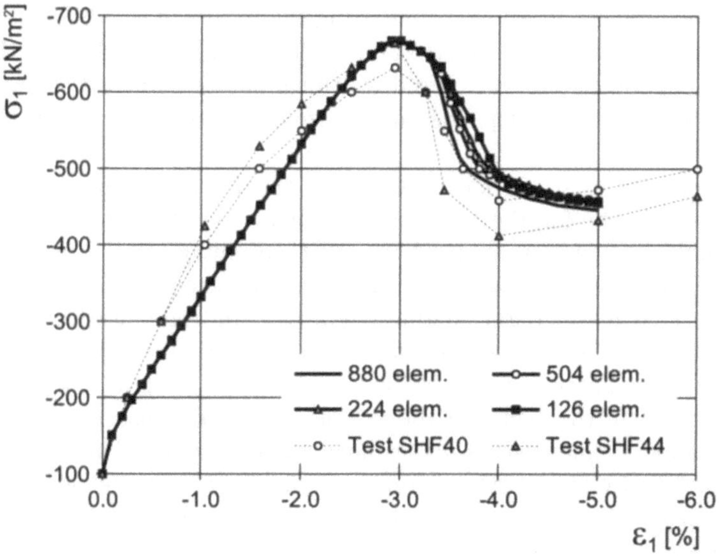

Fig. 15. Biaxial test, stress-strain curves for different meshes

In addition to the stress-strain curves obtained from different meshes, in Fig. 16 contour lines of the mobilised friction angle φ_m are presented for all analysed meshes. Only the mobilised friction angle at the "critical" plane, that is the plane with the largest accumulated shear strains, is shown. The contour plots describe the residual state of the biaxial test at a vertical strain ε_1 = 4.0%. At this stage the shear bands are fully developed and the mobilised friction angle has reached its residual value within the shear band

Table 1. Material parameters for simulation of biaxial test

Young's modulus	E	100 000	kN/m^2
Poisson's ratio	ν	0.20	-
Peak friction angle	φ	47	°
Residual friction angle	φ_{res}	36.1	°
Dilation angle	ψ	14.5	°
Cohesion	c	5	kN/m^2
Residual cohesion	c_{res}	5	kN/m^2
Peak shear strain at plane	γ_p^{vp}	0.28	-
Softening factor	F_{soft}	150	-

as indicated by the dark contours. A slight difference in the location of the shear bands can be observed if comparing the results of the different meshes, but the inclination of the shear band is the same independent of the mesh fineness. Obviously, the shear band width cannot be modelled correctly with this method because it depends on the size of the elements within the shear band.

The process of strain localisation and shear band formation is demonstrated in Fig. 17 for the mesh consisting of 504 elements. Contour lines of the mobilised friction angle at the "critical" integration plane where the largest amount of plastic shear strains has been accumulated are presented in a deformed mesh. Light shadings represent full mobilisation of the friction angle, black areas stand for the residual friction angle. The left figure represents mobilisation of the friction angle at the peak at a vertical strain of $\varepsilon_1 = 3.0\%$. In the centre of the sample the shear resistance has been exceeded and some integration planes have already switched into softening. In the second figure at a vertical strain $\varepsilon_1 = 3.4\%$ two symmetric shear bands start to grow. Due to the double biased mesh the shear band from the top right side to the bottom at the left side becomes dominant. The right figure displays the residual state at a vertical strain $\varepsilon_1 = 4.0\%$.

It has to be acknowledged at this stage that determination of the softening rate from biaxial tests as done here is not correct because geometric effects influencing the experimental results cannot be separated from the constitutive behaviour. In order to account for geometric effects as well, some sort of internal length scale has to be introduced which has not been done so far in the formulation discussed here. Therefore, the example serves only to show that results are not significantly influenced by the mesh configuration. However it is argued that geometric effects do not play a significant role when solving practical problems such as tunnel excavations. As in multilaminate models sliding on planes is the main mode of deformation it is believed that

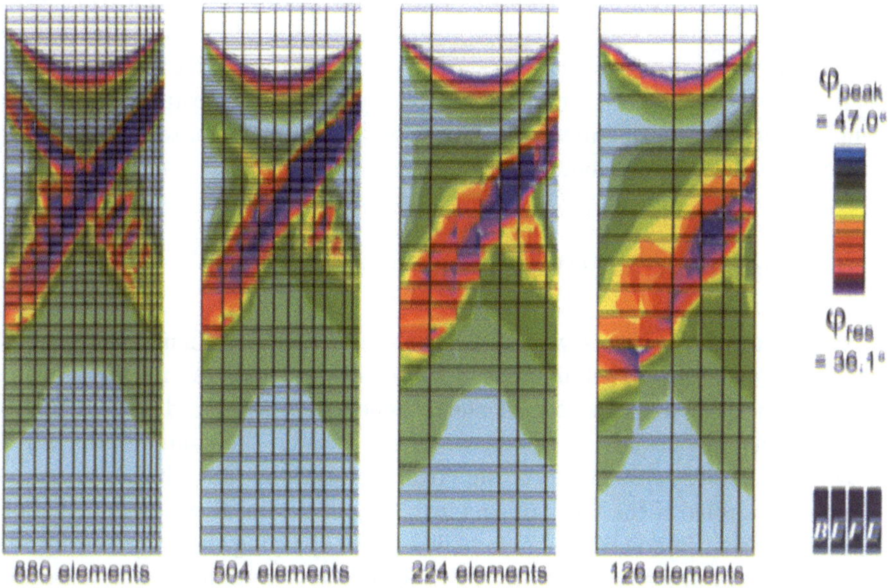

Fig. 16. Evaluation of mesh dependence: biaxial tests at residual state with fully developed shear bands ($\varepsilon_1 = 4.0\%$), contour lines of mobilised friction angle

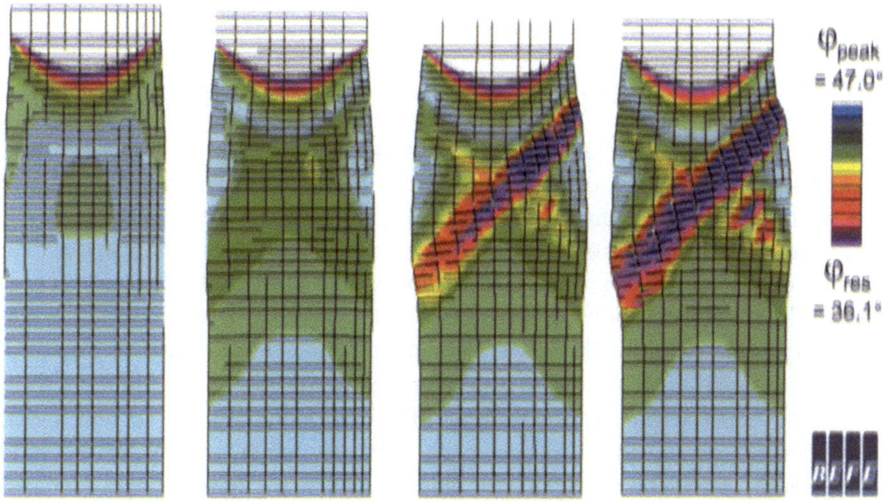

Fig. 17. Formation of shear band in biaxial test (504 elements), contour lines of mobilised friction angle in deformed mesh

the softening rate, required as input parameter, should be obtained by evaluating direct shear tests. This is convenient from a practical point of view also because results from these tests are more readily available than biaxial tests. However, further research is necessary in order to establish a consistent procedure for parameter identification for multilaminate models based on conventional experimental data.

6 Application to tunnelling

When excavating tunnels with low overburden under difficult geotechnical conditions it is of particular interest to identify possible failure mechanisms that involve the development of shear bands in the ground. These failure mechanisms may result in an increase of the load on the tunnel lining which may cause critical situations during construction (e.g. [39], [40]).

It was studied if the mechanisms involving strain localisation and the formation of shear bands could be modelled when applying the proposed formulation of the multilaminate model. For the reason of simplicity at first various studies have been performed by numerically simulating a tunnel where the full tunnel face is excavated. Then a staged tunnel construction according to the principles of the New Austrian Tunnelling Method (NATM) with excavation of the top heading, followed by bench and invert excavation, has been simulated.

6.1 Numerical model

A typical cross section of a shallow double-track railroad tunnel has been modelled. The overburden over the tunnel crown was chosen to be roughly two times the tunnel diameter (z/d = 2). Due to symmetry in the geometry only the right half of the tunnel has been discretised. The analysed domain has an extent of 75 m x 64.5 m with horizontal fixities at both sides of the mesh. The bottom nodes of the mesh are vertically fixed. Geometry and boundary conditions of the finite element meshes are presented in Fig. 18.

To evaluate possible mesh dependencies of results three different finite element meshes with a varying number of elements, namely 278 (coarse mesh), 546 (medium mesh) and 818 (fine mesh) finite elements, were analysed. In all calculations 8-noded, quadrilateral elements with parabolic shape functions and four integration points have been used. In Fig. 18 the medium mesh consisting of 546 finite elements is presented.

In the finite element analyses at first a drained, purely frictional material has been simulated, the chosen material parameters are included in Table 2. Although no 'real' material parameters obtained from laboratory tests have been utilised the parameters chosen would be typical for a medium dense sand.

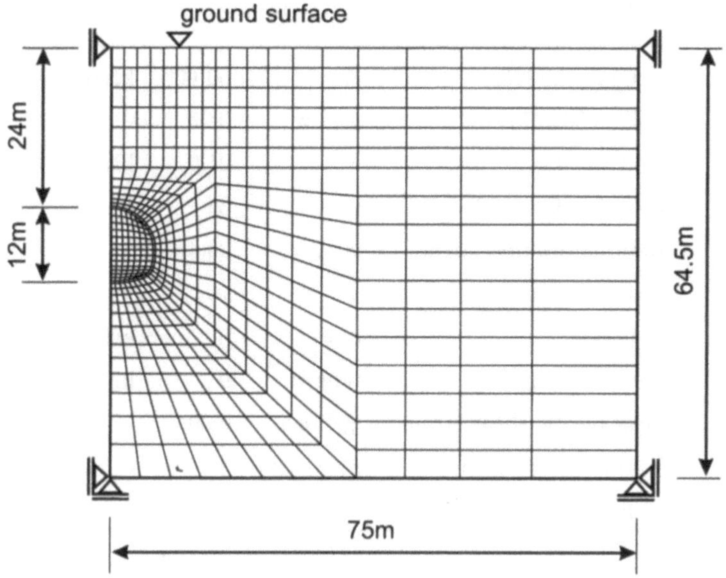

Fig. 18. Finite element mesh, geometry and boundary conditions for finite element analyses; medium mesh (546 elements)

In the first step of the finite element analysis the initial stress state in the ground prior to tunnel excavation has been calculated assuming a coefficient of earth pressure at rest $K_0 = 0.50$. According to the initial stress state the initial friction angle φ_0 is computed such that the stress state is located on the yield surface on the critical plane. The critical plane is the contact plane which experiences the largest amount of shearing. Prior to the excavation process the initial friction angle φ_0 is assumed to be isotropically distributed over all planes.

Excavation of the full tunnel face is modelled by gradually applying the nodal forces associated with excavation. In each load step the system is loaded with an additional load increment, the size of the load increments is 1% of the total excavation forces.

To be able to capture the mechanism leading to the collapse of the tunnel, no excavation support is installed in the numerical simulations. Thus, the shear resistance of the material surrounding the tunnel is increasingly mobilised until, at a certain load stage, a failure mechanism forms.

6.2 Mobilisation of shear resistance and formation of failure mechanism

The process of mobilisation of the friction angle during excavation of the tunnel is demonstrated in Fig. 19 for the load stages where 60%, 75%, 78%

Table 2. Material parameters for drained, purely frictional material

Young's modulus	E =	60 000	kN/m^2
Poisson's ratio	ν =	0.20	-
Unit weight	γ =	21.0	kN/m^3
Peak friction angle	φ =	35	^{o}
Residual friction angle	φ_{res} =	26.4	^{o}
Peak dilation angle	ψ =	10	^{o}
Cohesion	c =	0	kN/m^2
Tensile strength	tens =	0	kN/m^2
Peak plastic shear strain	γ_p^{vp} =	0.30	-
Softening factor for φ	$F_{soft,\varphi}$ =	2.0	-
Number of contact planes	n_{pl} =	64	-

and 79% of the excavation forces have been applied. For the right column of pictures, where 79% of the excavation forces have been applied, no converged solution could be obtained. Therefore, failure has been assumed and the finite element analysis has been terminated after a certain number of iteration steps arriving at the results presented. The pictures in Fig. 19 depict a part of the fine mesh consisting of 818 elements.

In the upper row of pictures the development of plastic shear strains is displayed. It has to be emphasised again that in the multilaminate model accumulated plastic shear strains vary over the contact planes in each integration point as a consequence of the stress history. In the diagrams always the critical contact plane containing the largest amount of plastic shear strains is chosen to represent an integration point. The contours in the diagrams are obtained by using the plastic shear strain increments that have been accumulated on the critical planes during the last load step (application of 1% of the nodal forces simulating excavation). Since plastic shear strain increments significantly increase when failure is approached the contour plots are presented in a logarithmic scale.

In the centre row of pictures in Fig. 19 the development of the mobilised friction angle is presented. Again, the values on the critical planes which primarily determine the material behaviour in softening are presented. Thus, the increase of the mobilised friction angle due to hardening in the pre-peak range and the reduction of the friction angle in post-peak softening can be displayed. In the lower row of pictures the volumetric strains are presented for the corresponding load stages.

With loading a considerable amount of plastic shearing takes place at first in a zone next to the tunnel sidewall. This leads to a mobilisation of the friction angle in this area. When 60% of the nodal forces associated with excavation have been applied the shearing resistance of the material is fully mobilised on the critical planes in the zone next to the tunnel sidewall and, consequently, dilatant material behaviour can be observed. With further loading shearing continues in the zone next to the tunnel sidewall and the critical contact planes switch into strain softening behaviour leading to a reduction of the friction angle on the respective planes.

At a load stage where 75% of the excavation forces have been applied, the strain softening zone starts to develop upwards indicating the initiation of shear band formation. Within the softening zone the friction angle is sharply reduced and the residual friction angle is reached in the ground next to the tunnel sidewall as indicated by the light grey shadings. At the same time dilatant material behaviour can be clearly observed in the strain softening zone.

At the load steps from 75% to 78%, plastic shear strains rapidly develop in a band reaching from the softening zone next to the tunnel sidewall vertically upwards to the ground surface. Accordingly, the friction angle is increased towards the peak friction angle in this band and the strain softening zone expands upwards, as well. Although the strain softening zone is relatively large now, the shear stresses can still be redistributed into the soil material away from the tunnel cross section and a converged solution is obtained from the numerical simulation.

When 79% of the excavation forces are applied, no convergence could be reached and the finite element analysis has been terminated after a certain number of iterations showing "snapshots" that indicate the formation of a shear band. In a relatively large zone of curved shape extending from the tunnel sidewall upwards to the ground surface large plastic shear strains have developed. The peak friction angle has been reached or exceeded in this zone. Next to the tunnel sidewall the friction angle has been reduced to its residual value as indicated by the light grey shadings, also leading to relatively large volumetric strains. The centre row of pictures showing results in a deformed mesh indicate the collapse of the tunnel. Significant surface settlements and large displacements of the tunnel sidewall directed towards the tunnel cross section are obtained.

The shape of the developing shear band should also be observed. Starting from the ground next to the tunnel sidewall it develops vertically upwards. However, when the ground surface is approached the band is curved slightly towards the tunnel axis.

6.3 Mesh dependence of results

To evaluate possible mesh dependencies of results three different finite element meshes consisting of 278 elements, 546 elements and 818 elements, respectively, were analysed. The geometry, the boundary conditions and the material parameters remained unchanged (see Fig. 18, Tab. 2). Again, in each

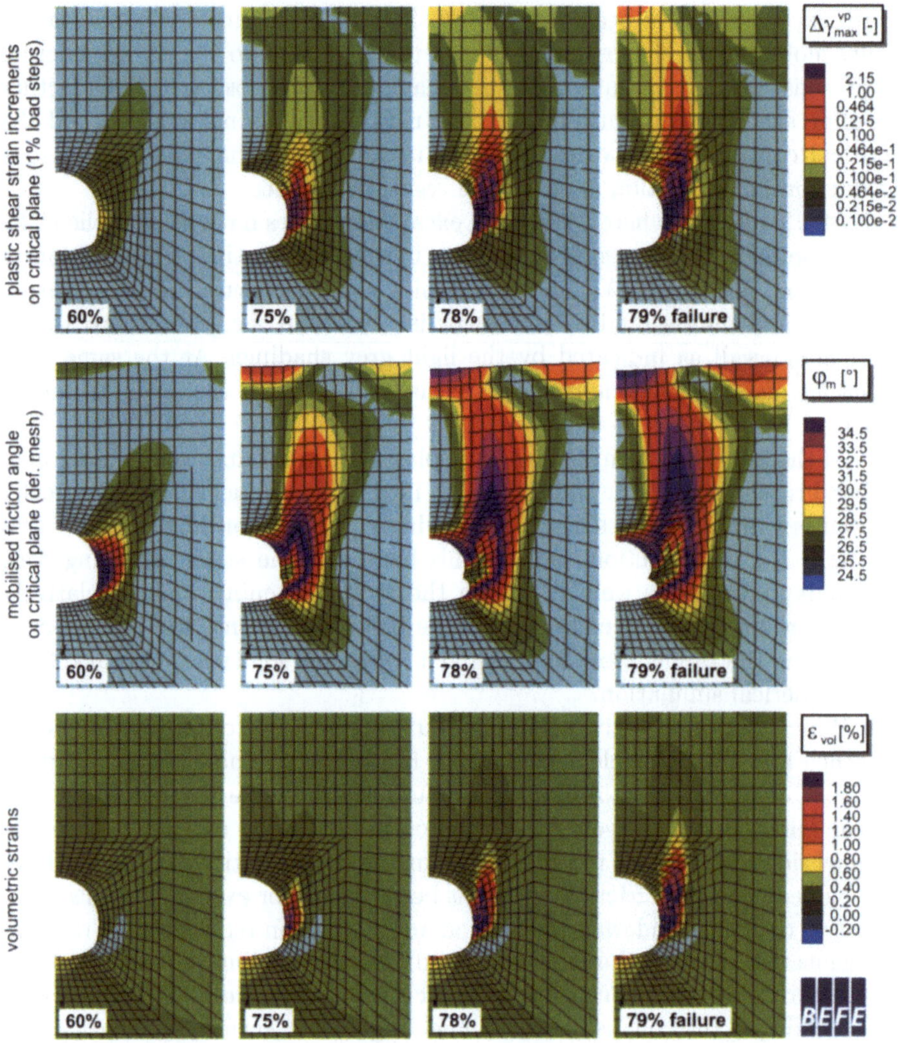

Fig. 19. Top row: development of plastic shear strain increments $\Delta\gamma_{max}^{vp}$ on critical plane
Centre row: mobilised friction angle φ_m on critical plane in deformed mesh
Bottom row: development of volumetric strains ε_{vol}

load step 1% of the nodal forces associated with excavation of the full tunnel face have been applied until no converged solution could be obtained and, consequently, failure has been assumed.

The development of tunnel crown settlements for all three meshes is presented in the diagram in Fig. 20. Crown settlements are plotted versus the percentage of the excavation forces that has been applied. In the diagram F_{exc} represents the total excavation forces and F_{appl} stands for the applied forces. It becomes clear that settlements significantly increase when approximately 70% of the excavation forces are exceeded, thus, indicating that failure is approached. Although the mesh fineness significantly differs, when comparing the three meshes nearly the same crown settlement curves are computed. No converged solution is obtained for the coarse mesh at 81%, for the medium mesh at 80% and for the fine mesh at 79% of the excavation forces. Thus, it can be assumed that the emerging progressive failure mechanism leading to the collapse of the tunnel is practically mesh independent in the presented finite element analyses. This has been shown also in more detail in Schuller and Schweiger ([41]).

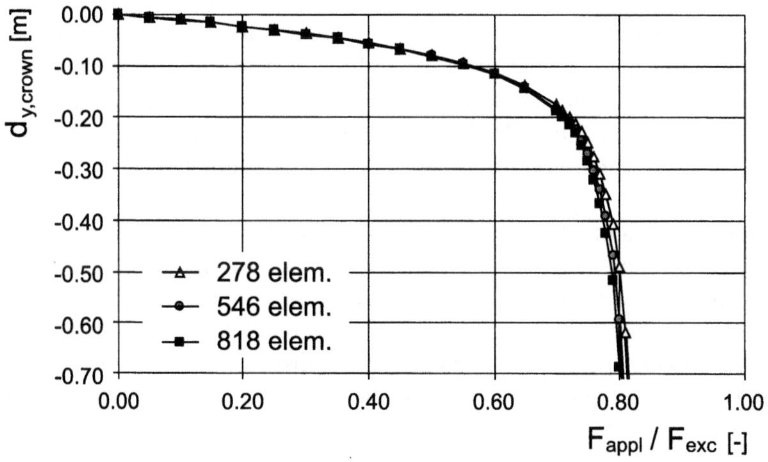

Fig. 20. Settlements of tunnel crown for different finite element meshes; evaluation of mesh dependence

6.4 Influence of initial stress state - K_0

The initial stress state in the soil prior to tunnel excavation depends on the soil material but is also a result of the loading history leading to different stress paths at various locations. The initial stress state is expressed by the coefficient of earth pressure at rest K_0 which is the ratio of effective horizontal stress $\sigma_h{}'$ to effective vertical stress $\sigma_v{}'$. As a consequence of the loading history of soils, a wide variation in K_0 values may occur under in-situ conditions. However, for normally consolidated soils the coefficient of earth pressure at rest can be assumed to be a function of the friction angle. According to an empirical criterion by Jaky it can be expressed as

$$K_0 \approx 1 - \sin \varphi. \tag{12}$$

All analyses presented below have been performed using the material parameters presented in Tab. 2 for purely frictional material. In this comparative study K_0 values of 0.35, 0.50, 0.80 and 1.20, respectively, have been taken. Assuming normally consolidated soil Jaky's criterion results for this material in $K_0 = 0.43$. Thus, the K_0 values of 0.35 and 0.50 are roughly within the range of normal consolidation for the employed material parameters. The higher values chosen in this study should demonstrate the effect of the horizontal in situ stress. It is acknowledged however that corresponding material parameters would be slightly different when considering an overconsolidated soil. This has not been done in order to isolate the influence of the K_0 value, and therefore only a value moderately higher than 1.0 has been considered.

The development of a failure mechanism for various K_0 values is depicted in Fig. 21. For each initial stress state contour lines of plastic shear strain increments for load stages prior to failure and at failure are shown. Only the plastic shear strains computed on the critical plane due to application of a load increment are included and plotted in logarithmic scale. In each load increment 5% of the nodal forces associated with tunnel excavation have been applied.

A low coefficient of earth pressure at rest of $K_0 = 0.35$ results in a relatively large increase of plastic shear strains already at a load stage of 60%. The strains are concentrated in a narrow zone ranging from the tunnel sidewall steeply inclined towards the ground surface. When 70% of the excavation forces have been applied large shear strains are computed next to the tunnel sidewall and in a zone ranging vertically upwards to the ground surface. At the load stage of 75% where no converged solution can be obtained a shear band forms in this zone.

For the higher K_0 value of 0.50 the development of significant plastic shear strain increments is restricted to a small zone close to the tunnel sidewall at the load stage of 65%. With further loading the shearing zone grows and finally, when 80% of the excavation forces have been applied, no convergence can be obtained. Compared to $K_0 = 0.35$ the plastic shear strain increments computed in the load steps prior to failure are distributed in a wider zone next to the tunnel sidewall. When failure is approached, the shear band forms in nearly the same location for both K_0 values.

A different failure mechanism is obtained for the large K_0 values of 0.80 and 1.20. Here, plastic shear strains develop in relatively large zones in the vicinity of the tunnel excavation without showing a concentration into distinct shear bands. In addition, the zone where the largest incremental shear strains are obtained at failure moves from the tunnel sidewall into the direction of the tunnel crown with increasing K_0 value.

6.5 Cohesive softening

When adopting the constitutive formulation proposed here strain softening can be modelled not only as a reduction of the friction angle but also by reducing the cohesion to a residual value or by doing both. The softening rate can be controlled for friction angle and cohesion individually by the softening factors $F_{soft,\varphi}$ and $F_{soft,c}$.

To show the influence of the softening type on the development of shear zones and the failure mechanism a cohesive frictional soil material has been modelled. The material parameters differing from the purely frictional material (see Table 2) are included in Table 3. For this material a peak friction angle of 20° and a cohesion of 40 kN/m^2 is assumed. The pre-peak range is characterised by strain hardening, however, when the peak friction angle is reached on certain contact planes the friction angle remains unchanged with further shearing. Softening is modelled by a reduction of the cohesion to a residual value of 5 kN/m^2. To be able to compare the results of the numerical analyses when using the cohesive frictional material to the purely frictional material, again the initial stress state prior to tunnel excavation is assumed corresponding to $K_0 = 0.50$. It is pointed out again that no particular soil is modelled in this study. It is presented in order to show the difference of behaviour when cohesive or frictional softening is dominating, respectively, although it is acknowledged that it may be very difficult to separate these effects experimentally. However it is argued that these numerical studies are still valuable in getting a better insight into the mechanical behaviour of the ground during tunnel excavation.

Table 3. Material parameters for cohesive frictional material – cohesive softening

Peak friction angle	φ =	20	°
Residual friction angle	φ_{res} =	20	°
Peak dilation angle	ψ =	0	°
Cohesion	c =	40	kN/m^2
Residual cohesion	c_{res} =	5	kN/m^2
Softening factor for c	$F_{soft,c}$ =	3.0	-

In Fig. 22, results of the numerical analysis using cohesive softening are displayed by showing contours of the cohesion in a deformed mesh. The peak shearing resistance of the material is reached first in the ground next to

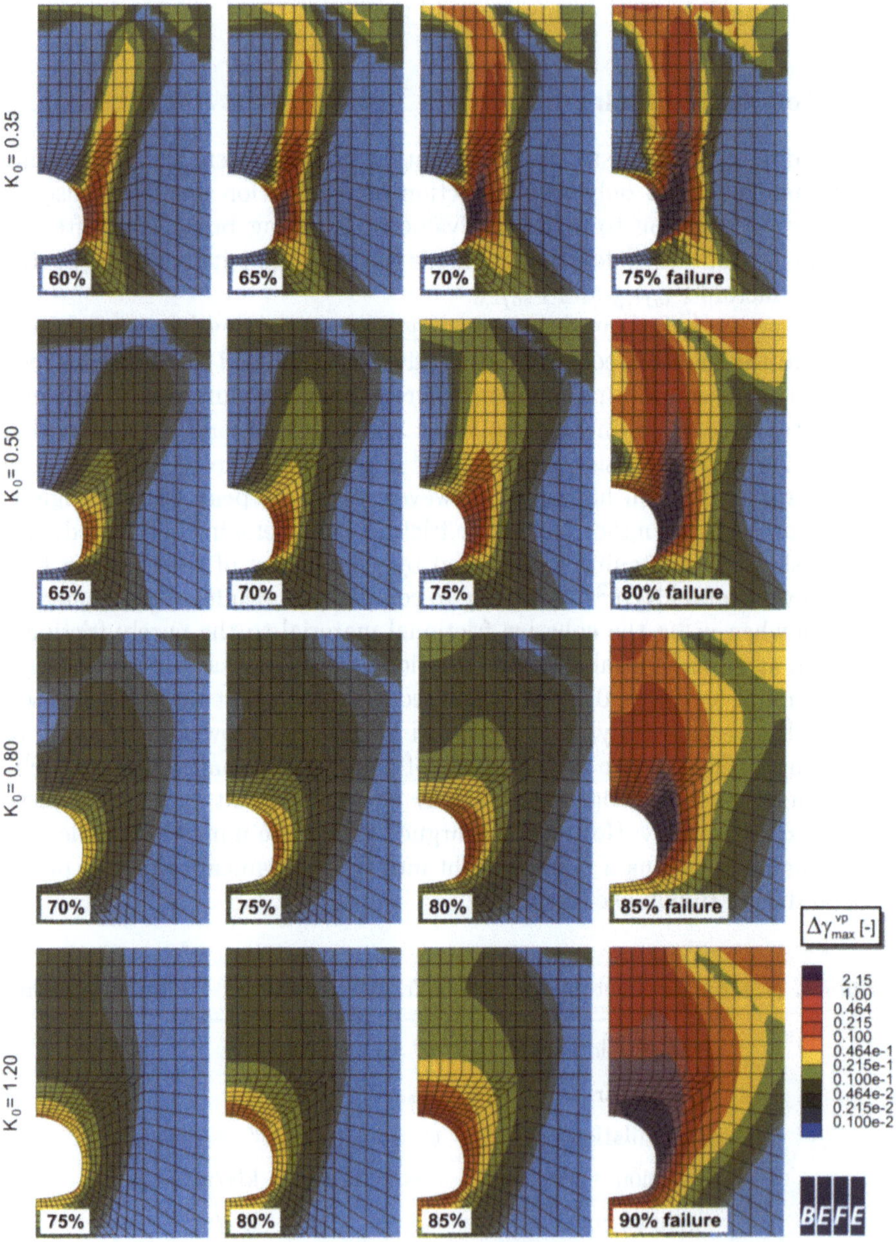

Fig. 21. Plastic shear strain increments due to 5% steps on critical planes for $K_0 = 0.35$, $K_0 = 0.50$, $K_0 = 0.80$ and $K_0 = 1.20$

the tunnel sidewall and strain softening can be detected by a reduction of cohesion on the critical planes of the respective integration points. Differing from the results of the purely frictional material here the strain softening zone grows further into the ground away from the tunnel at an angle of approximately 45° before a shear band develops. At a load stage of 67% no converged solution can be found and failure is assumed. The softening zone has developed into a curved band which starts at the tunnel sidewall at an angle of about 45° and slightly changes to vertical direction before growing towards the ground surface.

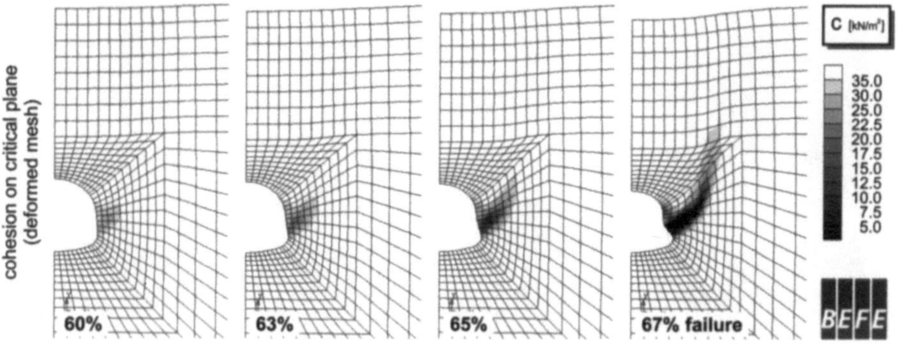

Fig. 22. Cohesive frictional material – reduction of cohesion during softening on critical planes

The development of the strain softening zone in both, the purely frictional material and the cohesive frictional material with cohesive softening, is depicted in Fig. 23. In these figures only a part of the mesh close to the tunnel is presented. Each line represents the critical plane of an integration point. The inclination of the line indicates the orientation of the plane and the length of the line stands for the magnitude of the plastic shear strain increment which has been accumulated on the respective plane during a load step of 1%.

In both materials, softening starts in the ground next to the tunnel sidewall as indicated by larger shear strain increments in this area at load stages of 70% (frictional material) and 60% (cohesive material). For the purely frictional material the critical planes are oriented at a higher angle to the horizontal direction. With further loading nearly all of the critical planes in the frictional material (75%) are orientated upwards at steep angles while in the cohesive frictional material (63%) two predominant directions of critical planes can be observed forming a shearing triangle that reaches horizontally into the ground next to the tunnel sidewall. When failure is approached in the frictional material (78%, 79%) large shear strains develop in a zone nearly vertically upwards from the tunnel sidewall at a distance from the tunnel axis of approximately 1.5 times the tunnel radius. In the cohesive frictional material (66%, 67%) a curved shear band develops out of the "shearing triangle" which turns gradually towards the ground surface. The orientations of the respective critical planes follow the shear band which finally grows vertically upwards at a distance from the tunnel axis of approximately 2 times the tunnel radius.

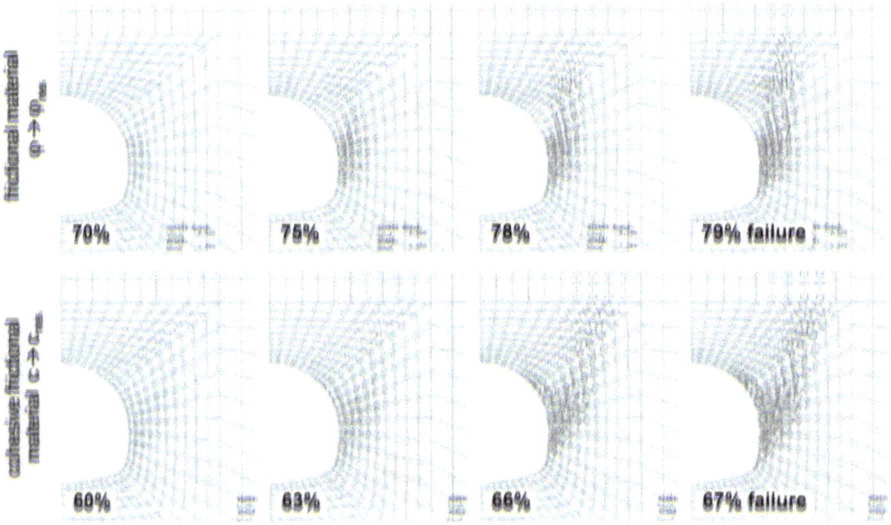

Fig. 23. Frictional softening – cohesive softening: Orientation and magnitude of plastic shear strain increments on critical planes due to 1% load steps

6.6 Tunnel – staged excavation sequence

To provide sufficient stability during construction, the tunnel cross section is frequently divided and excavated in stages. A typical construction sequence of NATM tunnels is to excavate the top heading in advance followed by bench and invert excavation. The development of a failure mechanism that involves shear band formation during excavation of the bench is numerically analysed for the tunnel cross section presented in Fig. 18. The material parameter set for the frictional material displayed in Tab. 1 has been used which is characterised by a peak friction angle of 35°, a residual friction angle of 26.4° and a peak dilation angle of 10°. The initial stress state was determined using a coefficient of earth pressure at rest of $K_0 = 0.43$ ($\sim 1 - \sin \varphi$).

Following common 2-D modelling technique, excavation of the tunnel top heading has been simulated in two computational steps. In the first step, 50%

of the nodal forces associated with excavation of the top heading have been applied. This should account for the deformations resulting from the three-dimensional stress redistributions at the tunnel face. In the second step, the "young shotcrete" lining has been put in place and, at the same time, the remaining 50% of the excavation forces have been applied. The shotcrete lining with a thickness of 35 cm was modelled using eight-noded solid elements. The complex material behaviour of shotcrete was simulated in a simplified way by using a linear elastic material and distinguishing only between "young shotcrete" (Young's modulus $E_1 = 10\,000$ MN/m^2, Poisson's ratio $\nu = 0.18$) and "hardened shotcrete" ($E_2 = 30\,000$ MN/m^2, $\nu = 0.18$). The "young shotcrete" in the lining of the top heading was replaced by "hardened shotcrete" at the begin of the excavation of the bench.

In Fig. 24, the mobilised friction angle φ_m and the development of plastic shear strain increments on the critical planes are presented for excavation of the top heading. Considerable mobilisation takes place in a zone above the tunnel sidewall and the shearing resistance of the material is already fully mobilised next to the footing of the shotcrete lining.

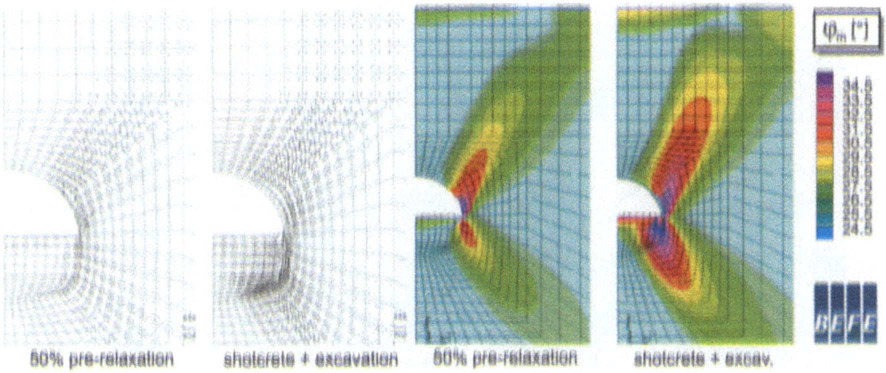

Fig. 24. Excavation of top heading; plastic shear strain increments on critical planes (plots on left side) and mobilisation of friction angle φ_m (plots on right side)

To study the development of a shear zone leading to a failure mechanism during bench excavation this construction stage was simulated in the same way as for the full tunnel face presented in the previous section. The nodal forces associated with excavation of the bench were stepwise applied in load steps of 5% without installing the shotcrete lining. At a certain stage shear zones with strain softening behaviour formed in the ground next to the footing of the shotcrete lining.

The development of plastic shear strain increments is demonstrated in Fig. 25. Contour lines are displayed for the largest strain increments that have been computed on the integration planes due to a load step of 5% of the nodal forces associated with excavation. When 55% of the nodal forces have been applied the strain softening zone in the soil above the tunnel sidewall increasingly grows towards the surface. However, it is still possible to redistribute stresses and a converged solution can be found. After the next load stage arriving at 60% of the excavation forces, no convergence can be obtained and a progressive failure mechanism forms. The calculation has been stopped after a certain number of iterations to obtain the contour plot displayed in Fig. 25. The formation of a vertical shear band starting from the ground close to the tunnel sidewall and advancing vertically upwards can be observed. When the ground surface is approached the shear band is slightly curved towards the tunnel axis.

It should be mentioned at this point that this construction stage is in fact a very crucial one from a practical point of view. A reasonably accurate assessment of deformations at this excavation stage is of considerable help in optimising the excavation procedure. Of course, 3D effects will also be playing a role and thus a 3D analysis with the proposed formulation would further improve the numerical model.

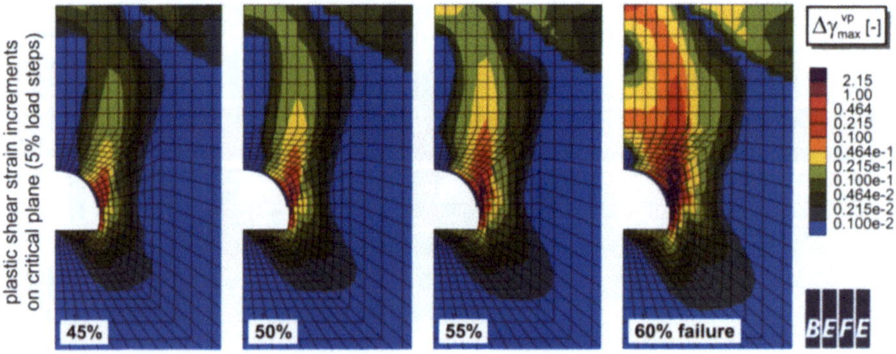

Fig. 25. Bench excavation; development of plastic shear strain increments on critical planes

7 Conclusion

A multilaminate model for soils has been presented. It could be demonstrated that the effect of principal stress rotation on the mechanical behaviour, which is well known from the literature, can be accounted for employing the multilaminate framework. Development of plastic shear strains on contact planes has been shown for various stress paths including rotation of principal stress axes. An enhanced strain softening formulation is introduced enabling the model to capture the development of shear bands and thus the post peak behaviour. A scaling of the softening rate, tailored especially for the multilaminate framework, guarantees solutions which are independent of mesh configuration, load increments and number of contact planes, at least to an accuracy relevant in practice.

The model has been applied to solve the problem of a shallow tunnel excavation constructed according to the principles of the New Austrian Tunnelling Method (NATM). It has been demonstrated that a realistic failure mechanism that involves the formation of shear bands is obtained for tunnel excavations in soil with low overburden.

It is acknowledged that with the proposed formulation the thickness of the shear band is not predicted realistically. However, this is not an important issue when solving practical boundary value problems as addressed here.

The response of the model to various initial stress states and when employing frictional or cohesive softening has been demonstrated by numerical simulation of a full face tunnel excavation. Finally, a possible failure mechanism due to the formation of a shear band has been numerically simulated for a tunnel constructed in a staged excavation sequence.

Acknowledgements

This research project has been supported by the Austrian Science Fund FWF (Project P12577-TEC) and was associated with the Joint Research Initiative "Numerical Simulation in Tunnelling" SITU.

References

[1] Belytschko T., Fish J. and Englemann B.E. (1988) A finite element with embedded localization zones. *Comp. Meth. Appl. Mech. Eng.*, Vol. **70**: 59-89

[2] Leroy Y., Ortiz M. (1989) Finite element analysis of strain localization in frictional materials. *Int. J. Num. Analyt. Meth. Geomech.*, Vol. **13**: 53-74

[3] Pietruszczak S., Mróz Z. (1981) Finite element analysis of deformation of strain-softening materials. *Int. J. Num. Meth. Eng.*, Vol. **17**: 327-334

[4] Brinkgreve R.B.J., Vermeer P.A. (1995) A new approach to softening plasticity. In: Pande and Pietruszczak (eds.) Proc. 5th Int. Symp. *Numerical Models in Geomechanics - NUMOG V*, Davos: 193-202

[5] Marcher T., Vermeer P.A. (2000) Macromodelling of softening in non-cohesive soils. *Proc. Int. Symp. Continuous and Discontinuous Modelling of Cohesive Frictional Materials - CDM 2000*: 89-108, Springer

[6] Vardoulakis I., Aifantis E.C. (1989) Gradient dependent dilatancy and its implications in shear banding and liquefaction. *Ingenieur Archiv*, **59**: 197-208

[7] Mühlhaus H.-B., Vardoulakis I. (1987) The thickness of shear bands in granular materials. *Géotechnique*, Vol. **37**, No. 3: 271-283

[8] de Borst R. (1991) Simulation of strain localization: A re-appraisal of the Cosserat continuum. *Eng. Comp.*, Vol. **8**: 317-332
[9] Tejchman J., Bauer E. (1996) Numerical simulation of shear band formation with a polar hypoplastic constitutive model. *Computers and Geotechnics*, Vol. **19**, No. 3: 221-224
[10] Oka F., Yashima A., Kohara I., Adachi T. (1994) Instability of a viscoplastic model for clay and numerical study of strain localisation. *Chambon, Desrues and Vardoulakis (eds) 3rd Int. Workshop Localisation and Bifurcation Theory for Soils and Rocks*, Grenoble: 237-247
[11] Zienkiewicz O.C., Zhu J.Z. (1987) A simple error estimator and adaptive procedure for practical engineering analysis. *Int. J. Num. Meth. Eng.*, Vol. **24**: 337-357
[12] Hicks M.A. (1997) An adaptive mesh study of localisation in a saturated soil. *Yuan (ed.) Proc. 9th Int. Conf. Computer Methods and Advances in Geomechanics*, Wuhan: 1853-1858
[13] Cividini A., Gioda G. (1992) A finite element analysis of direct shear tests on stiff clay. *Int. J. Num. Anal. Meth. Geomech.*, **16**: 869-886
[14] Sterpi D., Sakurai S. (1997) Numerical analysis of laboratory tests on a model tunnel. *Asaoka, Adachi and Oka (eds.) Proc. Deformation and Progressive Failure in Geomechanics*, IS-Nagoya '97: 211-216
[15] Sterpi D. (1999) An analysis of geotechnical problems involving strain softening effects. *Int. J. Num. Analyt. Meth. Geomech.*, Vol. **23**: 1427-1454
[16] Sterpi D., Cividini A. (1997) Numerical analysis of tunnels in strain softening soil. *Yuan (ed.) Proc. 9th Int. Conf. Computer Methods and Advances in Geomechanic*, Wuhan: 1389-1394
[17] Bazant Z.P., Oh B.H. (1985) Microplane model for progressive fracture of concrete and rock. *Journal of Engineering Mechanics*, Vol. **111**, No. 4: 559-582
[18] Beer G. (2001) BEFE user's and reference manual. CSS, Graz, Austria.
[19] Zienkiewicz O.C., Pande G.N. (1977) Time dependent multi-laminate model of rocks – a numerical study of deformation and failure of rock masses. *Int. J. Num. Anal. Meth. Geomech.*, **1**: 219-247
[20] Pande G.N., Sharma K.G. (1983) Multi-laminate model of clays – a numerical evaluation of the influence of rotation of the principal stress axes. *Int. J. Num. Anal. Meth. Geomech.*, **7**: 397-418
[21] Krajewski W. (1986) Mathematisch-numerische und experimentelle Untersuchungen zur Bestimmung der Tragfähigkeit von in Sand gegründeten, vertikal belasteten Pfählen. *Veröffentlichungen des Instituts für Grundbau, Bodenmechanik, Felsmechanik und Verkehrswasserbau der RWTH Aachen*, Heft 13.
[22] Sadrnejad S.A., Pande G.N. (1989) A multilaminate model for sands. *S. Pietruszczak and G.N. Pande (ed.) Proc. 3rd Int. Symp. on Numerical Models in Geomechanics*: 17-27, Elsevier
[23] Pietruszczak S., Pande G.N. (1987) Multilaminate framework of soil models - plasticity formulation. *Num. Analyt. Meth. Geomech.*, **11**: 651-658
[24] Karstunen M. (1999) Numerical modelling of strain localization in dense sands. *Acta Polytechnica Scandinavica* No. **113**, Espoo: The Finnish Academy of Technology
[25] Bazant Z.P., Prat P.C. (1988) Microplane model for brittle-plastic material: I. theory. *Journal of Engineering Mechanics*, Vol.**114**, No.10: 1672-1688

[26] Pande G.N., Yamada M. (1994) The multilaminate framework of models for rock and soil masses. *Proc. 1st Int. Workshop on Applications of Computational Mechanics in Geotechnical Engineering*, Rio de Janeiro: 105-123, Balkema

[27] Vermeer P.A. (1978) A double hardening model for sands. *Géotechnique*, **28**: 413-433

[28] de Borst R. (1986) Non-linear analysis of frictional materials. Delft University of Technology, Netherlands. Dissertation

[29] Rowe P.W. (1972) Theoretical meaning and observed values of deformation parameters for soil. *R.H.G. Parry (ed.) Stress strain behaviour of soils, Proc. Roscoe Mem. symp.*: 143-194. Henley-on-Thames: Foulis

[30] Spener U. (2000) Parameterstudien zur Anwendung des Multilaminate Models. Institute for Soil Mechanics and Foundation Engineering, Graz University of Technology. Diploma thesis

[31] Arthur R.F., Chua K.S., Dunstan T., Rodriguez del C. J.I. (1980) Principal stress rotation: a missing parameter. *J. Geotech. Engineering Div. Proc. ASCE*, **106**, No.GT 4: 419-433

[32] Ishihara K., Towhata I. (1983) Sand response to cyclic rotation of principal stress directions as induced by wave loads. *Soils and Foundations*, **23**, No.4: 11-26

[33] Joer H.A., Lanier J., Fahey M. (1998) Deformation of granular materials due to rotation of principal axes. *Géotechnique*, **48**: 605-619

[34] Schweiger H.F., Schuller H. (2000) New developments and practical applications of the Multilaminate Model for soils. *Smith and Carter (eds.) Proc. Developments in Theoretical Geomechanics, The John Booker Memorial Symposium Sydney*: 329-350. Balkema

[35] Wiltafsky C., Messerklinger S., Schweiger H.F. (2002) An advanced multilaminate model for clay. In: Pande and Pietruszczak (eds.) *Proc. 8th Intern. Symp. Numerical Models in Geomechanics*, Rom, Balkema: 67-73

[36] Akagi H., Saitoh J. (1994) Dilatancy characteristics of clayey soil under principal axes rotation. *Shibuya, Mitachi and Miura (Eds.), Proc. Int. Symp. Prefailure Deformation Characteristics of Geomaterials*, Vol. 1: 311-314, Sapporo, Balkema

[37] Neher H.P., Cudny M., Wiltafsky C., Schweiger H.F (2002) Modelling principal stress rotation effects with multilaminate type constitutive models for clay. *Pande and Pietruszczak (eds.) Proc. 8th Intern. Symp. Numerical Models in Geomechanics*, Rom: 41-47, Balkema

[38] Desrues J., Hammad W. (1989) Shear banding dependency on mean stress level in sand. *Dembicki, Gudehus and Sikora (ed.) Proc. 2nd Int. Workshop on Numerical Methods for Localization and Bifurcation of Granular Bodies*, Gdansk – Karlsruhe: 57-67

[39] Vavrovsky G.-M. (1994) Gebirgsdruckentwicklung, Hohlraumverformung und Ausbaudimensionierung. *Felsbau*, **12**, No. 5: 312-329

[40] Vavrovsky G.-M., Schubert P. (1995) Advanced analysis of monitored displacements opens a new field to continuously understand and control the geotechnical behaviour of tunnels. *Proc. 8th Congr. Rock Mechanics*, Tokyo: 1415-1419

[41] Schuller H., Schweiger H.F. (2002) Application of a Multilaminate Model to simulation of shear band formation in NATM tunnelling. *Computers and Geotechnics*. Accepted for publication.

Chapter 14
Advanced post-processing methods

Gernot Opriessnig, Gernot Beer

Institute for Structural Analysis, Graz University of Technology

Abstract. This chapter describes a visualisation system, which is specially designed for tunnelling. New visualisation-techniques for the improved perception of results of numerical simulations and measurements are presented. A requirement is that the characteristic properties of the data must not be changed by the display-methods used. In this chapter the implementation of a real-time system as well as the theoretical bases for visualisation is beeing described. Parameter studies are also presented to ascertain the display quality. The aim is to develop a system which is intuitive to use and that can display the results of a finite- or a boundary element calculation as well as data from measurements. To be applicable on a tunnel site, the system has to display the data in a way, that makes it easy to compare it to the measured data, which are available. The Tunnelling Visualisation System (TVS), which has been developed in the visualisation-project of the Austrian Joint Research Initiative "Numerical Simulation in Tunnelling" enables the user to perform a virtual walk through a tunnel. Navigation can be done using the mouse or a spaceball, which is a three-dimensional input device. With the help of this easy navigation is possible. Special hardware makes it possible to see a three-dimensional picture.

1 Introduction

The increasing computer performance and advances in simulation techniques make it possible to produce a large number of simulation results. More and more large scale 3D calculations can be done in an acceptable calculation time. However, large 3D-simulations place high demands on the visualisation software. This chapter describes novel visualisation techniques developed for the requirements of tunnelling and geomechanics. One of these methods is to display 3D-scalar-fields in a way, that enables the user to see important information clearly. Fuzzy geological structures and shear-bands or areas of plasticity can be pictured very well with this method. For the visualisation it is not important that the values are known on regular distributed points as it is the case in FEM (Finite Elements Method) or BEM (Boundary Elements Method) simulations. Therefore results from measurements can also be visualised. Real-time-rendering and user-friendliness increase the benefit of such techniques. Simulation or measurement results are seen immediately and changing display options is possible in real-time.

The use of numerical simulation on site to assist the tunnel engineer in often very difficult decisions is not very wide-spread. However, to be able to 'sell' this kind of technology to people working in tunnel construction one has to have very effective and easy to use software tools. Visualisation of data plays an important role here. There is a large market of visualisation software in the technical sciences. A large majority of this software however is concentrated on areas of mechanical engineering such as the aircraft and the automobile industry. Visualisation in the area of rock and soil mechanics has mostly been performed using software which has been adapted from other areas. However visualisation in tunnelling has special requirements with respect to the display of data and the ease of use of the system. The Tunnelling Visualisation Software (TVS) was developed, to fulfil these requirements.The visualisation system is currently available on a workstation with stereo option, on a PC running Windows NT using head mounted display and has undergone preliminary tests in the CAVE environment at Chalmers University in Gothenborg, Sweden.

Today a notebook, that can be seen almost as standard and has a performance, which seemed to be unreachable even in the beginning 90es. When the author started the research work in the field of scientific visualisation 1997, the hardware, that had to be used was a SiliconGraphics HighImpact workstation. One year later an SGI Octane MXI became affordable and was bought. At that time, this was the best and most expensive desktop computer hardware available. The most important feature, and the major reason for purchasing this was the graphics performance, OpenGL-compatibility [1], the support of hardware texturing and the ability of quad-buffering, which is necessary for smooth OpenGL-Stereo animations in real-time.

Now this features, and a much more, are available by a Dell Inspiron 8100 notebook, which is now the actual development environment for the Tunnelling Visualisation System – TVS (Fig. 1). The major advantage of this hardware is – without considering the price (less than 10% compared to the SGI-hardware) – the mobility. As mentioned above, a good visualisation is absolutely necessary to sell the novel simulation techniques to tunnelling engineers. Tunnelling has a very long history and a lot of people working in this area are very conservative, so it is not easy to receive acceptance, when novel methods are presented. Compared to a SGI-Workstation, which is very hard to transport and very difficult to handle without being damaged, a handy hardware, like the used notebook is an excellent argument for the usability to the described Tunnelling Visualisation System.

2 The Tunnelling Visualisation System - TVS

TVS allows the user to perform a virtual walk through a tunnel which exists in computer memory only.

During the virtual walk through the model the user may observe different results of numerical simulations and geological features. With the use of special hardware the model becomes 3D. Tunnelling engineers usually are

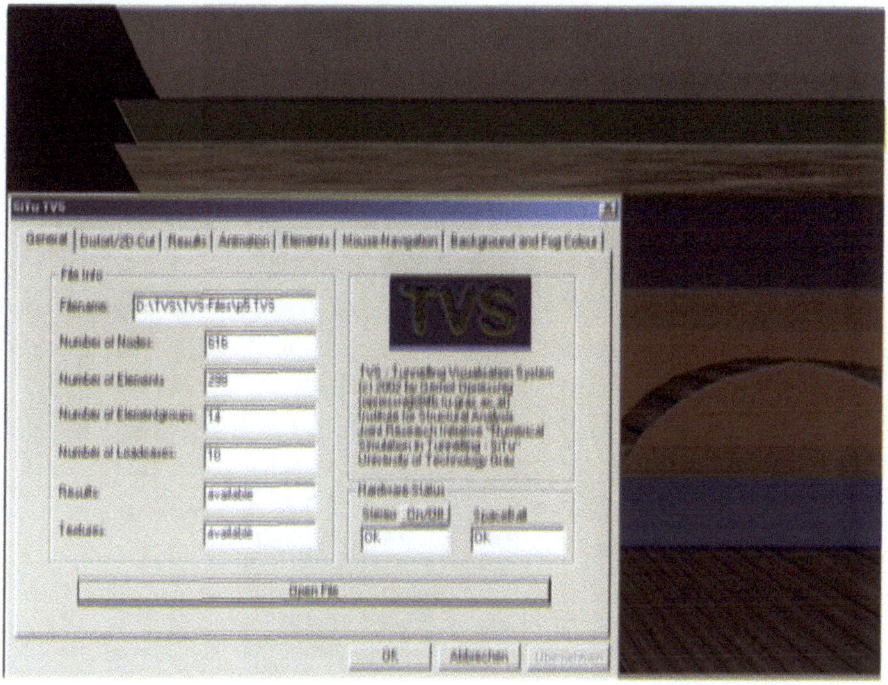

Fig. 1. Graphical User Interface of the TVS (Windows-Version)

not very familiar with the use of complicated computer-software, and they should not be required to study huge user-manuals, to be able to walk through the virtual tunnel The tunnel can be displayed with photo-realistic textures (Fig. 2). This textures are derived from pictures taken by a stereoscopic camera. Another way to display geological structures and groundwater-level is to paint surfaces in colours. The user is able to select, which part of the model should be visible, and in which way it should be displayed. Making the model translucent gives good results, if it is necessary to see more layers of the model.

During the virtual walk-through the position and the viewing-direction are displayed in the horizontal and vertical section of the model. To make the navigation easy a SpaceBall (Fig. 4) is used. Moving the ball of this treedimensional input device will change the position, turning it changes the viewing-direction.

A simple, linear transformation SpaceBall–movement to viewer – movement is not very efficient. The user should be able to determine his position very precisely, and, without moving his hand from the SpaceBall, move very fast. For this purpose a progressive scaling is used (Fig. 5). To improve the user-friendliness of the navigation, the user can decide, that only the

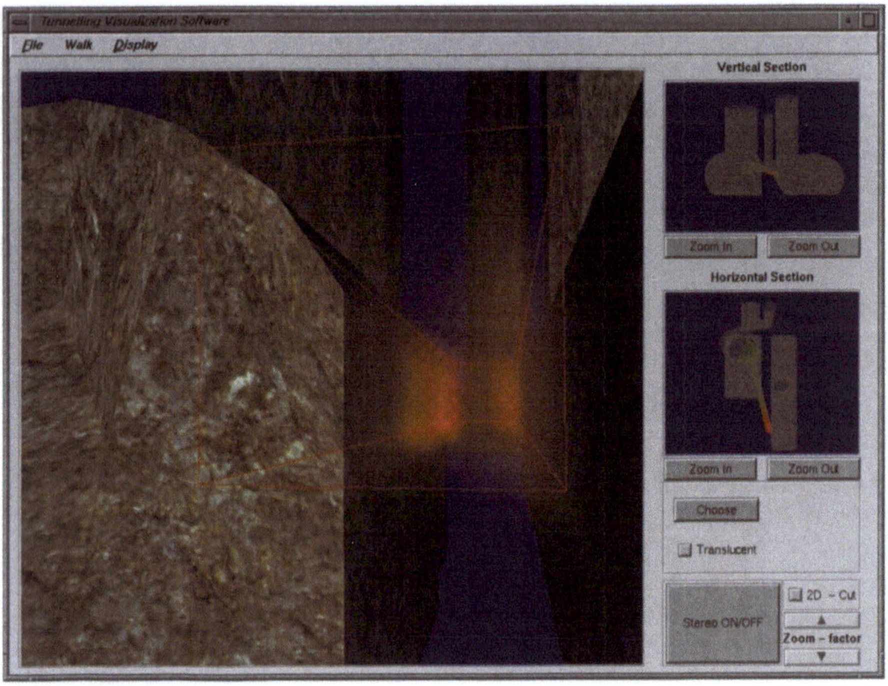

Fig. 2. Graphical User Interface of the TVS (SGI-Version)

dominant two directions of SpaceBall movement will be processed. It is also possible, to choose these directions manually [2].

2.1 Stereo viewing

Stereo viewing means that the user is able to see a threedimensional picture, like in reality. Many computer games use perspective distortion to get a 3D-effect. This method is unusable for scientific purposes. For real stereo-viewing it is necessary, to generate separate images for each eye. These two images show the model from different positions, like the pictures seen by the eyes in real life. With the help of these two pictures it is possible to see spatial. To get this effect on a computer a special hardware is necessary to ensure, that one eye can only see the picture generated for it. The Tunnelling Visualisation System supports three different technologies.

Electronical shutterglasses (Fig. 6) are one of the cheapest possibilities for stereo-viewing on a desktop-computer. Using this system the two pictures are displayed periodically on one screen. The infrared emitter sends a signal to the glasses. This signal contains the information, if the current picture is the right or the left one. The shutterglasses darken the view of that eye, that must not see the current image. To get a smooth animation, the change between

Fig. 3. Virtual Walk through the virtual Semmering Tunnel (Austria) with photo realistic textures, using a head mounted display (HMD)

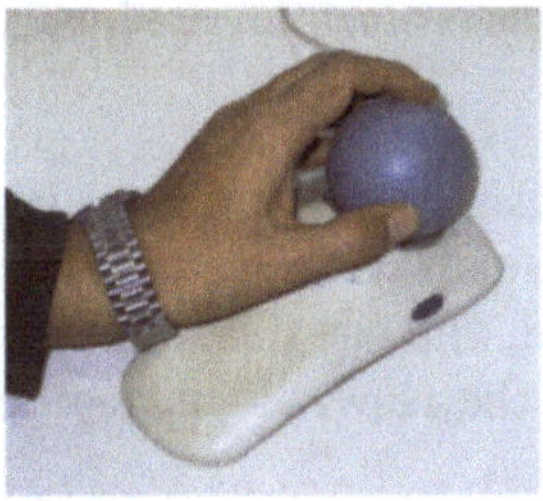

Fig. 4. Spaceball

left and right needs to be very fast. A frequency of 120 Hz is a good value, but generating 120 high quality frames in one second needs a good graphics hardware.

This is not necessary, if a headmounted display (HMD) is used. This device (Fig. 7) needs no extra computer monitor. In front of each eye is a small screen, showing the picture for "its" eye. There is no change between left and right necessary, so the frame-rate can be smaller than using shutterglasses. 30 Hz per eye is shown to be a good value, so the system has only to generate 60 pictures per second. The use of a HMD allows another possibility. With the help of a tracking system, the user can control the virtual viewing-direction by

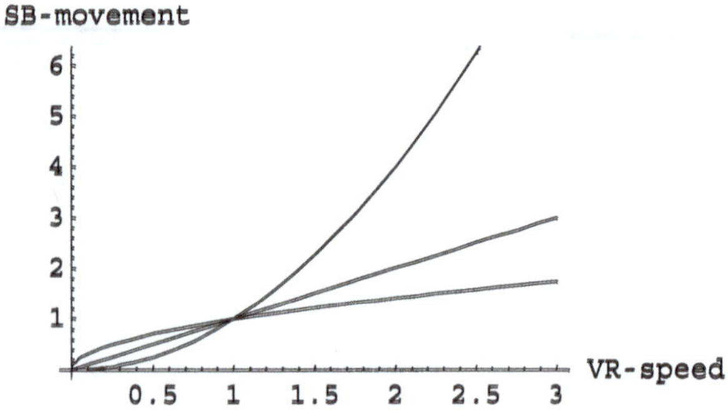

Fig. 5. SpaceBall (Progressive, linear and digressive SpaceBall-transformation)

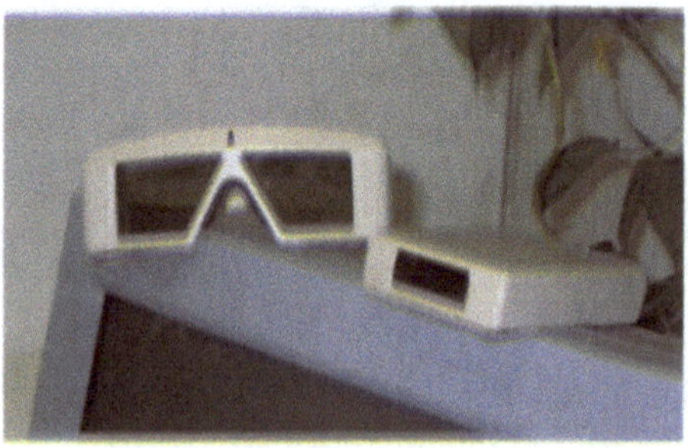

Fig. 6. Electronical shutterglasses incl. Infrared emitter

turning the head. The tracker localises the position of the head and submits this information to the visualisation system. So it is possible to detect the movement of the users head, and use this information for intuitive navigation. Turning the head changes the view, like in real life.

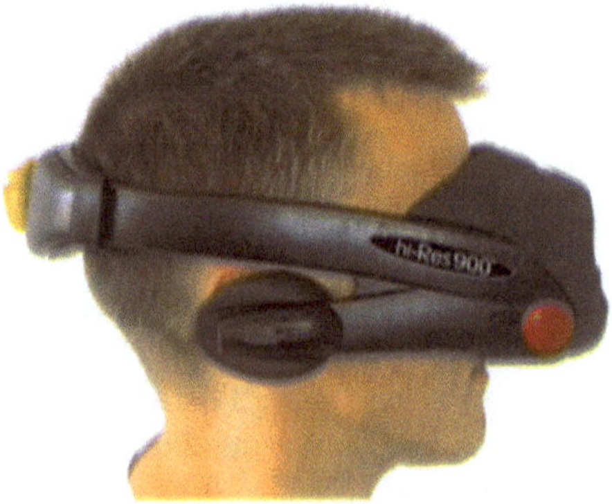

Fig. 7. Headmounted display

The third, most exclusive, possibility for real stereo is a CAVE (Fig. 8). The user is situated in a box with approximately $2*2*2$ m . Up to six sides of this box show different views of the model. For the stereo-effect shutterglasses are used. With the help of this environment it is possible to be in the virtual reality.

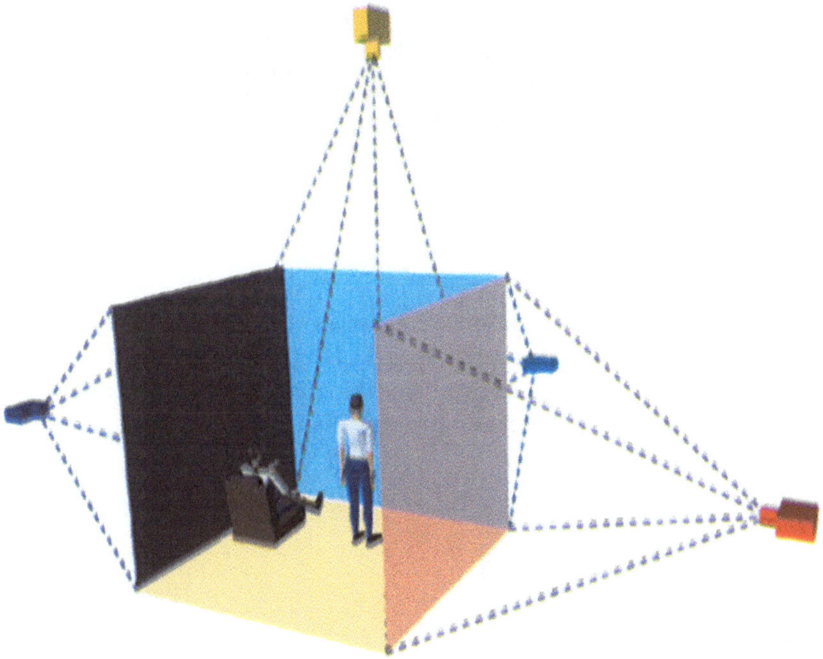

Fig. 8. Scematic display of a CAVE

2.2 TVS-features

In many cases it may be necessary to display results on 2D-cuts through the model. This is possible with TVS. The user can move a cutting-plane with the Spaceball and bring it in any position (Fig. 9). The plane is displayed in vertical and horizontal section

Tunnels usually are very long compared with their other dimensions. To display such models in a effective way it is possible to distort the geometry by defining the scaling-factor for each coordinate. While the tunnel is built, many physical and geometrical properties change. e.g. the amount of stress in the rock changes, new zones of damage are created, the level of groundwater moves and therefore the geometry of the tunnel is also a function of the

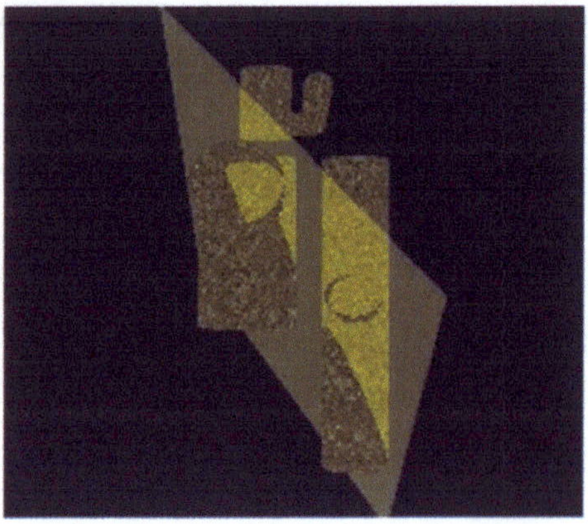

Fig. 9. Cutting plane

time. The Tunnelling Visualisation Software can display this as animation (Fig. 10). The geometrical change usually is very small, compared to the size of the tunnel, so the changes can be amplified with a user defined factor. The user is able to control many properties of the animation, like speed and direction.

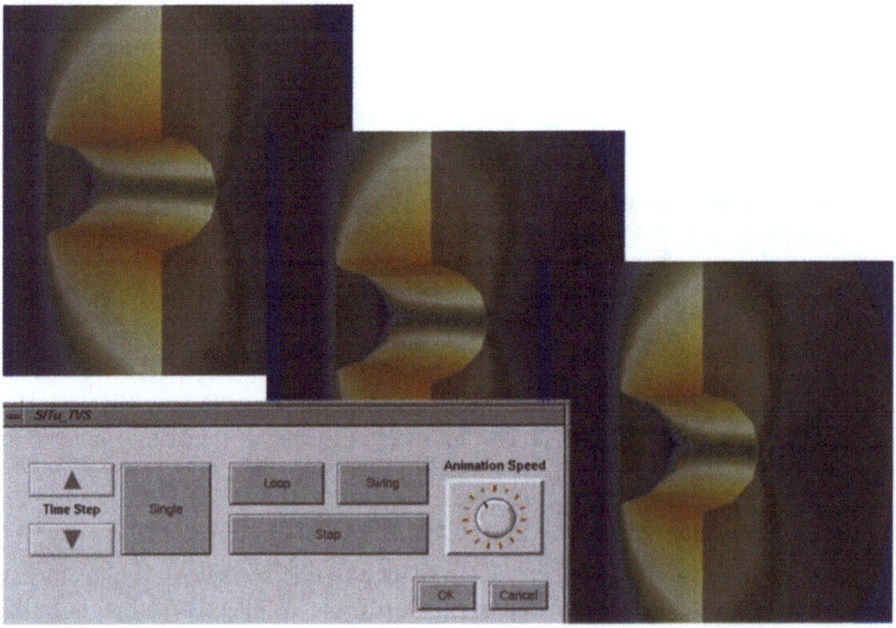

Fig. 10. TVS-Dialog box for animation adjustments and three animation steps

In a 3D model of the tunnel many properties are a function of three dimensions and should be pictured in a way, that takes account of these special three-dimensional properties. The method implemented in TVS is to display the data as fog [16] [3] with variable density depending on the value of the scalar being displayed (Fig. 11). This method gives good results. During the virtual walk through the model a high stress-value becomes visible as a dense cloud. The structures behind the fogged zone are still visible. Figure 12 shows the amount of stress in the rock surrounding a tunnel, displayed as fog.

Fig. 11. Display of 3D-scalar field as fog

The user can define different scaling-curves, e.g. to highlight maximum / minimum values. With this method it is also possible to draw "fuzzy" geological features also known as "shear bands". These are described as points clustered around a surface. The fogging technique is applied to scalar fields, if the values are vectored (e.g. principle stress) small arrows are drawn (Fig. 12).

The three-dimensional visualisation of spatially distributed values as fog makes great demands on the hardware used. If all pixels of the cloud have to be projected into the view level, a lot of calculation effort is necessary. The

Fig. 12. Surface-settlements, displayed as field of arrows

density values have to be integrated along all lines to the viewing point. In most cases, other objects will be visible and the calculation must take this into account. The projection has to be redone after each change of position or viewing direction. This would take too long for visualisation in real-time. If a fog scene is created as a cluster of points, the graphics system will be stretched to the limit. With a special procedure it is possible to create the fog in real-time [3] [4] [5].

The method is based on the following idea: Three-dimensional space is divided into planes on which the values are known. In these planes, the scalar values are represented by different densities to be displayed. If the planes are drawn at right angles to the viewing-direction and the borders are extrapolated to zero, the cloud appears to be three-dimensional (Fig. 13).

It is shown experimentally, that instead of at 90°, planes may be viewed at an angle between 45° to 135°. This means that it is enough to place the planes perpendicular to one of the three coordinate-directions. The values are known on fixed raster-points, so no calculation is necessary to define the planes. The addition of the density values is done by the graphics hardware. The depth-buffer ensures that only visible parts of the cloud are drawn. If the measured or calculated values are not regular it is necessary to interpolate. A good method to do this is to find a weighted average-value in each grid-point. To find the raster-values, a lot of calculation work is necessary, but this work has to be done only once. The only work, that has to be done during run-time is, to find the position of the planes and the best order to draw them.

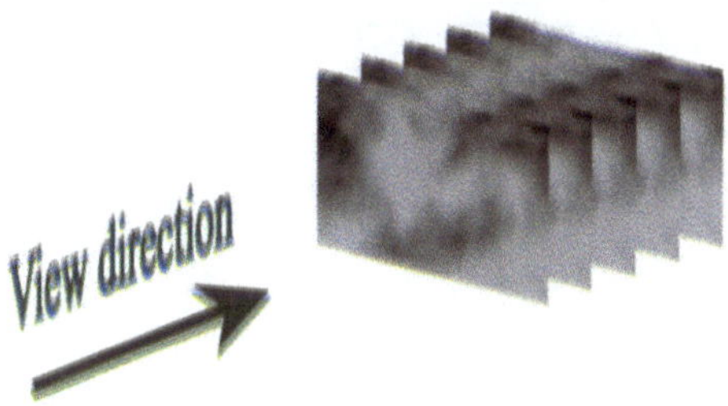

Fig. 13. Drawing of fog as parallel planes

For satisfactory results it is very important to find the right number of raster-values. If more values than necessary are calculated, they have to be stored and drawn. Graphics performance is very expensive, so the number of values should be as small as possible. But if the number is too small, information will be lost. In the field of communication engineering, such problems are very frequent (for example if analogue signals are digitalised) [6] [7]. But not only threedimensional results have to be displayed. In many cases the values on the surface are of interest. The data, to be displayed have different ranges of values, but to assign them to a colour scale a uniform range is needed. In TVS the input data is rescaled so that the minima and maxima correspond to 0 and 1. For technical purposes the relation between result value and the corresponding contour value should be linear. This is necessary to keep the typically characteristics of the values unchanged. A data set, in which the values change linearly, should appear with contours showing this properly, otherwise the data could be misinterpreted.

However in some cases a non linear scaling may be suitable. An example of this is when a measured value is out of range because of a fault in the meassuring device. Another case is where the user may wish to highlight values in a particular range. The Tunnelling Visualisation System offers a number of options to define a scaling curve with properties for each purpose. Fig. 14 shows the TVS dialog, in which these adjustments can be made. The user may also switch between sharp and smoothed contours can also be found in this window. The first group of scaling options produce a scaling curve, which is at least piecewise linear. The simplest is a linear amplification or attenuation. To get this, the data-value is multiplied by a constant factor.

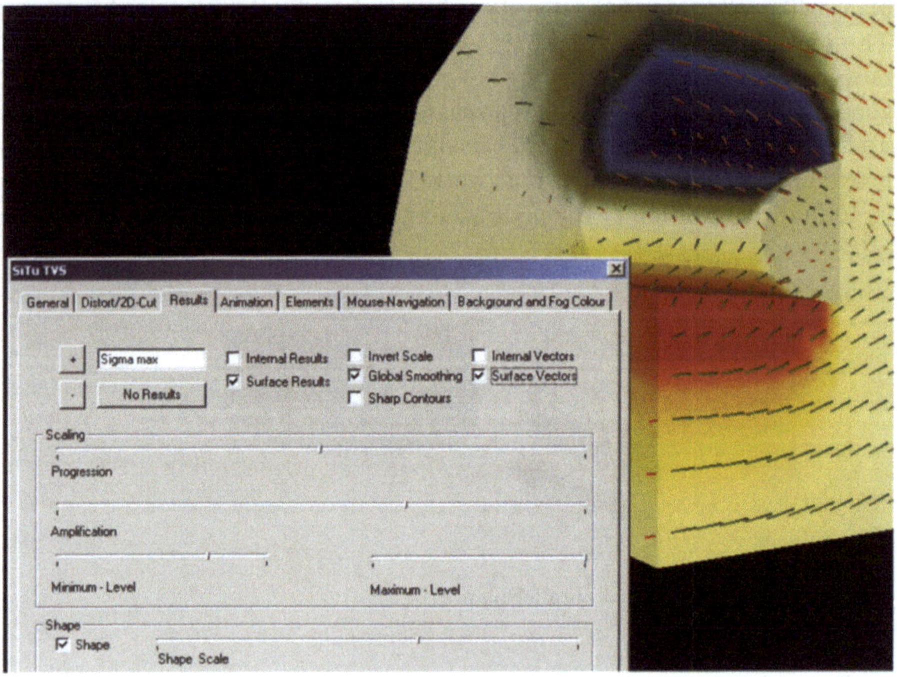

Fig. 14. TVS-dialog to set scaling options

This factor can be modified by the user. If it is greater than one, the values are amplified other-wise they are attenuated. The amplification factor can also be negative and this inverts the colour scale. For this purpose there is an extra button in the TVS-result dialog. Another possibility of linear result scaling is to cut off values outside a user defined range. This is useful if the data-set contains values, which are out of scale because of a measurement failure. In such a case the values of interest could not be interpreted from the contours. For this purpose TVS offers the possibility to define a range of values, for displayed values. This is also useful, if values lower or higher than a limit should not be displayed. A maximum stress for example is such a limit. Setting the display-minimum to this limit, produces an output of points, where the maximum stress is reached. All other points are drawn in the colour of the scale minimum. The user can determine a maximum and a minimum. Values outside this range are displayed in the same colour as the maximum respectively the mini-mum. Both limits can be set as a percentage of the available value-range. Non linear scaling is applied if it is necessary, to get a focus on peaks or throughs. In the field of geotechnics it is very frequently needed to display the concentrations of stress, because these zones are of significant interest in order to find out the risk of failure. For such cases a progressive scaling is necessary. Lower values have only a small part of the colour-scale, most of the scale is used for the larger values. The opposite of this is the declining scale. Using this, the major part of the colour-scale is reserved for values around zero. The resulting effect is, that there is no difference for the displayed colour if the values is 50% or 90% from the maximum, but there is a fine scaling for small values, so details in this range become visible.

The transformation between value and contour is made using the following equation:

$$C = A * \left(\frac{Value - Min}{Max - Min} \right)^{1*prog} \tag{1}$$

C : contour-value
A : linear amplification factor
Max : maximum value of scale
Min : minimum value of scale
$prog$: progression factor

2.3 Data processing

A typical 3D-simulation model consists of Boundary or Finite elements. If the calculation is done with finite-elements, the three-dimensional space is filled with brick-elements which are necessary for the simulation, but their geometry does not need to be drawn. Simulation results inside these elements are important for visualisation and they are drawn as fog. Only visible surfaces need to be displayed. If the tunnel-advance is visualised, elements become invisible, so this calculation must be redone for every loadcase. [8] To improve the smoothness of displaying elements with a midside node, additional points on the element edges are calculated. This calculation is done, using the "Newton-Interpolation algorithm". Usually elements are defined by their borders. To render surfaces, the midside node of the element has to be known. This position of this point can be calculated with the help of shape-functions [9]. OpenGL has options to use different kinds of lighting effects. The characteristic of a light source can be modified in a very large range. Position, direction, opening-angle, colour, attenuation-factors or the loss of intensity depending on the distance to the source may be varied. These adjustments are only effective if the normal-vector of the model-surface is known. If the model is drawn as a couple of triangles, this is no problem, because this vector can be found for each triangle. Using the mesh from numerical simulation, the quadrilateral elements usually consist of four or eight points. Such elements do not have a plane surface, so a normal-vector cannot be calculated exactly. Making triangles out of each non-triangular element will increase the necessary rendering effort. The Tunnelling Visualisation Software defines the normal-vector as perpendicular to both diagonals. This is not exact, but sufficient for good lighting effects.

2.4 Geometric data optimisation

Geometric data base for a FEM-calculation. For a calculation, using the finite element method, the space has to be described with threedimensional elements. To simulate the tunnel advance, it is necessary to also simulate the surrounding rock by modelling a block of material. To get realistic results, this block needs to be very large and sometimes, if the tunnel is not very deep must extend up to the surface. In this space geometrical or physical irregularity like faults or different materials, have to be considered. For the modelling of the tunnel, not only its shape is important, but it is also necessary to consider the tunnel support, such as shotcrete, and the sequence of tunnel-advance. The number of elements, which is necessary for a realistic model, is determined by the size of the modelled block and the amount of inhomogenities.

A mesh consists of nodes, and their position in space, and elements. Elements are defined by a list of node numbers, which are connected to it. Two different types of elements are commonly used. In the first type points are connected by straight lines. For the second group, this connection is a 2nd order curve. Therefore an additional point between the edges is necessary. Figure 15 shows an example of a 3D brick element with quadratic interpolation.

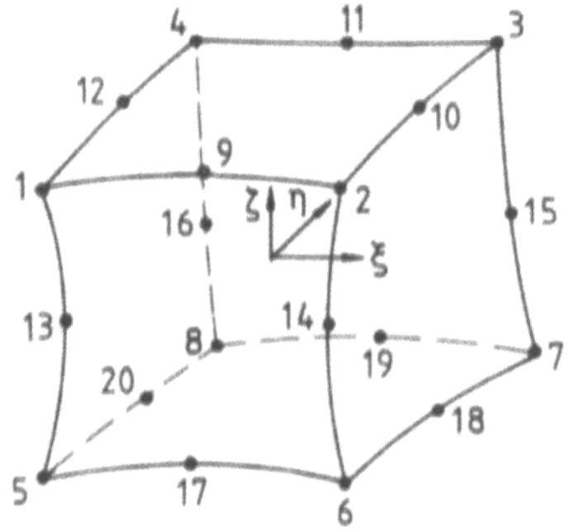

Fig. 15. Brick element

Other elements used in tunnelling are the shell element (Fig. 17) and the joint element (Fig. 16). The brick element is used to describe the rock/soil mass. Joint elements are used for describing faults ad for modelling shotcrete lining shell elements are used.

For each calculation with the finite element method, only one mesh is generated. If more than one excavation stage is simulated, it is necessary to

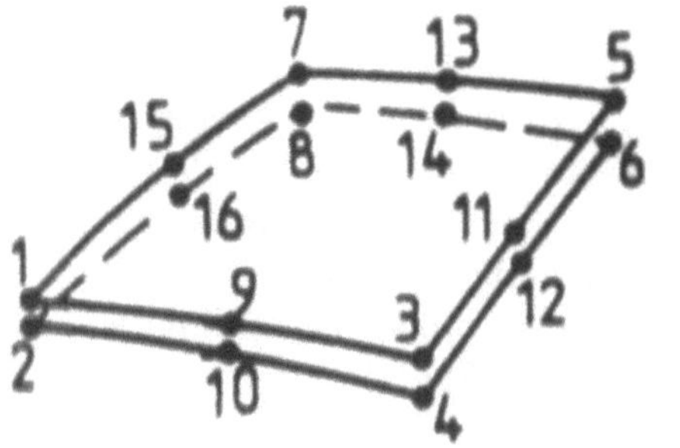

Fig. 16. Joint element

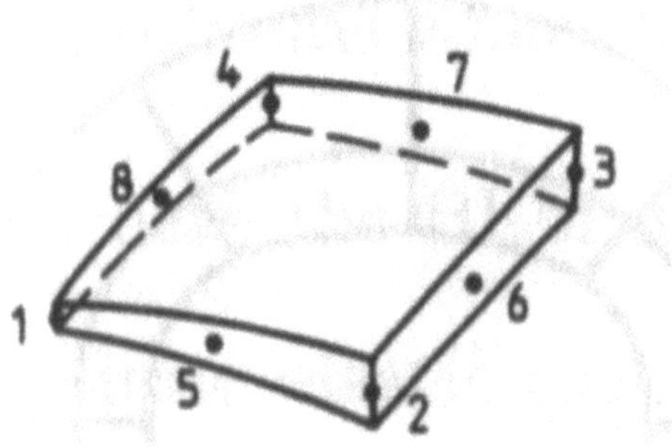

Fig. 17. Shell element

have information on which elements are active in each step. Loadcase 0 is the initial one. At this time, the tunnel-advance has not started, that means, the elements, describing the material inside the tunnel to be excavated have to be considered in the calculation, so they are active. In each excavation-step material is removed, so these elements are no longer of interest for the calculation. During the advance of the tunnel supports are introduced as shotcrete lining or rock-bolts To describe the tunnel-advance a list of active elements for each loadcase is needed in addition to information about nodes and elements.

Conversion of the geometry-data. For the visualisation of the geometry of a three-dimensional tunnel-simulation it is certainly possible to draw all three-dimensional elements. With the help of a good graphics-hardware, which includes a depth buffer, only the outside surfaces would then become visible.

There are two major reasons against this method. The first is that they represent a waste of expensive graphics-performance. Most of the elements, used for the simulation are never part of a visible surface, so they never become visible. The second reason is that having the three-dimensional space filled with brick-elements makes it impossible to display simulation results in that area. For an economic way of displaying the geometry it is necessary to find out, which part of the mesh is visible. The first step of doing this is reducing the element-order. Three-dimensional elements have to be divided into its six surfaces (Fig. 18) in order to extraxt surfaces whith can be displayed.

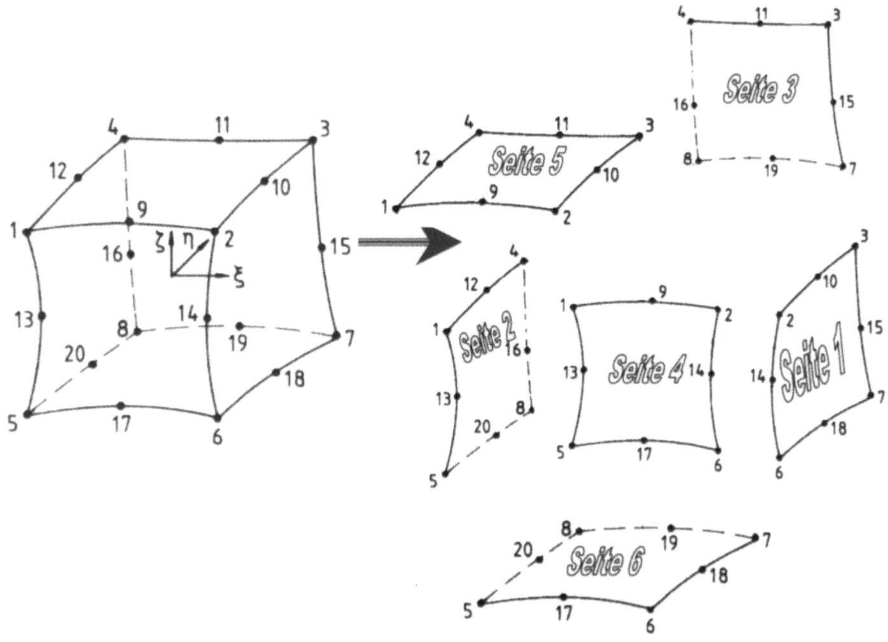

Fig. 18. From a three-dimensional brick element to its 2D-surfaces

The result of this procedure is a very large number of two-dimensional elements. Up to no simplification or reduction of the data has occured. The first step in the optimisation is to find out, if there are redundant elements. An element is redundant, if it describes the same surface. To find such elements it easier to check, if two elements are connect the same points. If one element consists of the four points N1, N2, N3 and N4, there are eight possibilities to define a regular element with the same points. An element is regular, if there is no intersection between two sides. Four possibilities can be found by a cyclic exchange of the node numbers, the other four ones by mirroring and cyclic exchange. The following combinations of elements are not redundant, even if they connect the same nodes.

- different element types,
- different materials,
- not active for the actual loadcase,
- joint-elements,
- shell-elements.

This search has to be redone for each loadcase. Elements, become visible during the excavation, form part of the tunnel-surface. Such elements are geometrically redundant, but they are not active at the same time. These elements are important for the visualisation, if only the tunnel is to be displayed. Without any optimisation this search is very calculation time expensive. It needs

$$n_c = n^2 * LC \tag{2}$$

compare operations to find all duplicates where n is the number of elements and LC is the number of loadcases. Having a mesh with about 8000 elements and 45 loadcases, this process can take up to one hour (Intel Pentium II, 366 MHz). The first optimisation step is to consider, that one element can only be identical to one other element. That reduces the number of necessary compare-operations n_c for one loadcase to

$$n_c = (n-1) + (n-2) + ...1 \tag{3}$$

or

$$n_c = \sum_{m=n}^{1} (m-1) \tag{4}$$

which becomes after simplification:

$$n_c = \frac{n^2}{2} - \frac{n}{2}. \tag{5}$$

The reduction of operations depends on n, but is always higher than 50% (see Fig. 19).

The next step of reducing the required number of operations is a pre-sorting of the elemet-types. This sorting is done while the data is read from the disk, so no additional effort is necessary. After this procedure, only compatible types of elements have to be compared to each others.

Without pre-sorting:

$$n_c = \frac{n^2}{2} - \frac{n}{2} \tag{6}$$

or

$$n_c = \frac{(n_1 + n_2)^2}{2} - \frac{n_1 + n_2}{2} \tag{7}$$

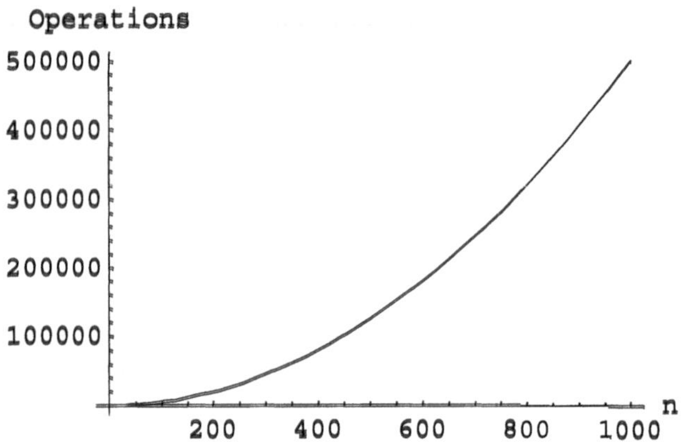

Fig. 19. Save of operations depending on the number of elements (n)

with

$$n = n_1 + n_2 \tag{8}$$

where $n_1 \, n_2$ ...number of elements type 1 and 2 respectively. With $(n_1 \neq 0)$ *and* $(n_2 \neq 0)$ it can be stated that:

$$(n_1 + n_2)^2 > n_1^2 + n_2^2 \tag{9}$$

that means, there is a decrease of required operations if the mesh consists of more than one type of elements, which is true for almost al meshes, used for a finite element calculation.

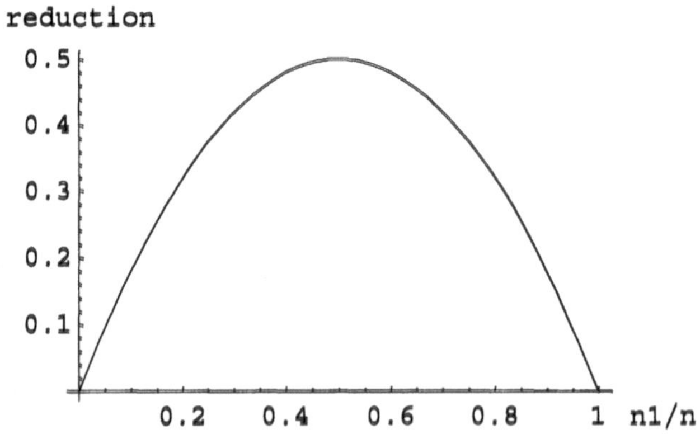

Fig. 20. Reduction of operations by pre-sorting the types of elements

Another possibility to save time is to prevent that 2D-elements with the same brick as source are compared to each other. If more loadcases are being displayed the following method can be used to save searching time. For the initial loadcase redundant elements are searched, independent of their activity-status. The result are elements, which can be redundant because of their geometry. These elements are put into a list, which is the base for all other searching operations. All other elements are excluded from the search, because it is impossible, that one of them becomes redundant. The number of necessary operations is no longer a quadratic function of n, it becomes linear to the number of elements in the list. Depending on the number of loadcases this method can reduce the necessary operations up to about 75% (tested with 8000 Elements and 45 loadcases) (Fig. 20).

2.5 Simulation data

Simulation results have special properties, which have to be considered by the visualisation tool. In some cases the values are not smooth, and discontinuous at element boundaries. This is caused by the discretisation procedure. The results of a finite element simulation usually are computed inside an element and not on the edges, where they would be necessary for drawing. With the help of shape functions, it is possible to find the values of the stress at the edge points by extrapolation. If one point belongs to more than one element, the extrapolated values for this node should be the same for each element. If there is a large difference, this indicates, that the mesh is too coarse, the results in error and the size or the type of the elements need to be changed. With a special global interpolation procedure it is possible to smooth differences between elements (Fig. 21) [10] [14] [15]. This is the same algorithm as used for displaying three dimensional scalar fields as fog.

Properties of the input-data. During the planning and building phase of a tunnelling project a lot of data are collected. The visualisation of this information and a user-friendly possibility to compare it to the simulation results must be possible with a state of the art visualisation system like TVS. The first and most important step to archieve this is common geometric data base for both types of data. As the mesh, generated for numerical simulation is the most accurate available description of the scene to be pictured, the Tunnelling Visualisation System uses this one. The mesh is not only a necessary part of the input data, it is also the geometrical data base of the simulation results. There are two principally different types of simulation results, deformations and different kinds of stresses. The difference between these two types is the location of the results. Deformations usually are calculated at the nodes, stresses in the gauss-integration points. These points are located in the element. Their exact position depends on the used element-type. In a local system of coordinates, where the edges have the position ± 1, the gauss-integration points have the coordinates $\pm 1/\sqrt{3}$ (Fig. 22).

Unfortunately the measurements on site, e.g. deformations at the excavation surface or the surface above the tunnel, are not performed on points

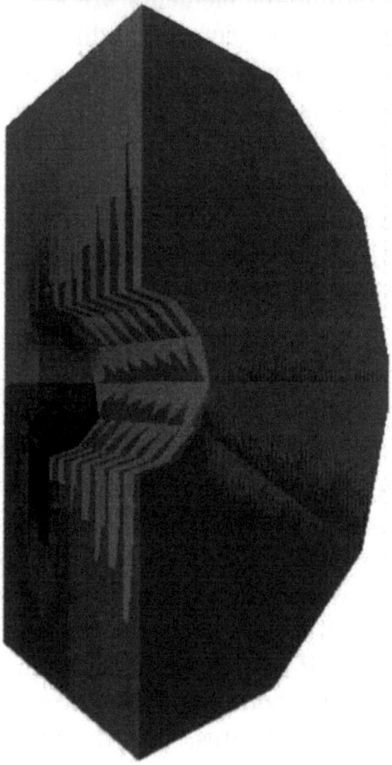

Fig. 21. Discontinuous contours between the elements

which are defined on the simulation mesh. However even simulation data are not always located in a position, where they would be needed for the visualisation.

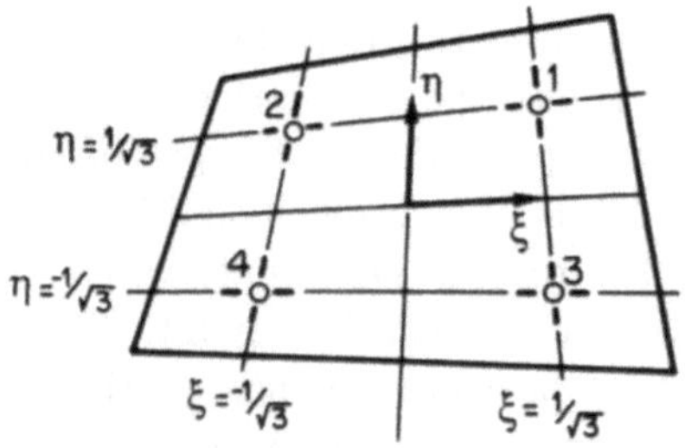

Fig. 22. Position of simulation results in an element

This has been the reason to develop the modified global smoothing algorithm, which has been implemented to the Tunnelling Visualisation System; a very essential feature for this visualisation tool.

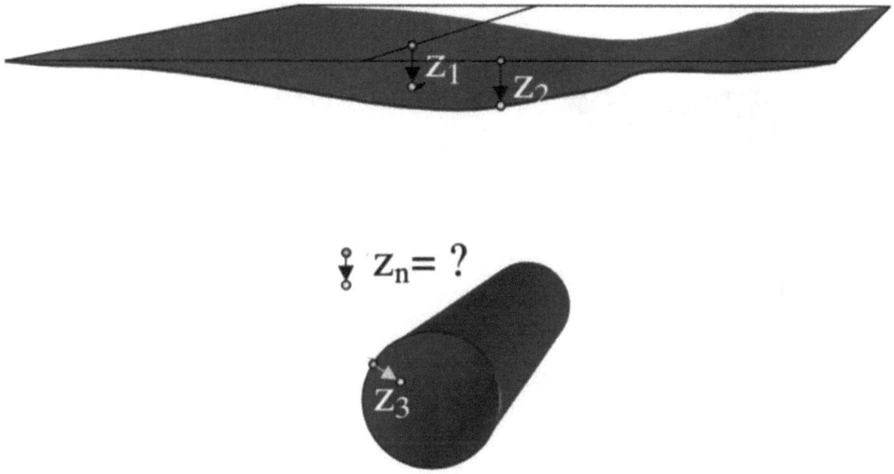

Fig. 23. Measured points (Z_1, Z_2 and Z_3) and a location (Z_n), where a value is searched

The task, to be solved by the algorithm is to find one unique value in a point in space, if a number of data points is known in the vicinity. In the case described the known values may be simulation or measurement results and the values to be calculated are located at mesh-points, which are used for the visualisation

Compared to simulation data, the amount of data produced by measurements is very small. But the requirements to the methods used are the same – no change should occur on the basic characteristics of the information. The number of known values in the mesh usually is much higher than the number of measured points. A circumstance, which must be considered during data processing. If less information is available, the results must become smoother. This is caused by the theorems of Shannon and Fourier [17], which state, that the content of information can never be increased by any procedure. That means that conversion algorithm, which makes mesh-data out of the measurement information must prevent a higher gradient of the resulting data which is more detailed than the distance and number of input-data-points can represent according Shannons-theorem, e.g. if there are five points available, the curve connecting them cannot have a grade higher than four.

The modified global smoothing algorithm [8] [11] has been tested extensively and after these tests it can be said that it fulfils the necessary requirements well.

2.6 Theoretical overview

The first part of the solution chosen is to take the average of all known values, to find the reqired one. A couple of parameter test have been performed to find out the best algorithm for determing the average.

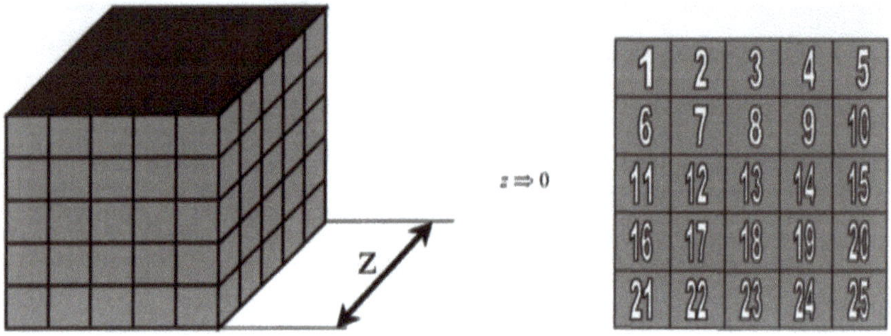

Fig. 24. Test configuration

The test have been based on a regular cube with 5*5*5 elements. The values in these elements were assumed to be constant in z-direction. Therefore it has been possible to decrease the depth to zero (Fig. 24). The algorithm is

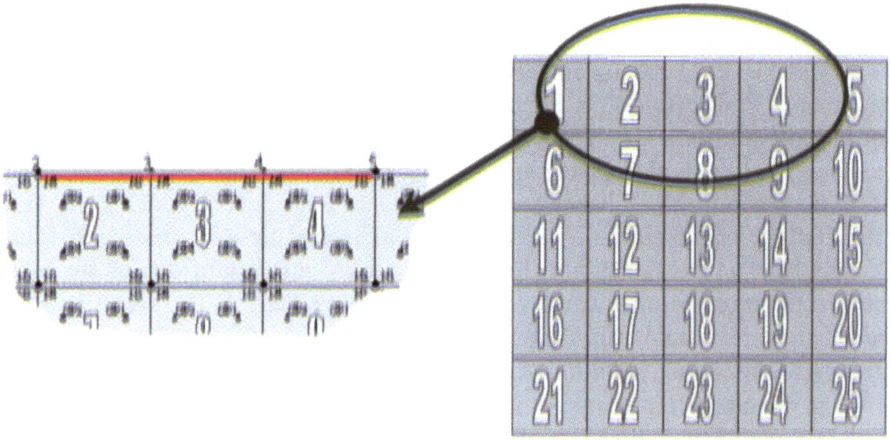

Fig. 25. Reduced test configuration

The resulting square has been used to perform several tests. To eliminate boundary-effects, which would occur because of the low number of elements, only the values on the marked line (Fig. 25) have been ecomputed. The extrapolated values have been compared to the averaged ones.

The first part was to choose the kind of averaging. In field of mathematics there are many average-types known. For this purpose four different ones have been tested:

$$Harmonic\,average:\ HM = \frac{\sum_{i=1}^{n} f(r_i)}{\sum_{i=1}^{n} \frac{1}{w_i} * f(r_i)}, \tag{10}$$

$$Geometric\,average:\ GM = \left(\prod_{i=1}^{n} w_i^{f(r_i)}\right)^{\frac{1}{\sum_{i=1}^{n} f(r_i)}}, \tag{11}$$

$$Quadratical\,average:\ QM = \sqrt{\frac{\sum_{i=1}^{n} w_i^2 * f(r_i)}{\sum_{i=1}^{n} f(r_i)}}, \tag{12}$$

$$Arithmetical\,average:\ AM = \frac{\sum_{i=1}^{n} w_i * f(r_i)}{\sum_{i=1}^{n} f(r_i)}. \tag{13}$$

For the mentioned types of average it can be stated that:

$$HM \leq GM \leq AM \leq QM \tag{14}$$

Figure 26 shows, that the arithmetic and the geometric average produce the best results. As the calculation effort for the first one is less and no prescaling of the values is needed, this one has been chosen.

Performing this calculation only will produce a correct output, if the distance between all known values and the searched one is the same.

In practical applications this model in unusable, because a configuration like this cannot be found very often. Therefore it is necessary, to consider the distance between each known point to the location, where we want to determine the value. Therefore a weighting function had to be included. Three different types have been tested in combination with the arithmetic average:

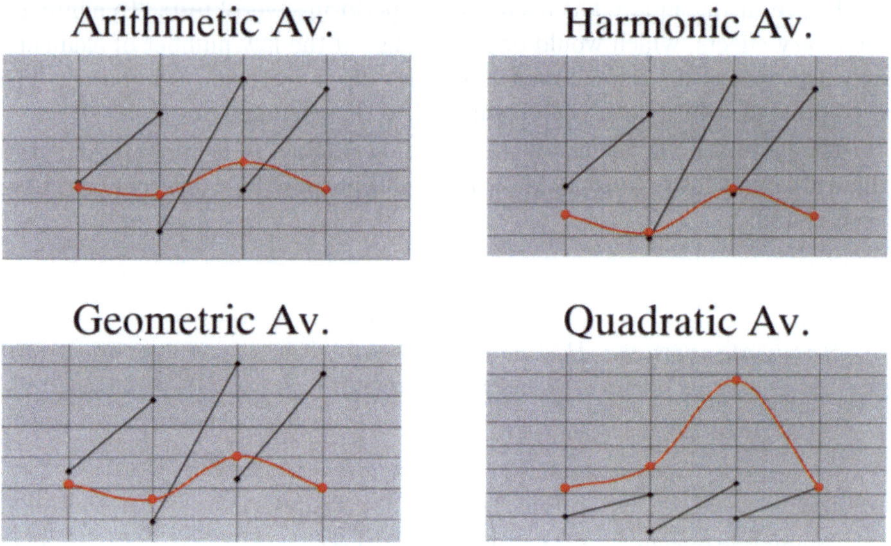

Fig. 26. Comparison of the tested averaging types

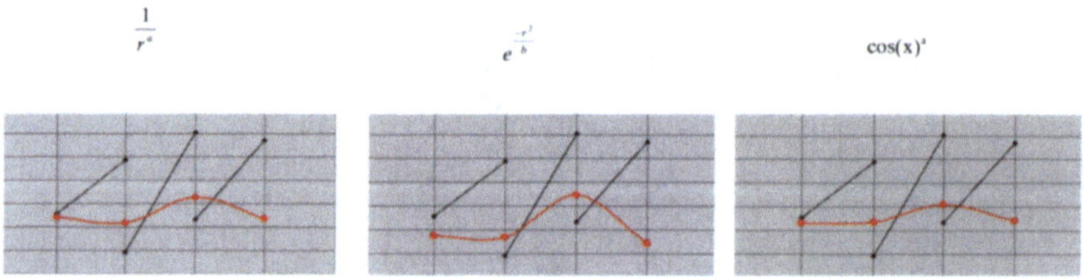

Fig. 27. Comparison between the tested weighting functions

The tests showed Fig. 27, that a weighting function $\frac{1}{r^2}$ and e^{-r^2}produced similar good results, but due to the circumstance, that the first one can be calculated much faster, this one has been chosen.

$$w = \frac{\sum_{i=1}^{n} w_i * \frac{1}{r_i^2}}{\sum_{i=1}^{n} \frac{1}{r_i^2}} \tag{15}$$

where n is number of known points and
r_i is distance between value to be computed and point i (see Fig. 28)

To ensure, that the chosen averaging method does not change the basic character of the source data, tests with regular data sets have been performed (Fig. 29).

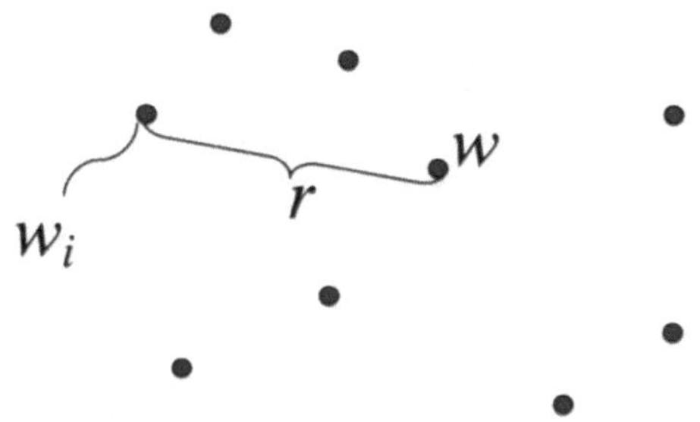

Fig. 28. Randomly distributed points and notation used

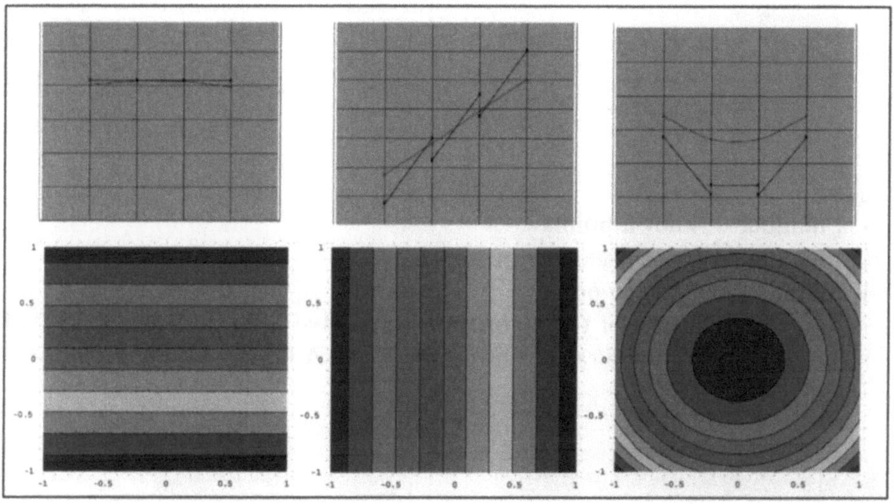

Fig. 29. Smoothing results of the shown regular data-sets

The third aspect, which has to be considered is the geometrical distribution of the known values. Ignoring this produces results, will become unacceptable, if the geometry of the mesh is distorted, or the distance between measuring points is irregular. The proposed solution of this problem is the following: To each point an area (in 3D a volume) is associated. It has been shown, how this parameter A can be found very easy: in [12] for measurement data and in [11] for simulation data. This value is added as an additional weighting factor. This means, that values, which are alone in their region have a greater influence on the result.

Figure 30 shows a cluster of points in 2D and their associated area.

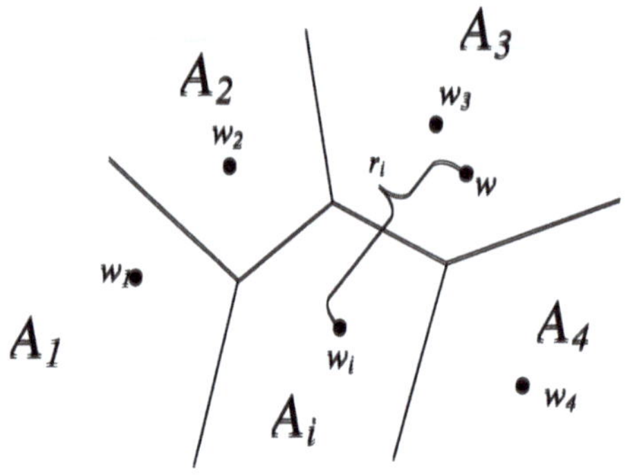

Fig. 30. Random distributed points and their associated area

$$w = \frac{\sum_{i=1}^{n} A_i * w_i * \frac{1}{r_i^2}}{\sum_{i=1}^{n} A_i * \frac{1}{r_i^2}} \tag{16}$$

where
n : number of known points
r_i : distance between searched w and w_i
A_i : Area associated to point i.

Also, for this part of the algorithm, tests have been performed. The following geometric configurations have been used: A finer mesh on the left side

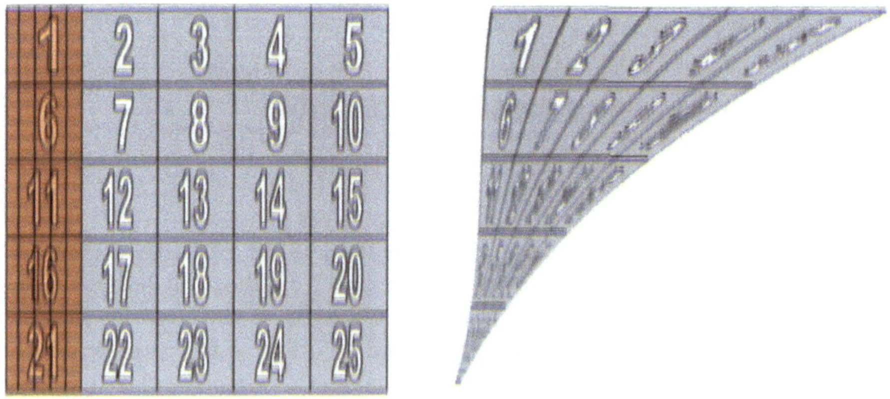

Fig. 31. Test geometry

If the, Modified Global Smoothing Algorithm [13] described here is applied to a FEM-calculation, no discontinuous contours appear. An additional applicable possibility to smooth the appearance of the contours is the replacement of the sharp contour scale by a blurred one (Fig. 32).

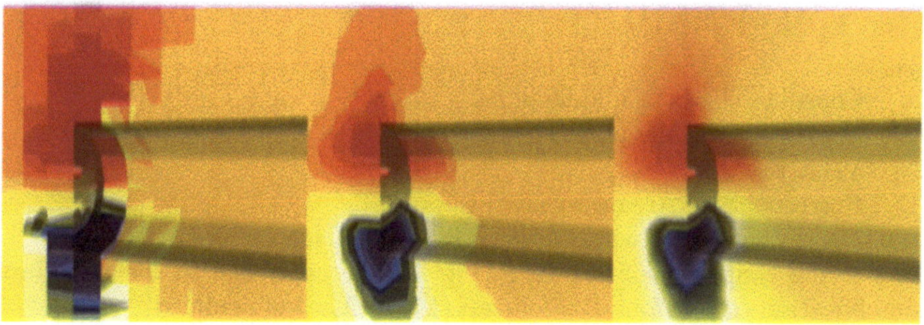

Fig. 32. Two steps of global smoothing (left: unsmoothed, middle: smoothed with sharp contours, right: smoothed with blured contours)

2.7 Runtime optimisation

The results of the described algorithm has been shown to be excellent, but if the amount of data becomes too high, the calculation becomes very time consuming, a circumstance that is a disadvantage in real-time visualisation. The best way to find the parameter A (the associated volume or area; (Fig. 32)) for measurement data is shown in [12], for simulation data this value is available due to the calculation method (Jacobian determinant of an element), so no additional afford is necessary. The distance is found using the theorem of Pythagoras, which cannot be optimised. So the only way to save time is by decreasing the number of necessary steps to compute one value. Theoretically the number of known points is equal to the number of terms to be calculated and added for each searched point. Experience has shown, that points, which are far away from the searched one have only a small influence on the results. But ignoring such values causes a theoretical error which becomes visible in specific cases, e.g. if a high number of high values is far away from the searched point. A tunnelling simulation usually has this property, because the area around the tunnel is described by a very fine mesh, compared to regions further away.

These circumstances cause, that all known values must be considered. The idea to fulfil these requirements and reduce the number of terms which have to be calculated is to subdivide the whole domain into a certain number of sub regions [13]. For each of these parts the represent value is calculated using the algorithm in [11]. This value is localised in one point. The coordinates are found using a similar algorithm:

$$\boldsymbol{x} = \frac{1}{n} * \sum_{i=1}^{n} A_i * w_i * \boldsymbol{x}_i \tag{17}$$

where:
$\boldsymbol{x}$: Coordinates of the middle point
$\boldsymbol{x}_i$: Coordinates of the point i
w_i : Value in Point i
A_i : Corresponding Area associated to point I
n : Number of points with known values in the sub domain.

For the configuration shown in Fig. 33 the error due to the described simplification can be found as follows: (w_i=1, A_i=1 is assumed)

Exact solution:	Using the simplification:
$\bar{r}_{exact} = \frac{1}{4} * \sum_{i=1}^{4} \frac{1}{r_i^2} = 125$	$\bar{r}_{sim} = 4 * \frac{1}{r_m^2} = 122$

Which gives an acceptable percentage error of 2,4%. This value decreases to zero for increasing distances that is

$$\lim_{r_{\min} \to \infty} (\bar{r}_{exact} - \bar{r}_{sim}) = 0 \tag{18}$$

r_{min} : Minimum of the involved distances

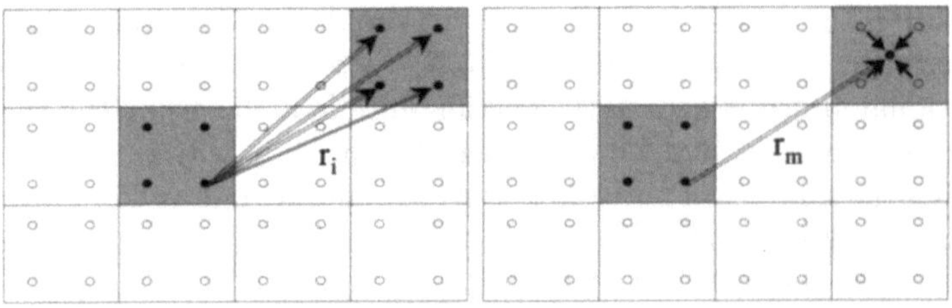

Fig. 33. Comparison between the exact solution (left) an the described simplification (right)

In 3D the domain is subdivided into cubic sub-domains. The one containing the searched value and those, that share one surface with it are considered exactly. The others are included using the simplification. In the worst case, if the searched point is not part of a sub-region on the surface, for one point seven sub-domains must be calculated exactly, all others in a simplified manner. This is an enormous reduction of calculation time (approx. 95% in practical examples), because, as mentioned above, this is being saved for each of the searched points.

Without subdivision ***m*n*k*** terms have to be evaluated and added. where:
m: number of sub domains
n: number of known points/sub domain
k: number of searched points

For the subdivided domain the following steps are necessary:

n*m calculations for the representing points of each sub domain (the time for calculation for the location of the representing point is very small compared to the one for one averaging term, so it is ignored here)

In the worst case, if the searched point is an internal one ***k*(7*n+m-7)*** steps are needed.

Therefore the time factor Δt_{rel} saved by defining the sub domains is given by:

$$\Delta t_{rel} = \frac{k*(7*n+m-7)+n*m}{k*n*m}. \tag{19}$$

This term goes to zero (Fig. 34) for large values for k, m and m, that means that the larger the data set is, the better the increase of performance is.

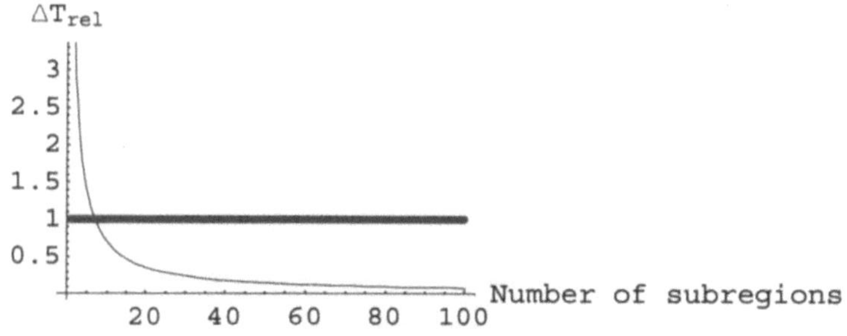

Fig. 34. Decrease of time due to subdividing

2.8 Parameter tests in 3D

The tests have been performed using the following configuration: A couple of values in a cubic domain, containing 10^6 distributed known values are calculated one time without any sub dividing, and another time using 125 subdivisions. The smoothed values are calculated along the line shown in Fig. 35. Three sets of source data are used linearly and parabolically changing ones.

Exact Solution					
	P 1	P 2	P 3	P 4	P 5
T1	2,01	4,03	6,05	8,07	10,09
T2	11,10	11,10	11,10	11,10	11,10
T3	6,47	5,44	5,05	5,44	6,47

Smoothed Values					
	P 1	P 2	P 3	P 4	P 5
T1	2,00	4,01	6,05	8,09	10,10
T2	11,00	11,00	11,00	11,00	11,00
T3	6,40	5,36	4,95	5,36	6,40

Error					
	P 1	P 2	P 3	P 4	P 5
T1	0,47%	0,58%	0,00%	0,29%	0,09%
T2	0,90%	0,90%	0,90%	0,90%	0,90%
T3	1,11%	1,54%	1,97%	1,54%	1,11%

$T1$: linear increasing in x-direction
$T2$: linear increasing in y-direction
$T3$: value equals to the distance from the centre
Pi : Point Nr. i.

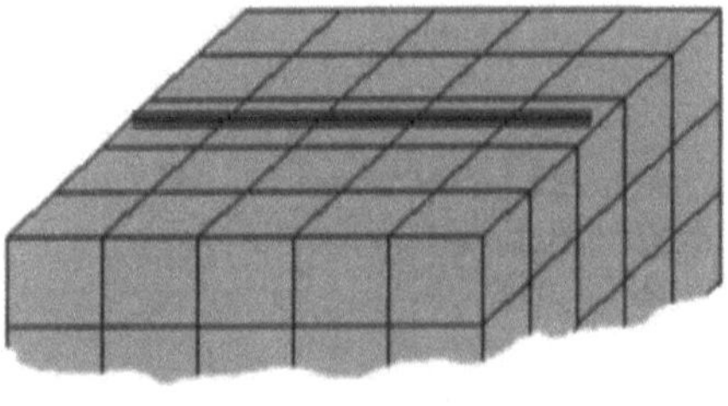

Fig. 35. Test configuration and the exact values in the centre of the sub-domains along the line

The performed tests show an impressive quality of the smoothed values. If the theoretically values are compared to the smoothed results like shown in Fig. 36, this can be seen.

These tests have also been performed along other lines crossing the cube and different distributions and numbers of known values and subdivisions, all with the same accuracy of the results. An error of less than 2% for the purpose of visualisation is a value, which can be accepted for technical and scientific

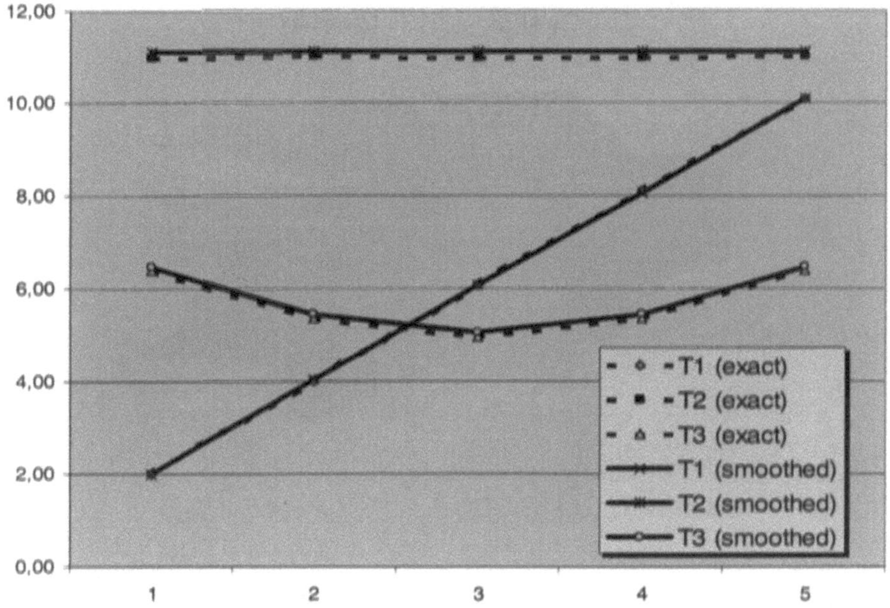

Fig. 36. Smoothed values; comparison to exact ones and failure percentage

applications. The theoretical time for the tested configuration compared to the one, needed for not optimised smoothing is approximately 5%, that means the tested subdivision into 125 parts with 10^6 known values gives almost the same results 20 times faster than without the modification, described here. This saving in necessary calculation time also could be observed during the tests. The original global smoothing algorithm took about 7 seconds on an Intel Pentium™III 1000 MHz, the optimised one less than a half second.

3 Examples

Shown here are some examples of visualisation taken for the SiTu-projects.

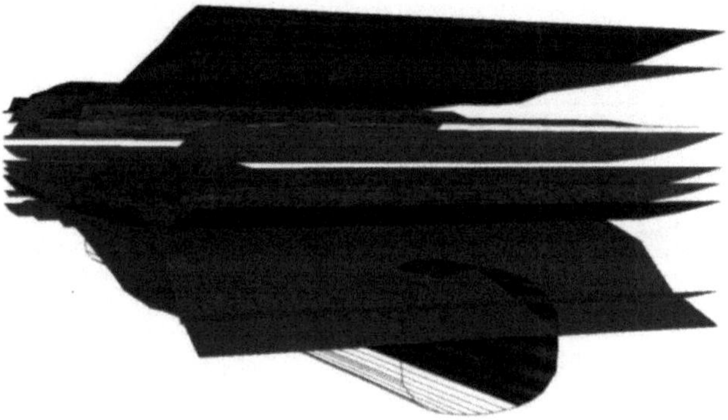

Fig. 37. Display of geological structures above a tunnel (Chap. 11)

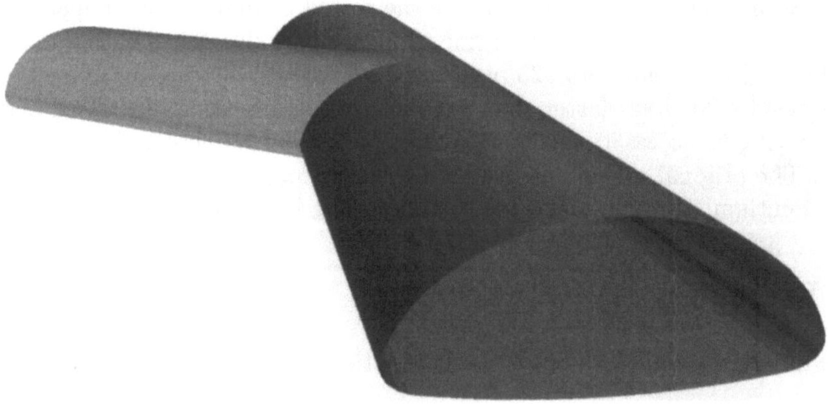

Fig. 38. Intersection of two tunnels

Fig. 39. Display of a sequential tunnel advance with deformation vectors (Chap. 15)

Fig. 40. Display of a cutting-plane (total displacement in the rock)

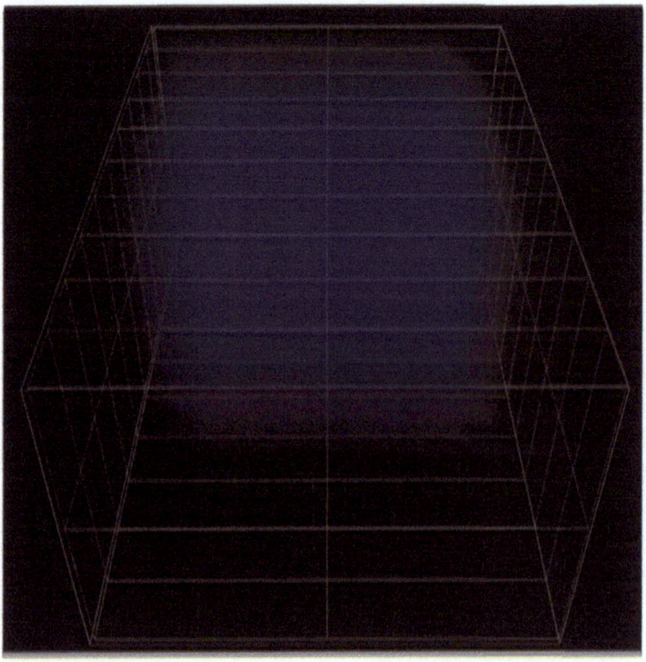

Fig. 41. Display of the water saturation as fog. The picture shows a sand pilar in which the water saturation is decreasing in time. (Chap. 12)

4 Conclusion - futural outlook

Numerical Simulation, advanced techniques of measurement and high performance real-time visualisation are inextricably parts in modern engineering. Specifically a user friendly system to visualise the data is necessary for the usability and for the acceptance of this techniques in conservative fields of engineering like rock-mechanics or tunnelling. But it is at least just as important for this purpose, that the system is portable. On a tunnel-site the offices are usually not so comfortable and spacious so a large workstation for visualisation is a strong argument against the usability of these technologies on site, particular if the excessive sensibility of such an equipment concerning dust is considered. The Tunnelling Visualisation System - TVS, presented in this chapter is running on a notebook. Only for stereo-viewing there is an additional equipment necessary, but even if all supported hardware is included (spaceball and head mounted display), a brief-case is enough to transport it. That means a mobile high-end visualisation system is available on a platform, on which also the numerical simulation (e.g. using the Boundary and Finite Elements simulation software BEFE) can run with an excellent performance. This platform is a Dell Inspiron 8100 notebook, GeForce2 GO graphics-board, Pentium III processor running with 1000 MHz. External devices are a Spacetec Spaceball 3003 and Cybermind I-glasses 800*600. The operating system is Windows 2000 professional. This configuration makes it possible to use the system, where it is needed: on a tunnel site or at any other place e.g. if you want to provide a customer with a guided tour through a future tunnel. This fact and the novel and very innovative visualisation-techniques used by TVS sum up to a system, which will be a an important step on the way to the application of simulation and visualisation in almost all fields of engineering as a matter of course.

References

[1] Neider J., Davis T., Woo M. (1995) OpenGL Programming Guide – The Official Guide to Learning OpenGL, Release 1. Addison-Wesley Publishing Company, ISBN 0-201-63274-8

[2] Noble R.A., Clapworthy G. J. (1998) Improving Interactivity within a Virtual Sculpting Enviroment. *Proceedings IEEE Conference on Information Visualization*

[3] Opriessnig G., Beer G. (1998) Visualisation in Tunnelling. *Proceedings IEEE International Conference on Information Visualisation. IEEE Computer Society*, Southbank University London, UK

[4] Jobard B., Lefer W. (1997) The Motion Map: Efficient Computation of Steady Flow Animations. Laboratoire d'Informatique du Littoral, Calais France

[5] Yuanxian Gu, Xiaosong Yang, Cangzhou Yuan, Yungpeng Li (1998) Advanced Visualization Techniques for FEM-Computing-Scan-Buffer based direct Volume Rendereing for 3D irregular meshes. CIMNE, Barcelona Spain

[6] Freyer U. (1981) Nachrichtenübertragungstechnik, Grundlagen – Komponenten, Verfahren – Systeme. Carl Hanser Verlag München Wien, ISBN 3-446-13093-4
[7] Herter E., Lörcher W. (1987) Nachrichtentechnik–Übertragung-Vermittlung-Verarbeitung. Hanser ISBN 3-446-14593-1
[8] Opriessnig G., Beer G. (2000) Data Optimisation for the Visualisation of FEM-Results. *Proceedings IEEE International Conference on Information Visualisation. IEEE Computer Society*, Southbank University London, UK
[9] Beer G., Watson J. (1992) Introduction to Finite and Boundary Element Methods for Engineers. John Wiley and Sons
[10] Opriessnig G., Beer G. (1999) Visualisation in Tunnelling - New Developments. *Proceedings IEEE International Conference on Information Visualisation. IEEE Computer Society*, Southbank University London, UK
[11] Opriessnig G., Kettler M., Beer G. (2001) Elimination of the Influence of the Mesh Geometry to the Quality of Global Smoothing. *Proceedings IEEE International Conference on Information Visualisation. IEEE Computer Society*, Southbank University London, UK
[12] Feigl W. (2001) Entwicklung und Implementierung von Algorithmen zur Visualisierung von FEM-Daten und Messresultaten in einem Modell (in German). Masters Thesis, Institute for Structural Analysis, Graz University of Technology
[13] Opriessnig G., Beer G. (2002) Improving the Modified Global Smoothing Method for spatial distributed and large Data-Sets to use it in the Real-Time Visualisation of Simulation and Measurement Information. *Proceedings IEEE International Conference on Information Visualisation. IEEE Computer Society*, Southbank University London, UK
[14] Duarte C. A., Oden J. T. (1992) H-p Clouds-an h-p Meshless Method. TICAM – Texas Institute for Computational and Applied Mathematics, The University of Texas at Austin Taylor Hall 2.400 Austin, Texas, 787712, USA
[15] Belytschko Organ Krongauz (1993) A Coupled Finite Element Free Galerkin Method. Department of Civil Engineering Northwestern University Evanston IL 60208-3109 USA
[16] Haas W. et.al. (1997) Visualisierung und Animation in der Geomechanik für Strömungs-, Transport- und Spannungsberechnungen. Institut für Informationssysteme Joanneum Research Graz
[17] Lenze B. (1997) Einführung in die Fourier Analyse. Logos Berlin

Chapter 15
Application of numerical simulation at the tunnel site

Harald Golser[1] and Wulf Schubert[2]

[1] GEOCONSULT ZT GmbH / GEOCONSULT Consulting Engineers
[2] Institute for Rock Mechanics and Tunnelling, Graz University of Technology

Abstract. Since the tunnel engineer often has to respond quickly to unexpected ground conditions, rapid results from numerical simulations performed on site could serve as a tool, which assists important decisions. One object of the project was to evaluate the efficiency and applicability of 2D and 3D finite element methods (FEM) and boundary element methods (BEM) to simulate tunnel advance under different geotechnical conditions. Another task to be performed during this project was to make site data available to other projects within the JRI.

The finite element and boundary element code BEFE was used ([1]). The simulations were occasionally performed on site in order to gain experience with the practical applicability and efficiency of the numerical tools. Modelling techniques for excavation sequence and support were investigated, as well as the model influences on the results. The modelling approach was evaluated using criterion such as the time spent for model set-up, calculation and processing the results. Improvements in methods of data evaluation of results for a direct comparison with monitored data are presented. The calculation time and disk storage requirement using a Conjugate Gradient Solver (CG) are compared to a Frontal Solver for different model sizes. The development of longitudinal displacements and the orientation of displacement vectors are presented for homogenous and heterogeneous ground conditions. The influence of different primary stress conditions on the displacement pattern was investigated as well. Finally, the practical application of the numerical models is demonstrated by case histories.

1 Introduction

Numerical simulations are routinely performed during tunnel design, especially for tunnels with low overburden such as subway tunnels or tunnels in built-up regions. The knowledge of the geological conditions in such projects is generally good, allowing the ground characteristics to be determined to a satisfying degree. The main target of the computations, about from proving the stability of the tunnel is the determination of displacements to be expected in order to assess the influence of tunnelling on surface structures. To a limited extent, the stresses in linings can be approximated under such boundary conditions. Due to limited knowledge of the real ground mass conditions, detail design of tunnel excavation and support has to continue during excavation. Continuous adaptation of excavation sequence and support design to the ground behaviour and the performance of the tunnel are required to ensure safety and efficiency. Due to lack of appropriate modelling tools and time, this optimisation process is often based on experience. Experience will remain important in tunnelling also in the future, but it would be desirable to supplement this process with numerical simulations. Numerical simulations on site, continuously updating the model and using experience from construction presently are not performed routinely. Since the tunnel engineer often has to respond quickly to unexpected ground conditions, numerical simulations on site in future could serve as a tool to assist in important decisions.

Currently forward modelling techniques may not be sufficiently accurate for detailed design. This is because frequently the models used are not adequate for the problem and the knowledge of the actual ground condition is limited. Because these inaccuracies mainly occur during the design phase, numerical simulations during construction can enhance the on-site decision process if they can be performed in time. Large computational times, as well as deficiencies in the models have limited the application of routine numerical simulations on site.

There are numerous mathematical models and simulation methods available, but they are not equally suitable to simulate different geotechnical conditions. The decision for using a certain simulation method is not only based on respective geotechnical conditions, but also on the time spent for model preparation, the calculations and processing of the results. The effort involved in generating the model should provide an equal return in accuracy and cost savings. The application of routine numerical simulations on site will be increased only if the results of such simulations are realistic and available to be used for daily decisions on excavation sequence or support efficiency.

The number of numerical simulations in the tender and detail design phases has increased. At the same time considerable progress has been made in observation methods. Huge amounts of data have been collected. Only in a few cases have these data been used to critically review the results of the previously prepared computations. This lack in feedback to the engineers in charge of computations has hindered the improvement of computation tools. The general case of nonconformity of the calculated results with the observed behaviour has led to a loss of reputation of numerical simulations on the side of the engineers involved in construction.

An important task of this project was to narrow this gap between designers and site engineers by applying numerical simulations directly on site. Two sites in Austria, the Sieberg railway tunnel, and the Spital highway tunnel were selected for this purpose. The results of the simulations should also make site engineers familiar with simulation techniques, value of calculation results and limitations in the accuracy.

In the course of the work it was shown, that the time requirement for pre- and post processing, as well as for the computation itself were too high. Appropriate tools were developed to speed up the process.

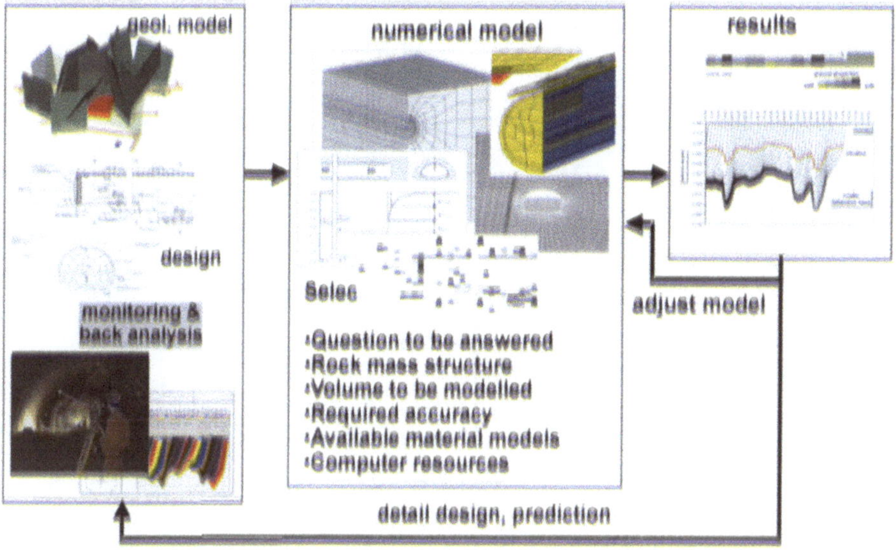

Fig. 1. Modelling scheme

2 Modelling strategies

In numerical simulations one attempts to model the very complex construction process. In most cases the ground through which the tunnel is driven will be very heterogeneous and an-isotropic. In addition the supports used in tunnel construction (especially rock anchors and shotcrete) have a strong time dependent behaviour. Efficient modelling requires simplifications. The allowable simplifications to still produce realistic results with acceptable effort strongly depends on the ground characteristics and the construction procedure, as well as the environment, in which the underground structure is located [2].

2.1 Pseudo 3D models

Temporal and spatial sequencing of the excavation influences the redistribution of the stresses, the load developing in the support, and the displacements of the tunnel. Accordingly initial relaxation of the rock mass and consequently the final load of the support is partially governed by the excavation process (Fig. 2).

With pseudo 3D models, time dependent and three-dimensional effects can be considered with 2-dimensional models. For this purpose, modelling methods were developed to consider the pre relaxation of the rock mass. Two methods are commonly used, the load reduction method and the stiffness reduction method [4].

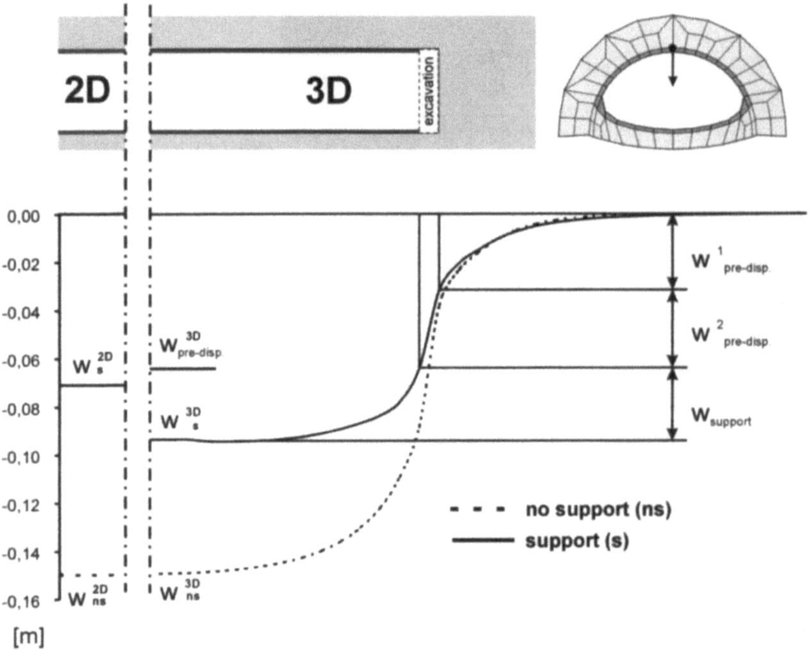

Fig. 2. Comparison of the magnitude of settlements of the crown point from top heading excavation for a 2D and 3D finite element calculation with and without shotcrete lining

A 3D FE model was used to show the complete settlement history of a line along the crown. The results were compared to 2D computations. $W^1_{pre-disp.}$ and $W^2_{pre-disp.}$ occur before the support is installed. It can be seen, that about 20% to 35% of the final displacements occur ahead of the tunnel face. The effect of a support on the reduction of the displacements can be clearly seen.

With a corresponding 2D model, the excavation and the simultaneous installation of the shotcrete lining causes the crown settlements $W_s{}^{2D}$, leading unrealistically low displacements and high stresses in the shotcrete lining. Results from the simple 2D model without relaxation prior to supprt installation result in a conservative design for the lining thickness and strength, while providing too low displacements.

To arrive at more realistic results, the 3D effect can be simulated with methods such as the "Load reduction Method".

Load reduction method. The load reduction method is based on the idea of decreasing the internal pressure p_0 in several steps ([3], [4]). The excavation process can be split into several steps Fig. 3a-c. The stiffness of the support can be increased stepwise by updating corresponding material parameters before each load case. With the β factor the magnitude of the internal pressure is controlled.

The β factor can be obtained by evaluating the measured d). It can be assumed, that the settlement magnitudes are proportional to the corresponding (1-β) values.

$$u_1 : u_2 : ...u_i = (1 - \beta_1) : (1 - \beta_2) : ..(1 - \beta_i) \tag{1}$$

The measurements can be taken from crown settlements or surface settlements and they must include pre-displacements (Fig. 4). In the case of shallow tunnels, the displacement data can be obtained by

- surface levelling,
- vertical extensometers from the surface,
- horizontal inclinometers from the tunnel face.

For deep tunnels, horizontal inclinometers and extensometers are the only way to obtain pre displacement data.

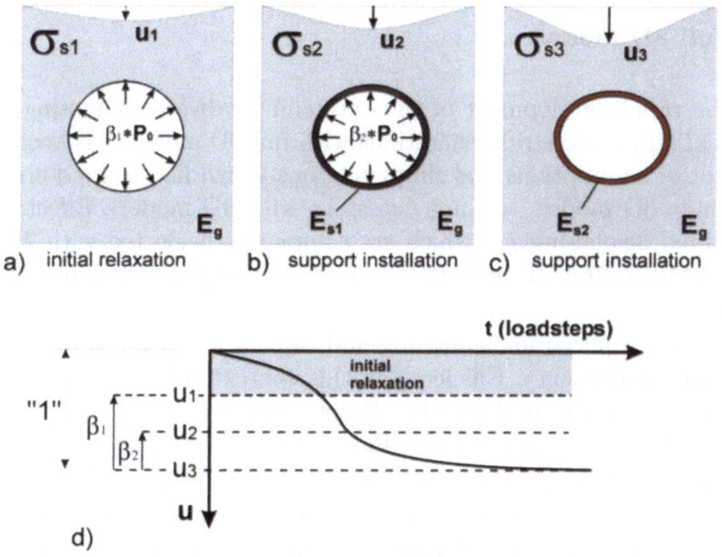

Fig. 3. Principle of the load reduction method

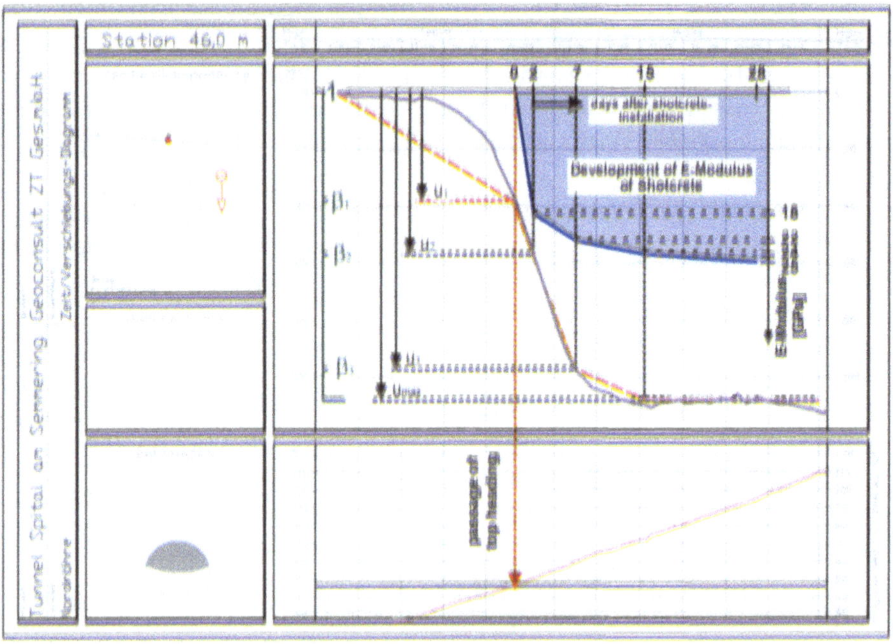

Fig. 4. System for estimation of load factors out of measured surface settlements during tunnel approach

2.2 Full 3D models

With the rapid development of software and hardware simulating three dimensional stress and strain conditions with full 3D models has become feasible. A lot of assumptions and simplifications which had to be done with 2D and pseudo 3D models, are not necessary with 3D models. Effects, such as longitudinal displacements, which are completely neglected with 2D models can be obtained from 3D models. The stress redistribution can be realistically simulated.

The time required for modelling and calculation requires models with a high degree of efficiency. Efficiency in this context stands for minimal effort to obtain calculation results with sufficient accuracy. Only models with high efficiency allow a day-to-day modelling on site.

For site applications there are several limitations to the use of 3D computer models. 3D models are usually avoided because of the difficulties of mesh generation, storage requirements, long run times and evaluation and visualisation of the results. For mesh design and data visualisation interactive packages are being developed and are available. To reduce run time and the storage problem, equation solvers, such as the PCG (Preconditioned Conjugate Gradient) are used.

There are several computational methods, which can be practically used for 3D analysis. The most important ones are the following ([5]):
Finite Element Method (FEM)
Boundary Element Method (BEM)
Coupled FEM/BEM
Finite Difference Method (FDM)
Distinct Element Method (DEM)

Finite element method. For 3D applications, the finite element method is the most widely used method for analysing tunnel problems. The main advantages and disadvantages are summarised as follows.

+ Nonlinear and time dependent material behaviour as well as the smeared effect of closely spaced parallel joints can be considered.
+ Excavation sequences and the timely installation of support such as rock bolts and shotcrete can be simulated.
+ The method has been extensively applied in geotechnics and thus a lot of experience is already available.
- The volume of the infinite domain surrounding the tunnel to be analysed has to be completely modelled. Dependent on the code used, excessive run time and disc storage may be required.
- Modelling of distinct fault zones is time consuming and cumbersome.
- Not suitable for geotechnical conditions, where the rock mass behaviour is mainly governed by joints.

Boundary element method. The boundary element method is young compared to the finite element method, but significant advances have been made recently in the development of this method. Under certain circumstances, this method is an alternative to the finite element method ([6]).

+ Only surface discretisation is required and this causes less effort for pre- and post-processing. Smaller equation systems require less disc storage requirements and computation time.
+ Distinct faults can be modelled very efficiently as interfaces and non-linear behaviour of the fault can be included in the analysis ([7]).
+ Results can be obtained anywhere in the domain, and not only at predefined (Gauss)points using postprocessing
- Except for interfaces (faults), only elastic material behaviour can be considered.
- A detailed modelling of excavation sequences and stepwise activation of support measures is not possible at this stage but is under development
- The method is not suitable for highly jointed rock mass.
- Up to now, this method was not used very often for geotechnical problems, so less experience is available than with the finite element method.
- Results can only be obtained on pre-defined dummy elements.

Coupled finite element /Boundary element method. By coupling of FEM and BEM, the advantages of both methods can be utilised. With this method, only the near field region around the tunnel, where nonlinear material behaviour is expected, is discretised with finite elements. The far field, where linear elastic material behaviour can be assumed, is represented by boundary elements. The main advantage of this method is the low discretisation effort especially with shallow tunnels, where the shape of the surface and surface settlements are a matter of interest. In this case, the surface can be modelled with boundary elements and updated independently from the tunnel and the surrounding rock mass. Considering a new surface topology only requires an update of the discretisation of the surface. Support measures such as shotcrete and rock bolts as well as sequential tunnel excavation can be simulated as with the conventional finite element method (Fig. 5, [8]).

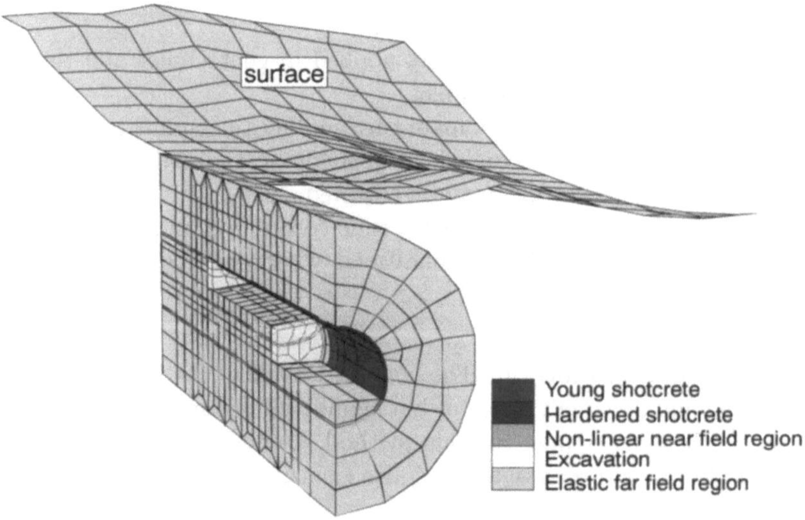

Fig. 5. Model of coupled boundary elements and finite elements

3 Modelling effects

3.1 Element types and sizes

BEFE offers a number of element types with different number of nodes, degrees of freedom and shape functions. Principally, elements with both linear and parabolic shape functions can be used in the same model. Element type and sizes greatly influence the behaviour of the model. These influences are investigated for 2D plane strain and 3D models and the results are discussed in the following sections. Calculations are carried out with different mesh refinement and with different element types. A hydrostatic stress condition is used and the overburden ranges from 100 to 600m. The tunnel diameter is 10m and no support is installed. The results from FE simulations are compared with analytical solutions after Feder ([9]).

Fig. 6 shows the crown settlements carried out with a 2D plane strain model with linear elastic material behaviour for meshes with different refinement. The model behaviour using 4 node elements with a linear shape function are compared to that of 8 node elements with a parabolic shape function. The results are very similar for all cases investigated. The divergence from the analytical solution by Feder ([9]) is generally very low.

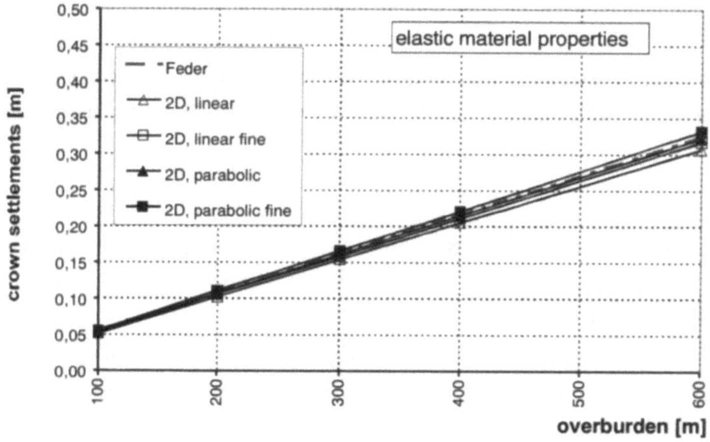

Fig. 6. Crown settlements for different overburdens carried out with meshes of different refinement and element types and elastic material behaviour

In Fig. 7, the same calculations are carried out with plastic material properties. Thereby essential differences in the results occur with different mesh refinements and element types. Using linear elements, the calculated displacements for an overburden in excess of 200m are too small. This effect is progressive with increasing overburden. A better agreement between the results is obtained with parabolic elements.

Mesh refinement effects the displacements equally whether elastic or plastic material behaviour is assumed. Independently from the element type, refined meshes behave "more flexible" and thus cause slightly higher displacements. With increasing overburden and consequently larger plastic zones around the tunnel, results obtained with linear elements increasingly differ from the exact (analytical) solution.

The difference in the displacements when using linear elements in contrast to parabolic elements results from the smaller plastic zones around the tunnel.

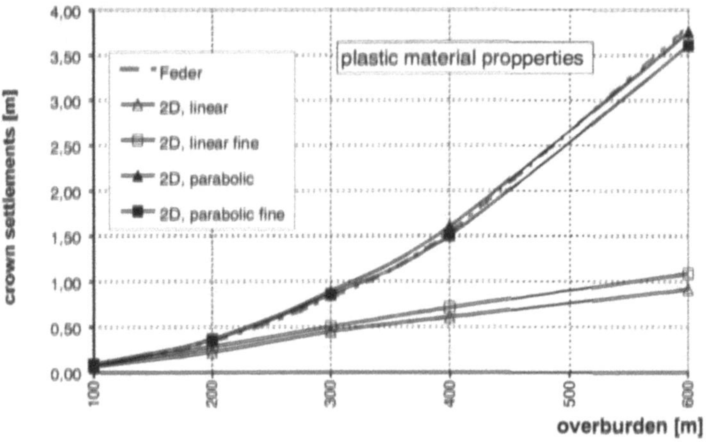

Fig. 7. Crown settlements for different overburden height (100-600m) carried out with meshes of different refinement and element types and plastic material behaviour

This can be seen in Fig. 8, where the development of the plastic zones is plotted for different element types and mesh refinement. For this example a 400m overburden was used.

The mesh refinement minimally effects the development of the plastic zones. With parabolic elements, wider plastic zones and a higher plastic strain are calculated.

The results suggest that, at least with non-linear material behaviour, parabolic elements should be used. But the use of parabolic elements implies a correspondingly high number of unknowns, which have to be solved and thus long calculation times.

The most effective way is to use parabolic elements only, where plastic material behaviour is expected during tunnel excavation. The development of plastic zones around the tunnel depending on the primary stress state is well known and can roughly be estimated with numerical and analytical methods. The efficient use of both linear and parabolic elements reduces the number of unknowns to be solved and thus calculation effort without reducing the accuracy of the result.

An example is shown in Fig. 9, where the displacements along a vertical profile above the crown are shown. The approach with a mesh of combined parabolic and linear elements leads to almost the same displacements as a mesh existing only of parabolic elements.

With the refinement of the mesh, the number of unknowns increase extraordinarily, even though the accuracy of the result are not improved that much.

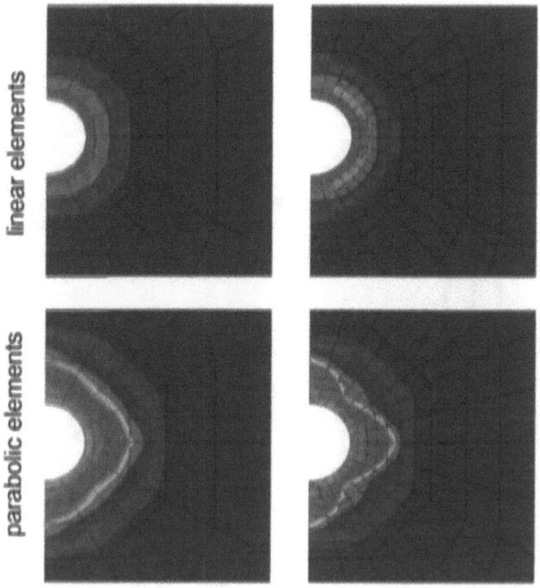

Fig. 8. Extension of plastic zones around the tunnel for different element types and mesh refinement

3.2 3D application

The statements made above for 2D simulations are valid for 3D applications as well. For a step-by-step excavation, the influence of the thickness of the mesh slices on the results was investigated.

With displacement trend lines, these model effects can be easily detected. If material parameters do not change and the excavation sequence and support installation remain uniform the displacement trend lines should be horizontal. Any trend line divergence from the horizontal indicates a mesh influence.

A study showed that in the area where reliable results are required the size of the mesh slices should be kept constant. As shown in Fig. 10, different mesh slice thickness strongly influences displacements in the vicinity of the boundary between smaller and larger elements. This influence increases rapidly with frequently changing slices and shows in strongly undulating trend lines.

Figure 11 shows the settlement trend line X+0 (referring to the tunnel face) at the transition from a zone of 5,0m mesh slices (zone A) to a zone of 1,5m mesh slices (zone B). Additionally, the final settlements are displayed over the same range. The calculations are carried out with linear elements and elastic-ideal plastic material conditions.

The crown settlements are considerably higher in zone B, caused by the thinner mesh slices. Comparative calculations with linear elastic material

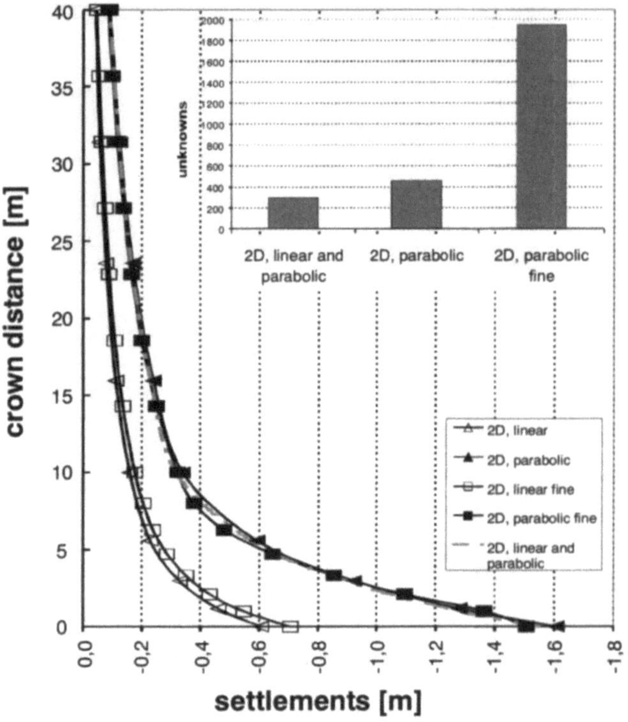

Fig. 9. Settlements for different elements and mesh refinement; comparison of number of unknowns

behaviour showed almost no effect due to changes in the dimension of mesh slices.

The only way to eliminate such mesh-induced problems is to use constant element sizes in the area of the model, where a high accuracy of the results is required. Comparative calculations should always be done with the same mesh geometry. Different round lengths must not be modelled by varying the corresponding mesh slice thickness in longitudinal direction. This would superpose the effects of round length and changing mesh stiffness ([10]).

3.3 Model geometry

Another model effect is the ratio between model width and the length of the model. This influence only effects longitudinal displacements. In Fig. 12 the deflection curves and trend lines for the longitudinal displacement are displayed for a calculation carried out with different meshes. The models differ in their ratio between the cross section area and the model length. With a ratio of W/L smaller than 0,4, trend lines show strong variations. Settlements and lateral displacements are not effected by this model effect.

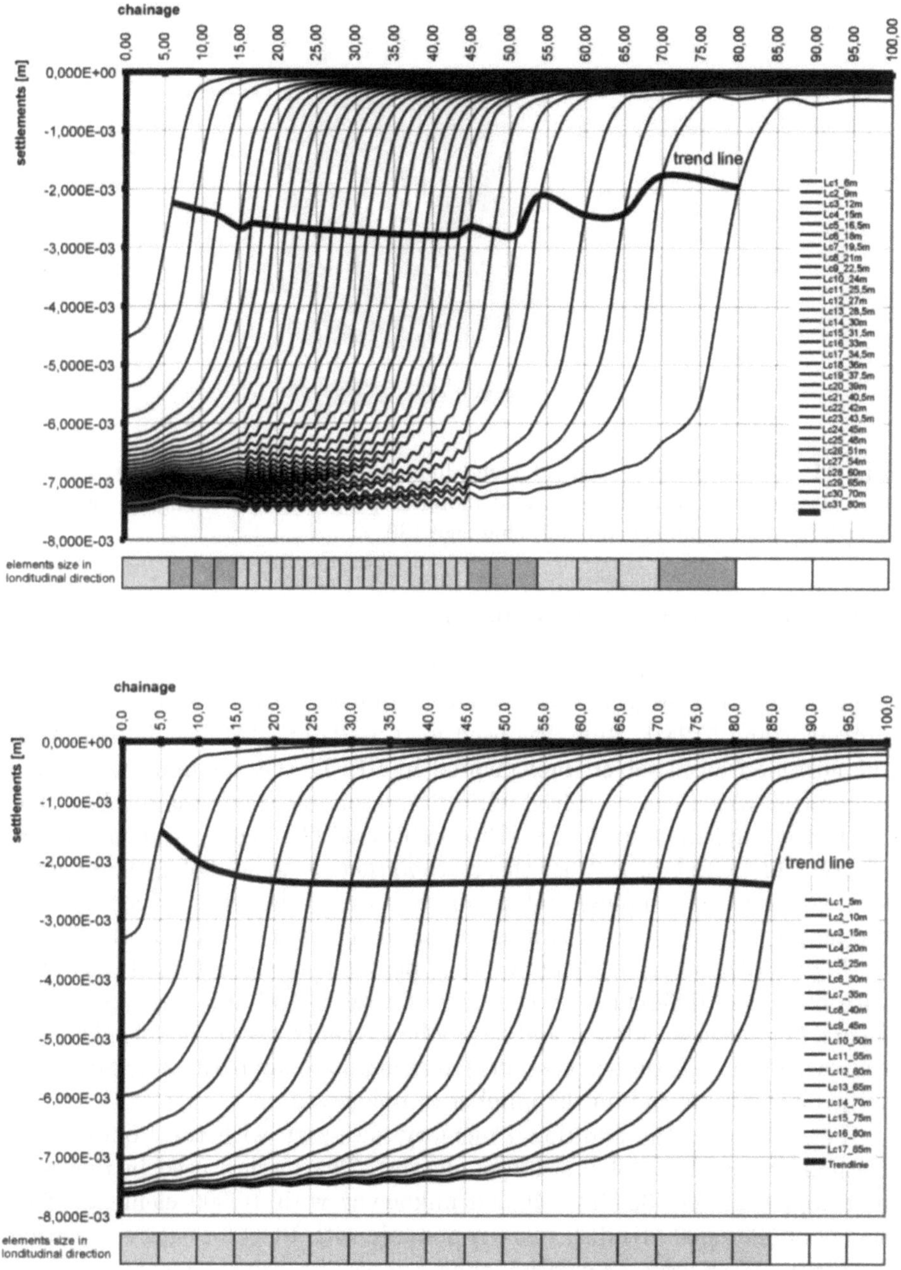

Fig. 10. Graphs of crown settlement curves; top model consists of elements with different sizes; bottom model consists of elements of uniform size

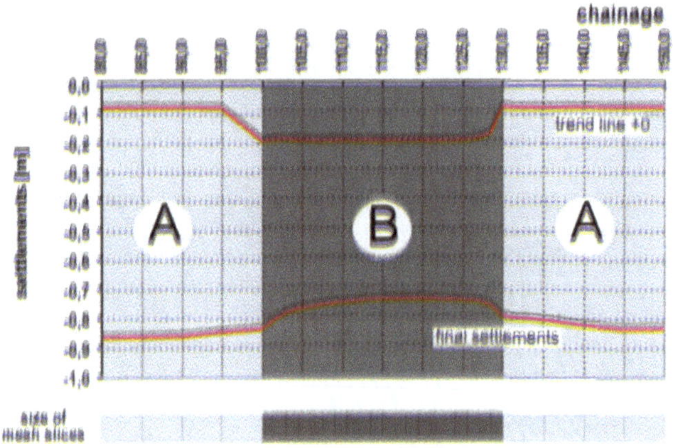

Fig. 11. Crown point settlement trend line (x+0) and final settlement line for a tunnel excavation with ideal plastic material properties for changing mesh slice length between chainage 100 and chainage 130

3.4 Improvement of the equation solver

Large meshes with up to 10.000 elements and 100.000 unknowns have been used during this project. A Personal Computer with a Pentium 450 MHz processor and 512MB Ram was used for the 2D and 3D calculations. With the current frontal solver implemented in BEFE, the maximum number of unknowns, which can be solved within a reasonable time, was reached. This led to the implementation of a "Preconditioned Conjugate Gradient solver" (PCG) ([11]) into BEFE. With this method there is no need of assembling and storing the global stiffness matrix for each excavation step, which occupies the major part of the total storage requirements and calculation time.

Figure 13 shows the calculation time per load step for a tunnel excavation with meshes of different sizes (number of mesh slices) and two types of elements. For the case of linear and parabolic elements, the saving of calculation time with the PCG per excavation step amounts to approximately 80%. This value does not depend on the number of unknowns.

Fig. 14 shows the additional memory necessary with the frontal solver method (assembled global stiffness matrix). The meshes used in this work had approximately 30.000 - 40.000 unknowns with 10.000 elements. So a saving of 600 - 800 MB disk space is possible with PCG for each calculation.

Figure 15 shows a calculation time comparison using a mesh with 9.804 linear elements and about 31.000 unknowns. This mesh size and the number of unknowns correspond with most of the meshes used in this work. In most cases 70 excavation steps were calculated. With the frontal solver, at least 55 hours per calculation were needed. With the PCG solver, results could be obtained after 8 hours. The calculation time depends on the convergence speed i.e. the number of iteration steps required to satisfy the convergence

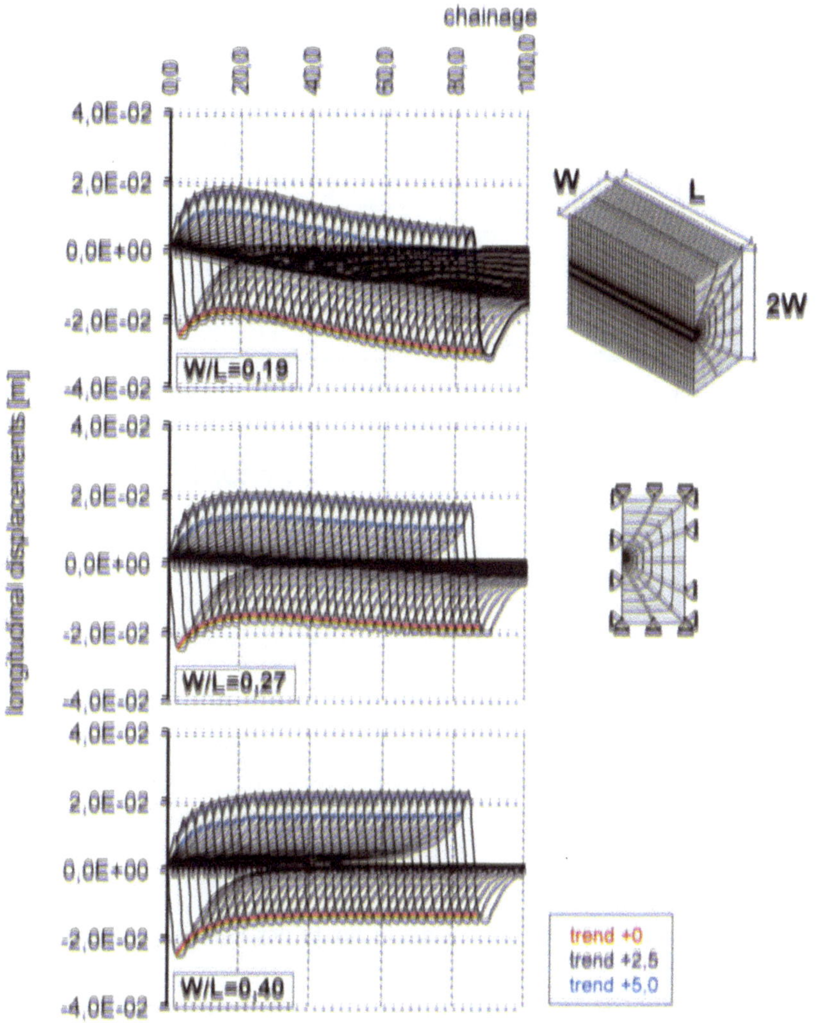

Fig. 12. Longitudinal displacements (deflection curves and trend lines) with different ratios between model width and model length

criterion. With a larger plastic zone around the tunnel the number of visco-plastic iterations might increase and correspondingly the calculation time increases. But nevertheless, most of the calculations with up to 70 excavation steps could be handled within 24 hours with PCG.

3.5 Summary

The influence of element size and types on the calculated displacements was investigated. For both elasticity and plasticity, double refined meshes do not

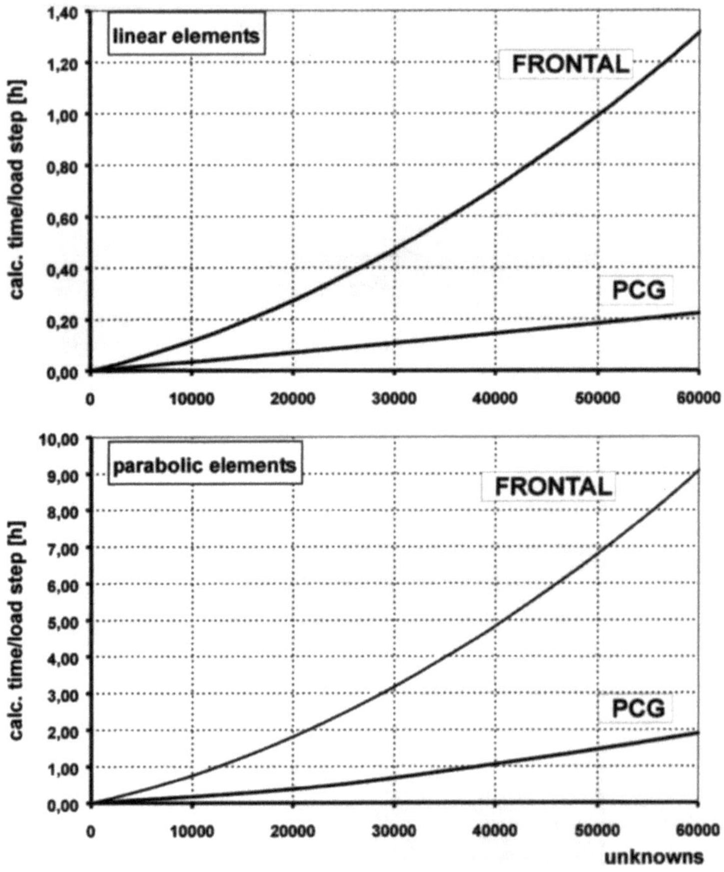

Fig. 13. Comparison of calculation time in hours per load step carried out with different equation solvers

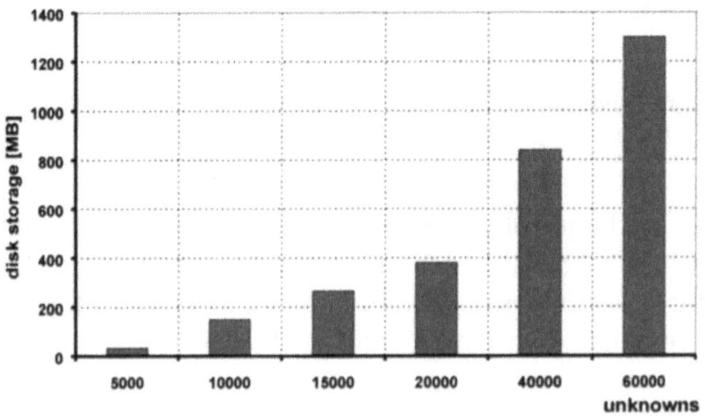

Fig. 14. Additional disk storage requirement with frontal solver

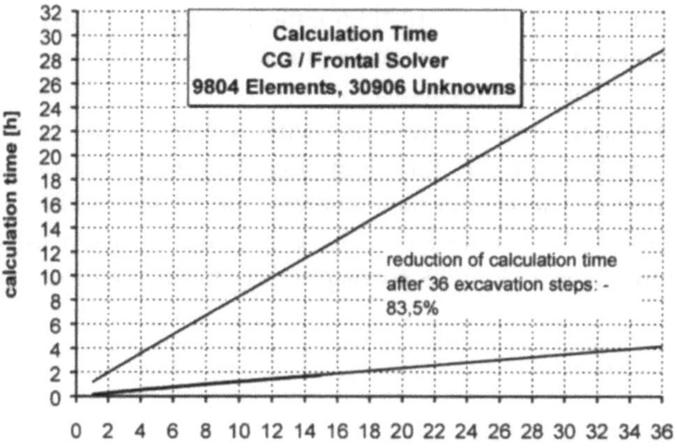

Fig. 15. Calculation time for simulation of a top heading excavation with a mesh existing of 9.800 linear elements and 31.000 unknowns

significantly increase the results accuracy. Parabolic and linear elements behave equally in elastic cases. Parabolic elements are much more accurate in zones with non-linear material behaviour. Using both parabolic and linear elements in one model, whereby the parabolic ones only cover those zones with expected non-linear material behaviour, turned out to be most efficient (maximum accuracy with least number of unknowns).

For sequential excavations with 3D models, the size of mesh slices must be kept constant to avoid mesh-induced influences on the displacements. Influences of the model geometry on the longitudinal results were investigated.

With a Preconditioned Conjugate Gradient (PCG), calculation time and disk storage requirement could be reduced appreciably in comparison with to the frontal solver. Models with 10.000 elements and up to 70 excavation steps could be handled within 24 hours.

4 Data evaluation

4.1 General

For a practical application of numerical simulations at tunnel sites, it is necessary to make calculation results directly comparable with monitoring data. Thus the applicability of numerical models to predict tunnel performance can be assessed quickly and objectively. Applying the same visualisation and evaluation methods used for displacement monitoring to the calculation results makes the comparison easier.

4.2 Deflection curves

In Fig. 16, it is shown how deflection curves can be used for comparison. The data structure of the deflection curves and trend lines is shown in a spreadsheet. The trend lines are then superimposed on the model geometry for visualisation and to demonstrate the comparability between numerical results and monitored performance.

Each deflection curve corresponds with a row in the data matrix. The trend lines result from the corresponding diagonals. "X+0" indicates the trend measured at the tunnel face whereby for "X+2" is the trend taken at nodes 2m behind the tunnel face. The trend can be taken for any distance from the tunnel face as long as corresponding mesh nodes are available.

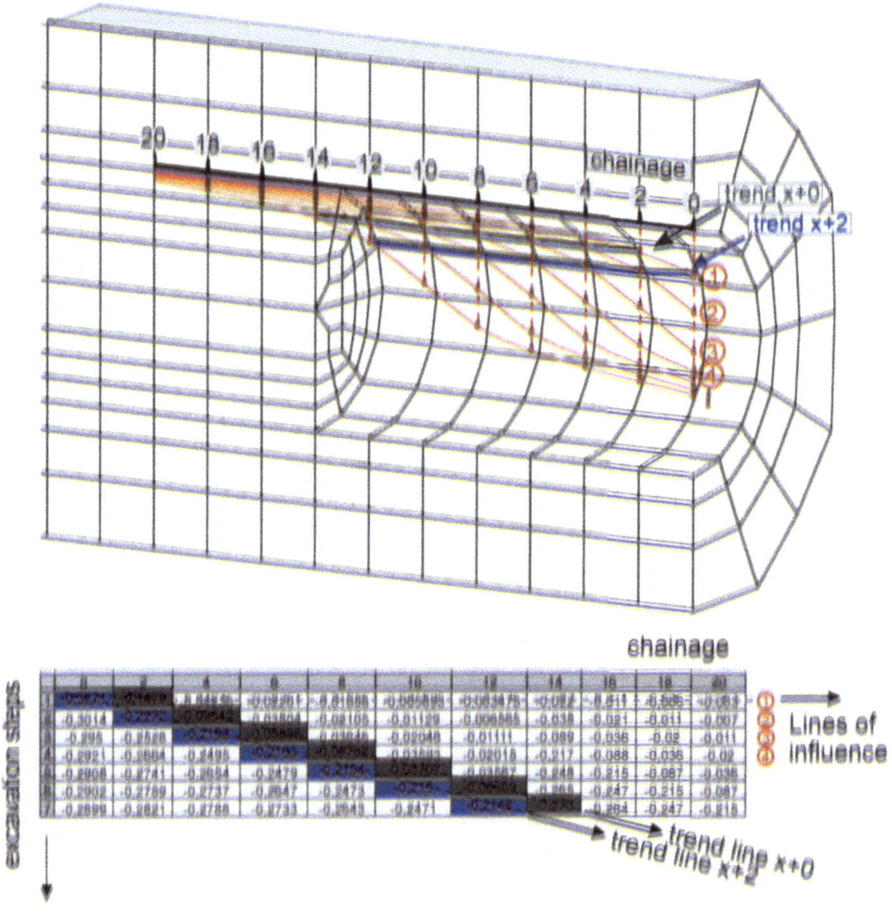

Fig. 16. Principle of "deflection curves" and trend lines for the crown point

Computed deflection curves of crown settlements in a practical layout are shown in Fig. 17. A situation in which the tunnel approaches a weak rock mass is used for demonstration. The transition from stiff rock to weak rock is at chainage 102,5m. From there on, the increased space between the deflection curves, which is the effect of a weaker rock mass, can be clearly observed. The deflection curve for a face position at chainage 90,0 (LI 90,0) is shown in pink colour in this diagram. Trend line x+0 corresponds with the displacements of the crown points at the tunnel face. It shows the magnitudes of the pre-settlement ahead of the tunnel face.

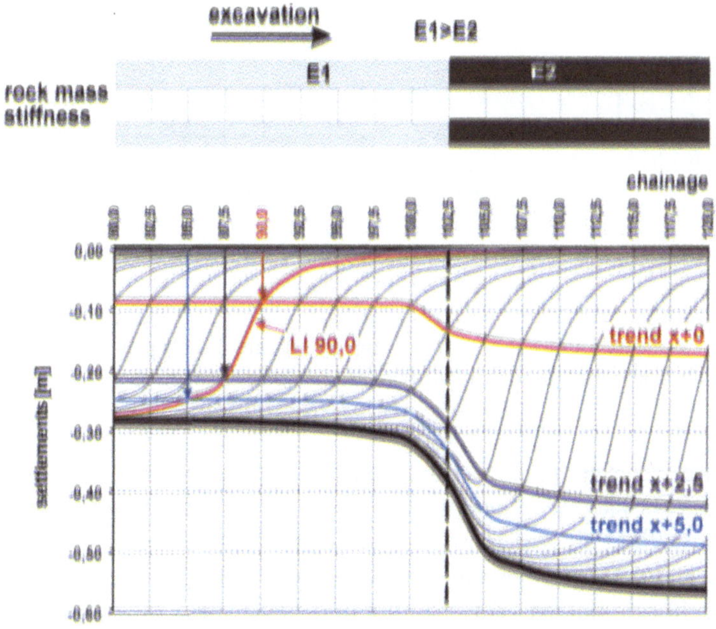

Fig. 17. Computed deflection curves and trend lines for top heading excavation when approaching a weak region

Figure 18 shows the deflection curves and trend lines for crown settlements of a simulated sequential top heading/bench excavation for a 30m long section. The complete model has a length of 200m. The trend lines refer to the crown points at the tunnel face respectively 5m behind the face and are horizontal as long rock mass conditions do not change. Settlements caused by the top heading excavation are about 33mm, while the bench excavation induces an additional settlement of 7mm. With the influence of top heading and bench excavation, total settlements amount to 40mm. With this display, the spatial influence of each excavation step can be visualized.

4.3 Displacement history plots

With displacement history plots, the columns of the result data matrix in Fig. 16 are plotted. The difference to the plots of monitoring data is, that dis-

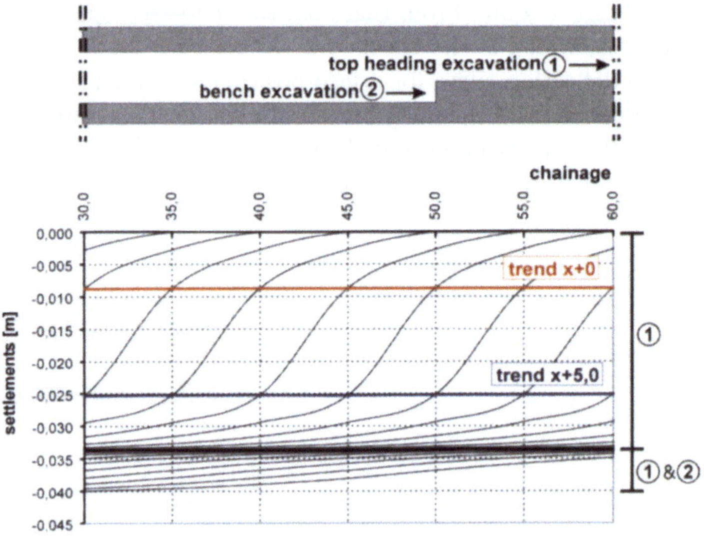

Fig. 18. Computed deflection curves and trend lines for top heading/bench excavation with homogenous rock mass conditions

placement data refer to excavation steps but not to real time. Time dependent displacements like creep are not considered with this numerical simulation, thus the results only consider the face advance effect.

In Fig. 19, the settlements of the crown point of the cross section at chainage 40 are plotted. Top heading and bench pass the cross section with excavation step 8 and 29, respectively. Settlements that occur before step 8 are pre-settlements, which cannot be easily monitored in reality.

4.4 Displacement vector orientation

In Fig. 20, calculated trend lines of displacement vector orientation of a top heading excavation approaching weak rock mass conditions are displayed. The round length is 2,5m. The trend lines refer to 2,5m, 5,0m, and 7,5m distance from the tunnel face. The calculated vector orientations basically match the measured data. A fundamental difference is in the absolute orientation of the vector. In measurements, the vector orientation within homogeneous rock mass conditions (normal state) is approximately 10 ° against the direction of excavation. The calculated vector orientations tilt in direction of excavation.

This deviation results from the fact, that displacement monitoring can start only after the excavation has been made, while the calculation shows the whole history also ahead of the face. To be able to compare the calculated results with the measured ones, a normalization is required.

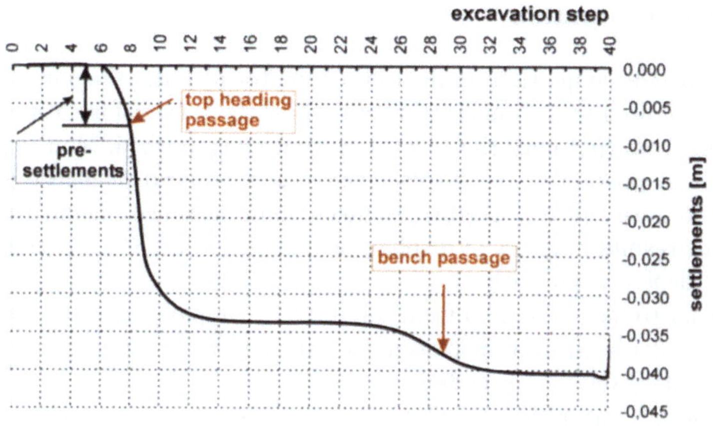

Fig. 19. Computed settlement history diagram of sequential top heading/bench excavation

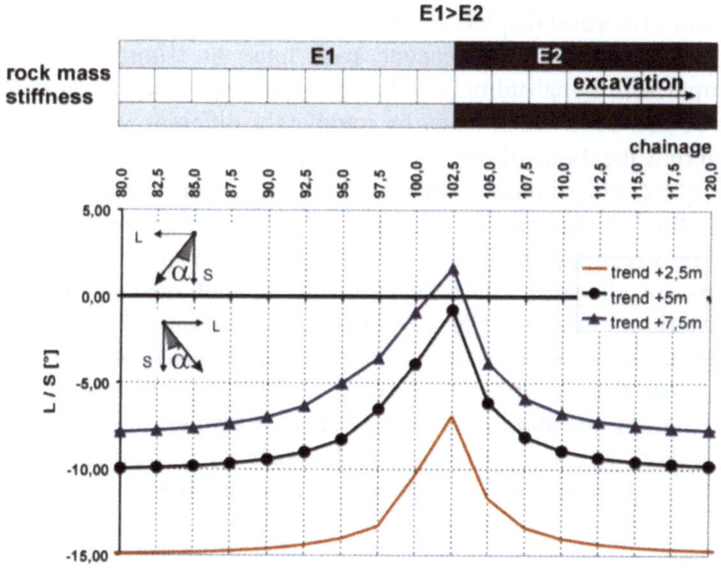

Fig. 20. Calculated trend lines of displacement vector orientations of the crown point when approaching a weak region

4.5 Data calibration

To compare the simulation results to the monitoring data, the numerical data have to be edited first. Numerical results show the total deformation process, beginning from (p_o) until the excavation process is finished. Monitoring data only show a portion of the displacements as a reduced displacement path. This is because the zero reading is taken some distance behind the tunnel face. This results in a loss of displacement information and a corresponding reduced displacement path.

In order to make the data from simulations comparable to the monitoring data, the calculated total displacement path has to be reduced according to the monitoring data zero reading.

In Fig. 21, the influence of the zero reading location on the evaluated displacement vector orientation is shown.

A "zero reading" at the tunnel face is achieved by subtracting the displacements of step 1 from the total displacements. For a zero reading, taken one excavation step behind the face, displacements of step 1 and 2 are subtracted from the total displacements. The later the zero reading is made, the more of the calculated displacement path must be trimmed to allow a direct comparison. Longitudinal displacements change orientation twice in the vicinity of the face, which leads to completely different vector orientations with different locations of the "zero reading".

In Fig. 22, longitudinal displacement history lines referring to different "zero reading" locations are displayed. The evaluation approach referring to the "zero reading" behind the tunnel face corresponds best with measurements. In all cases, the relative trends do not depend on the respective evaluation method or the "zero reading" location. Settlements increase when approaching weak material, longitudinal displacements increase relatively against the direction of excavation. This is confirmed by measured trends. Focusing on absolute values requires a proper determination of the "zero reading" location, especially with longitudinal displacements, which tend in opposite directions with different "zero readings".

5 Displacement vector orientation

5.1 General

Attempts to verify the observations on site by 3D numerical simulations were initiated by Budil ([12]). He used the boundary element method (BE) included in BEFE for his work. The simple model preparation and low computation time with BE in contrast to FE was the motivation to use this method. But the main disadvantage was, that only elastic material behaviour could be considered. Further, the simulation of step-by-step excavation was not possible.

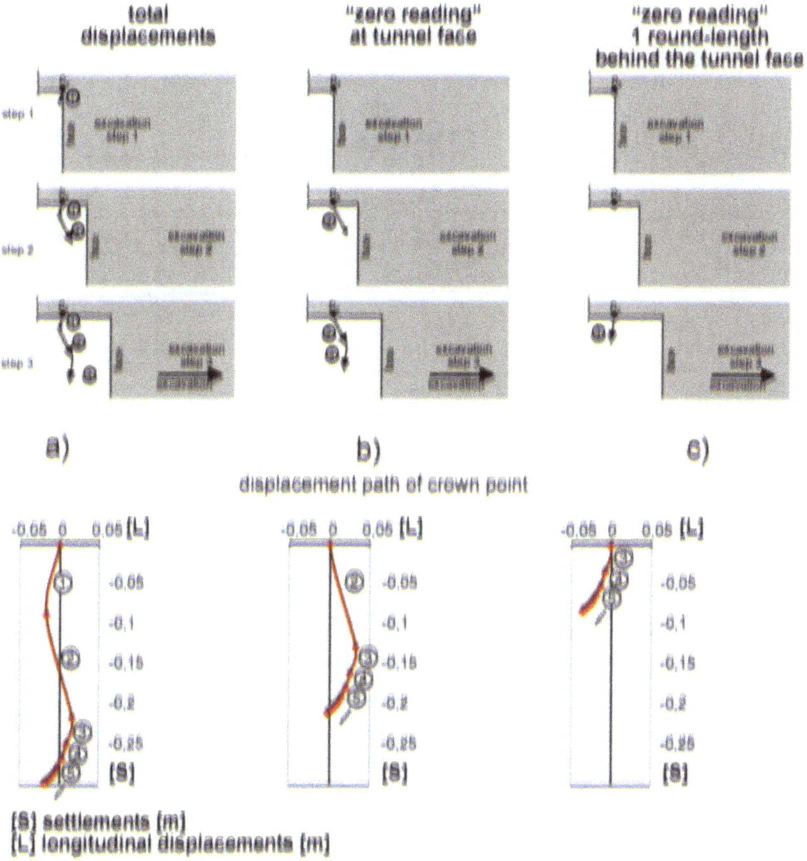

Fig. 21. Influence of the location of the "zero reading" on the evaluated displacement vector orienta-tion.
a) total displacement path
b) monitored displacement path in the case of "zero reading" at the face
c) monitored displacement path in the case of "zero reading" 5m behind the face

Steindorfer ([13]) carried out additional calculations with the BE method focusing on the deformation behaviour and stress rearrangement in the vicinity of the tunnel face.

With the implementation of the PCG solver in BEFE, the calculation of finite element models with up to 10.000 elements became efficient. This was essential for the investigation of longitudinal displacements, which requires large meshes to avoid boundary influences. With the step-by-step excavation method and the ability to consider non-linear material behaviour, tunnel excavation in heterogeneous rock mass conditions can be more realistically simulated.

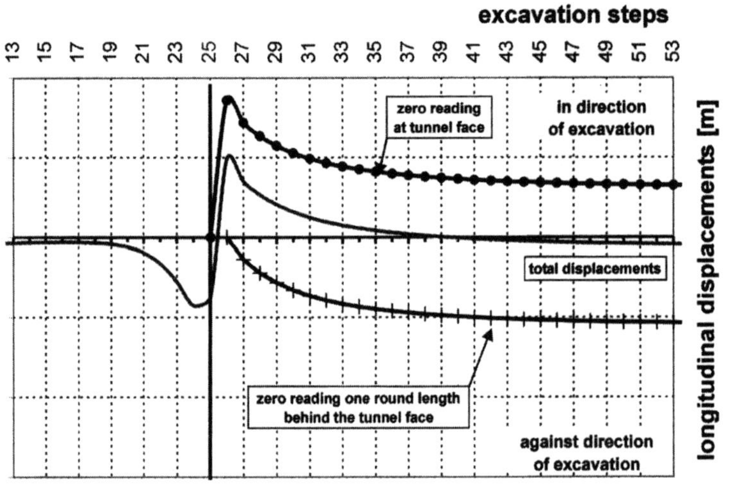

Fig. 22. History diagram of longitudinal displacements referring to different locations of zero readings

In the following section, the development of longitudinal displacements and the orientation of displacement vectors are studied for homogenous ground conditions with elastic and ideal-plastic material behaviour. Further the influence of changing rock mass conditions on the displacement vector orientations are studied as well as the influence of different primary stress conditions.

5.2 Finite element model

The 3D model contains approximately 10 000 linear isoparametric finite elements. The boundary conditions are displayed in Fig. 23. Assumptions include plane strain conditions at the model boundaries and a vertical symmetry plane along the tunnel axis. The model consists of 2,5m and 20m slices with a total length of 240m. The tunnel is circular with a diameter of 10m. The tunnel is excavated full face step-by-step, with a total of 70 excavation steps (load cases) until station 192,5. The maximum primary stress is generally taken to 40 MPa, the K_0 –value is 0,5. To arrive at sufficiently accurate results the following model conditions are determined:

- Element size and thickness of the slices are constant along most of the model.
- The minimum ratio between the cross section area and the model length is maintained.
- Because of the longitudinal boundary conditions, reliable results cannot be obtained within 2,5 tunnel diameters (in this case 25m) from the boundary of the model. This section can be excavated in one step using a thick slice without affecting the accuracy of the results.

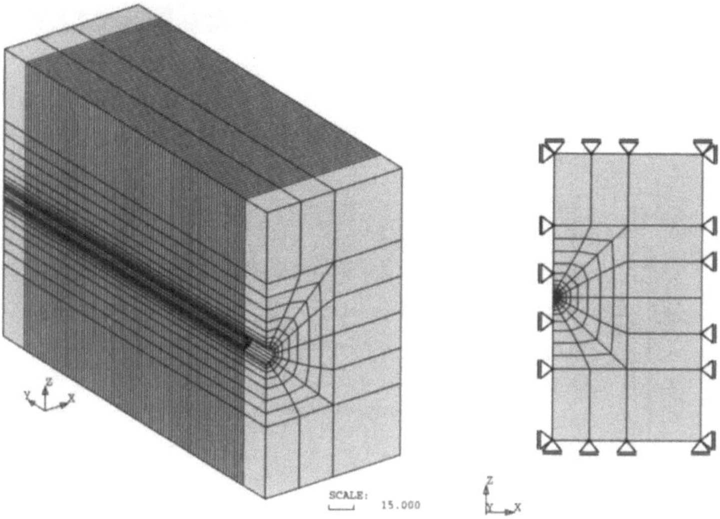

Fig. 23. 3D Finite Element Model

5.3 Influence of relatively "soft" rock ahead of the face

To simulate a tunnel approaching a "soft" rock mass, the model is split into two regions of different material stiffness, 2.000 MPa for E_1 and 400 MPa for E_2, and with the Poisson ratio of 0,25 for both regions. The boundary between "stiff" and "soft" rock is at station 102,5.

Fig. 24 shows the normalised displacement vector orientation trend lines. Trends for the crown points 2,5m (1/4 D), 5,0m (1/2 D) and 7,5m (3/4 D) behind the tunnel faces are displayed. Before station 80 a "normal" displacement vector orientation can be observed. In homogenous ground conditions the distance from the face to the measuring point only influences the trend line magnitude while the shape of the curve is unaffected. Approximately 2 tunnel diameters (20m) in front of the transition the vector orientation deviates from the "normal", pointing more against the excavation direction. The vector orientation returns to the "normal" position approximately 2 diameters behind the transition. It is interesting to note, that the trend taken at 7,5m behind the face shows a larger deviation from the "normal" (20 °) than the trend 2,5m behind the face (17 °).

The "normal" vector orientation is the same in both the "soft" and "stiff" material, although the radial displacements in the "soft" material are a multiple of those in the "stiff" material.

Figure 25 shows a comparison between the settlement trend lines and the displacement vector orientation. It can be seen that the change in rock mass stiffness can be recognised much earlier by evaluating the vector orientation trend than the radial displacements only.

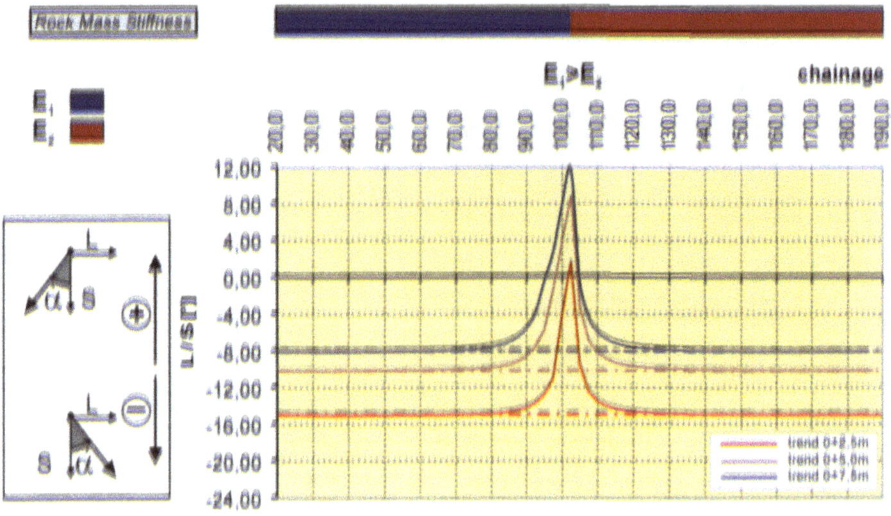

Fig. 24. Trend lines of displacement vector orientation of the crown point, trends shown 2,5m, 5,0m, and 7,5m behind the face when approaching comparatively "soft" rock

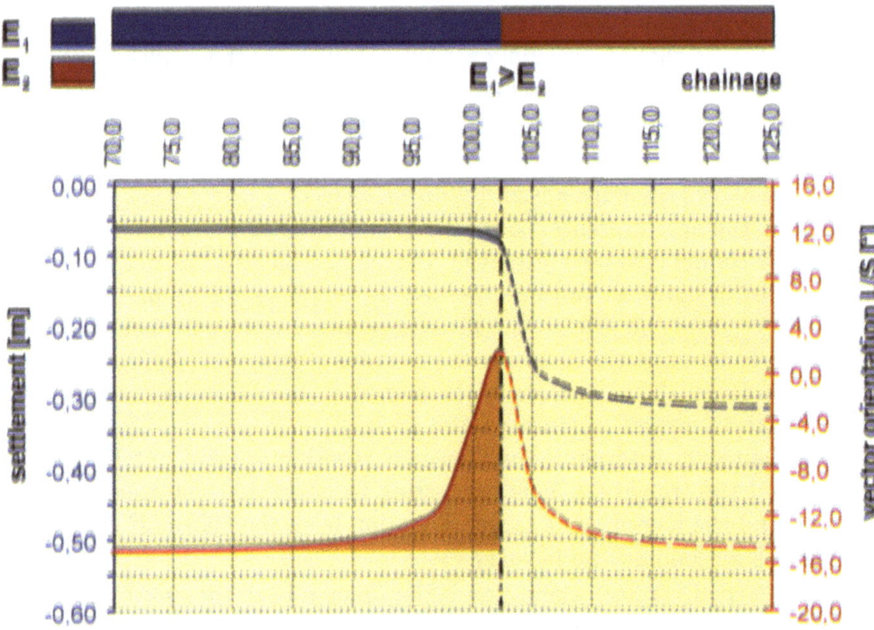

Fig. 25. Comparison between trends of settlement and displacement vector orientation (L/S), when excavation approaches relatively "soft" rock

5.4 Influence of relatively "stiff" rock ahead of the face

The influence of a stiffer rock mass ahead of the face on the displacements was studied with the same mesh as used for the previous case. The E-modules for the two regions are taken to E_1=2.000 MPa and E_2=10.000 MPa, boundary conditions and primary stress situation are unchanged. The results are shown in Fig. 26.

Compared to the previous model, the vector orientation trend is opposite to that observed when the excavation approaches the "stiff" rock mass. Approaching the boundary to the "stiff" region, the vectors point more in the direction of the excavation. The zone of influence in which the divergence of the vector orientation from the "normal" is observed is the same as in the previous case.

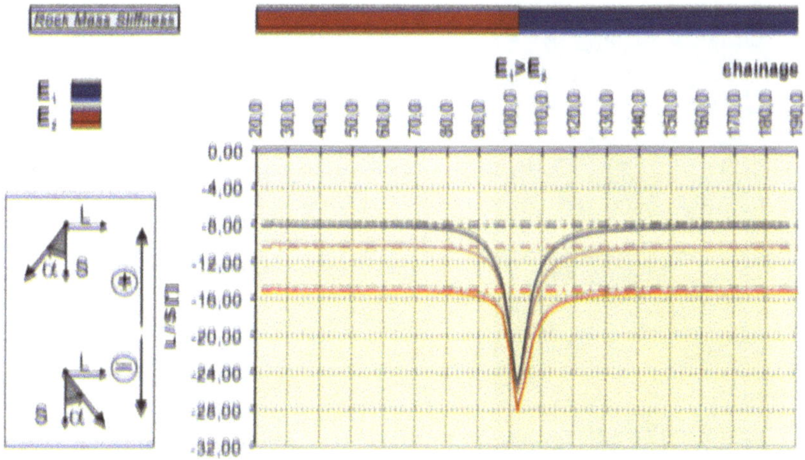

Fig. 26. Trend lines of displacement vector orientation of the crown point, trends shown 2,5m, 5,0m, and 7,5m behind the face when approaching comparatively "stiff" rock

5.5 Stress conditions in the vicinity of the face

In Fig. 27a the contours of the maximum principal stresses are shown for cases when the excavation approaches a "stiff" rock mass, for a homogeneous rock mass and a transition to a "soft" rock. The tunnel face in all cases displayed is two round lengths (half a tunnel diameter) ahead of the transition. When excavation approaches a "soft" rock mass, stress concentrations develop between the tunnel face and the region with "soft" rock. When the excavation approaches a "stiff" rock a stress increase is observed in the "stiff" material, with lower stresses between the face and the stiff region. The different stress conditions influence the spatial displacements around the excavation.

In Fig. 27b, the corresponding maximum principal stresses along the tunnel sidewall and trend lines for the excavation up to the transition are shown. Starting from around two excavation steps ahead of the transition, differences in the stress trend (red lines), especially in front of the face can be observed. The trend lines show a stress increase when approaching weak rock and correspondingly a decrease when approaching stiff rock. The same trends are more pronounced with a further excavation towards the transition.

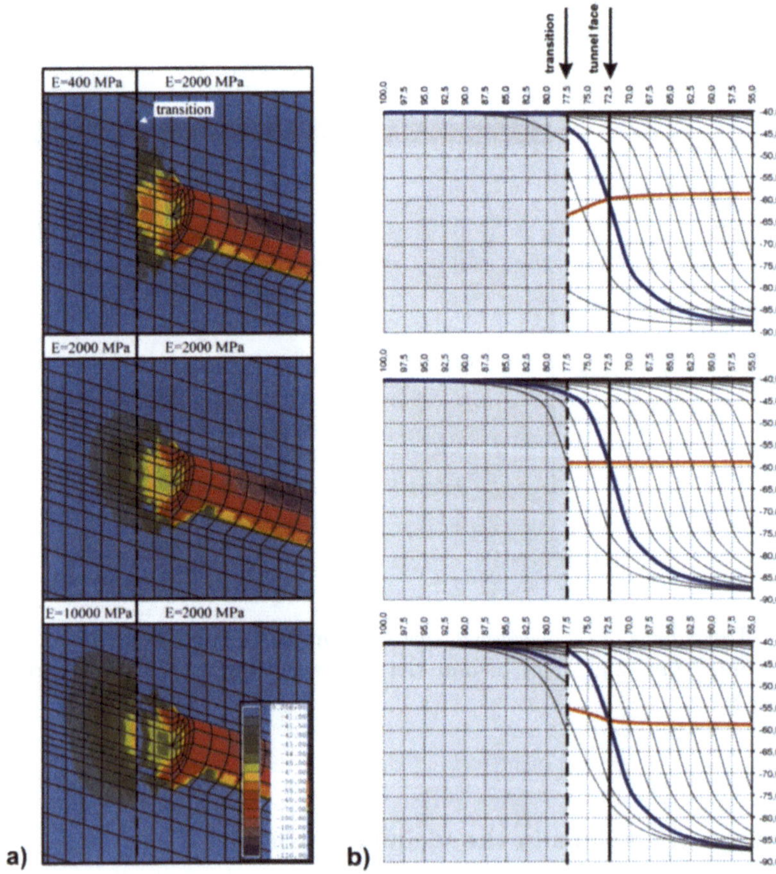

Fig. 27. a) Contours of maximum principal stresses
b) Deflection curves for the maximum principal stress in the side wall when approaching "stiff" and "weak" rock respectively in comparison with homogeneous rock mass conditions

5.6 Tunnelling through rock mass with frequently changing stiffness

The next case studied is a sandwich type rock mass with a sequence of sections with different stiffness and of different length. The E-moduli of the different sections are: E_1=1.000 MPa, E_{12}=800 MPa and E_2=500 MPa. Other conditions were the same as in the models described before. Figure 28 shows the structure of the model and the trend lines of displacement vector orientation. As shown in the previous computations, the heterogeneity of the rock mass significantly affects the displacement vector orientations. The relative difference in stiffness between two regions influences the magnitude of the displacement vector orientation deviation from the "normal". The diagram also shows, that the vector orientation within longer homogeneous regions normalises, while in short regions the trend changes without an intermediate normalisation.

This finding can contribute to more accurate decisions on site. When encountering better rock mass conditions within a fault zone a decision has to be made to what extent excavation and support shall be modified. This decision depends on the length of the zone with better rock mass quality and on the rock mass quality itself, as well as organizational and economical considerations. Using the trend of the displacement vector orientation one can determine if the better region is short or more extended.

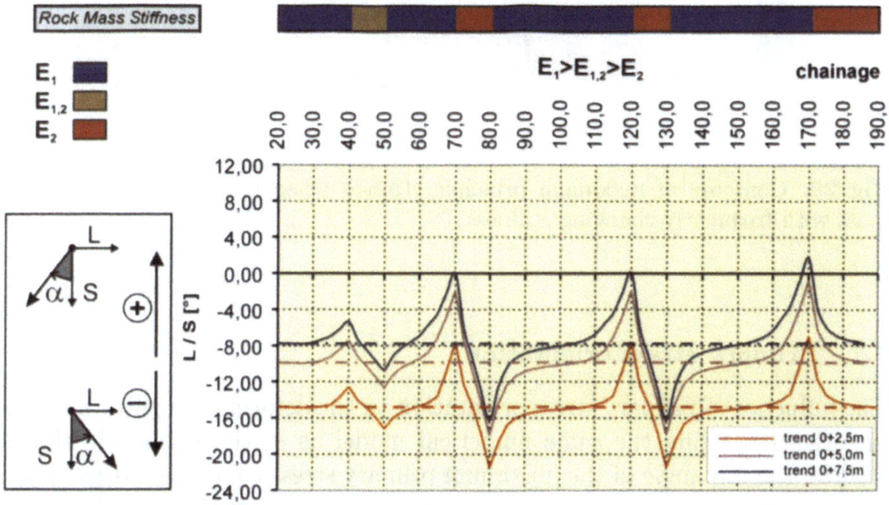

Fig. 28. Trend lines of displacement vector orientation of the crown when tunnelling through rock mass with frequently changing stiffness; trends shown 2,5m, 5,0m, and 7,5m behind the face

Figure 29 shows the maximum principal stresses, with concentrations in the "stiff" rock mass at the boundary between "stiff" and "soft" rock, and a reduction in the "soft" section. This effect is the more pronounced, the bigger the difference between the stiffnesses of neighbouring sections is.

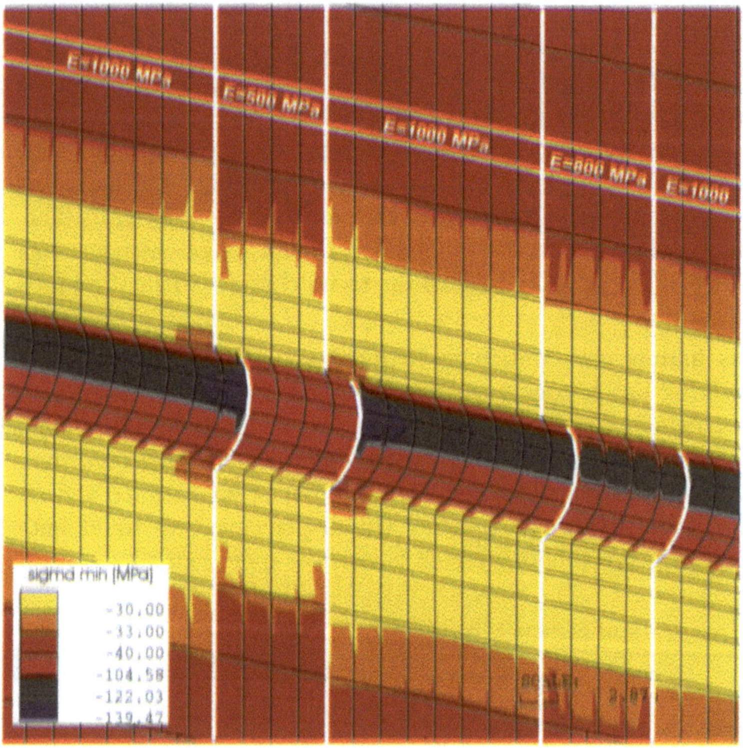

Fig. 29. Contours of maximum principle stresses when tunnelling through rock mass with frequently changing stiffness

5.7 Primary stress orientation

The influence of primary stress orientation on the deformation pattern is also investigated. Using the same numerical model as in the previously shown studies, the dip angle of the maximum primary stress component σ_{I}gs varied from 60 ° from the horizontal (against direction of excavation) to 120 ° from the horizontal (towards direction of excavation). The values of $\sigma_{\mathrm{II}g}$nd $\sigma_{\mathrm{III}g}$ are chosen to be 0,5 σ_{I}.

To evaluate the results, data points around the tunnel perimeter at a distance of 5 m (1/2D) behind the face are selected. The normalised displacement vectors are shown in stereographic projection ([14],[15]). The intersections of the displacement vectors with the upper and lower hemisphere were connected. Fig. 30 shows the intersections of the normalised displacement vector orientations for the upper and lower half of the tunnel with the lower, and the upper hemisphere respectively.

The results show a significant influence of the orientation of the primary stresses on the crown and invert, while the orientation of the displacement vectors at the sidewalls is nearly unaffected.

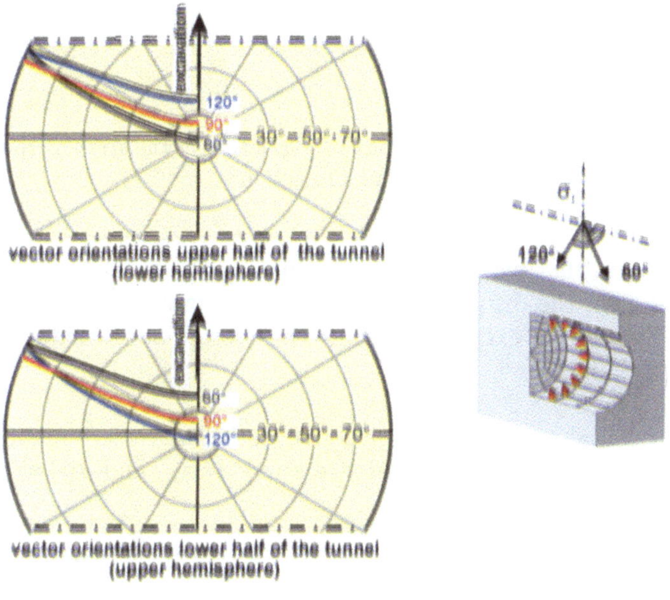

Fig. 30. Normalised displacement vector orientation for different primary stress orientations shown in stereographic projection ($K_0 = 0,5$)

5.8 Variation in far field stress

Variations of the ratios between the principal stresses have been studied. Following cases were investigated, and the results shown in Fig. 31 to Fig. 33:

⇒ Stress in vertical direction (σ_v) same as in longitudinal direction g(σ_L), horizontal stresses perpendicular to the tunnel (σ_T) axis double and triple of the vertical stress ($\sigma_v = \sigma_L$, and $\sigma_T = 2$ and 3* σ_v, respectively).

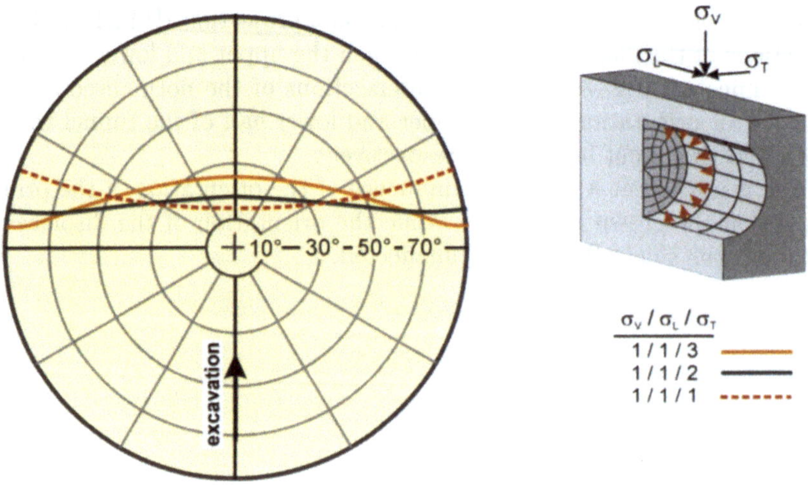

Fig. 31. Displacement vector orientation for a variation in the stress perpendicular to the tunnel axis shown in stereographic projection

It can be seen, that with an increase of horizontal stresses perpendicular to the tunnel axis, the displacement vector in the crown increasingly tends to point in direction of excavation, while the vectors at the sidewalls show the opposite trend.

⇒ Same horizontal stresses in longitudinal and transverse direction being half, double or triple the vertical stress ($\sigma_T = \sigma_L$, and $\sigma_V = 1$, 2, 1/2 and 1/3* σ_L, respectively).

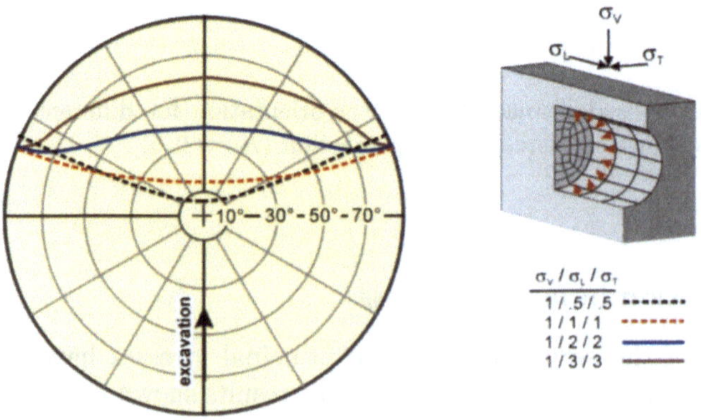

Fig. 32. Displacement vector orientation for a variation in the horizontal stresses shown in stereographic projection

In this case the study shows an increasing trend of the displacement vector orientation at the crown in direction of excavation with increasing ratio $\sigma h/\sigma v$, while the orientation of the vectors at the sidewalls is very similar in all cases.

⇒ Same stresses in vertical and transverse direction, longitudinal stress increased to double and triple of the vertical stress ($\sigma_V = \sigma_T$) and $\sigma_L = 2$ and 3* σ_V, respectively).

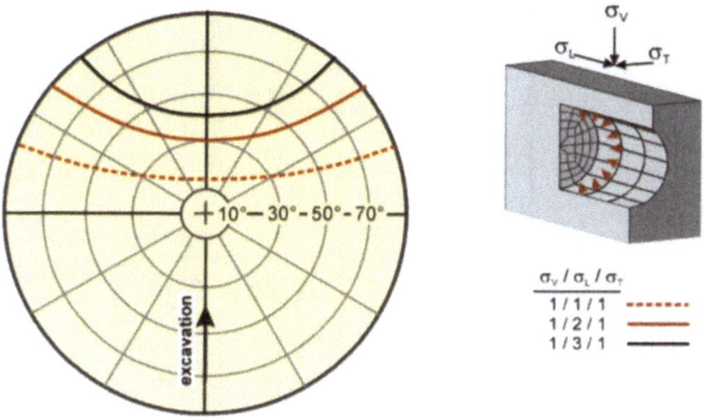

Fig. 33. Displacement vector orientation for a variation of longitudinal stresses shown in stereographic projection

With increasing longitudinal stresses all displacement vectors tend to increasingly point in the direction of excavation

The effect of varying K_0 was also investigated. K_0values in this study ranged from 0,5 to 0,9. The total displacement paths are shown in Fig. 34. It is observed that for high K_0 values the change in longitudinal displacements in the vicinity of the face is much higher, than for small K_0 values (step 1 and 2).

6 Case study S6

6.1 General

The tunnel Spital is part of the S6 Semmering Expressway. (Fig. 35) shows a plan view of the project. The twin tunnel system is sequentially excavated with top heading, bench and invert excavation (New Austrian Tunnelling Method). The overburden height varies from 7 to 87m, with an average of about 40 m.

In the first half of the tunnel sheared phyllites with isolated lenses of quartzite, rauhwacke and marble prevail. Half of the advance is predominantly

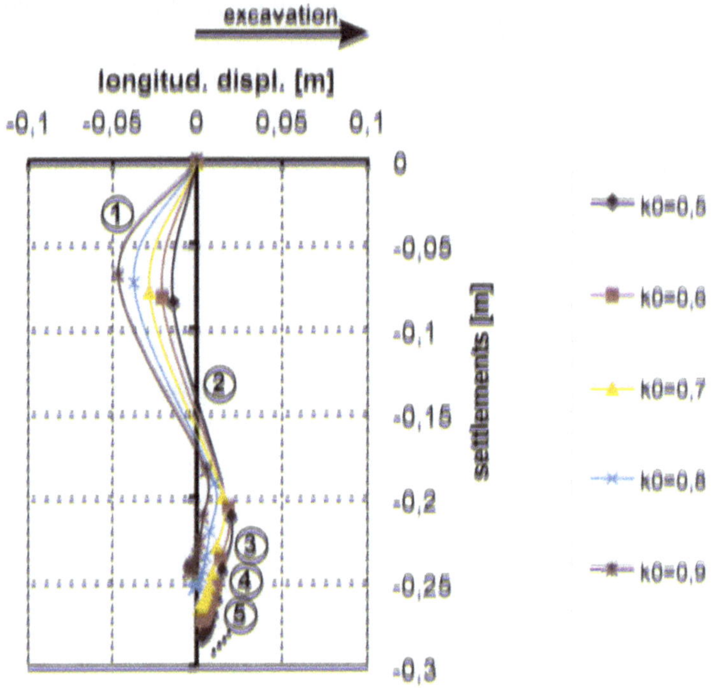

Fig. 34. Total paths of the displacement vector orientation of the crown with increasing K_0 values

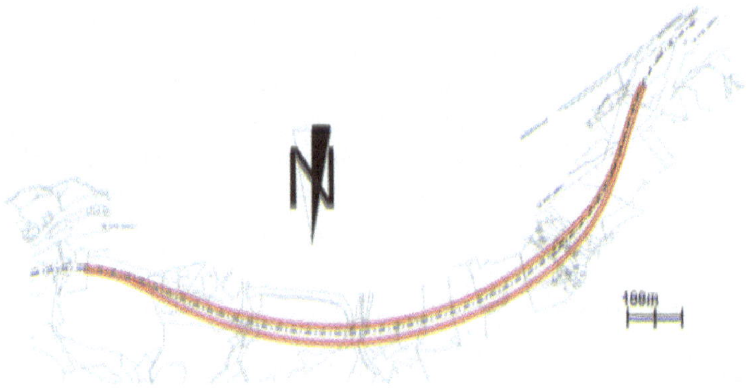

Fig. 35. Plan view of tunnel site S6 / Tunnel Spital

located in quartzphyllites. Squeezing ground conditions and local water inflow were expected for most of the tunnellength.

The support measures and excavation sequences for the squeezing sections are designed as follows.

- Face support by shotcrete and rock bolts.
- 25 - 30cm shotcrete lining and temporary lining for top heading invert
- Lattice girders, forepoling, and systematic pattern of self-drilling, post grouted rock bolts.
- Ring closure at a maximum of 80m behind the face

6.2 Problem statement

Around 500m from the west portal the tunnels underpass several houses with an overburden of approximately 18m. To avoid damage to the houses, the maximum tolerable surface settlement was determined to be 4cm. Experience from the first 200m of the tunnel excavation showed, that surface settlements in average reached 9cm above the tunnel, with a maximum of approximately 16cm. As the rock mass conditions in the area of where the houses were located was expected to be at least as poor as in the previous excavated sections, methods to reduce surface settlements were examined.

One of the options discussed was the division of the top heading into two parts with a staged excavation and a temporary lining of the sidewall. Finite element simulations were applied to study the effect of the staged excavation and the influence of the distance between the faces.

An alternative option discussed was a full top heading excavation with a "pipe roof". A finite element model and a coupled finite element / boundary element model were used to study the effects on the displacements of this option.

6.3 Numerical simulations

The predicted geological condition in the vicinity of the houses is shown in Fig. 36. For the FE simulation the ground model was simplified by using one material only instead of considering the rauhwacke layer and alluvial material overlaying the tunnel. The Geological Strength Index (GSI) (Lit 16) was used to determine rock mass properties from laboratory tests.

The goal of the simulations was to evaluate the effect of the staged top heading on the surface settlements.

6.4 2D numerical simulation

Model generation. The mesh consists of 964 isoparametric, 8 node elements. The nodes at the lateral and bottom boundaries were fixed. To eliminate boundary effects the distance from the tunnel axis to the vertical boundary is 8 tunnel diameters and to the bottom boundary 4 tunnel diameters

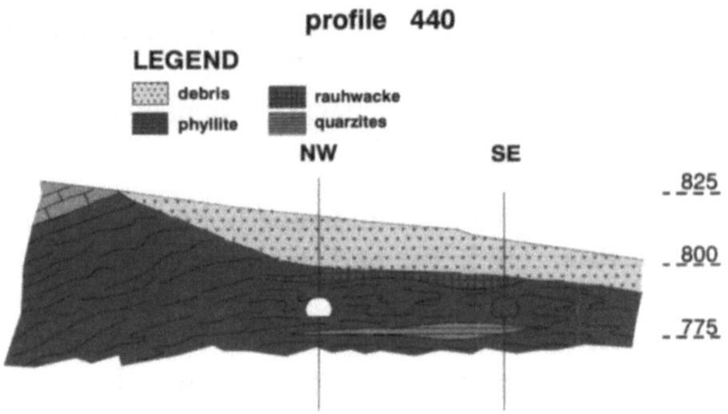

Fig. 36. Geological cross section at station 440

(Fig. 37). The upper mesh boundary represents the ground surface. The overburden is about 23m.

The vertical component σ_z of the primary stress was assumed to be the overburden pressure with a specific gravity of $25KN/m^3$. The horizontal primary stress was assumed to be 50% of the vertical stress.

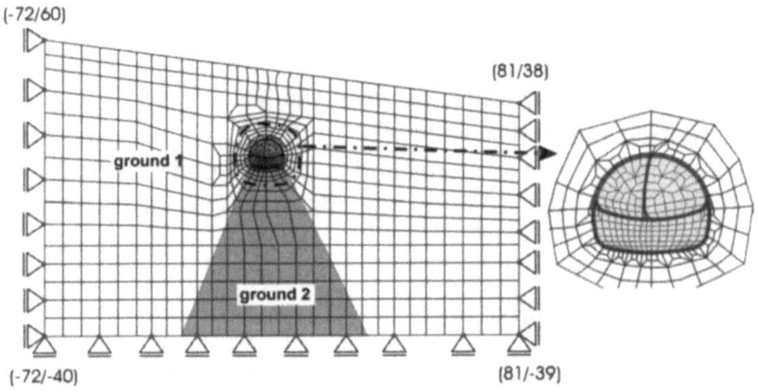

Fig. 37. 2D finite element model; material types for ground; boundary conditions

Load steps. In this simulation, the load reduction method was applied. The excavation steps and the "back-fill" sequences for staged top heading are shown in Fig. 38. Each section was excavated in two steps. With each step, "young" shotcrete is installed, and previousely installed shotcrete is replaced by hardened shotcrete.

The load factors were obtained by evaluating the surface settlements along the tunnel axis from station 82 to 250. The ratio of the pre-settlements and

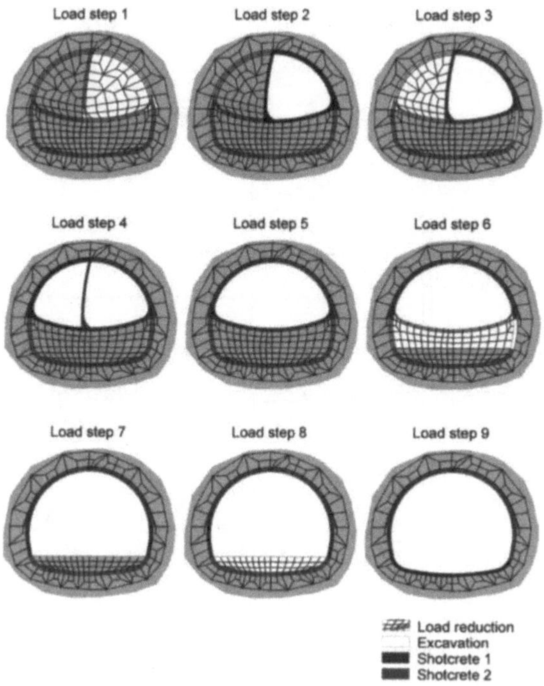

Fig. 38. Load steps for the staged excavation

the total settlements from top heading excavation and bench excavation are identified from monitoring results (Fig. 39). The load factors correspond to the ratios between the pre-settlements and the total settlements. At cross section 105, pre-settlements amount to 34% of the total top heading settlements and 55% for bench excavation respectively. The load factors for the top heading and bench excavation were taken proportional to the settlement components

Calculation results. The calculation results are plotted as pseudo time histories (displacements vs. load case) for different points on the tunnel perimeter. This kind of visualisation is considered to be the best way for practical use, especially when the direct comparison with measuring data (time histories) is necessary. Figure 40 shows the complete settlement paths calculated for full top heading and staged excavation. The evaluation is made for a surface point, the crown, the upper and lower sidewall points. The position of these points approximately corresponds to the measuring points in the tunnel cross section.

Monitoring data of tunnel displacement do not show the whole settlement paths, but only the part after the zero reading while for the surface point the

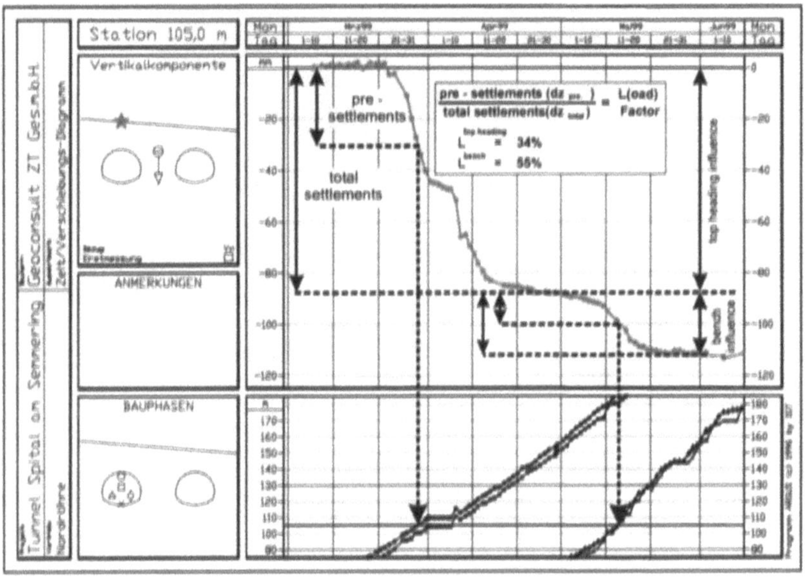

Fig. 39. Data sheet of monitored settlements at cross section 105

whole path is available (Fig. 41). Fig. 42 shows the calculated "displacement" histories without predeformation.

It was observed that the calculated shape of the surface trough did not agree with the measured one (Fig. 43, Fig. 44). The excavation sequence showed minimal effects on the trough shape. The measured width of the settlement trough is about one tunnel diameter. Outside this area, the measured settlements are negligible (Fig. 44).

The calculated surface troughs have a lateral extension of more than 5 tunnel diameters at each side of the tunnel axis. Also, the surface uplift in the vicinity of the model boundaries does not correspond with the observations. This effect is well known and cannot be avoided as long as the simplified material model (elastic, ideal plastic with a Mohr-Coulomb failure criterion) is used. It has been shown [7] that for shallow tunnels good agreement with measured values is obtained if the unloading modulus of soil is higher than the loading modulus. Without considering unloading conditions under the invert (assigning a higher Young's modulus to soil material), this uplift effect is even greater.

The shape of the surface trough is influenced by different material parameters. To investigate these effects the rock mass strength and the type of material model were varied. Fig. 45 shows different surface troughs developed with a full face model after two load steps (top heading excavation). In a) and b), ground 1 is assigned Mohr-Coulomb material behaviour with different friction angles or different cohesion respectively. In c), a material

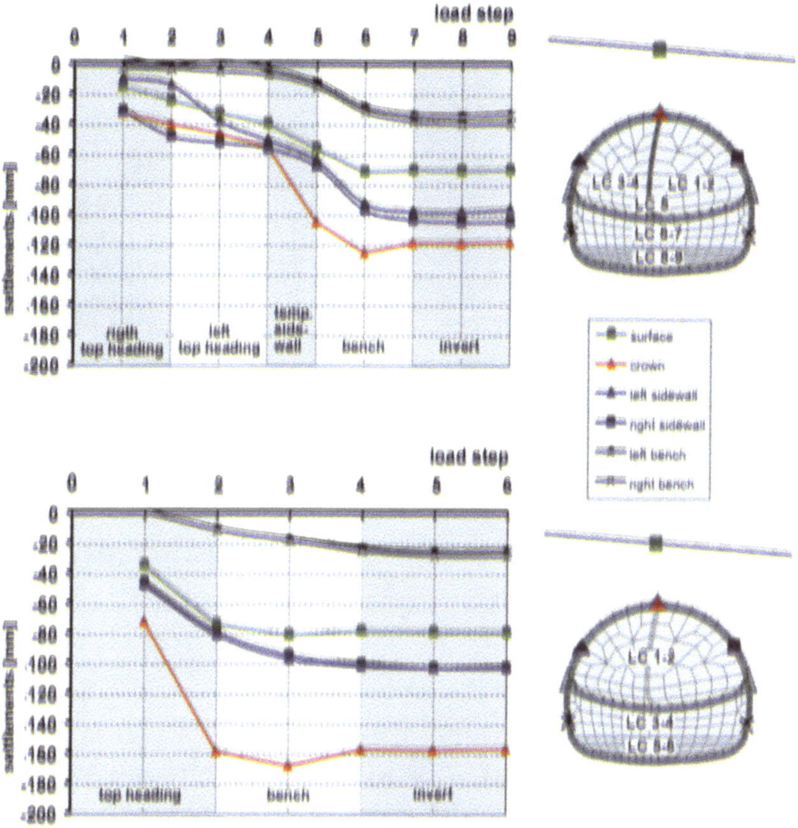

Fig. 40. Calculated displacement histories for full excavation and staged excavation

model is used, where the friction angle and cohesion of horizontal smeared joints are varied. The diagrams show, that with a slight variation of certain parameters, completely different surface trough shapes can be obtained. The shape of the calculated trough with the multi laminate material model agrees best with the measured profile.

6.5 Alternative method using a "pipe roof system"

In the following section, the amount of a surface settlement reduction by a "pipe roof system" is investigated with both a finite element model and a coupled finite / boundary element model. Tubes with 15m length which are installed after every 12m of excavation are used. The pipe roof extends approximately over half of the top heading. The pipe roof is modelled as elastic shell with solid finite elements.

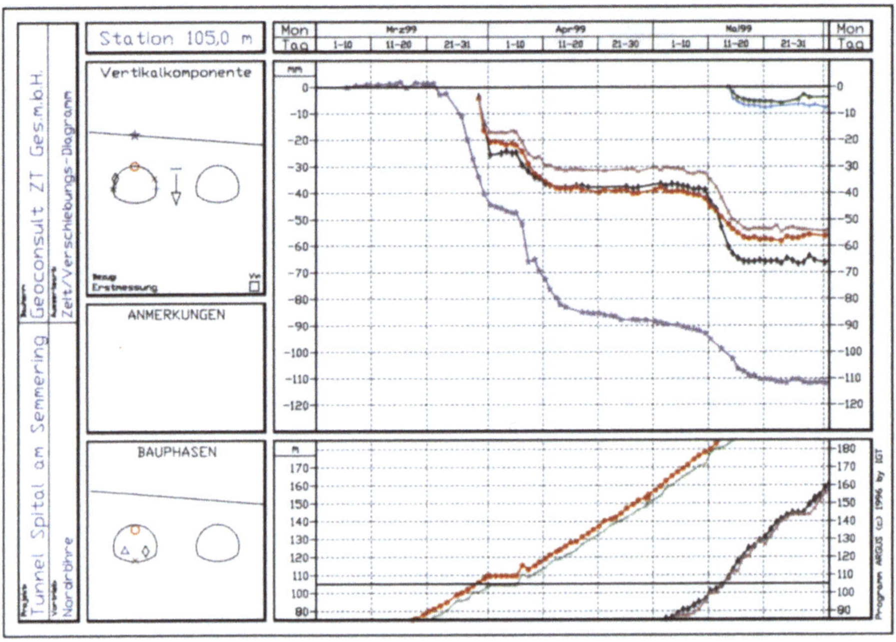

Fig. 41. Monitored settlements at cross section 105

Finite element model. The mesh consists of 4.560 isoparametric, linear and parabolic finite elements, with about 22.500 unknowns (Fig. 46). Symmetry is assumed for the plane along the tunnel axis. Each slice corresponds to the excavation length of 1,0m.

In contrast to the 3D models used so far, this mesh could not be developed by just copying equal slices in excavation direction. Due to the inclination of the pipe roof, its position is different for sequencing slices. This required the separate modelling of most of the slices. The overlap zone additionally increased the discretisation effort. The discretisation effort of this mesh was considerable in comparison to meshes developed by slice copying.

The effect of the pipe roof system was modelled as follows:

Initially, all finite elements are assigned rock mass parameters. During the excavation the pipe roof is sequentially activated by assigning corresponding material parameters. For the rock mass, elastic ideal-plastic material behaviour is assumed, whereas the pipe roof is considered to behave elastically.

Coupled finite element / boundary element model. The coupled model consists of 2.900 finite elements and 500 boundary elements with 15.800 unknowns. The excavation area (core) and pipe roof system were separately generated with finite elements. One mesh slice (core) and one solid shell re-

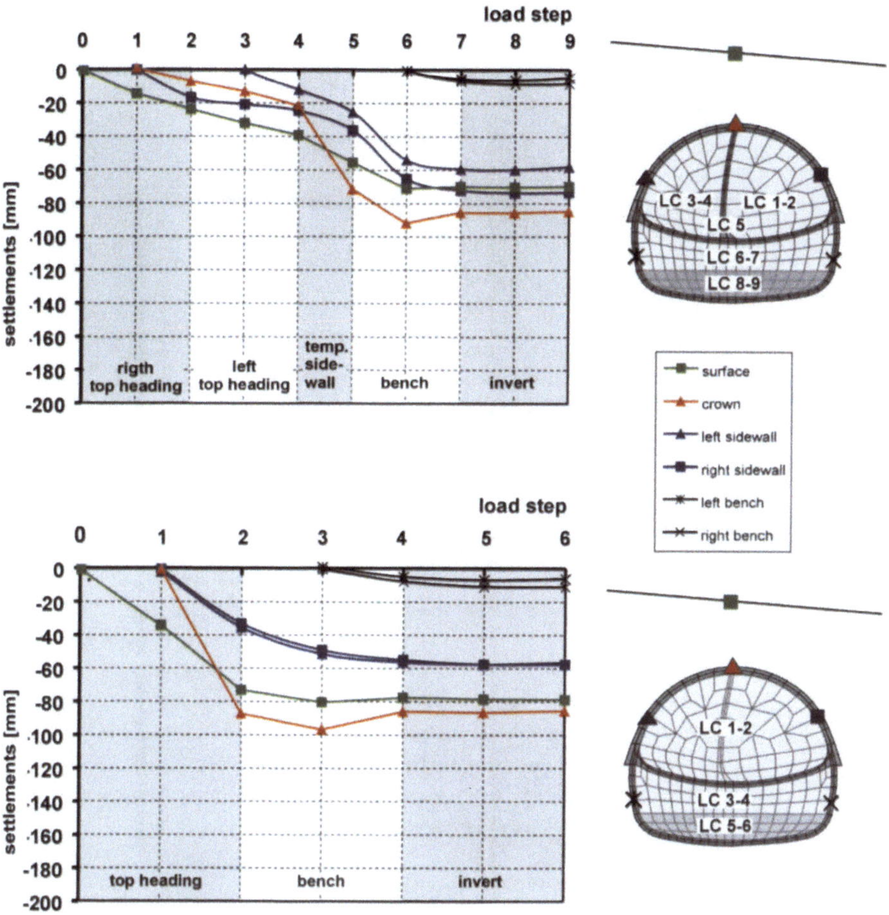

Fig. 42. Calibrated load case histories for full excavation and staged excavation Surface settlements

spectively had to be discretised and were then copied in longitudinal direction to develop the model.

The finite elements are embedded in an elastic far field region, which is bordered by the surface elements at the upper side and by a plane at the bottom of the model (Fig. 47). At the bottom plane, vertical displacements are restrained, preventing unrealistic model uplifts. The discretisation effort for this model was extremely low. A further benefit is the possibility to vary the overburden just by moving the surface elements correspondingly.

Results. Figure 48 shows the crown settlement lines and trend lines for excavation with and without the pipe roof calculated with the finite element

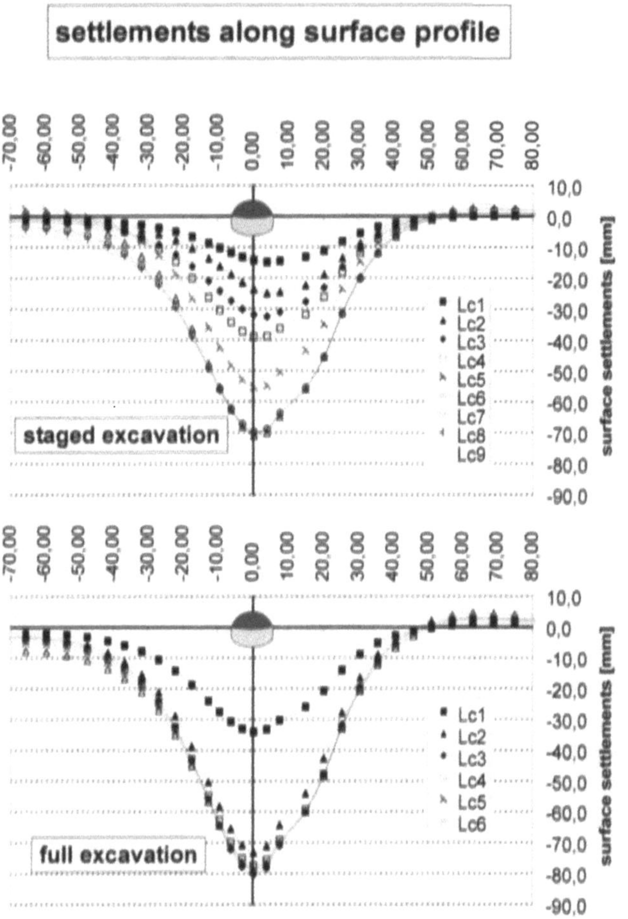

Fig. 43. Surface settlements in a cross section for staged and full excavation

model and the coupled model respectively. With both models the crown settlements decreased by approximately 20% with the pipe roof support. The wavy shape of the deflection is caused by mesh influences due to the transition between linear and parabolic elements. The trend lines at the tunnel face are barely affected by the pipe roof except for a slight uplift tendency within the whole model.

With the same input parameters, lower settlements are obtained with the coupled models in comparison to the finite element models.

Displacement results calculated with finite element models are approximately the same as with coupled finite / boundary element models. In cases, where 3D meshes cannot be developed by slice copying, the coupled method provides considerable less effort for preparing the model.

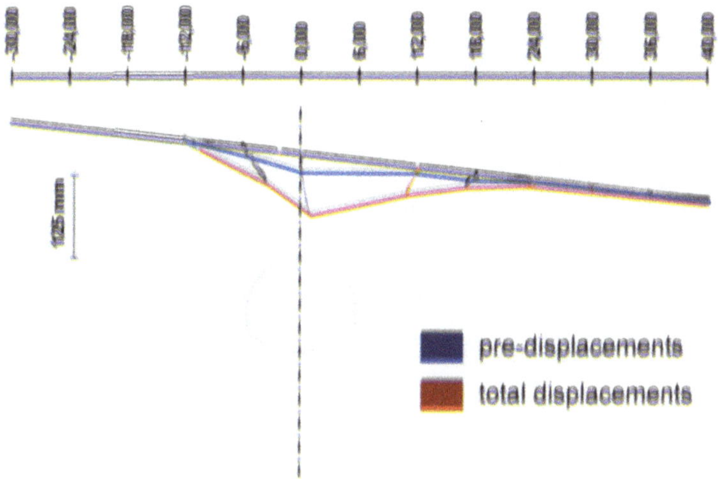

Fig. 44. Measured settlements along the surface profile at station 130

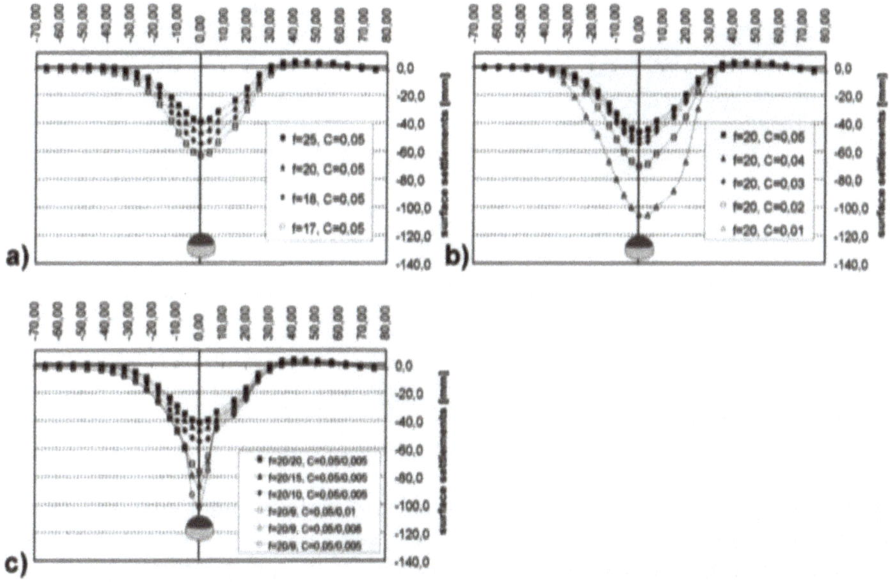

Fig. 45. Calculated surface troughs after load step 2 (full excavation of top heading) by variation of the friction angle and cohesion of ground 1.

a) variation of the friction angle from 25 ° to 17 °

b) variation of the cohesion c from 0,05 [MPa] to 0,01 [MPa]

c) material model where the friction angle and the cohesion of the smeared joints are varied

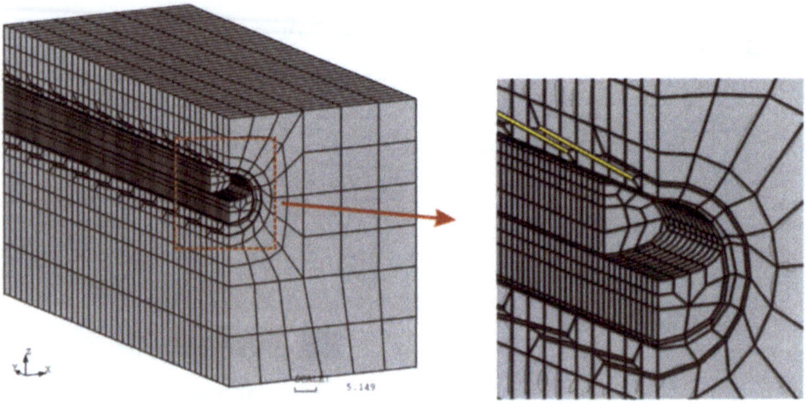

Fig. 46. Finite element model for the simulation of the pipe roof system

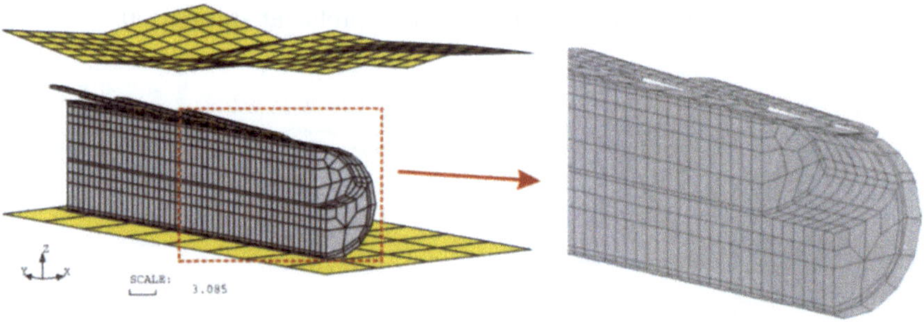

Fig. 47. Coupled finite element / boundary element model for the simulation of the pipe roof system

7 Conclusion

With the experience gained from numerical simulations under site conditions, basics for numerical modelling, calculation and result data evaluation were obtained. The **B**oundary **E**lement **F**inite **E**lement code **BEFE** was used to investigate modelling techniques with respect to practical suitability and efficiency.

Result data evaluation with deflection curves, trend lines and load-step (time) history lines were demonstrated. It was shown how result data calibration has to be done to allow a comparison with monitored data.

The influences of element size and types on the calculated displacements were investigated. Using both parabolic and linear elements together in a model, whereby the parabolic only cover those zones with non-linear material behaviour, turns out to be efficient.

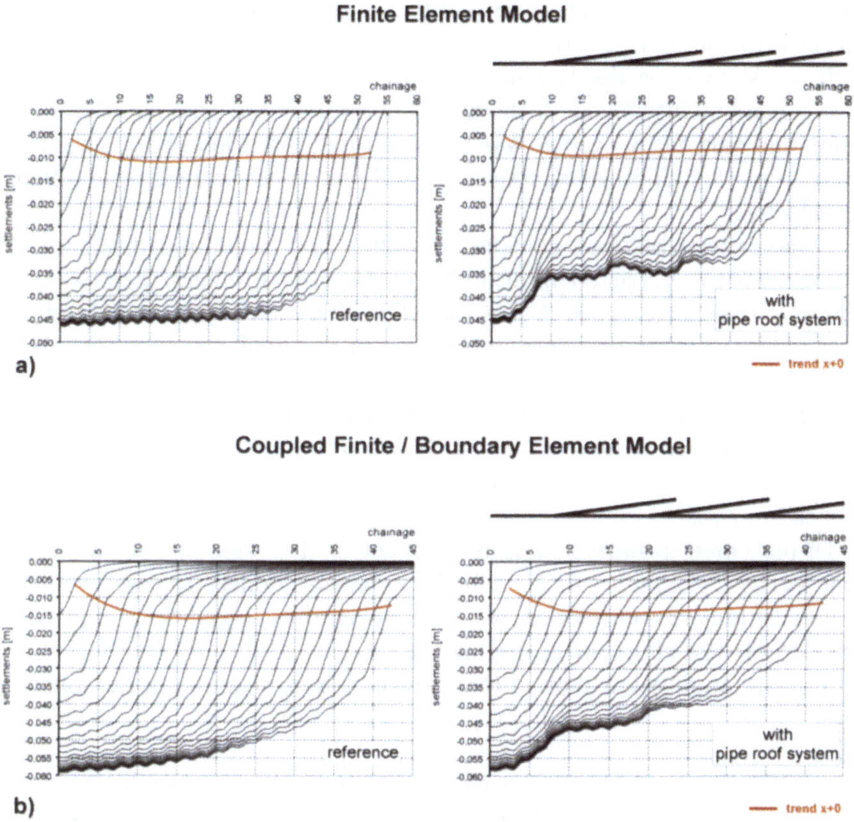

Fig. 48. Crown settlement deflection curves and trend lines for sequential excavation with and without pipe roof system for
a) Finite Element Model
b) Coupled Finite / Boundary Element Model

For sequential excavations with 3D models, the size of mesh slices must be kept constant to avoid mesh-induced influences on the displacements. Influences of the model geometry on the longitudinal displacements were investigated.

With a Preconditioned Conjugate Gradient (PCG) solver, which has been incorporated into BEFE, calculation time and disk storage requirement can be reduced appreciably in comparison to a frontal solver. Non-linear models with 10.000 elements and up to 70 excavation steps could be handled within 24 hours.

Longitudinal displacements were investigated for elastic and elastic-ideal plastic material behaviour and for different primary stress conditions. The direction of the longitudinal displacement changes twice in the vicinity of the face. Utmost care therefore has to be exercised in monitoring and in interpreting monitoring data.

It was shown how displacement vector orientations correlate with changes in rock mass quality ahead of the face. The effect of the primary stress orientation and varying K_0 values was demonstrated for several basic situations. These numerical findings corresponded well with observations from different tunnel sites.

Discrepancies between calculation results and monitored displacements demonstrate the restricted applicability of the Mohr-Coulomb material model for many rock mass types. For a realistic simulation of shotcrete and rock bolts the material models described earlier in this book were not available at the time of the work.

A main shortcoming of plane strain models is the difficulty to realistically determine the load-reduction factors. These factors generally are roughly estimated. The considerable influence on the results demonstrates the importance of back-analyses.

Displacement results from finite element models were approximately the same as these of coupled finite/boundary element models. In cases where 3D meshes could not be developed by slice copying, the coupled method provided considerable less effort for preparing the model.

The site application showed that advances were required in the following areas:

- constitutive models considering loading and unloading conditions
- improvements in the assessment of rock mass parameters for modelling
- constitutive models for shotcrete and rock bolt systems
- mesh generation tools with higher automation degree; adaptive mesh refinement
- improvements in data visualisation following monitoring data evaluation practice

References

[1] Beer G. (1985) BEFE C.S.S User's Manual, Computer Software and Services International, Austria

[2] Beer G., Golser H. (2002) Application of Numerical Simulation at the Tunnel Site. Institute for Structural Analysis; Graz University of Technology

[3] Laabmayer F., Swoboda G. (1978) Beitrag zur Weiterentwicklung der Berechnung flachliegender Tunnelbauten im Lockergestein. *Moderner Tunnelbau bei der Münchner U-Bahn*, Springer Verlag: 55-71

[4] Jedlitschka G. (1997) Vergleich Stützlastverfahren- Stützkernverfahren. Geotechnical Project, Institute for Soil Mechanics and Foundations, Graz University of Technology

[5] Beer G., Schweiger H.F. (1996) An Overview of Different Computational Models with Emphasis on Practical Applicability. *Felsbau*, 14

[6] Beer G., Sigl O., Brandl J. (1997) Recent developments and application of the boundary element method in geomechanics. Numerical Models in Geomechanics, Pietruszczak and Pande (eds), Balkema

[7] Beer G., Poulsen B.A. (1994) Efficient Numerical Modelling of Faulted Rock Using the Boundary Element Method. *Int. J. Rock Mech. Min. Sci. and Geomech.*, Vol. 31, No. 5: 485-506,

[8] Beer G., Golser H., Jedlitschka G., Zacher P. (1999) Coupled Finite Element / Boundary Element analysis in rock mechanics – Industrial applications. *Proceedings of the 37th U.S. Rock Mechanics Symposium*: 133-140

[9] Steindorfer A. (1995) Excel spreadsheet for Feder Formulas. Institute for Rock Mechanics and Tunnelling, Graz University of Technology

[10] Golser H. (1999) Application of Numerical Simulation Methods on Site. *Felsbau*, 17: 21-25

[11] Beer G. (1999) Institute for Structural Analysis, Graz University of Technology

[12] Budil A. (1996) Längsverschiebungen beim Tunnelvortrieb. doctoral thesis at the Institute of Rock Mechanics and Tunnelling, Graz University of Technology

[13] Steindorfer A. (1998) Short Term Prediction of Rock Mass Behaviour in Tunnelling by Advanced Analysis of Displacement Monitoring Data. doctoral thesis at the Institute of Rock Mechanics and Tunnelling, Graz University of Technology

[14] Schubert W., Steindorfer A. (1996) Selective displacement monitoring during tunnel excavation. *Felsbau*, 14/2, VGE: 93-97

[15] Steindorfer A., Schubert W., Rabensteiner K. (1995) Problemorientierte Auswertung geotechnischer Messungen. *Felsbau*, 13/6, VGE: 386-390

[16] Hoek E., Brown E.T. (1997) Practical estimates of rock mass strength. *Int. J. Rock Mech. and Min. Sci. and Geomech.*, 34(8): 1165-1187

Chapter 16
Summary and outlook

In this book the main results of the Austrian joint research initiative Numerical Simulation in Tunnelling have been presented. The main aim of this research work was to advance the reliability and efficiency of existing numerical simulation methods. To achieve an increase in reliability of the methods, i.e. to ensure that predictions made agree well with actual behaviour a significant amount of development had to be made with respect to material models for rock/soil and shotcrete as well as with respect to the material parameters used. To increase the efficiency of the methods and to make the methods usable for tunnel engineers the input of data to the simulation program has to be made automatic to a great degree. Meshes that are necessary for the numerical simulation need to be generated completely automatic. An essential part of this development was improvements in the acquisition of geological data. The new method of data acquisition developed means that the information can be obtained in digital form and fed directly into the pre-processing software for the numerical simulation program.

There has been a general consensus within the tunnelling community that material models based on plasticity theories do not present the actual behaviour of a jointed rock mass very well. Also, it has been found that the classical Mohr-Coulomb material law does not represent well the behaviour of soil and this affects the quality of results from simulation of shallow tunnels to a greater degree. Therefore some effort was spent in developing the material models for rock and soil further. The determination of material parameters has presented a great challenge for tunnelling engineers because often parameters obtained from the laboratory can not be used directly in the numerical simulation. Many times some factors are applied quite arbitrary to obtain "rock mass" parameters. The two approaches presented in this book to determine material parameters use two completely different methods: one based on the minimisation of the error the other on concepts of soft computing.

Shotcrete is frequently used as support in tunnelling and forms an integral part of the NATM. The time elapsed until the support becomes active is an important aspect in this tunnelling method, since some deformation has to be allowed in order for the ground arch to become fully effective. Indeed in the diagrams originally conceived by Pacher (as shown in Chap. 2) curves that describe the response of the shotcrete appear but their accurate determination has eluded engineers so far. In this book we believe that a model has been presented for the first time that allows to accurately consider the thermal, chemical and mechanical processes that occur in the shotcrete after placement.

Compressed air is often used in tunnelling in order to reduce the inflow of ground water and to reduce the surface settlements for shallow tunnels.

Up to now it was difficult to estimate the amount of compressed air required. Theoretical considerations together with experiments form the basis of a method of accurate prediction of the air losses and the effect of compressed air on the surface settlements.

The most sophisticated theoretical developments are useless unless one is able to transfer these to the tunnel site. For this to be successful one has to have a system that can be used by tunnel engineers with no previous experience in numerical simulation. Essential for this is a good end efficient pre-processing and an effective post-processing. There should be a minimum of effort required to input a problem and most of the data should be taken form existing geometrical and geological data bases. The results should be presented in such a way that tunnelling engineers can immediately see what is going on. As a good example we mention here the prediction of the degree of utilisation that has been presented in Chap. 10 and whose visualisation is shown in Chap. 14. This system is currently being used at an Austrian tunnel site to warn tunnel engineers of excessive degrees of utilisation of the concrete shell during tunnel advance. This is also a good example how numerical simulation and measurements interact. Further examples of site application have been presented in Chap. 15, where the visualisation of data and comparison with measurements have also been shown.

We hope that with this book we have made a significant contribution to the advance of numerical simulation in tunnelling. Some of the developments discussed here are already some way towards commercialisation. This applies for example to the system developed for the acquisition of geological data and the system for determining the degree of utilisation based on displacement measurements. Others, such as the pre- and post-processing software have great potential for commercial exploitation.

The trials with numerical simulation software reported in Chap. 15 have shown that improvements in efficiency have to be made in order to make it possible for 3-D numerical simulations to be carried out at the tunnel site to provide assistance to the tunnel engineers. Using a conjugate gradient solver, finite element analyses can be carried out in a short time but the effort in generating the mesh is still considerable. Even if this problem is solved using efficient automatic mesh generators, there is still the question of errors introduced by mesh truncation and mesh coarseness.

It has been shown that boundary element analyses (BEM) offer a significant reduction in effort as compared with finite element analyses (FEM). The effort is basically reduced by an order of magnitude because only surface instead of volume discretisation is needed. Also, there is no need for mesh truncation and it has been proven that errors introduced by the coarseness of the mesh are also smaller than with the FEM. Another advantage of the BEM is that results can be obtained accurately anywhere in the domain and not only at Gauss points as a post-processing exercise. Up to now the use of the BEM in tunnelling has been restricted because of the lack of software that allowed to consider sequential excavation/construction and non-linear material behaviour. It has been recently shown that sequential excavation can be modelled with the multi-region BEM (see Figure 1).

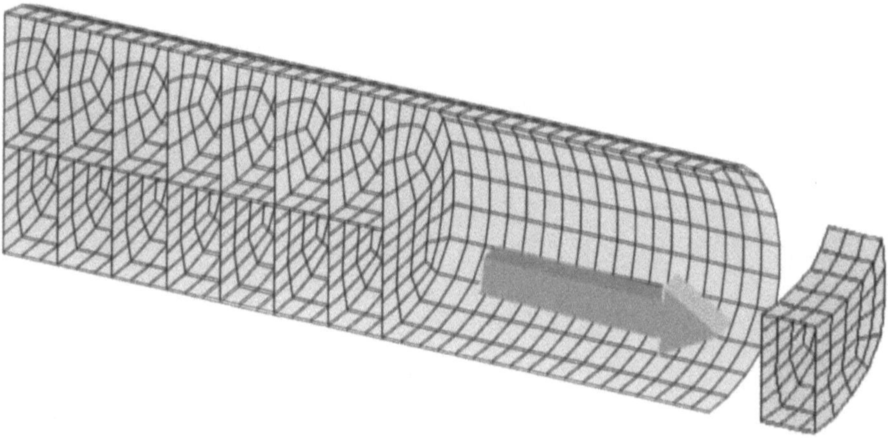

Fig. 1. Boundary element mesh enabling the simulation of sequential excavation. Only half of the tunnel surface is discretised and symmetry conditions applied. The results of this mesh will be more accurate than ones obtained for the FEM mesh as for example shown in Chap. 15. The effort in getting the results however will be an order of magnitude smaller.

The theory for dealing with non-linear material behaviour in the BEM is well known and only needs to be implemented efficiently. Progress on this development is underway in a follow on project funded by the Austrian national science fund. It is hoped that by 2005 software based on the BEM will be available for the tunnelling engineer, that allows accurate 3-D analyses of tunnels to be carried out in a very short time. This should make numerical simulation an easy to use tool for any tunnelling engineer and will be able to assist in the often difficult decisions.

Finally the acceptance of numerical simulation in the design office and on the tunnel site will only be achieved by a well executed transfer of technology. This book is meant to serve as the first step in this transfer and we hope that it will be successful in its purpose.

Springer-Verlag and the Environment

WE AT SPRINGER-VERLAG FIRMLY BELIEVE THAT AN international science publisher has a special obligation to the environment, and our corporate policies consistently reflect this conviction.

WE ALSO EXPECT OUR BUSINESS PARTNERS – PRINTERS, paper mills, packaging manufacturers, etc. – to commit themselves to using environmentally friendly materials and production processes.

THE PAPER IN THIS BOOK IS MADE FROM NO-CHLORINE pulp and is acid free, in conformance with international standards for paper permanency.

MIX
Papier aus verantwortungsvollen Quellen
Paper from responsible sources
FSC® C105338

If you have any concerns about our products,
you can contact us on
ProductSafety@springernature.com

In case Publisher is established outside the EU,
the EU authorized representative is:
Springer Nature Customer Service Center GmbH
Europaplatz 3, 69115 Heidelberg, Germany

Printed by Libri Plureos GmbH
in Hamburg, Germany